Rehm/Espig
The Cultivated Plants of the Tropics and Subtropics

The Technical Centre for Agriculture and Rural Co-operation (CTA) operates under the Lomé Convention between member States of the European Community and the African, Caribbean and Pacific (ACP) States.

The aim of CTA is to collect, disseminate and facilitate the exchange of information on research, training and innovations in the spheres of agricultural and rural development and extension for the benefit of the ACP States.

Headquarters: "De Rietkampen", Galvanistraat 9, Ede, Netherlands
Postal address: Postbus 380, 6700 AJ Wageningen, Netherlands
Tel.: (31) (0) (8380) - 60400
Telex: (44) 30196 CTA NL
Telefax: (31) (0) (8380) - 31052

# The Cultivated Plants of the Tropics and Subtropics

## Cultivation, Economic Value, Utilization

Sigmund Rehm
Gustav Espig

IAT
Institute of Agronomy in the Tropics
University of Göttingen

101 Figures
and 55 Tables

1991

CIP-Titelaufnahme der Deutschen Bibliothek

**Rehm, Sigmund:**
The cultivated plants of the tropics and subtropics:
Cultivation, economic value, utilization / Sigmund Rehm;
Gustav Espig. CTA. [Transl. by George McNamara; Christine Ernsting]
- Weikersheim: Margraf, 1991
Einheitssacht.: Die Kulturpflanzen der Tropen und Subtropen <engl.>
ISBN 3-8236-1169-0
NE: Espig, Gustav



Title of the original German edition:
Die Kulturpflanzen der Tropen und Subtropen

English edition by:
Verlag Josef Margraf, 1991
Mühlstrasse 9, D-6992 Weikersheim
West Germany

Translated by:
George McNamara, Christine Ernsting

Illustrated by:
Ulrike Griese

Cover design by:
Katrin Geigenmüller

Typesetting by:
Atelier Niedernjesa: Konrad Vielhauer, Hartmut Bremer

Printed by:
Priese GmbH, Berlin

Printed in Germany

ISBN 3-8236-1169-0

# Foreword

The exponential growth of the world population calls for a proportionate increase in agricultural production, particularly in the tropics and subtropics where the largest population increase is taking place. Topography, soils, climate, and socio-economic conditions vary widely in these regions with the consequence that several thousand plant species are used to provide food for man and his animals, to supply beverages, spices, medicinal plants, and raw materials for clothing and a multitude of technical requirements, and to assist sustainable land use practices. The wealth of useful plants is far from being fully exploited and harbours genetic resources for industry and fuel on a much larger scale than today. More than 1,000 plants are discussed in this pocket manual; forest and ornamental plants had to be omitted.

Each chapter begins with an introduction to the particular properties of the plant group, giving an overview on the economics, production trends, nutritional aspects, chemistry, and technological features. Only the major crops could be treated in detail. With regard to these, we sought to cover all essential points: production, botany, breeding, ecophysiology, cultivation practices, diseases and pests, processing and utilization. The numerous minor or only locally important crops were collected in the tables; these give the valid botanical name of each plant, a selection of its vernacular names, and indicate its distribution, economic importance and uses. The drawings should help to identify the plants, and depict important morphological peculiarities. The diagrams illustrating the production during the last ten years are intended to offer visual information on the relative importance of a crop and on current trends.

The necessarily concise form of our presentation will need complementation from other sources in many instances. As a key to available information we give a large number of references to all the species included. In selecting the quotations we aimed at covering all aspects of production and utilization, and all regions of the tropics and subtropics. Therefore, we also quote some publications in foreign languages, particularly in French and Spanish. Works published before 1970 are included only in exceptional cases since we felt the current literature to be of the greatest value to the reader. All information on the plants dealt with in this book is available off-line from a continuously updated data bank for cultivated plants and relevant literature at the Institute of Agronomy and Animal Health in the Tropics at Göttingen University.

The original German edition of this book was based on the lectures of the senior author (S.R.) at Göttingen University and on the documentation on tropical crops collected by G.E., assisted by Mr. E. von Hippel. The drawings are the work of Mrs. Ulrike Griese. The English translation was done by George McNamara, supported and advised by Dr. Harold Gough, both of Epsom, Surrey. The authors thank Mrs. Chr. Ernsting for proofreading the English text; they remain responsible for the technical correctness of all the data and for the opinions expressed in the book. Further, they wish to thank Mr. Walter

Eisenberg and Mr. Konrad Vielhauer for their help with the production of the tables, the list of references and the subject index, and also the publisher for the pleasant cooperation throughout the preparation of this edition.

Göttingen, September 1990

Sigmund Rehm
Gustav Espig

# Contents

# Introduction

Of the surface area of the earth which is used for agriculture, two thirds are in the tropics and subtropics. Almost 4 billion people live in these regions, three quarters of the world's population. Most of the countries in the tropics and subtropics are developing countries, with little industry. In them, agriculture is the basis of their economy. It must not only supply the nutritional needs of their own populations, but also must earn the majority of the export income. Without the foreign exchange gained by exports, the development of their national industries is not possible. Thus, the transition from developing country to industrial country depends primarily on the performance of agriculture, and, within agriculture, mainly on the agronomic sector, which provides far more export income for warm countries than animal production.

The greatest problem for the agronomist in the developing countries, however, is the supply of food for the increasing population (194). In the year 2000, it is predicted that the population in the developing countries will number 5 billion. Without a substantial improvement in the methods of cultivation, this challenge cannot be met. A further spread of cultivation is possible only to a small extent into the areas which have not yet been used, without endangering the permanent productivity of the soil by erosion. On the contrary, in some regions, the cultivation of plants has already extended into areas of land which are unsuitable for it, especially in arid regions (e.g. the Sahel), which should only be used carefully as pasture land, and it has spread to steep hillsides in the mountains, which should only be forested, or not used for economic purposes at all. The solutions to the problem can and must be found in:

- the regionalization of cultivation (limiting it to areas which are able to permanently supply high yields),
- in multiple cropping (the cultivation of two or three field crops in one year),
- by the use of all inputs to increase the yield per unit area (improved cultivation and maintenance techniques, choice of cultivars, fertilizers, plant protection, irrigation),
- and in making sure of the basic supply of fodder for animal husbandry (pastures and fodder crops).

The number of plants which are cultivated in the tropics and subtropics is very large. ZEVEN and DE WET (2026) name 2,500 species, excluding ornamental and forest plants; this number includes the cultivated plants of the temperate zone, and some close relatives of the species cultivated. This abundance has a number of causes:

a) The climatic conditions, especially the absence of long periods of frost, lead to less acute selection in the tropics than in the temperate zone; on the other hand, the conditions are so widely varying (arid to humid regions, highlands) that it has been necessary to develop a great variety of cultivated plants.
b) Plant cultivation in the tropics and subtropics is 10,000-12,000 years old. All primary centres and regions of the prehistoric development of agriculture lie

in these zones (699, 2026). The rich floras of Asia, Africa, and America, which were not impoverished by ice ages, were available for the work of the primitive agriculturalists.

c) The agriculture of many of the countries of the tropics and subtropics remained unaltered from these days. Adaptation to new techniques, which might have led to an elimination of unsuitable species, was not necessary. Definite improvements in breeding particular species, which would have led to the replacement of species which perform less well, were hardly ever made.

World trade, the drive for exports, and the transition to rational production procedures are nowadays the causes of rapidly progressing changes in plant cultivation in the tropics and subtropics. High-yield potential cultivars of maize, sorghum and pearl millet have replaced the cultivation of the weaker millet species; only hevea remains as a source of rubber which is worth cultivating; the development of animal husbandry in the tropics has induced the breeding of new fodder plants; the chemical industry has forced the disappearance of some raw materials from the world market (indigo), but has been the driving force behind the development of new sources of raw materials (*Dioscorea* as a supplier of diosgenin for the synthesis of sex hormones). Our main concerns in this book are to comprehend these changes, to exclude obsolete plants, to indicate new developments, and to consider the economic importance of each plant. Completeness cannot be our goal in the space available. There are listings of further agricultural plants (126, 200, 239, 242, 258, 341, 371, 447, 866, 1023, 1095, 1138, 1146, 1211, 1444, 1445, 1493, 1581, 1779, 1789, 1846, 1947, 1948); there are works which deal with the products of particular regions (Australia (712), Bolivia (269), Brazil (1518), India (19, 376, 999), Indonesia (684), Colombia (1378), Malaysia (259), East Africa (7), Oceania (145), Philippines (241), West Africa (258, 262, 397, 587, 843, 1477), Central America (91, 1039, 1960)); and, there are specialized works for the individual plant groups which are cited in the relevant chapters. For the general aspects of agriculture in the tropics and subtropics, we refer to a range of works (44, 62, 92, 666, 683, 1297, 1492, 1743, 1911, 1949, 1990).

Agronomy is an integrating science, which is based not only on the various natural sciences (botany, climatology, soil science, phytopathology, breeding, technology), but has to consider also the socio-economic aspects. We have aspired to deal with all of these factors for the important crop species.

With regard to the scientific nomenclature of plants, we have endeavoured to use the names which are valid according to the International Code (603, 878, 1728). Where plants are still frequently cited in the literature under a name which is no longer valid, we have given the most used synonym, and if necessary, two synonyms. Our sources have been especially (1611, 1637, 1789, 2026). We have deviated from these authors where it has seemed to us that it is more useful to gather together several lesser species under one name for the convenience of agriculturalists and breeders. Cultivars could only be named as examples; mostly we have had to limit ourselves to mentioning the cultivar groups.

It seemed desirable to give the common names of plants also in several of the world's most important languages, because the scientific names of the plants are not always given in the foreign literatures. Where frequently used, we have also given the common names in the local languages.

We have limited botanical particulars (morphology, anatomy, physiology) to the features which are important for the agronomist. Complete botanical descriptions for many of the plants dealt with are found in (341, 1297, 1444, 1445) and other handbooks.

The cytogenetics of the most important tropical cultivated plants have been widely clarified in the last 50 years, and have provided the basis of the planned developments in breeding which have led to a prodigious improvement of cultivars for many species (402, 1657). This applies not only to the high-yielding cultivars of rice, wheat and sorghum, but also to sugar cane (1737), papaya (535), cotton (784), and banana (1654). In this field, it must not be forgotten that some species have hardly yet been worked upon by the breeders, and with many other species, their genetic potential is far from being exhausted (563, 1258, 1513). Breeding can only be dealt with in this book by indicating its most important goals, and by quoting the specialist literature, even for the most important species. For further information the relevant reviews should be consulted (535, 781, 1413, 1558, 1698).

The treatment of agricultural procedures is difficult within the space of a concise book, because of the extraordinary diversity of cultivation conditions. Among the cultivated plants of the tropics and subtropics, there are not only field and garden crops, but also plants which are cultivated in water, and tree crops whose cultivation is more in the nature of forestry. Propagation can be either sexual (by seed) or asexual (vegetative). Problems of maintaining productivity are far more serious than in the temperate zone, i.e. protection against erosion by wind and water, and these can require diverse precautions. Special difficulties are the extraordinary differences in technical equipment; over large areas, the only tool of the cultivators is the hand hoe. On the other hand, nowhere else are such complicated and expensive machines used as in the mechanized harvesting of cotton or sugar cane. The same applies to plant protection, which over vast areas, is limited to crop rotation and mixed cultivation, and perhaps the collection of harmful insects by hand; in other places, the normal routine is to apply chemical agents by spraying from an aeroplane. The same multi-faceted picture is found with the use of fertilizers.

Obviously, it is impossible for us to present the multitude of agricultural methods and possibilities. We must therefore limit ourselves to emphasizing the most basically important and generally valid aspects. In addition we have attempted to include the procedures which require no great financial means or special technical knowledge.

Detailed advice about fertilizers has been omitted because of the extraordinary differences in soil types found in the tropics and subtropics; such advice is found especially in (595) for all important crops. Special techniques of fertilizer usage for dryland and irrigated soils are dealt with in (677). The monographs about the individual cultivated plants usually discuss the problems of fertilizer usage.

Diseases and pests can only be reviewed in as much as they cause severe damage and are of more than regional importance. In this area, much has been done in recent years. For most of the major cultivated crops, there are nowadays plant protection handbooks (see the literature references in the relevant chapters), and there are also works covering the whole world (83, 567, 734, 735, 749, 982, 1007, 1017, 2001), books which deal with particular groups of insects (707, 1008, 1367), and handbooks for particular regions (Africa (200, 265, 1604, 1802, 1917), America (64, 1916), Asia (108, 629, 801, 925, 1488, 1517)).

Remarkable ideas about the general problems of plant protection in warm countries are presented in (251, 869, 951).

Perhaps the most important aspect of plant cultivation in the warm countries is the control of weeds; this, too, can only partially be dealt with. Weed control requires the greatest input of labour, and in spite of this, weeds cause the greatest yield losses (98, 750, 1097, 1355). The importance of the weed problem is widely acknowledged, and there are handbooks for almost every country and for every region, which describe the most important weeds and their control (Australia (110, 1769, 1944), Brazil (1077), Malaysia (144), New Guinea (729), East Africa (846, 1790), South Africa (727), Sri Lanka (680), Taiwan (323), USA (754, 1847), Near East (5, 184, 311), West Africa (847), West Indian Islands and Central America (9, 582), Zambia (1860)). Several publications have also appeared which deal with tropical weeds as a worldwide subject (22, 221, 271, 424, 751, 752, 937, 982, 1203, 2029). The parasitic weeds deserve special attention (Loranthaceae, *Orobanche* spp., *Striga* spp., etc.) (118, 991, 1238, 1356, 1908). Mankind is still far from solving the problem of weeds on a worldwide level. Herbicides are definitely a help, but they can only be used to a limited extent. In primitive agriculture, the solution in the first place must be sought with improved agricultural methods (crop rotation, timing of hoeing, mulching, ground cover plants, etc.).

With the climatological information, we follow the division into 5 large regions made by WALTER (1893):

1. permanently humid tropics, with 0-2.5 months dry periods;
2. summer rainfall regions of the tropics and subtropics, with 2.5-10 months dryness;
3. deserts and semideserts, with 10-12 dry months;
4. permanently humid subtropics (sometimes also classified as warm-temperate zone);
5. and, winter rainfall regions of the subtropics, with 2.5-12 months of dryness.

To these must be added the highlands of the tropics, which offer special and sometimes particularly favourable conditions for plant cultivation. Irrespective of the elevation, the tropics are defined by the uniformity of temperature throughout the year; the difference in temperature between the warmest and coldest month is smaller than the mean daily temperature fluctuation between day and night. The subtropics, on the other hand, are characterized by considerable differences in the temperatures prevailing in summer and winter. As a consequence of these definitions, the climatological borders between the tropics and subtropics show considerable deviations from the Tropics of Cancer and Capricorn (Fig. 1). In the tables, we use the following symbols: Te - temperate zone, S - subtropics, Tr - tropics, TrH - tropical highlands. From the viewpoint of plant cultivation, the criteria chosen by TROLL (1010) are the most suitable, and thus the world climate map by him and PAFFEN also corresponds very well with the distribution of the various plant cultivation systems (Fig. 1). For a quick orientation, the climatic diagrams (1892, 1893) are very useful. Local conditions (slopes, wind) can, however, be more decisive than the broad climatic region, and a detailed analysis of the climatic factors is then often necessary. The plant cultivator has many possibilities to alter the local and micro-climate (irrigation, shade and windbreak trees, mulch) (621), and many

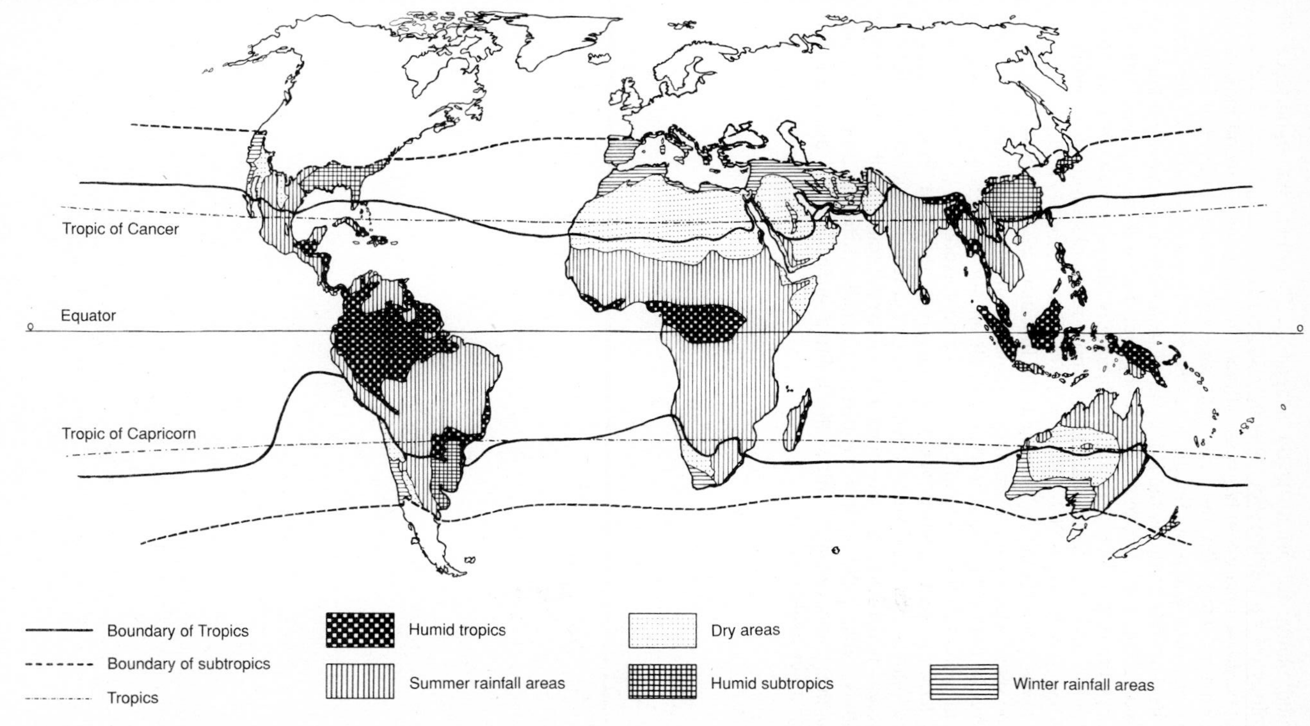


Fig. 1. Climatic regions of the tropics and subtropics (1010)

cultivated plants are astonishingly adaptable. Therefore, the importance of climatology for agriculture should not be exaggerated.

The pedologic data given here can only be limited to very general indications (pH, structure, etc.). On the one hand, the soils within individual climatic regions, and also within a climatic area can be extremely varied. On the other hand, the demands made on the soil by most plants are not narrowly defined; in each individual case, an exact knowledge of the soil in question can be extremely important, but it is not important for the overview that we have attempted to give. For more exact information, the books on tropical pedology should be referred to (18, 207, 256, 924, 1004, 1196, 1256, 1577, 2022).

The statistical information is taken in the first instance from (514, 518). In so far as no other sources are given, they originate from these publications, and are taken for the year 1988. The uncertainty of all statistical data is well known, and unfortunately, even the collected figures in the publications of the FAO are quite incomplete for some plants. We have therefore abstained from presenting detailed figures. The total figures generally give a realistic picture for most of the species which play an important role in world trade. With others (fruit, vegetables, pulses) the figures are certainly widely different from the actual productions; however, they allow the relative importance of the individual products to be evaluated.

# Starch Plants

In all parts of the world, starch is the most important and inexpensive source of energy in human nourishment. To satisfy the average daily human energy need of 6.7-8.4 MJ (1359), 400-500 g of starch are necessary (1 g starch provides about 17 kJ). In nutrition, the carbohydrates starch and sugar have no function physiologically other than as energy providers.

A considerable part of the starch production is utilized industrially (956, 1460). Pure starch is produced from cereals or roots by fine milling in a watery medium, then separation by sedimentation or by centrifuging (moist starch has a density of 1.5), followed by drying. Pure starch is used as a foodstuff, but mostly industrially, for sizing threads and cloth finishing, as an inert carrier in pharmaceutical preparations, as a powder in cosmetics and medicines, as a drying material, as well as for glues (pastes), and as a filler in paper manufacture. It finds further uses in the synthetic chemicals industry (with an increasing significance in biodegradable products) and in the production of ethyl alcohol. A significant amount of starch is converted into dextrin and sugar, either by acid hydrolysis, or enzymatically. Since sugar from starch is cheaper than sucrose, it is widely used in the food industry; also pure, crystallized glucose (dextrose), malt sugar (maltose), and fructose syrup are all produced from starch.

Starch is built up in almost all the higher plants, and it is stored in seeds, fruits, stems or roots. Agriculturally, the most important storage organs are seeds (cereals and pseudocereals) and roots or tubers. Grain and root crops are so different in cultivation that we present them here in separate sections, as is usual, and we have also included information on several other starch-producing plants.

## CEREALS

Cereals are the fruits of graminaceous plants which can be milled to flour (955, 1041, 1965). They account for by far the greatest amount and value of all agricultural products. The total world production of the most important types for the last 10 years is set out in Fig. 2. The figures for wheat, rice and maize show considerable increases, as in previous decades (a rise of about 20%), while all the other species show little change. The proportion of world grain production in the tropics and subtropics amounts to 100% for rice, about 75% for maize, and more than 90% for millets. Even for wheat, 42% of the world production is grown in the tropics and subtropics, particularly in those areas with a Mediterranean climate. Barley, oats, and rye are predominantly products of the temperate zone. Because of its resistance to drought and salt, barley is of considerable importance in some subtropical countries. Triticale (see p. 401), a cross between wheat and rye, may also become important for the warmer regions in the future (648, 665, 769, 1031, 1216, 2028).

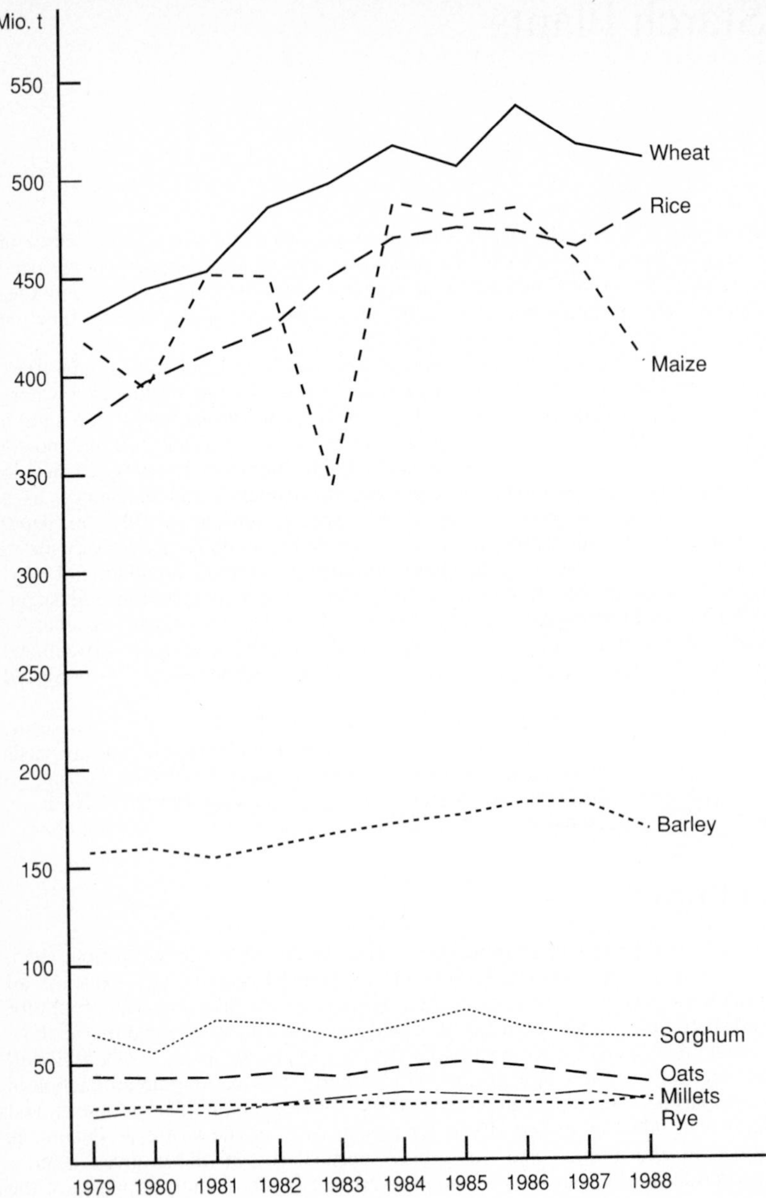


Fig. 2. World production of cereals, 1979-1988

Table 1. Grains: nutrients in 100 g of air-dried material (309, 310, 376, 1408)

| Species | Energy | Water | Crude protein | Starch | Crude fat | Crude fibre | Ca | P | Fe | Thi-amine | Ribo-flavin | Niacin |
|---|---|---|---|---|---|---|---|---|---|---|---|---|
| | kJ | g | g | g | g | g | mg | mg | mg | mg | mg | mg |
| Rice | | | | | | | | | | | | |
| unpolished | 1500 | 13 | 8 | 77 | 1.5 | 0.5 | 15 | 100 | 2.0 | 0.35 | 0.10 | 2.5 |
| polished | 1520 | 12 | 7 | 80 | 0.5 | 0.3 | 10 | 80 | 1.0 | 0.08 | 0.04 | 1.0 |
| Maize | 1570 | 12 | 10 | 71 | 4.5 | 2.0 | 12 | 220 | 2.5 | 0.35 | 0.13 | 2.0 |
| Wheat | 1450 | 13 | 12 | 70 | 2.0 | 2.0 | 40 | 350 | 3.5 | 0.40 | 0.10 | 5.0 |
| Barley | 1450 | 12 | 11 | 70 | 1.8 | 4.0 | 35 | 320 | 4.0 | 0.45 | 0.15 | 5.5 |
| Sorghum | 1500 | 12 | 12 | 70 | 3.4 | 2.0 | 32 | 300 | 4.5 | 0.40 | 0.12 | 3.5 |
| Pearl millet | 1570 | 10 | 11 | 72 | 5.0 | 2.0 | 30 | 320 | 4.0 | 0.33 | 0.15 | 2.1 |
| Finger millet | 1420 | 12 | 7 | 76 | 1.4 | 2.8 | 350 | 250 | 5.0 | 0.35 | 0.05 | 1.3 |
| Foxtail millet | 1490 | 11 | 10 | 72 | 3.0 | 2.9 | 30 | - | 4.0 | 0.60 | 0.10 | 0.8 |
| Proso millet | 1490 | 12 | 12 | 70 | 2.4 | 2.2 | 20 | 330 | 5.0 | 0.78 | 0.10 | 1.0 |

Wheat is the most important grain in the world export markets: 21% of the total production is exported, compared to 14% for maize, and only 3% for rice.

Cereals play a vital role in the nourishment of both humans and animals, not only because of the amount produced, but also because they are excellent for both storability and transportability. Unfortunately, the storage losses in the developing countries are often very high (up to 50%). Improved storage methods are one of the most important steps that could be taken to ensure an adequate supply of food for all the world's people (687, 1260). Cereals supply not only energy, but also protein (901, 1293, 1406), minerals and vitamins (Table 1). About 50% of human protein needs are provided by cereals - severe protein deficiency occurs only in those countries where starch plants other than cereals form the basic diet (see p. 38). From a quality viewpoint, cereal protein is often unbalanced; there is frequently a deficiency of the essential amino acids lysine and tryptophan. Of the mineral contents, the potassium and also the calcium contents are important. With the vitamins, thiamine is especially important, and some species of grain contain a useful amount of niacin. Vitamins A and C have been omitted from the table, since they are not found in grains, or only in slight amounts. Grains with a yellow endosperm (such as maize, sorghum, pearl millet and finger millet) have a certain value as a source of vitamin A, but they do not contain large amounts (about 200 IU/100 g); the yellow colour is caused predominantly by carotenoids without vitamin A properties (zeaxanthin and others).

As a foodstuff, cereals are eaten in a variety of forms, including pastes, noodles, different types of bread, etc. The food value is greatly influenced by the fineness of milling, and by other processing (e.g. see the figures for polished rice in Table 1). A large part of the production of barley and sorghum is used in making beer and pombe, and some other types of grain are likewise used to a small extent in the making of fermentation products.

Bran is a by-product from milling, which can be used unchanged for animal feed. From the bran, and from the germs which remain after starch extraction, oil can be extracted, especially from maize, but also from rice, sorghum and wheat (Table 15). The residues after oil extraction, and also the gluten which is recovered from the aqueous phase in starch production, are especially useful as animal food and for the culture of microorganisms. From the husks of rice and sorghum, wax can be extracted (see p. 394).

## Rice

fr. riz; sp. arroz; ger. Reis

The genus *Oryza* is the most important representative of the graminaceous subfamily Oryzoideae, characterized by uniflorous spikelets with hard lemma and palea which enclose the grain when ripe, and a hermaphrodite flower with 6 anthers (Fig. 3). The genus is found in the warm parts of all continents (1855). Of the cultivated species, *O. sativa* L. is found in Southeast Asia and is distinguished by its long ligule (8-20 mm), and *O. glaberrima* Steud. is found in West Africa and has a shorter ligule (3-6 mm). Both types are perennial, but are mostly grown for only one harvest, and both comprise many cultivars; even the well known deep-water form of *O. sativa*, the floating rice, is also found for *O. glaberrima*. (103, 305, 1273, 1298, 1305, 1314, 1835). Nowadays, the cultivation of *O. glaberrima* is restricted to the flood-plains of the Niger basin; everywhere else it is being replaced by the more productive cultivars of *O. sativa*.

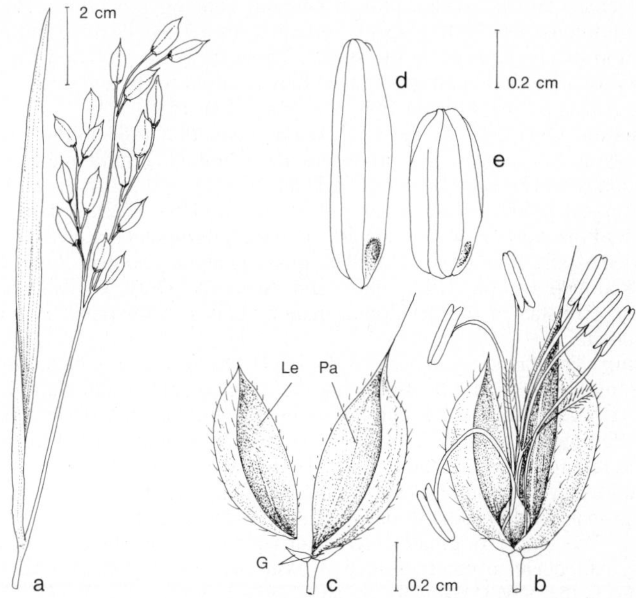


Fig. 3. *Oryza sativa*. (a) mature panicle, (b) flowering spikelet, (c) the bracts of the spikelet: G = glumes, Le = lemma, Pa = palea. Hulled grain of (d) Indica, and (e) Japonica types

*O. sativa* embraces many thousands of cultivars (1510), which have been developed, since rice is self-fertilizing and for thousands of years the harvest has been carried out by hand, cutting off each individual panicle, selecting the best for the next crop. The two most important cultivar groups are Indica and Japonica. Indica cultivars are predominantly tropical, with thin, long grains (length/width index more than 3, Fig. 3d) and most are short-day plants, with wide variability in growth height, panicle size, hairiness, grain pigmentation, and tillering capability. Cultivars in the Japonica group are generally found in the subtropics, have grains which are round to oval (length/width index of 1-3, Fig. 3e), are either unaffected by day-length, or react to it only mildly, and have modest growth heights and small panicles. Hybrids between Indica and Japonica strains are sometimes to a large extent sterile.

Apart from the different characteristics of the groups, there are differences between the cultivars, such as:

- adaptation of the rice to wet and dry growing conditions. The dryland (upland) cultivars mostly also grow well in water, but many wetland rices are unsuitable for dry cultivation,

- vegetation period (80-240 days),
- potential yield, and response to fertilizers,
- dormancy period of the seeds (0- several months),
- tendency to shed the seeds too early,
- quality characteristics (taste, protein content, cooking properties. The East Asian glutinous rices form a special group, with 3.7-5% dextrin, and a high proportion of amylopectin in the starch. These rices are preferred for many uses in East Asia, and in other countries they are used for puffed rice),
- disease resistance (61, 80, 103, 300, 653, 963, 1403, 1510).

**Production**. Over 90% of the total world production (483 million tons) is grown in Asia. The biggest rice producers are: China (172 million tons), India (102), Indonesia (42), Bangladesh (22), Thailand (21), and Japan (12). Outside Asia the largest producers are Brazil (11.8) and the USA (7.2). A fairly large proportion of the world production is grown in the subtropical regions.

Total rice exports amount to 12 million tons, and about a sixth of this is grown in industrial nations, the USA, Italy, and Australia. Only Thailand, China, Burma and Pakistan of the developing nations play a comparable part in the export markets.

**Breeding**. The breeding of rice (836, 1414) has received a great stimulus since the middle of the 1960s, following the development of the high-yielding varieties (HYVs) which were introduced by the International Rice Research Institute, Philippines (396, 832, 1151). They have short straw, and therefore are resistant to lodging even with high N fertilization; their leaves are short, stiff and erect, thus using the sunlight with optimal efficiency; they have a short to medium growing time (100-130 days); and they thrive in a variety of climatic conditions. The HYVs originally had an unsatisfactory resistance to diseases, and were inadequate in qualities such as taste and cooking characteristics; in these respects the newest cultivars are substantially improved (73, 1419, 1624). Meanwhile, since the 1960s, most of the rice producing countries have carried out their own breeding programmes using the original material developed in the Philippines, and good cultivars, well adapted to local conditions, are now available. The latest development is the breeding of hybrids. Hybrid rice was first grown in China, with a yield increase of 20-30 per cent. Today, hybrid rice breeding is carried out in many countries (464, 720, 1768, 1873).

**Ecophysiology**. Rice demands a high growing temperature; during its growth 30-32°C is optimal. The minimum temperature for germination is 18°C for tropical varieties and 10-12°C for subtropical types. Rice cannot endure frost at any stage of growth. Because of this need for warmth, rice is seldom grown higher than 1200 m above sea level, even in the tropics (44, 834, 1292).

While wetland rice can thrive on a very low $O_2$ content in the ground, it needs a rich supply of $O_2$ during germination, especially the Indica cultivars. If rice must be directly sown into water, then pregerminated seeds should be used, or, alternatively, pelletizing with $CaO_2$ improves germination under water.

Rice gives the best yields in regions with long hours of sunshine, such as in the Mediterranean area or in California. In the tropics, yields are reduced in cloudy localities or during the rainy season, compared to the yields in areas or seasons with a lot of sunshine. Pronouncedly short-day cultivars flower fastest when the days are extremely short (10 hours or less); even when there is little more than 12 hours of daylight, their flowering is delayed. For some cultivars, a thermophase precedes the photophase, and for these, the need for warmth must be satisfied before the reaction to day-length has any effect (1738). All cultivars have a juvenile phase of 4-6 weeks, in which they do not react much to the day length. Modern cultivars are mostly day-neutral, or have only a slight reaction to the photoperiod (495, 1861).

Agriculturally, the most important characteristic of rice is its ability to thrive in water, and this is shared by very few other cultivated plants (e.g. *Colocasia esculenta* - Table 12). The extensive flood-plains of the great Southeast Asian rivers could hardly be used for agriculture if it were not for rice. However, rice is not a water plant, and it grows very well in dry conditions. The water needs of dryland cultivars are also quite high. The rainfall during the growing period should not be less than 800 mm, evenly spread throughout this time. The optimum is 1250-1500 mm. Sprinkle irrigation can replace insufficient rainfall; with it, however, the high yields of wetland rice cannot be obtained (1224). The disadvantage of permanent flooding is the high loss of water by evaporation and seepage; climatic conditions and soil permeability determine the extent of water losses in wetland rice.

Rice makes few demands on the soil type (835). It grows in a pH range of 4.5-8, the optimum pH is 6-7. A heavy soil which can hold a lot of water is preferred by upland types. Wetland rice can be grown on all types of soil, as long as they are not too permeable. In spite of the low $O_2$ requirement, the $O_2$ level should not fall too low, otherwise the uptake of nutrients will be disturbed, leading to problems such as autumn decline and red wilting. These dangers are mostly caused by high water levels and high temperatures. The $O_2$ availability in the root zone can be improved by regulating the water level, and in some situations by temporary complete drainage of the water from the field (44, 844).

The nutrient removal by 1 ton of rice harvest is set out in Table 2. Apart from the elements mentioned in this table, $SiO_2$ is also important for the healthy growth of rice plants; from 800-1200 kg/ha are needed. In many rice-growing soils, there is a problem due to insufficient $SiO_2$ (595, 942).

Table 2. Mineral removal by 1 ton of rice as harvested (853)

| Element | In 1 t grain kg | In 1.7 t straw kg | Total kg |
|---|---|---|---|
| N | 22 | 6.6 | 28.6 |
| P | 2.3 | 0.8 | 3.1 |
| K | 3.4 | 15 | 18.4 |
| Ca | 0.4 | 2.3 | 2.7 |
| Mg | 0.1 | 2.1 | 2.2 |

**Cultivation**. An important part of the world's rice harvest is produced without irrigation (for example over 80% in Brazil, 67% in Sarawak, probably about 50% in Africa, and 20% in the Philippines) (505, 833, 857, 1819, 1820). Like all other cereals, upland rice can be sown broadcast (the seed rate is then about 110 kg/ha), or it can be drilled (seed rate about 80 kg/ha). The main problem is weed control, which demands continual hand weeding. This can be made easier by sowing in rows, and it can be practically eliminated through the use of herbicides (66). With sufficient fertilizer and good care, the harvest may be not much less than that for wetland rice (over 3 t/ha), but in practice, due to insufficient care, harvests of more than 1.5 t/ha are seldom achieved.

For wetland rice (506, 653, 1747, 838), the surface of the soil inside the basin should be exactly horizontal, so that the water level is the same everywhere. Before sowing or planting, the ground should be compacted, so that water and fertilizer losses through seepage are minimized. The top 10 cm should be homogeneous fine soil or mud. The classical method is to raise the young plants in seedbeds. The area of these should be 5-8% of the size of the eventual field, and 40-50 kg seed are needed per hectare of field. Planting out takes place 18-45 days after sowing, (dependent on the temperature), when the plants are 15-25 cm tall. With the Dapog method (1329), seeds are first germinated, then sown in a thick layer placed on an impermeable material (banana leaves, plastic sheet), and they can then be planted out 10-14 days later. A 20-25 $m^2$ seedbed is enough for 1 ha. The planting out is done mostly by hand, but it can also be done mechanically. Transplanting creates a lot of work, but it has the advantages that for the first few weeks little land and water are needed, the seedlings do not have to compete with weeds, and that the weak plants are eliminated. The planting distances are usually 10 cm in the row, and 10, 15 or 20 cm between rows, and usually 3 (but also up to 5) seedlings are planted at each position. Where workers are scarce and costly, wetland rice is sometimes directly sown; when it is sown broadcast, the use of herbicides is usually necessary. The fully mechanized growing of rice has been developed in the USA (1041, 1896) and Australia; it is now spreading to other countries.

As N fertilizer, ammonium sulphate or urea are used. Nitrates are unsuitable, as they are reduced in the soil, and losses occur in the form of $N_2$. For HYVs (see p. 12), for which up to 120 kg N/ha are given, half should be applied before the time of sowing or transplanting, and the remainder as a top-dressing at the time of panicle initiation; in this event the water is drained off before the application of the fertilizer. Further data on rice fertilization are given in (230, 412, 595, 829, 844, 1088, 1843, 2008). In low-input systems of agriculture, the supply of rice with N from blue-green algae or *Azolla* is of increasing importance (76, 1498, 1531, 1532).

After fertilizer usage, the most important factor in cultivation is the battle against weeds, for which nowadays a range of selective herbicides are available (66). However, most of this work is still done by hand, by hoeing, or using small, motorized machines.

The grain is ripe 30-40 days after flowering. To make the harvest easier, and to achieve an equal degree of ripening, the water is drained off 2-3 weeks before the harvest date. When the grain is ripe, the stalks and some of the leaves are still green. With enough humidity and warmth, the growth of the plants is restimulated after the panicles are cut off, and a second or even third harvest is possible (ratoon crops) (837, 1240, 1570, 1571). To achieve this, the rice straw is mown just above the surface of the ground, so that good new tillering takes place. With careful cultivation, the first ratoon ripens 10-20 days faster than newly planted rice, and gives a 20-30% better yield. The utilization of the water is about 100% improved in the ratoon.

Wetland rice is the only arable crop which has been grown in monoculture for thousands of years without harmful consequences. Rice can also be farmed in crop rotation with sugar cane, pasture, cotton, or legumes. The use of rice fields for fish farming offers the possibility of improving the protein intake of the local populations (342).

**Diseases and Pests**. The most important diseases are summarized in Table 3, and some of the insect pests in Table 4. Worldwide, the worst losses are caused by the fungi *Pyricularia* (830) and *Cochliobolus*, and by the stem-borers and bugs; locally, other organisms can be the main cause of damage. Other vermin, such as rats, birds and storage pests have to be taken equally seriously. But, of all the control measures for rice crops, the most directly valuable are the agronomic precautions that can be taken (the choice of healthy seed, disposal of all debris after the harvest, a balanced use of fertilizers), and these are measures which can be carried out easily and with little cost. With these improved practices alone, the crop failures which are so common in Asia could be avoided.

**Harvest**. Rice is harvested when the panicles have turned yellow, that is, before full ripening, so that the losses of grain in harvesting, and breakage of dry grains during threshing are minimized. With a manual harvest, the grain has a moisture content of about 20%, and for combine harvesting, between 25 and 30%. After harvesting by hand, the grains are usually threshed several days later, when they are dryer, or immediately after the harvest using one of the small Japanese threshing machines. Usually further drying is necessary, as grain should not contain more than 14% moisture when it is put into store. The yield (rice in husks, "paddy") is still under 2 t/ha in many countries and the global average is 3.2 t/ha, although with HYVs (see p. 12), 6-10 t/ha is attainable, even in the tropics.

**Processing and Use**. The best rice qualities are obtained by hulling and polishing of otherwise untreated, dried rice (white rice) (85, 334, 547, 1088). After the weight loss from husks (about 20%) and bran (8-10%), a further loss of 2.5-5% occurs through broken rice, even in the most favourable circumstances. In general, one can expect a yield of Head A rice of 65-67% of the original harvested weight. The amount of broken rice is reduced, and the total yield from milling improved by 5-10% if the rice, before shelling, is soaked, rapidly cooked, and then dried again (parboiled rice). This procedure was introduced into many countries during the Second World War (583, 938). The rice loses some of its flavour, but it has a higher vitamin content due to the diffusion of vitamins from the aleurone layer and the embryo into the centre of the grain. The grains of parboiled rice appear yellowish rather than white, and they are semi-translucent. They are cooked more quickly and with less water than white rice. Parboiled rice is sold more cheaply than good white rice.

From the remains left behind in the rice mill, oil (Table 15) (368), wax (see p. 394), furfurol and protein-rich animal fodder can be made. The husks can be used for building panels, as a polishing material, or used for burning. The ashes, which are rich in silicates, find uses in fireproof bricks, cement, and as an additive to rubber (156, 747).

Rice straw has greater usefulness than that of other grains. In rice growing countries, it is usually the basic fodder in animal rearing (852). During and after the rice harvest, the stems and leaves are not dead, so they have a correspondingly high fodder value (see p. 401). In Southeast Asia, large volumes of rice straw are used in the cultivation of the padi straw mushroom (*Volvariella volvacea*, Table 25). Because of its fineness and durability, rice straw is locally used in basket-making, for bags, mats and hats. In some countries, there are

Table 3. The most frequent parasitic diseases of rice (830, 963, 982, 1062, 1326, 1674)

| Name and agent | Carrier | Main symptoms | Distribution | Control |
|---|---|---|---|---|
| **Viruses** | | | | |
| Stripe virus | Delphacidae (e.g. *Laodelphax striatellus* Fallen) | chlorotic streaks, stunted growth, new leaves twisted and drooping | Japan, Korea | sanitation, breeding for resistance |
| White leaf Hoja blanca | Delphacidae (*Sogatodes* spp. and others) | leaves with white streaks, stunted growth, normal leaf forms | N and S America | sanitation, breeding for resistance |
| Tungro | Cicadellidae (e.g. *Nephotettix* spp.) | stunted growth, little tillering, chlorosis | SE Asia, India | sanitation, breeding for resistance |
| **Mycoplasms** | | | | |
| Yellow dwarf | Cicadellidae (*Nephotettix* spp.) | severe stunting, over-tillering, chlorosis | Asia | sanitation, breeding for resistance |
| **Bacteria** | | | | |
| *Xanthomonas oryzae* Uyeda et Ishi. Bacterial leaf blight | wind, water | grey, dead spots on points and edges of leaves, withering and dying back | Japan, SE Asia | water regulation, sanitation, breeding for resistance |

| | | | | |
|---|---|---|---|---|
| **Mycoses** | | | | |
| *Pyricularia oryzae* Cav. Rice blast | spores, wind | spots on leaves, browning and breaking off at stalk nodes and panicle base | all rice areas | correct use of fertilizers, sanitation, seed dressing, spraying with fungicides, breeding for resistance |
| *Cochliobus miyabeanus* Drechsler Brown spot | spores, wind | numerous small brown spots on leaves, later with grey/white centres | all rice areas | correct use of fertilizers, crop rotation; chemical and hot-water treatment, breeding for resistance |
| *Leptosphaeria salvinii* Catt. Stem rot | sclerotia, cultivation, irrigation, | spots on the leaf sheaths, broken stalks | all wetland rice growing areas | early drainage, sanitation, breeding for resistance |

Table 4. The main groups of the insect pests of rice (654, 831, 963, 982, 1342, 1876)

| Group | Representatives | Symptoms | Distribution | Control |
|---|---|---|---|---|
| Stem borers | *Chilo* spp., *Sesamia* spp., *Tryporyza* spp., | dying of the youngest leaves (dead heart), chlorosis of panicles (white heads), holes eaten into stems and leaf sheaths | all areas, particularly Asia | sanitation, crop rotation, fallow period; insecticides |
| Rice bugs | *Leptocorisa* spp. | puncture sites on glumes, shrivelled spikelets and grains | especially in S, SE, and E Asia and Australia | sanitation, crop rotation, seeding time, contact insecticides |
| Leafhoppers Planthoppers | *Sogatodes* spp., *Nephotettix* spp. | chlorosis, brown staining (hopper burn), viruses (see Table 3) | especially Asia and America | light traps; sanitation, fallow period or crop rotation, contact insecticides; systemic insecticides |

cellulose and papermaking factories which use rice straw as a raw material (747).

## Maize

fr. maïs; sp. maíz; ger. Mais

Maize, *Zea mays* L., originated in Central and South America, and in early times spread to other parts of the world, probably even before the time of Columbus (242, 782, 880, 889). It is the most extreme instance of a plant which has been developed by man, adapted to his needs, and which can only survive through man's work (796, 1117, 2026). It is the cereal with the highest yield potential, and therefore it is invaluable for the nourishment of both men and animals. In its classification, it belongs to the small tribe Maydeae of the Gramineae, subfamily Andropogonoideae.

In its area of origin, there are thousands of primitive types of maize; in East Asia, wax-maize has been developed (starch only in the form of amylopectin). The modern, highly-selected cultivars can be put into the following groups:
- flint corn, the most important for human consumption,
- dent corn, which has the highest yields,
- floury corn (soft corn), especially suitable for starch production.

There are also groups for more specialized uses, such as popcorn and sweet corn (which contains 30% glycogen in the kernels), however the yields of these are substantially lower than those for the major groups.

**Production**. Although maize has been grown for hundreds of years outside the New World, the largest production is still in the Americas (185 million tons, of which 125 million are grown in the USA alone). In Asia 107 million tons are grown, in Europe 66 million, and in Africa 30 million.

**Breeding**. The breeding of maize (687, 918, 1416, 1558, 1717, 1886) has achieved the greatest increases of yield from the introduction of hybrid cultivars, which have been grown on a large scale since about 1940. They can also greatly increase the yields in the less developed countries, but the hybrids are more difficult to introduce there, as the farmers cannot breed their own seed, so that each year new seed-corn must be bought. In such circumstances, synthetic cultivars are to be preferred.

In the forefront of development in the last 20 years has been the improvement of protein content (there are now forms with more than 20% protein in the grain, and forms with several rows of cells in the aleurone layer), and the improvement of protein composition (higher lysine and tryptophan contents, genes *opaque*-2 and *floury*-2). The amylose-rich forms are of interest for human consumption and for starch extraction (starch contents up to 80% as amylose, compared to only 21-24% in normal maize, with the remainder as amylopectin). Recently, cultivars are being bred which are earlier ripening, smaller growing, and fairly drought tolerant, so that maize can be grown in less favourable climatic zones.

**Ecophysiology**. For good growth, maize needs a lot of sunshine and warmth. Average temperatures of 20-24°C are ideal, and at night the temperature should not fall below 14°C. Temperatures over 26°C accelerate the growth too much, so that yields fall, and with temperatures over 30°C, the susceptibility to diseases increases. Most cultivars have little sensitivity to photoperiod, however short days accelerate their development; at high temperatures, the influence of day-length is slight (495, 777, 1292).

Maize is not drought tolerant; dry periods during the flower formation and fertilization can lead to a total loss of the harvest. The vegetation period of the high-yielding varieties is long, 140 days or more. At least 500 mm of rain should fall during this time. Dryness at the time of harvest is hoped for - it makes the harvesting easier, and reduces the danger of spoiling the crop by fungi (*Gibberella* spp., *Diplodia zeae* (Schw.) Lev.).

Maize grows on many different types of soil, but is best on sandy loam, in the pH range 5.5-8. It fails in acid soil. In regions with high rainfall, the soil should be light and permeable, as waterlogging leads to yellowing (chlorosis), and a reduction of the yield. With good climatic conditions, maize develops very quickly, and needs a good supply of nutrients during the months until flowering. To produce 1 ton of corn, the plants take from the soil 24 kg N, 4 kg P, and 23 kg K (93, 921, 1197).

**Cultivation** (175, 930, 1041, 1717). In modern operations, maize is always dibbled in rows. The row-spacing depends most of all on the rainfall, and then on the cultivar and the fertility of the soil. In dry areas, the normal spacing is 90 cm between rows, and 20-30 cm within the row, which gives a seed usage of 20-30 kg/ha. In extreme cases, the row spacing is increased to 2 m in regions with very low rainfall, so that the plants have a greater volume of soil in which their roots can find water, and the seeds are sown deeply (8-10 cm). Sowing in furrows is another aid to the development of a deep-reaching root system, and it helps to improve drought tolerance. These furrows are filled by later working, such as hoeing the weeds. With a wider spacing between rows, it is advisable to place mineral fertilizers (especially phosphates) in bands at the side of and several centimetres below the seeds (1191).

Where herbicides are not used, the most important task for ensuring a good yield is early weeding, thus eliminating the competitors for water and nutrients.

In many areas, mixed planting is carried out, i.e. maize with beans, pumpkins or watermelons. This is a very successful technique when there is correct care and sufficient rainfall, as it minimizes the danger of soil erosion (which can be very great for maize, as for all other row crops), prevents the compaction of the soil surface by the impact of raindrops, and shelters the soil from overheating from direct sunlight; moreover, it can also diminish the effect of infections by diseases and pests. In many countries, there have been very good experiences with the use of minimum tillage systems for maize farming (1397).

Maize yields show an extraordinary variation. The global average of 3.6 t/ha is alarmingly low. Normal yields for well-cared-for maize, grown without irrigation in a warm country, are 5-8 t/ha. Where a plentiful supply of water is available, and the top priority is the production of basic foods, maize is very successfully grown with irrigation. If sufficient fertilizer is used, and the usual plant hygiene measures are taken, then harvests of 20 t/ha and more are attainable.

**Diseases and Pests**. The most important diseases of maize are the leaf blights *Helminthosporium maydis* Nis. et Miy., and *H. turcicum* Pass., downy mildew (*Sclerospora philippinensis* Weston), corn smut (*Ustilago maydis* (DC.) Cda.), ear rots (*Diplodia*, *Gibberella*, etc.), and streak virus (carried by leafhoppers, also on wheat). The most important insect pests are stem-borers (*Busseola fusca* Hmpsn., and other types) and the European corn borer (*Ostrinia nubilalis* Hbr.);

both of these can be easily controlled by contact insecticides. Genetic resistance can be an answer to the diseases, otherwise the damage can be minimized by choice of location, planting time, plant spacing, crop rotation, etc. (48, 982, 1717).

**Harvesting and Use**. When the ripening period coincides with the end of the rainy season, the harvest can be postponed for weeks or even a month, as the maize grains will not be shed, and they are covered by the husk leaves which protect them from infestation by insects. Maize is picked by hand in most of the warm countries, but in large-scale agriculture, it is usually harvested mechanically using special machines which also shell the cobs at the same time. In Central America and in some African countries, maize is the most important foodstuff of the population, but in the USA and some other countries it is mostly used for animal feed. A large part of the harvest is industrially processed into starch, of which a considerable proportion is used for high fructose corn syrup, and the by-products from this, fodder protein and vegetable oil (Table 15), find a good market (814). Worldwide, cooked or roasted unripe corn cobs from hard maize are an enjoyable and nourishing vegetable (see p. 126). Silaged green-maize is an especially valuable animal food (see p. 398).

## Wheat

fr. blé; sp. trigo; ger. Weizen

Common wheat, *Triticum aestivum* L. (*T. vulgare* Vill.), is by far the most important of the cultivated species of wheat. Durum wheat, *T. turgidum* (L.) Thell. var. *durum* (Desf.) Mac Key, also plays a major role because of its high protein content, and it is grown particularly in the Mediterranean countries, where wheat flour is often processed into farinaceous pastes such as macaroni. The other species of wheat have only local importance, although they are now becoming useful as carriers of resistance genes. In the tropical and subtropical mountain regions, the following can still be found in cultivation:

*T. monococcum* L. - einkorn;
*T. turgidum* var. *carthlicum* Nevski - Persian wheat,
var. *dicoccon* (Schrank) Thell., emmer,
var. *polonicum* (L.) Mac Key, - Polish wheat;
*T. aestivum* var. *compactum* (Host) Mac Key, - club wheat
var. *sphaerococcum* (Perciv.) Mac Key, - shot wheat (498, 1094, 1389, 1657, 2026).

Because of their different gluten qualities, the cultivars of *T. aestivum* are divided into several groups: hard wheat, semi-hard wheat, and soft wheat (1041). In the tropics and subtropics, nearly all the wheat grown is soft wheat, with some semi-hard wheat also. An exception is Pakistan, where a large volume of white hard-wheat cultivars are grown, as far as they have not been replaced by HYVs (see p. 12) in the years since 1964/5.

**Production**. The world's total wheat production (510 million tons) exceeds that of any other grain (Fig. 2). The largest producers are (in million tons): China 88, USSR 85, USA 49, India 45, France 30, Turkey 21, and Canada 16. A total of 40 million tons are grown in the subtropical parts of Europe (the Mediterranean and Balkan areas).

Wheat accounts for 120 million tons of exports, 50% of the total grain exports worldwide. The major exporters are (in million tons): USA 42, Canada 20, France 17, and Australia 12. The relatively large amount of wheat in storage worldwide makes it the most important foodstuff for famine relief operations.

**Breeding**. The new high-yielding varieties of wheat (HYVs) have been bred by crossing short-straw Japanese soft wheat cultivars with American, European and Australian cultivars (1558). This work has been done in the USA, and especially in Mexico, at the Centro Internacional de Mejoramiento de Maís y Trigo (CIMMYT). Apart from their short, firm stalks, they are noted for their strong awn formation, and their erect leaves. Further breeding programmes involve crossing these with native cultivars to try and improve disease resistance and grain quality (light colour, taste, suitability for local cooking customs). In general, the top priority in wheat-breeding is the development of resistance against the new races of rust fungi which are always appearing. To secure a wide genetic basis for its resistance, multiline varieties are preferred in many of the warmer countries. In many places, the breeders are also giving their attention to the preservation and improvement of the local strains, which possess qualities of disease resistance, are less demanding, are adapted to higher temperatures, are resistant to dryness and salinity, have high protein content and quality, good grain colour and taste. The breeding of protein-rich cultivars is especially important for the developing countries (genetic starting material 'Atlas 66') (295, 396, 422, 782).

**Ecophysiology**. Wheat is grown in the tropics and subtropics in three different climatic zones:

- in regions with winter rain, grown during the cool period of the year (sown in autumn or early winter, harvest in early summer),
- in subtropical areas with summer rain, grown in the dry, cool part of the year (mostly with irrigation, or with particularly favourable conditions also making use of the water stored in the soil during the rainy season),
- in high regions of the tropics, above 1800 m above sea level.

In all three climatic zones, it is cool at least up to the time of flowering, but not so cold as to provide the cold conditions needed by winter wheat. The decrease in temperature with height means that for every 100 m increase in elevation, the vegetation time is increased by about 5 days (7). At greater heights, frost during the time of flower differentiation, or at flowering, can cause empty ears. An average temperature of 18.5°C is ideal; higher temperatures overstimulate the development and lead to reduced yields. At temperatures over 27°C, the likelihood of diseases is increased, and, with some cultivars, the pollen becomes sterile (14, 17, 63, 495, 975).

The water requirement is dependent on temperature and cultivar. Cultivars with a modest need for water can manage with 250-300 mm rain with cool weather, high-yielding varieties generally need 400-900 mm during the vegetation period. HYVs are especially well adapted for cultivation with irrigation. They must be sown at a shallow depth, and therefore need their first supply of water soon after germination. Their yield potential will be fully exploited only if they are given a sufficient supply of water till shortly before the harvest. Most HYVs require two waterings more than other cultivars.

Wheat prefers neutral to slightly alkaline soils which are well provided with easily available nutrients. The removal of nutrients needed for the production of

1 ton of wheat is: 20-30 kg N, 6-8 kg P, 20-28 kg K (these are the total mineral contents in the grain plus the straw) (283, 953).

**Cultivation**. Wheat is still sown broadcast in many countries, but it is more often drilled, with a row distance of 18 cm. For the cultivation of 1 ha, 80-110 kg of seed are required. When wheat is irrigated, the border strip or corrugations systems are usually used, but it also thrives well with sprinkling. Weeds are not a great problem with good soil preparation, except when the temperature is high, and then the cultivation of wheat is only possible with the use of herbicides (2,4-D, MCPA, etc.). The vegetation time depends very much on the temperature (height above sea-level, time of year); most HYVs are relatively early ripening, and need 120-150 days from sowing to harvest.

**Diseases**. The rust diseases,

*Puccinia graminis* Pers., black stem rust,

*P. striiformis* West., stripe rust,

*P. recondita* Rob. ex Desm. (*P. triticina* Eriks.), leaf rust,

are the greatest problems in wheat growing in warm countries, above all because new strains of the fungi are always developing, against which the existing cultivars of wheat have no resistance. The most important counter-measures are the growing of different cultivars in an area, or the use of multiline varieties (see p. 22) (812). Also leaf-spot disease (*Septoria* spp.) and foot-rot diseases (mostly *Fusarium* spp.) can cause damage to the harvest, and suitable precautions must be taken (crop rotation, seed disinfection, care in sowing time, choice of cultivars) (982, 1433).

**Harvesting and Use**. The full yield potential is in fact about 10 t/ha, but, for high-yielding varieties with irrigation, only 3-6 t/ha are harvested in practice. Low-yielding varieties grown in dry conditions produce an average of only 0.6-1 t/ha. The world average is about 2.3 t/ha.

Wheat is used primarily for human nourishment, and also as animal fodder (1389, 1417). In countries with long, cool winters, the wheat fields are used for grazing (before ear emergence); the grain yield is only reduced a little, and there is the benefit of valuable feed for the animals.

## Barley

fr. orge; sp. cebada; ger. Gerste

Barley, *Hordeum vulgare* L., belongs to the subtribe Triticinae of the Poaceae (tribe Triticeae, subfamily Pooideae), as do wheat and rye, and like them, it originated in the Middle East (233, 1480, 1657). In the tropics and subtropics, it is grown mostly for human consumption (376, 809), not mainly for brewing and animal fodder as in the temperate zone. The total production in the warm regions amounts to around 25 million tons. Its particular usefulness is that it can be grown under the most extreme conditions (at the edge of the desert in areas with winter rain, high places in the tropics, up to 4700 m high in the Himalayas), even though the yields can be very small (in Libya, the average yield is 0.7 t/ha). Its cultivation in such places is made possible by its very short vegetation time (55 days in dry places), its tolerance of both cold and heat, and its resistance to salinity (some varieties can stand up to 1% salt in the soil) (92, 482, 483, 1993).

Its quick ripening in dry and warm weather makes it possible to have a small yield even with rainfall of only 150-200 mm. From a nutritional viewpoint, barley is valuable because of its high protein content (over 15%), the good digestibility (83%) and the relatively high biological value of its protein (70%). Modern breeding has developed types of barley with improved amino-acid composition (4.1% of the protein as lysine) (782, 1467). With the widening popularity of European beers, there is now interest in the warm countries in the cultivation of low-protein barley for brewing (in Australia, etc.). For brewing, it is out of the question to grow barley in marginal areas - it must be grown in fertile fields with good rainfall to get a good formation of grain.

Barley is similar to wheat with regard to cultivation techniques (see p. 23), and 50-70 kg/ha seed is needed.

As with wheat, barley suffers from a number of types of rust, and damage is often caused by mildew (*Erysiphe graminis* DC), covered smut (*Ustilago hordei* (Pers.) Lagerh. et al.), and foot rot (especially *Helminthosporium sativum* P.K. et B.) (1041, 1755).

Besides its uses for brewing, for flour, and for animal fodder, barley is also processed into pearl barley (the manufacture of this is similar to that for hulling and polishing rice to make white rice); germinated barley is used to make malt for beer-brewing, but also for making malt syrup or malt sugar (1041). When it is cut green, barley is a valuable animal fodder, and can also be made into hay (Australia, California).

## Sorghum

fr. sorgho; sp. sorgo; ger. Sorghum

Sorghum includes a profusion of different forms (434). Its growth height can vary between 0.6 and 7 m, the stems can still be juicy when the grain is ripe (kafir corn), or they can be totally dry and woody (kaoliang). The panicles can be loose, with drooping branches, or they can form compact, round heads. As with maize, there is the normal starch composition (21-28% amylose), but in East Asia there is also wax-sorghum (entirely amylopectin), and cultivars in India with sweet grains, which contain 30% glycogen, as with sweet corn. Then there are the types grown purely for animal fodder (Sudangrass, etc.), and the broom sorghum.

The classification into botanical species (376, 1700) has not been found useful in plant breeding (700, 1890). All forms can be crossed with each other, and "species" differences sometimes depend on only a single gene. Therefore in the agricultural literature, the general name *Sorghum bicolor* (L.) Moench is used in the wide sense to apply to the whole complex of forms, and then a subdivision into subspecies and groups is carried out within this species (1625, 1920). In the latest classification (700), the many forms are divided into five groups, classified on the basis of the morphology of the ripe spikelet; while this may help the non-specialist, it can not replace a full classification. The distinguishing features of the most important groups for agriculture and for breeding are compiled in Table 5.

In sorghum, the spikelets are arranged in pairs, as is usual for all members of the Graminaceae subfamily Andropogonoidae (Fig. 4). The outer glumes of the fertile sessile spikelet persist, and more or less enclose the ripe grain. Fig. 4. illustrates two stages of the development of a type from the kafir group.

**Production**. Sorghum is an important food for humans and animals in tropical Africa, in India and China. The main producers are the USA (17 million tons, mainly for fodder), India (11), Mexico (5.5), Nigeria (4.9), Sudan (4.6), and Argentina (3.2). The total production in the African continent is 15.3, and in Asia 18.4 million tons. Sorghum plays no significant part in the export markets.

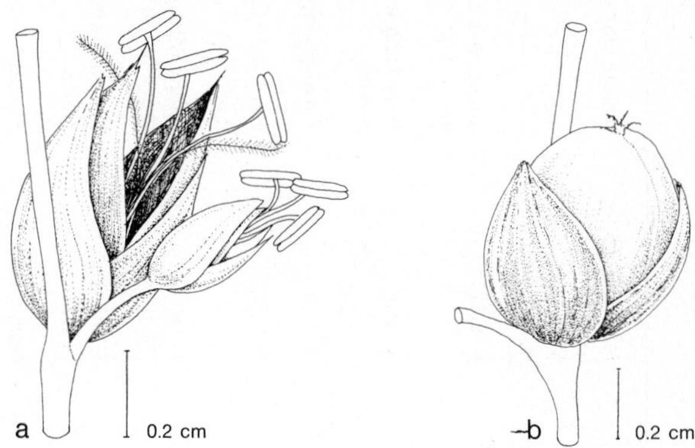


Fig. 4. *Sorghum bicolor*. (a) pair of spikelets - lower spikelet pedicellate, male; upper spikelet sessile, hermaphrodite. (b) ripe fruit with persistent glumes

**Breeding**. Sorghum has been established in the USA as a fodder material, and has been developed into a highly productive type of grain (1540). The new hybrid cultivars are high-yielding, large-grained, low-growing, and ripen all at the same time, so they can be harvested with a combine harvester (1890). In the past decades they have been successfully grown in Africa and Asia, and have already been responsible for a considerable increase in sorghum production there. In individual countries, the breeders are seeking to combine the high yield potential of the American cultivars with the resistance qualities of their local cultivars (resistance to bird damage, *Striga* (1558, 2005), and parasitic diseases and insects (757, 824)). The breeders are also giving attention to quality factors (protein content, amino acids, taste, etc.) (770, 815), and to adaptation to extreme conditions ,e.g. low-growing cultivars which ripen in 80-90 days for poor, dry soils (1475). They are also developing varieties based on the Guinea group, which ripen in 120-240 days in extremely humid conditions, and which can survive several weeks of flooding, and varieties which have open panicles, which dry up quickly after rainfall.

**Ecophysiology**. Sorghum is a typical representative of the plants of the tropical summer-rainfall areas; it needs high temperatures for good growth (optimum temperature 27-28°C), and possesses better heat resistance than any other grain.

There are great differences in water requirements among cultivars. For the highest yields, 500-600 mm rain are necessary. Sorghum withstands wet soil better than maize. On the other hand, many cultivars are extremely drought

Table 5. The most important types of *Sorghum bicolor* (L.) Moench s.l.

| Groups after HARLAN & DE WET (700) | Common names | Species after SNOWDEN (1760) | Origin | Remarks |
|---|---|---|---|---|
| Bicolor | Sudangrass | *S. sudanense* (Piper) Stapf | E Sudan | Forage grass |
| | Broomcorn | *S. dochna* (Forsk.) Snowden var. *technicum* (Koern.) Snowden | Mediterranean region | |
| | Kaoliang | *S. nervosum* Bess. ex Schult. | E Asia | Small, dark-brown grains; tall, dry, woody stem |
| | Sorgo, sweet sorghum | *S. saccharatum* Nees | Africa | Juicy stem with high sugar content |
| | | *S. bicolor* (L.) Moench s.a. | Africa | Small-grained, primitive forms |
| Guinea | Guinea corn | *S. guineense* Stapf. em. Snowden | W Africa | Hard grain, tall, for humid regions |
| Caudatum | Feterita | *S. caudatum* (Hack.) Stapf | E Sudan | Large, white, flattened grains; important for cross breeding |
| Kafir | Kafir corn | *S. caffrorum* (Retz.) Beauv. | E and S Africa | Round, white, brown or red grains |
| | Hegari | *S. caffrorum* (Retz.) Beauv. | Sudan | Round, white grains |

| | | | | |
|---|---|---|---|---|
| Durra | Durra | *S. durra* (Forsk.) Stapf | NE Africa, W Asia, India | Flattened, yellow grains; compact, often drooping panicle; stem covered with wax |
| | White durra | *S. cernuum* (Ard.) Host | W Asia | Large, round, white grains |
| Bicolor x Caudatum | Kaoliang | *S. nervosum* Bess. ex Schult | E Asia | As kaoliang above |
| Bicolor x Kafir | Kaoliang | *S. nervosum* Bess. ex Schult. | E Asia | As kaoliang above |
| Guinea x Kafir | Shallu | *S. roxburghii* Stapf. | India | Flattened, white grains; loose panicle |
| Bicolor x Durra | Milo | *S. subglabrescens* (Steud.) Schweinf. ex Aschers. | E and NE Africa, India | Large, flattened, yellow or white grains; important for cross breeding |

tolerant, and can be cultivated with substantially less rainfall than maize (824, 886, 1292, 1748, 1749). The drought tolerance is based on a low transpiration rate, on the dense and deep-reaching root system, and on its ability to stop all growth and metabolism in times of severe dryness (dormancy, abiosis). Early ripening cultivars have the least need for moisture, and can be grown with only 200-300 mm rain. Sorghum is a short-day plant, but most of the new cultivars will also flower in long-day periods, although short days accelerate their development. The demands on the soil are slight, as sorghum is unusually efficient at absorbing mineral nutrients from the soil. It grows in a pH range of 5.0-8.5, withstands salt and alkali well, and it also grows on badly drained soils. For high yields, the soil should be deep and not too light, and contain enough nutrients. The nutrient removal for 1 ton of grain is about 50 kg N, 9 kg P, and 45 kg K, calculated from the above-ground plant mass (921) .

**Cultivation**. In modern agriculture, sorghum is drilled, with a seeding depth of 2.5-5 cm, a row-spacing of 25-110 cm, and a distance within the rows of 6-25 cm (1474). These dimensions depend above all on the rainfall and cultivar. The smaller distances are only applicable to land which is irrigated, and to cultivars which tiller weakly. For 1 ha, 2-15 kg seed are necessary. Because the development of the plants in the first weeks is slow, the soil should be well prepared, to eliminate competition from weeds as much as possible. In mechanized cultivation, soil herbicides are employed (atrazin, propazin, etc.). Sorghum is mostly grown as an annual, but it is originally a perennial plant. Some cultivars, including some of the high-yielding hybrids, respond very well to being harvested more than once (ratoon cropping) (448). Because of its high absorptive power for nutrients, the field must be heavily fertilized for each succeeding crop, and for each ratoon crop. The danger of soil erosion is less with sorghum than with maize.

**Diseases and Pests**. Sorghum suffers from many parasites. Fungus diseases are especially found in humid areas of cultivation (1961). The most important are the kernel smut (*Sphacelotheca sorghi* (Link.) Clint et al., controlled by seed disinfection) and downy mildew (*Sclerospora sorghi* West. et Up., *S. graminicola* (Sacc.) Schroet.) (376, 1782). The greatest damage is caused by insects. The sorghum midge *Contarinia sorghicola* Coq. appears in all cultivation areas; it can almost completely destroy a harvest, and is controlled by the use of insecticides at the beginning of flowering. In the first 4-6 weeks the seedlings are very vulnerable to the shoot fly, *Atherigona varia* var. *soccata* Rond., however control is possible with systemic and contact insecticides, and the breeding of resistant varieties looks promising. The insects which are well known for their damage to other grain crops (stem-borers, bugs, aphids, grasshoppers, etc.) also affect sorghum, and can cause considerable damage (376, 982, 1687). In tropical Africa, the red-billed weaver (*Quelea quelea*) is one of the most serious pests to deal with.

**Harvest and Yields**. The harvesting and threshing of sorghum presents no problems worth mentioning. The yield depends on the growing conditions: with good fertilizer and water quantities, similar yields to those of maize can be obtained (20 t/ha and more), but in the driest areas, the yield is usually less than 1 t/ha, and the global average is about 1.3 t/ha. Remarkably high yields from ratoon crops have been reported from India and Taiwan, where, with three harvests a year, 20 tons per year per hectare are harvested (376).

**Processing.** In Africa and Asia, mash and flat-breads are made out of sorghum flour and groats. An important part of the harvest in Africa is used solely for the brewing of beer. By-products from the manufacture of starch are fodder protein, oil (Table 15), and wax (see p. 394). In some countries, dyes are extracted from the glumes and used for the colouring of foodstuffs and textiles (see p. 386).

The stems and leaves of some varieties are still partly green and juicy at the time of harvest, and make a valuable animal fodder. The dried stems and roots are an important fuel in some parts of India and China, the strong stems of kaoliang are used in house-building, fibres from the stems are used in basket-making, and the threshed panicles are used as brooms (not only those of broom sorghum).

## Further Cereals

Rye (*Secale cereale* L.) and oats (*Avena sativa* L.), both of which are widespread in the temperate regions, are seldom grown as cereals in the warmer countries, but limited amounts are grown for animal fodder in the subtropics and in mountainous areas of the tropics. In contrast, the mostly fine-grained species of millet (Table 6) play a major role in Africa and Asia (152, 1041), and the annual production can be estimated at 30 million tons. In some areas, they are more important than rice or sorghum. Their cultivation as a cereal has remained limited to the Old World from which they originated.

Most millets (770, 784, 1455) belong to the graminaceous sub-family Panicoideae, in which the lemma and palea tightly enclose the ripe grain. Canary grass belongs to the Phalarideae tribe of the Pooideae, tef and finger millet to the Eragrostoideae, and Job's tears to the Maydeae tribe of the Andropogonoideae. The structures of spikelets and their arrangement are shown in Figs. 5 and 6, illustrating some of the more important representatives of the Panicoideae. The unusual inflorescence of Job's tears is shown in Fig. 7.

All species of millet are heat-loving. Some, such as proso, barnyard and little millets, can ripen in two months, thus yielding a harvest even with very little rainfall (180-250 mm), though the yield may then be very small. Some grow well on very poor soils, which could hardly be used for any other field crop (little, barnyard and ditch millets, and fonio). On the other hand, proso, pearl and foxtail millets can produce harvests of 3-5 tons of grain per hectare, under favourable conditions. For these three types, breeding over the last 30 years has brought about considerable improvements in yield and quality (538, 1456). The prospects are especially good regarding further improvements in the breeding of pearl millet, a species which is cultivated in many of its types from Senegal to the Transkei in Africa, and also in India (52). Splitting up pearl millet into many botanical species (1733) is of little value to the breeder, as there is so much variability and cross-fertilization between the groups. As with *Sorghum bicolor*, a collective name for all the types is usually used in the agricultural literature. It had been agreed that the valid name will be *Pennisetum americanum* (L.) Leeke, but recently, DE WET (1921) has emphasized that *P. glaucum* (L.) R. Br. is the correct name. Apart from its use as a cereal, pearl millet is one of the best yielding and most adaptable fodder plants in the tropics and subtropics (see p. 398).

Table 6. Millet species

| Botanical name | Vernacular names | Distribution | Remarks and literature references |
|---|---|---|---|
| *Brachiaria deflexa* (Schumach.) C.E. Hubb. ex Robyns | signalgrass, kolo rassé | Burkina Faso, Mali | Only cultivated locally (152, 262, 783) |
| *Coix lacryma-jobi* L. | Job's tears, adlay, larmes de Job, lágrima de San Pedro, capim rosarie | SE Asia, Tropical Africa | (Fig. 7) Perennial. Thin-shelled forms used as a cereal, thick-shelled for 'pearls'. Green fodder (3, 94, 152, 262, 617, 782, 1051) |
| *Digitaria cruciata* (Nees) A. Camus | raishan | Khasi hills, India | Only cultivated locally, also as a fodder plant var. *esculenta* Bor. (152, 376, 581, 1668) |
| *D. exilis* (Kipp.) Stapf | hungry rice, fonio, fundi, acha | W Africa | Annual production estimated at about 100,000 tons (152, 262, 587, 783, 815, 843) |
| *D. iburua* Stapf | black fonio, iboru, iburu | Nigeria, Niger, Togo, Benin | Only cultivated locally (152, 262, 783) |
| *Echinochloa frumentacea* (Roxb.) Link | barnyard millet, billion dollar grass, sawa, sanwa | India, SE Asia | Quick ripening, salt tolerant. Also as a fodder plant (152, 376, 427, 1462) |
| *E. utilis* Ohwi et Yabuno | Japanese millet, Japanese barnyard millet | Japan, China | As for *E. frumentacea*, but suitable for cooler climates (152, 1225, 2004, 2026) |

| *Eleusine coracan* (L.) Gaertn. | finger millet, dagusa (Ethiopia), ragi (India), kurakkan | Central and E Africa, India to Japan | Eaten as gruel and flat-breads. Used for beer production. Keeps well in storage (19, 152, 376, 738, 766, 1455, 1457, 1650, 1799, 1924) |
|---|---|---|---|
| *Eragrostis tef* (Zuccagni) Trotter (*E. abyssinica* (Jacq.) Link) | tef (Amhari), tafi (Galinya), paturin d'Abessinie | Ethiopia | Used for beer-brewing and flat-breads, but outside Ethiopia only used as a quick growing fodder plant (see p. 398) (152, 374, 766) |
| *Panicum miliaceum* L. | common millet, true millet, mijo común, proso | Mediterranean countries to E Asia, USSR | Used for beer-brewing and flat-breads. Exported as bird-seed (19, 152, 376, 581, 1041, 1455, 1871) |
| *P. sumatrense* Roth ex Roem. et Schult. (*P. miliare* Lam.) | little millet, samo, samai, kutki (India) | India, Sri Lanka | Very low-demanding, drought tolerant, also tolerant of wetness (19, 152, 376) |
| *Paspalum scrobiculatum* L. | bastard millet, ditch millet, koda (India) | India, China, Japan | Low-demanding, drought tolerant, also cultivated as a fodder plant (19, 152, 376, 1925) |
| *Pennisetum americanum* (L.) Leeke (synonyms, see p. 29) | pearl millet, bulrush millet, millet à chandelle, mijo perla, dochan, kala sat, bajra (India) | Africa, India | (Fig. 5) Salt tolerant, drought tolerant. Exported as bird-seed. In the Sahel, there are cultivars with a vegetation time of 60 days (52, 246, 376, 531, 815, 875, 1292, 1345, 1456, 1476, 1862, 1921, 1961) |
| *Phalaris canariensis* L. | canary grass, alpiste | Mediterranean countries (Morocco), Argentina | Bird-seed, also green fodder (152, 1871) |

Table 6. Millet species cont'd

| Botanical name | Vernacular names | Distribution | Remarks and literature references |
|---|---|---|---|
| *Setaria italica* (L.) Beauv. | Italian millet, foxtail millet, millet des oiseaux, moha, painço | Mediterranean countries to Japan. Particularly China, Central Asia, India | Less need for warmth than other millets. Eaten as gruel and flat-breads. Bird-seed, green fodder (19, 152, 376, 1041, 1455, 1871) |

Because of their tiny seeds, millets can only be sown shallowly. The best results are achieved when, after sowing broadcast, the field is lightly harrowed and then rolled, or, when the seed is drilled, if the seedbed is rather firm.

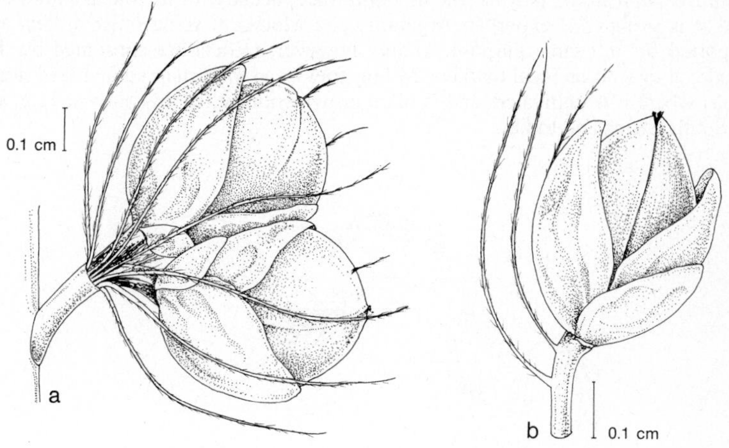


Fig. 5. (a) *Pennisetum americanum*, group of spikelets surrounded by numerous bristles. (b) *Setaria italica*, few bristles underneath each spikelet

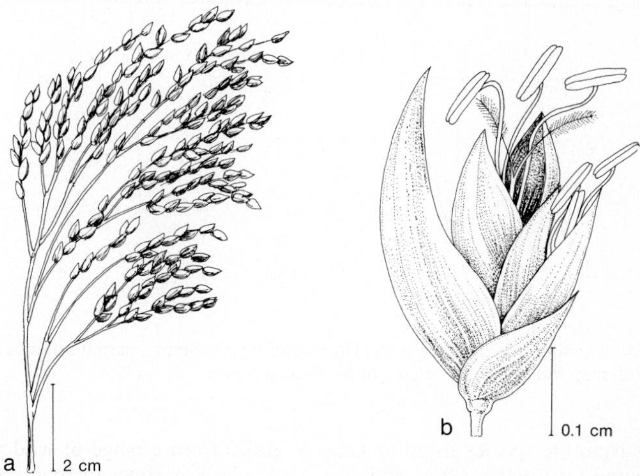


Fig. 6. *Panicum* sp. (a) panicle; (b) two-flowered spikelet, lower flower male or sterile, upper flower hermaphrodite

The majority of millets have only local importance. Some species were grown even in prehistoric times, and they are still important food crops, such as finger, proso, pearl, and foxtail millets. Millets are consumed in the forms of flour, porridge, groats and flat-bread, and they are also widely used in beer-making. *Phalaris canariensis* plays a role in world trade, because of its role as bird-seed, and it is grown for export in Argentina and Morocco; some other millets are exported for the same purpose. Mainly however, millets are consumed by the producer or sold on local markets. Millet straw is an important animal feed in all areas where it is cultivated, and is often grown entirely for this purpose (e.g. tef in South Africa and India).

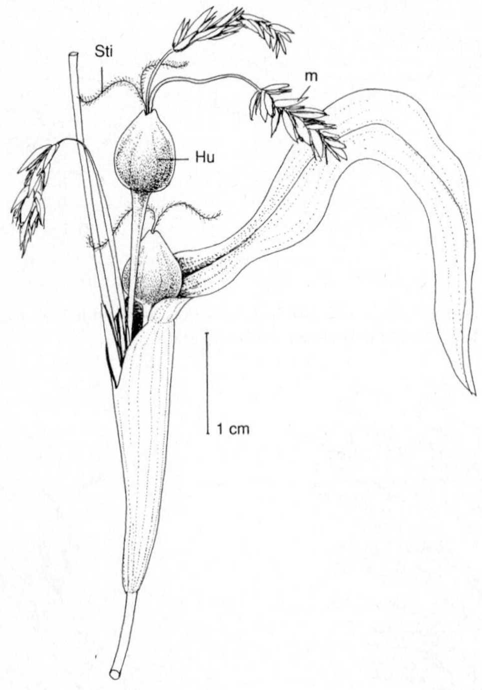


Fig. 7. *Coix lacryma-jobi*, inflorescence, Hu = solid husk around a partial inflorescence, m = stalk with male flowers, Sti = stigma of the female flower

Apart from the species listed in Table 6, grains from a range of wild grasses are gathered for food in Africa and Asia, particularly in times of food shortage. Some of these might occasionally be cultivated, but they can hardly be considered cultivated plants in the usual sense. The following examples are known: *Echinochloa colona* (L.), shama millet; *E. crus-galli* (L.) Pal. Beauv., barnyard grass; *Brachiaria ramosa* (L.) Stapf, browntop millet.

# PSEUDOCEREALS

Apart from the Poaceae, a few dicotyledons are cultivated for their starch-containing seeds; the term "pseudocereals" has been adopted for them.

Of the three genera in Table 7, the *Amaranthus* and *Chenopodium* species have been in cultivation for a very long time, and there are many hundreds of types of them (152, 287, 1258, 1513, 1655, 1657, 1667). Apart from the species in Table 7, the following species of *Amaranthus* are sometimes used as grains: *A. dubius* Mart. ex Thell., *A. hybridus* L., and *A. lividus* L. (242, 782, 1513, 1657). For *Chenopodium*, *C. nuttaliae* Safford is probably only a type of *C. quinoa*, or a subspecies of *C. berlandieri* Moq. (1789, 2026).

Fig. 8. *Chenopodium quinoa*. Fruiting plant

Fig. 9. *Fagopyrum esculentum*. Flowering plant

*Chenopodium* and *Fagopyrum* are plants for the cool mountain regions. Quinoa (Fig. 8) is grown on the high plateaux of the Andes mountains, and, in fact, only at heights where no other grain can ripen. The annual production in Ecuador, Bolivia and Peru amounts to 25,000 tons. The cultivation of buckwheat (Fig. 9) is limited to the Himalaya region and the mountains of southern India (1667).

Table 7. Pseudocereals

| Botanical name | Vernacular names | Distribution | Remarks and literature references |
|---|---|---|---|
| **Amaranthaceae** | | | |
| *Amaranthus caudatus* L. | Inca wheat, trigo del inca, quihuicha | Central and S America, India, Iran, China | Also used as a vegetable and ornamental plant (152, 361, 376, 1268, 1667) |
| *A. cruentus* L. | purple amaranth | Central America, India, China | White seed, also used as leaf vegetable (152, 1268, 1667, 2026) |
| *A. hypochondriacus* L. | princess feather, alegria, guautli, huauthli | N India, China | White, tasty seed, also used as vegetable and ornamental (152, 1268, 1667, 2026) |
| **Chenopodiaceae** | | | |
| *Chenopodium album* L. | lamb's quarters, pigweed, ansérine blanche, bethu sag (India), | India (Himalaya) | Grain types are up to 3 m high, cultivation slight. The leaves are eaten as a vegetable in all continents (376, 1667) |
| *C. pallidicaule* Aellen (*C. canihua* Cook) | cañahua, cañihua | S American Andes, Peru, Bolivia | The ashes of the plant used instead of lime for chewing coca leaves (572) |
| *C. quinoa* Willd. | quinoa, quinua | S American Andes | (Fig. 8) (see p. 35) Also as a leaf vegetable (146, 152, 1258, 1406, 1584, 1585, 1657, 1667, 1964) |

| | | | |
|---|---|---|---|
| **Polygonaceae** | | | |
| *Fagopyrum esculentum* Moench (*F. sagittatum* Gilib.) | buckwheat, blé noir, grano saraceno | Central Asia, India, Nepal | (Fig. 9) (see p. 35) Also as a vegetable, chicken feed, green fodder, green manuring, ground cover (see p. 435) and as a bee plant (see p. 402) (299, 376, 1251, 1657, 1667) |
| *F. tataricum* (L.) Gaertn. | Tatarian buckwheat, sarrasin de Tartarie, ku-chiao-mai | Central Asia, USSR (Siberia), India (Himalaya) | Grains are slightly bitter. Also green fodder and green manuring (376, 581, 1251) |

*Amaranthus* can stand higher temperatures, and grows on lowlands as well as in mountainous regions.

Even in cool climates, the development of pseudocereals is quick - about 3 months. They are generally cultivated on poor soils, and the yields are correspondingly small; when there are sufficient nutrient supplies, all three genera yield about 3 tons or more per ha (1034). Their nutrient content is similar to that of the cereals; quinoa and amaranths contain 14-16% protein, which has good digestibility and high biological value (298, 1667). Amaranth has particularly high P and Ca contents (520 mg P, 185 mg Ca/100 g.). The starch of amaranth exists almost entirely as amylopectin, as in the waxy types of cereals. When heated the grains puff up, and in this form, mixed with honey or syrup, amaranth is most appreciated in India ("laddoo") and Central America. Quinoa contains bitter, poisonous saponins in the grains, which must be washed out with water before eating.

The leaves and young shoots of all types are used as vegetables. *Fagopyrum* is an important plant for green-manuring and for ground cover because of its rapid growth (see p. 435).

# ROOT AND TUBER CROPS

Many plants have an underground storage organ which accumulates mostly starch as reserve material; of these, there are only four species or genera of widespread importance: cassava, sweet potato, yam, and potato. All of them originate from the tropics, the first three from the lowlands and the potato from the mountain areas of the South American Andes. The potato is mostly cultivated in the temperate zones; only about 25% of the world production comes from the tropics and subtropics. Fig. 10 shows that worldwide only minor changes have occurred in the production of root and tuber crops in the last decade.

Roots and tubers contain up to 80% water when freshly harvested (Table 8), so that transport over long distances is hardly worthwhile, except for the most expensive products such as early potatoes from the Mediterranean region to Western Europe and seed-potatoes. Additionally, the tropical species in particular can only be stored for a limited period of time. The total exports of fresh roots and tubers remain at less than 1% of the total production, with some unimportant exceptions limited to potatoes. The export of dried products is of greater importance, for example dehydrated sweet potato for human consumption, dried cassava and sweet potato chips for animal fodder (840), and also starch for human foodstuffs and industrial uses. This is extracted from cassava, sweet potato, potato, and from some of the other species listed in Table 12 (e.g. arrowroot starch).

Where roots and tubers are the main source of nourishment, the local population often suffers from protein deficiencies (see p. 120), although the individual species differ greatly in protein contents (Table 8). It is clear that the protein content of fresh tubers cannot give a correct picture for the comparison of root and tuber crops with each other, or with other sources of nutrition. On the basis of dry weight (Table 9), the protein content is extremely low only in the case of cassava (of the other starch plants similarly low figures are found only in the case of plantains). All of the other species have a protein content of over 5%,

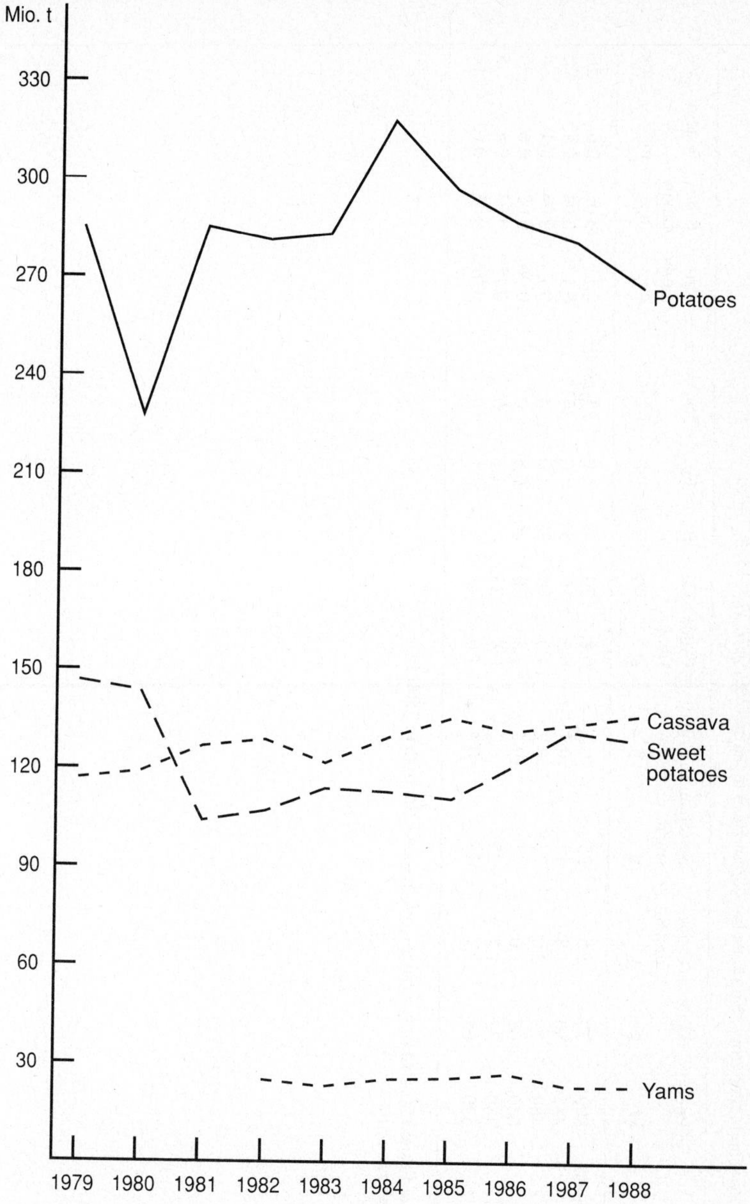

Fig. 10. World production of the most important root and tuber crops, 1979-1988

Table 8. Root and tuber crops. Nutrients in 100g of fresh material (309, 310, 376, 1315, 1408)

| Species | Energy | Water | Crude protein | Starch and sugar | Crude fat | Crude fibre | Ca | P | Fe | Vit. A | Thi-amine | Ribo-flavin | Niacin | Vit. C |
|---|---|---|---|---|---|---|---|---|---|---|---|---|---|---|
| | kJ | g | g | g | g | g | mg | mg | mg | I.U. | mg | mg | mg | mg |
| Sweet potato | 500 | 72 | 1.5 | 25 | 0.2 | 0.8 | 30 | 42 | 1.0 | 25-2500* | 0.1 | 0.05 | 0.6 | 20 |
| Potato | 335 | 80 | 1.8 | 17 | 0.1 | 0.5 | 10 | 51 | 1.0 | 40 | 0.1 | 0.04 | 1.4 | 20 |
| Yam** | 460 | 72 | 2.0 | 24 | 0.2 | 1.0 | 22 | 88 | 1.0 | 0 | 0.1 | 0.03 | 0.4 | 5 |
| Cassava | 630 | 62 | 1.0 | 35 | 0.2 | 1.3 | 30 | 40 | 0.8 | 0 | 0.06 | 0.02 | 0.6 | 30 |
| Taro | 480 | 72 | 1.7 | 25 | 0.2 | 0.8 | 23 | - | 1.1 | 0 | 0.15 | 0.03 | 0.9 | 5 |
| Tania | 570 | 65 | 2.1 | 32 | 0.3 | 1.0 | 13 | - | 1.1 | 0 | 0.09 | 0.03 | 0.6 | 10 |

* White fleshed 50, intensely orange-coloured even more than 2500 I.U.

** *Dioscorea alata*

but this still does not provide a sufficient supply to fulfill the normal protein requirement. Yam approaches the figures for rice and maize (1715). When the yield of protein per hectare is considered, it is found to be no less for roots and tubers than for cereals or peanuts, due to the high total yields that are possible (678). With respect to the other nutrients, there are considerable differences between the species. Sweet potatoes with yellow flesh provide a lot of vitamin A, and the B-vitamins are found in amounts similar to the cereals (on a dry weight basis). Sweet potato, potato and cassava are good sources of vitamin C, but only yam contains enough vitamin C to provide the amount needed daily by a human being (379).

Roots and tubers differ from the cereals in that there are no by-products of any great value from the starch extraction, apart from small amounts of animal fodder.

Table 9. Basic nutrient contents in 100 g dry matter of various starch plants (average values calculated from Tables 1, 8, and 26).

| Species | Crude protein g | Starch and sugar g | Crude fat g |
|---|---|---|---|
| Rice (polished) | 8.1 | 91 | 0.6 |
| Maize | 11.3 | 80 | 5.1 |
| Wheat | 13.8 | 80 | 2.3 |
| Potato | 9.0 | 85 | 0.4 |
| Sweet potato | 5.4 | 89 | 0.7 |
| Yam (*Dioscorea alata*) | 7.2 | 86 | 0.7 |
| Cassava | 2.6 | 92 | 0.5 |
| Taro | 6.1 | 89 | 0.7 |
| Plantain | 3.8 | 90 | 1.6 |

Some of the roots and tubers with long vegetative periods are among the species which provide the largest supplies of usable energy (1881). Cassava, with good care, can produce 20 tons of starch per hectare and per year, which represents about 340 million kJ energy. This performance is only surpassed by that of the sago palm (1732). For this reason, the interest in using root crops in tropical agriculture has increased in recent years. Species with a short vegetation time (sweet potato, tuber-forming labiates, potato) are particularly important in regions with a short rainy season, and for multiple cropping (380, 946, 1023, 1201, 1315, 1881).

## Cassava (Manioc)

fr. manioc; sp. mandioca, yuca; ger. Maniok

All cultivated forms of cassava belong to the species *Manihot esculenta* Crantz (*M. utilissima* Pohl), Euphorbiaceae. The allocation to different botanical species has been abandoned in the agricultural literature, as the types are not easily distinguished from each other (1533, 1534). The wild forms are found between the Amazon region and southern Mexico. It is probable that they were first brought under cultivation in Central America, and, perhaps independently of this, in the Amazon region as well. Soon after the discovery of America, cassava spread to other parts of the world.

The cultivars show remarkable differences in leaf form (1039), in the length, thickness and colour of roots, and in the linamarin content. Two types are distinguished for cultivation: early-ripening cultivars, which are harvested after 6-9 months, have a low linamarin content, and are usually used fresh (sweet cassava); and late-ripening, high-yielding cultivars, which mostly have a high linamarin content, are usually harvested 18-24 months after planting, and are mostly used for the production of flour and starch, or chipped for use as animal fodder (343, 1880). The importance of cassava is that it gives the highest yields of all root crops with relatively simple cultivation, and that the roots can stay in the ground for several years without deteriorating, so that it often serves as a reserve for times of need (242, 345, 843, 882, 1202, 1315, 1653, 1675).

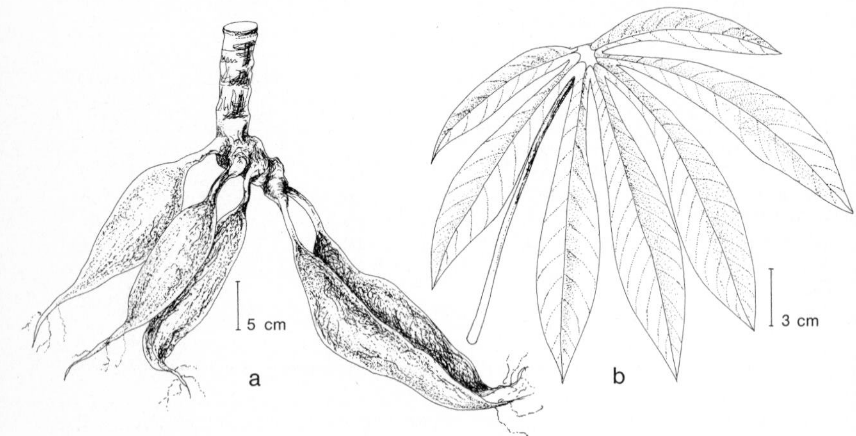


Fig. 11. *Manihot esculenta*. (a) roots, (b) leaf

**Production**. Cassava takes 6th position among the world food crops and it is grown everywhere in the tropics. Africa produces 57 million tons, Asia 52, and Latin America 29. The countries with the largest production are Brazil with 22, Thailand with 22, Zaire with 16, Indonesia with 15, and Nigeria with 14 million tons.

Cassava starch as well as dried chips play a major role in exports; the EEC imports 5-6 million tons of this important fodder material each year.

**Morphology and Anatomy**. Cassava is a perennial shrub, up to 4 m high, with deep-lobed spirally arranged leaves. The stems are very woody, with swollen nodes (insertion region of leaves). The roots (Fig. 11), like all parts of the plant, contain the bitter tasting glycoside linamarin, from which hydrocyanic acid is released by the action of the enzyme linamarase (1280). Linamarin is found in all cells of the tuber, not only in the latex (244). In the tuberous roots, the linamarin content is higher in the rind than in the pith; in the sweet cultivars there are less than 50 mg/kg in the pith, for the bitter cultivars, the corresponding figure is over 100 mg/kg (215, 616).

**Breeding**. The breeding of cassava is concerned above all with the development of types which are resistant to mosaic virus and bacterial blight. Next in importance are the following:

- root form (short thick roots with high starch content, this being especially important for mechanical harvesting, but it also makes manual harvesting easier);
- the position of roots in the soil (as far as possible horizontal and near to the soil surface);
- quality (low fibre content, little linamarin, higher protein content);
- tolerance of dry conditions.

Breeders in some regions are also selecting plants with lower growth height (which gives stability and resistance against storms) and with better branch formation (not too close to the soil, but with an extensive branch system, quickly forming a full canopy of leaves). The breeding procedure is based mainly on the selection and combination of suitable local strains, but crossing with other *Manihot* species is also possible (7, 290, 343, 345, 1253, 1254).

**Ecophysiology**. Cassava is a plant of the humid tropics. Because of its need for high temperatures, it is seldom grown above 1400 m, and at the northern and southern boundaries of its range (about 30 degrees in latitude), it is grown only near sea-level. Optimally, it receives 1000-2000 mm rainfall per year. Since it is moderately drought tolerant (in dry periods, the majority of the leaves are shed), its cultivation is possible with only 750 mm rainfall, and even with only 500 mm in cooler subtropical places. However, these conditions certainly do not lead to high yields. With well-drained soil, it can also be grown in the wettest places, with rainfall up to 5000 mm (842, 1089).

For high yields, cassava needs an open position with no shade and with as much sunlight as possible (776, 1907). The highest weights of roots will be attained under short-day conditions (1080, 1292). Cassava makes few demands on the soil, as long as it is deep soil, free of stones, well aerated, and well drained. On heavy soils in very rainy regions, it is usually planted on ridges, which also makes the harvest easier. A good supply of minerals, especially K, is necessary for a good yield. The absorption of P is assisted by a very active VA mycorrhiza (931, 1651). For the production of 1 t of fresh roots, the whole plant will remove from the soil about 3 kg N, 1 kg P, 7 kg K, 2 kg Ca, and 1 kg Mg (595, 759, 921, 1858).

**Cultivation**. Cassava is vegetatively propagated by cuttings. Ripe wood is used, with 4-6 eyes (a piece about 25 cm long, diameter 2.0-3.5 cm). The stems from 1 ha are enough to plant 5-6 ha. The planting distance varies from 80 x 80 to 150 x 150 cm, depending on the cultivar and the soil fertility. The planting distance has little influence on the yield per hectare, especially with late-ripening cultivars. The cuttings are planted differently in individual countries: horizontal, inclined, or vertical; probably inclined planting, at an angle of 45°, is best. The planting depth can be between 5-15 cm; deeper planting can be useful in drier areas if there is a danger of the upper layers of the soil drying out, however it means that the tubers develop deeper in the earth, and are therefore more difficult to harvest.

Considerable increases in yield have been achieved in Indonesia (up to 150 tons/ha) with plant material where *M. glaziovii* has been grafted on to *M. esculenta* (245).

The first roots appear within a few days, and leaf formation begins after about 14 days. In the first 3-4 months, the land must usually be hoed twice to keep the soil free of weeds, and in this way the soil can also be piled around the stems. After this, the canopy of leaves is so dense that the growth of weeds is generally suppressed.

Because of its high intake of nutrients, cassava is used in a shifting cultivation system preferably as the last crop before the bush fallow (562). In permanent systems of agriculture, it is best if cassava is followed by a green manure plant (see p. 429).

**Diseases and Pests**. About 30 diseases of cassava have been described, of which the most dangerous are African mosaic disease (AMD, a virus), and cassava bacterial blight (CBB), which is caused by *Xanthomonas campestris* (Pammel) Dowson pv. *manihotis* (Arthaud-Berthet) Starr; both of these are especially devastating in Africa. The mosaic virus is spread by *Bemisia* spp., and to counteract it, resistant cultivars can be grown, and virus-free plant material should be chosen for cuttings. Bacteria-free cuttings are the most important measure which can be taken against CBB (193, 251, 1083, 1084, 1386, 1794). The most important animal pests are the leaf-eating insects - in America, the hornworm, *Erinnyis ello* L., and in Africa the grasshopper, *Zonocercus elegans* Thunb.; also to be feared are thrips, green spider mites, and the cassava mealybug, *Phenacoccus manihoti* Matile-Ferrero (164, 284, 1083, 1794, 1979). Great damage can also be caused by herbivores such as rats, swine, monkeys, and hippos, which do not eat the bitter varieties with as much enthusiasm; in some circumstances, the field must be fenced.

**Harvest and Processing**. For fresh consumption, cassava is best harvested before the roots are too hard and fibrous, but it is harvested later for starch production (see p. 42). The starch content is higher after several months of dry conditions than during the rainy season. Before harvesting, the stems are cut off and taken away. The roots can be removed by hand on light soils, or when the crop has been grown on ridges. On heavy soils, and with cultivars with long and deep-reaching roots, the use of hoes, spades or ploughs is necessary. For large-scale agriculture, cultivars with short roots are needed, so that they can be harvested by ploughing. For human consumption, and also for the production of starch or dried chips, the further processing should be done as soon as possible, and it should certainly follow within 2 or 3 days, because the roots decay quickly after harvesting. With fresh roots, the danger of poisoning is decreased, as hydrocyanic acid is released by the linamarase only after several days (the undecomposed linamarin has less acute toxicity).

Experiments to improve the storage quality of the roots are being carried out in a number of countries (291, 359, 946, 1398).

The maximum yields possible are 60-80 t/ha, while commercial operations usually achieve 30-40 t/ha, but the global average is only 9-10 t/ha.

For consumption, the rind is usually peeled off as far as the cambium, and only the pith of the roots is utilized. With bitter varieties, the linamarin is removed by crushing or by soaking. In West Africa, the cassava meal can be fermented into "gari", a product which has a higher protein content than the meal (946, 1009).

For processing in factories (86, 463, 628, 817, 818), only the outermost cork layer is removed, so that the starch in the skin layer is also utilized. The extraction of starch takes place in a wet process, after the root tissue has been ground as finely as possible. In this process, the bitter compounds and the hydrocyanic acid are washed out. When dried chips are produced, 98% of the linamarin is destroyed during the drying process. The importance of cassava as an animal fodder has greatly increased over the last 10 years (615, 1279, 1399).

Tapioca is a special product made from cassava starch, produced when the moist starch is heated to about 70°C. At this temperature, the starch gelatinizes and becomes more easily soluble and digestible. Tapioca comes in the form of sago-like small (seeds) or large (pearls) pellets, and it can also be bought in the form of flakes.

The liquid which is a by-product of starch extraction is often boiled down and used for sauces in Latin America (West Indian pepper-pot "cassaripo", and "tucupí" in Brazil). The young, protein-rich leaves are an important vegetable in many countries (see p. 127, Table 23).

## Sweet Potato

fr. patate douce; sp. batata, camote; ger. Batate

The sweet potato, *Ipomoea batatas* (L.) Poir., Convolvulaceae, originates in the part of the Andes from northern South America to Mexico (111, 1289, 1675). It is hexaploid (2n = 90), and it is assumed that the ancestors were tetraploid and diploid species found in this region. Sweet potato was known in the Pacific area in the time before Columbus, and taken from there to Southeast Asia and to Africa (2016). Nowadays it is grown everywhere in the tropics and subtropics (242, 461, 843, 946, 1315, 1411, 1869), and in some areas it is the major source of nourishment (7, 219). There are thousands of cultivars:

- with white, light yellow, to deep orange-coloured flesh,
- with light ochre, brown, red, to deep violet-blue skins,
- with short, almost round, oblong roots, or up to 1 metre long roots,
- with short (3 months), medium (4-6 months), or long (9 months or more) growing times,
- and with entire heart-shaped leaves, to leaves with narrow lobes (with some cultivars, individual plants can show a variety of leaf forms).

**Production.** The total world production amounts to 130 million tons. By far the largest producer is China, with 110 million tons. The other main producers are also situated in East and Southeast Asia (Vietnam 2.1 million tons, Indonesia 2.0, India 1.5, Japan 1.4, the Philippines 0.8, and the Korean Republic 0.7). The production in Africa totals 5.9, and in the Americas 2.8 million tons. Dehydrated sweet potato and starch produced from it play a very small role in the export markets.

**Morphology and Anatomy**. The roots (Fig. 12), when stored, produce adventitious shoots. All parts of the plant, including the roots, contain latex tubes with a neutral-tasting milky juice. The sweet potato is one of the root crops with a relatively high nutritive value (Tables 8 and 9, the cultivars with yellow flesh containing a lot of vitamin A). Their cultivation can be completely mechanized. The roots can be stored for several weeks, and with special treatment for many months.

**Breeding**. The root shape and size are considered to be most important, after which come the other quality factors (starch and sugar content, fibre content, vitamins, taste, etc.). For sale in the local markets, oval roots with a weight of 250-500 g are most suitable. The breeder must bear in mind that roots with a particular skin colour may be preferred in the local markets. Harvesting will be made easier if the roots are formed near to the centre of the plant. Quite often, the breeder has the challenge of developing new, virus-free cultivars when yields from long-established cultivars have decreased due to virus attacks. Because of the frequent self-sterility of the plants, cross-breeding is necessary as a rule; the parent plants are easy to bring into flower at the same time in any particular climate when they are grafted on to an annual *Ipomoea* species (1141, 1869, 1897).

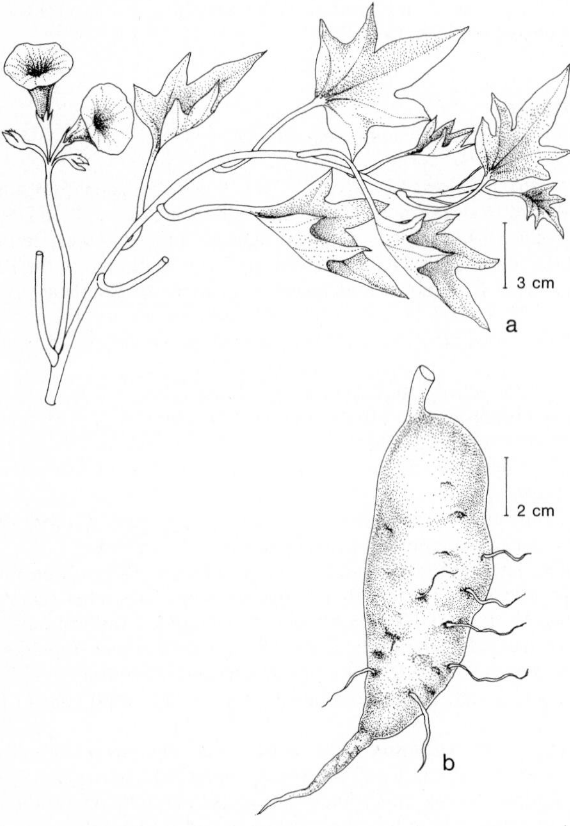


Fig. 12. *Ipomoea batatas*. (a) shoot and flowers, (b) single root

**Ecophysiology**. Sweet potato grows well only when there is an average daytime temperature of more than 18°C; at 10-12°C, growth comes entirely to a standstill, and the slightest frost kills all the parts above the ground. However, the sweet potato's need for warmth is less than that of most of the other root and tuber crops. They are grown in the tropics from the lowlands up to 2500 m above sea level, which is substantially higher than cassava can be grown. Distinct differences between day and night temperature encourage the formation of tuberous roots. Long days delay flowering, and reduce the formation of tuberous roots (1165, 1292, 1315). Cultivars with short vegetation times grow well in the subtropics, up to 40 degrees in latitude. Cultivation is essentially limited to regions with 750-1250 mm rainfall. The plants survive long dry periods, but for high yields and for good tuber quality, a water supply which is evenly spread throughout the growing period is necessary. When roots are produced for the fresh market in areas with irregular rainfall, irrigation can be economic. Dryness is desirable during the ripening period, for then the roots will be firm and keep well. In subtropical climatic conditions, the tubers mostly contain more starch and less sugar than those grown in the wet tropics (Table 10). High yields are only achieved in areas with a lot of sunshine. As for cassava, the sweet potato needs well-aerated and well-drained soil above all. With heavy soils, and where there is a danger of water-logging, it is grown on ridges or on mounds. Very acid soils (pH under 5) will not be withstood, nor saline soils (44). The nutrient needs are similar to those for cassava, and 1 ton of roots contains about 4 kg N, 1 kg P, 7 kg K (921, 1833).

Table 10. Starch and sugar contents of sweet potato from sites in the tropics and subtropics.

| Site | g/100 g fresh weight | | g/100 g dry weight | |
|---|---|---|---|---|
| | Sugar | Starch | Sugar | Starch |
| Subtropics | 2-4 | 18-30 | 6-10 | 60-80 |
| Tropics | 6-10 | 14-20 | 20-34 | 45-65 |

**Cultivation**. Sweet potato is vegetatively propagated, in most countries by stem cuttings about 30 cm long, which can be taken from any part of the shoots; the leaves are stripped off the cutting before planting. In moist, warm soils, the cuttings take root after 2 or 3 days. They cannot be planted into dry soils. In some subtropical countries (USA), the roots are placed in warm beds after winter storage, and the shoots which are formed are used as the planting material. A normal planting distance is 90 cm between rows, and 30 cm within the row. Weeding is only necessary in the first 6 weeks (by hoeing or herbicides, linuron or diphenamid). When the leaf-mass is well developed, it suppresses the further growth of weeds; sweet potatoes have a very positive influence on weed control, as they can repress troublesome perennial weeds such as nutsedge (*Cyperus rotundus* L.). Sweet potatoes are self-compatible, and are planted in New Guinea by some tribes in monoculture (965, 1318), but they are mostly grown in crop rotation with other field crops.

**Diseases and Pests**. The greatest damage to the yield is caused by a variety of virus diseases (1285, 1869). These can make it necessary to grow less susceptible cultivars, or to use newly bred virus-free cultivars. Losses caused by fungus diseases are generally slight, particularly since cultivars with resistance against these diseases are available in most regions. With carelessness, infestation is possible by beetles which eat the leaves, and whose larvae hollow out the roots, and by butterfly caterpillars, which feed on the leaves (284, 1794, 2001); any of these can cause serious losses. To prevent severe damage, useful measures include: harvesting at the right time, the use of insecticides, and crop rotation.

**Harvesting and Processing**. As already mentioned, the harvesting of sweet potatoes can be fully mechanized, as with potatoes. In small-scale farming, they can be dug up with a hoe or fork. For fresh marketing, sweet potatoes are packed in sacks where they keep well for about three weeks without any special treatment. For longer storage, they must be held for the first 12-14 days in conditions of high temperature (30-35°C) and in a humid atmosphere (85-90% relative humidity) (curing), to promote the suberization of the periderm; after this they can be kept for many months at a temperature of about 13°C. This method is mostly used in subtropical regions for the preservation of the planting material. The majority of the harvest is consumed fresh, but some suitable cultivars can be dehydrated, and in Japan, 45% of the harvest is cultivated for commercial starch extraction (220, 461, 946, 1754).

For good eating cultivars, with a 5-month growing time, 20 t/ha is a successful yield. With excellent conditions, and especially with the high-yielding cultivars which are used for animal fodder, yields of 40-50 t/ha can be achieved. With early cultivars, or with bad conditions for cultivation, the yields are much lower. The global average is 14.6 t/ha. All parts of the plant are valuable: the leaves are consumed in many countries as a protein-rich vegetable (Table 23) (1870); leaves, foliage, unsaleable tubers (too small, too large, or damaged), and all remnants from processing can be used raw for animal food, especially for pigs, but also for sheep, goats, and cattle.

## Yam

fr. igname; sp. ñame; ger. Yam

The genus *Dioscorea*, Dioscoreaceae, has hundreds of species, and is distributed throughout the tropical and subtropical world. Almost all are climbing plants, and most of them have very thin stems, and storage organs under the surface of the earth (Fig. 13). *D. bulbifera* grows large tubers in the axils of the leaves. Some of the wild species taste good, and can be used for human consumption without any further processing. People have been using these as a food source since primeval times. Since the settlement of people into villages, yams have been taken into cultivation in all parts of Africa, Asia and America. Even the bitter species have been made use of, as the bitter principle, the poisonous alkaloid dioscorine, is destroyed by cooking. About 13 of the 40 or so species in cultivation can be designated as important on a regional or worldwide basis (Table 11). The others are of only local importance, and some are only eaten when there is a shortage of other food sources (262, 284, 378, 413, 815, 946, 1186, 1315).

The common name of all *Dioscorea* species is the West African Mande word "niam"; the anglicized version of this - yam - originated in America. Yam is still the staple food in some countries of tropical West Africa; the yam production in Nigeria amounts to 16 million tons (main species *D. rotundata*). West Africa supplies about 95% of the world production of yams, but yams are also an important foodstuff in some parts of Southeast Asia.

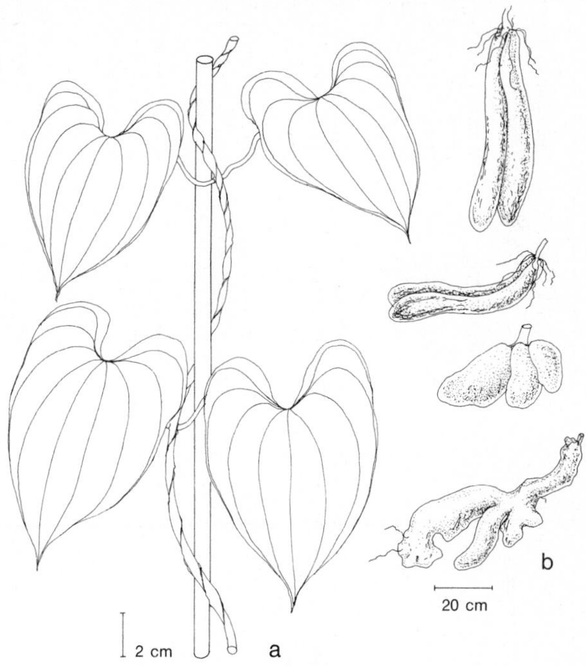


Fig. 13. *Dioscorea alata*. (a) part of shoot, (b) different root forms

The cultivation of yams necessitates more manual work than other tuber crops. Propagation is carried out vegetatively using small daughter tubers or pieces of larger tubers. For breeding purposes, yams can be propagated using stem cuttings, or small pieces of tuber. A more economical propagation method (microsetts) has also been developed for cultivation (793, 926, 927, 1186). Yams are usually planted on mounds or ridges. For each plant, a stake which is strong and tall enough must be erected, on which the high-climbing shoots can grow; only in this way can yields of 20-40 t/ha, and even up to 60 t/ha be achieved. Improved methods have been proposed (679), but a full mechanization of the cultivation of present-day cultivars is not possible, and in fact will only be achieved with types having small tubers, formed near to the surface (e.g. *D. esculenta*). The systematic breeding of yams is still in its earliest stages, and many potentialities are open to the breeder.

Table 11. The most important yam species (*Dioscorea* spp.) (259, 376, 378, 946, 1186, 1315, 1611, 2026)

| Botanical name | Vernacular names | Origin | Distribution | Remarks |
|---|---|---|---|---|
| *Dioscorea abyssinica* Hochst. | rikua (Chaga) | Ethiopia | Ethiopia, E Africa | Important in this area, never the staple diet |
| *D. alata* L. | water yam, greater yam, winged yam, white yam, igname ailée, ñame blanco | SE Asia | India to Polynesia, Africa, W Indies | With *D. rotundata*, the most important yam species. Tubers have many shapes (Fig. 13), and are often very large. Growing time 8-10 months |
| *D. bulbifera* L. | aerial yam, potato yam, igname bulbifière, batata de aire | Tropical Asia and Africa | SE Asia, N Australia, Africa, Central and S America | Mostly only the aerial tubers are edible. Some African forms must be de-toxified. Tubers keep well |
| *D. cayenensis* Lam. | yellow Guinea yam, igname jaune, ñame amarillo | W Africa | W Africa, W Indies, Northern S America | Does not keep well due to short dormancy. Growing time 10 months |
| *D. dumetorum* (Kunth) Pax | cluster yam, African bitter yam, three-leaved yam, ñame amargo | Africa | Africa between 15°N and 15°S | Widespread cultivation of both poisonous and non-poisonous forms. Growing time 8-10 months |
| *D. esculenta* (Lour.) Burk. | Asiatic yam, lesser yam, potato yam, igname des blancs, ñame de China | Indochina, Oceania | India to Polynesia, W Africa | Only known in cultivation, no bitter forms. Tubers taste good, similar to sweet potato, but do not keep well. Growing time 7-10 months |

| | | | | |
|---|---|---|---|---|
| *D. hispida* Dennst. | Asiatic bitter yam, intoxicating yam, karukandu, | India, SE Asia | India, SE Asia | Mostly gathered wild, sometimes cultivated. All forms are poisonous, important food in times of need |
| *D. japonica* Thunb. | Japanese yam, shan-yu-tsai (China) | Japan | China, Japan | Important food plant in Japan, medicinal plant in China |
| *D. nummularia* Lam. | kerung (Java), ubing basol (Philippines) | SE Asia | Indonesia, Oceania | Cultivated in a wide area, tubers large, deep-lying, mostly harvested 2-3 years after planting |
| *D. opposita* Thunb. (*D. batatas* Decne.) | Chinese yam, cinnamon yam, igname de Chine ñame de China, sain-in (China) | China | E Asia | Very tolerant of cold, cultivated in a wide area, growing time 6 months |
| *D. pentaphylla* L. | sand yam, buck yam, ubi passir (Java) | SE Asia | Indonesia, Oceania | Tubers slightly poisonous. Cultivated in a wide area |
| *D. rotundata* Poir. | white Guinea yam, eboe yam, ñame blanco | W Africa | W Africa, W Indies | Most important species in W Africa, many cultivars. Only known in cultivation. Large tubers, good taste, keep well. Growing time 6-10 months |
| *D. trifida* L.f. | cush-cush yam, couche-couche, yampi, mapuey, aja | Northern S America | Central America, W Indies, Sri Lanka | The only one of the S and Central American species which is much cultivated and has reached greater importance. Depending on the cultivar, the tubers have a variety of colours and tastes. Growing time 9-10 months |

The majority of yam species are most at home in the wet, hot tropics, but *D. abyssinica*, *D. alata*, and *D. esculenta* can also be grown in regions with a dry season of several months (dormancy in the dry months), and *D. japonica* and *D. opposita* are good species for those parts of the subtropics which are always wet (1292).

With careful handling and well-aerated storage, several yam species can be stored for a few months (e.g. *D. alata*, *D. bulbifera*, *D. rotundata*) (359); these play a considerable part in local and regional trade (429). England annually imports about 10,000 tons of *D. alata* tubers. However, the major part of the crop is grown for local consumption. Compared to its fresh use, the production of dried yam products plays a subordinate role (for example:- dried flakes, "gari" (see p. 44), meal, and starch) (620, 1508).

Other species of *Dioscorea* containing large amounts of the saponin dioscin have become of great importance in the pharmaceutical industry (see p. 308).

## Potato

fr. pomme de terre; sp. patata; ger. Kartoffel

Of the many tuber-forming species of *Solanum*, Solanaceae, which are cultivated or are found wild in the South American Andes, only the species *Solanum tuberosum* has world-wide importance (242, 260, 294, 425, 706, 946, 1179, 1656, 1657).

**Production**. The cultivation of the potato in the tropics and subtropics has been greatly increased because of the eating habits of the Europeans who live in those areas. There has been an extraordinary increase in encouragement from the governments of many developing countries in recent times on account of its great physiological importance in nutrition (Tables 8 and 9), because of its good ecological adaptation (for example, short growing time), and because of its good qualities for storage and transportation. The development of techniques of plant propagation in the regions of cultivation has been an important factor in the increased efficiency of cultivation (1447). The major producers in the tropics are India 14.1 million tons, Colombia 2.5, Brazil 2.3, Peru 2.0, and Bolivia 0.7 million tons. The export of early potatoes to the industrialized countries is important for many countries, and about 1 million tons are exported each year from the European Mediterranean and Balkan regions. The major exporters of early potatoes are Italy (275,000 tons), Egypt (166,000 tons) and Cyprus (133,000 tons).

**Breeding**. The resistance capabilities of South American wild species and of local varieties play an important role in modern potato breeding (1049, 1558, 1845). For the breeding of cultivars which are suitable for cultivation in tropical mountain regions, the frost resistant species *S. acaule* Bitt., *S. ajanhuiri* Juz. et Buk., and *S. curtilobum* Juz. et Buk. are of increasing importance. Breeding cultivars suitable for local conditions is only now beginning in hot countries; at the moment medium-late and late European cultivars are the most important (1049). A new development is the breeding of homozygous cultivars which can be propagated by true seed (614, 1111).

**Ecophysiology**. The optimal temperature for the growth of tubers is 16-18°C; the upper limit is about 30°C. Under short-day conditions, the tuber formation is generally not dependent on the temperature; the vegetative period is shortened, the growth of the tops is slight, and the stolons remain short (495, 993, 994, 1053, 1179, 1656).

**Cultivation**. Potatoes are planted much closer together in the tropics (e.g. 50 x 20 cm) than in the temperate zones because of the more rapid formation of tubers. Because of this, the ground is covered more quickly, and irrigation water is more efficiently used. The amount of seed potatoes needed is large, 3-4 t/ha. When the seed potatoes have to be planted immediately after harvesting, then the dormancy period must be broken using chemicals, such as carbon disulphide or rindite (120).

**Diseases and Pests**. Potatoes can be severely infested by diseases and pests in the hot countries, especially when they are cultivated soon after a previous crop of potatoes, or, where they are grown all year round in a region without a natural break in cultivation. As in Europe, viruses play a major role (1285), and the fight against them can only be successful by planting certified virus-free seed. Such seed potatoes can be produced in cool, high places, or in hot, dry areas, both of which are unsuitable for the aphids which are the carriers of the viruses (137, 1447).

*Phytophthora infestans* (Mont.) de Bary, late blight, can destroy the whole crop in cool, high regions, and also when the weather is wet, but it is not important at temperatures over 25°C; above this temperature, *Alternaria solani* (Ell. et Mart.) Sor., early blight, takes over. Frequent application of fungicides is necessary. On lowland sites, *Pseudomonas solanacearum* E.F. Smith, bacterial wilt, is widespread, and so far this can only be controlled by phytosanitary measures, such as crop-rotation, and the strict control of the seed potatoes used (292, 293, 1509).

The cyst-forming nematodes (*Globodera rostochiensis* Woll., golden nematode, and *G. pallida* Stone) appear in the cool highlands of many countries, and can endanger the profitability of potato growing. *Meloidogyne* spp. (root-knot nematodes) appear in warm areas. Inspection of seed materials is strictly carried out in many countries, especially in relation to the nematodes which have been named here (982).

Of the insect pests, *Agrotis* spp. (cutworms) and the potato tuber moth (*Phthorimaea operculella* Zell.) must especially be mentioned. The *Agrotis* species can be controlled by insecticides. Strict attention must be paid in the case of the potato tuber moth that the tubers in the field and after the harvest are always kept covered with soil.

**Harvesting and Processing**. Yields are small in the tropics, seldom over 20 t/ha, and in most developing countries the yield is even under 10 t/ha. The potatoes produced in hot countries are almost invariably sorted and sold immediately after harvesting. Their high market value justifies long-distance transport to customers, who regard the potato as a valuable vegetable. Because of the high growing costs in the tropics, using potatoes for animal fodder is out of the question, as well as for starch or alcohol production.

## Further Root and Tuber Crops

Of the many less important root and tuber crops (946) (Table 12), only *Colocasia esculenta*, *Maranta arundinacea* and *Xanthosoma sagittifolium* (Figs. 14 and 15) are found throughout the tropics and are of economic interest.

Table 12. Further root and tuber crops (for explanation of symbols see footnotes)

| Botanical Name | Vernacular names | Climatic region[1] | Distribution[2] | Econ. value[3] | Remarks and literature references |
|---|---|---|---|---|---|
| | | | **Dicotyledonae** | | |
| **Basellaceae** | | | | | |
| *Ullucus tuberosus* Lozano | ulluco, melloco, papa lisa, kipa uljuko, chugua | TrH | r: Colombia, Peru, Bolivia | + | Production in Peru about 30,000 tons. When cooked, the tubers taste similar to potato (242, 946, 970, 1201, 1378, 1544, 1864) |
| **Cruciferae** | | | | | |
| *Lepidium meyennii* Walp. | maca, chijura | TrH | l: Peru, Bolivia | + | Cultivated at heights of 3800-4200 m. Tubers cooked to prepare a sweet gruel. Cultivation is decreasing (946, 1038, 2026) |
| **Labiatae** | | | | | |
| *Plectranthus edulis* (Vatke) Agnew (*Coleus edulis* Vatke ) | gala dinich, oromo diniche | TrH | l: Ethiopia | + | Fields of this cultivated in the highlands of Ethiopia. Several cultivars (1221, 1650, 2026) |
| *P. esculentus* N.E.Br. (*Coleus esculentus* (N.E.Br.) G. Tayl.) | kafir potato, dazo, ndazu, rizga | S, Tr | r: Central and W Africa to S Africa | + | Widely cultivated in the drier parts of tropical Africa. Tubers eaten raw or cooked; elongated, often branching, good tasting (200, 1306, 1611, 2026) |

| | | | | | |
|---|---|---|---|---|---|
| *Solenostemon rotundifolius* (Poir.) J.K. Morton (*Coleus rotundifolius* (Poir.) A.Chev. et Perr., *C. parviflorus* Benth.) | Hausa potato, Chinese potato, country potato, fra-fra-salaga, koorkan | Tr | r: Africa, S India to SE Asia | + | Only known in cultivation, many cultivars, generally small tubers. Vegetation time 3-4 months (259, 262, 347, 376, 731, 946, 1221) |
| **Leguminosae** PAPILIONOIDEAE | | | | | |
| *Flemingia vestita* Benth. ex Barker (*Maughania vestita* (Benth. ex Barker) O. Kuntze) | sòh-phlong | TrH | l: N India | + | Cultivated in the Khasi and Jaintia hills. Vegetation time about 7 months. Yields up to 10 t/ha (376, 581, 1261, 1597, 1669, 2026) |
| *Pachyrhizus ahipa* (Wedd.) Parodi | yam bean, ahipa, jíquima, jícama, ajipa | Tr | l: Bolivia, N Argentina | + | Old cultivated plant of the Indios. Leaves and seeds poisonous, utilized as for *P. erosus* (1039, 1846, 1864, 2026) |

[1] Te = temperate zone; S = subtropics; Tr = tropics; TrH = tropical highlands
[2] w = worldwide; r = regionally; l = locally
[3] + = low, only locally traded; ++ = medium, locally sometimes of high value; +++ = high, important for export

Table 12. Further root and tuber crops, cont'd (for explanation of symbols see footnotes page 55)

| Botanical Name | Vernacular names | Climatic region | Distribution | Econ. value | Remarks and literature references |
|---|---|---|---|---|---|
| *P. erosus* (L.) Urb. | potato bean, yam bean, patate cochon, jícama, sin kama, fan-ko | Tr | w: mainly Asia | ++ | Propagated by seed. Young tubers up to 8 months after sowing are sweet and watery, and can even be eaten raw. Tubers over 1 year old are mainly used for starch production. Yields up to 95 t tubers/ha. Young pods used as a vegetable, leaves and ripe pods poisonous. The seeds contain rotenone, and are used as an insecticide and fish-poison (145, 376, 444, 731, 946, 1039, 1201, 1411, 1672) |
| *P. tuberosus* (Lam.) Spreng. | yam bean, jícama, fejião yacatupé | Tr | r: S America, W India, China | + | Similar to *P. erosus*. Tubers larger (39, 444, 946, 1039, 1864) |
| *Sphenostylis stenocarpa* (Hochst.) Harms | African yam bean, kutonoso, roya | Tr | r: Africa | + | Mostly grown from seed. Growing time 8 months. Roots taste similar to potato. Young pods and seeds are also eaten (200, 262, 444, 480, 946) |
| **Nyctaginaceae** | | | | | |
| *Mirabilis expansa* Ruiz et Pav. | marvel of Peru, mauka | TrH | l: Bolivia, Ecuador | + | Cultivated in a limited region because of its edible roots (242, 1201) |

| **Oxalidaceae** | | | | | |
|---|---|---|---|---|---|
| *Oxalis tuberosa* Mol. | oca, oka, ibia, apillia | TrH | r: Colombia, Bolivia, Chile, New Zealand | + | Tubers with high sugar and starch contents, but with sharp taste, only edible cooked. Yields up to 20 t/ha possible (242, 946, 950, 970, 1201, 1378, 1735, 1864) |
| **Tropaeolaceae** | | | | | |
| *Tropeaolum tuberosum* Ruiz et Pav. | añu, cubio, isañu, mashua | TrH | r: Andes, Colombia to Chile | + | Tubers with mild taste, only eaten cooked. Yields of 20-30 t/ha (242, 946, 970, 1201, 1864) |
| | | **Monocotyledonae** | | | |
| **Alismataceae** | | | | | |
| *Sagittaria sagittifolia* L. ssp. *leucopetala* (Miq.) Hartog ex van Steenis | Chinese arrowhead, muya muya, ubi keladi | S, Tr | r: SE and E Asia | + | Cultivated in ponds. Rhizomes are not edible raw, prepared by peeling, grinding and cooking; also used for pig food and for starch production. Leaves sometimes eaten as vegetable (145, 259, 376, 731, 946, 1411) |
| **Araceae** | | | | | |
| *Amorphophallus konjac* K. Koch | elephant yam, konjaku (Japan) | S, Tr | r: Japan, China | + | Area cultivated in Japan 17,000 ha. Flour used for the production of vermicelli and cakes (259, 301, 946, 1300) |

Table 12. Further root and tuber crops, cont'd (for explanation of symbols see footnotes page 55)

| Botanical Name | Vernacular names | Climatic region | Distribution | Econ. value | Remarks and literature references |
|---|---|---|---|---|---|
| *A. paeoniifolius* (Dennst.) Nicolson (*A. campanulatus* Bl. ex Decne.) | oroy elephant foot, whitespot arum, telinga potato, suran, pongapong | Tr | r: Asia, Oceania | + | Old tubers up to 25 kg, some forms contain large amounts of oxalate, which is removed by soaking and cooking. Young leaf stalks cooked as a vegetable. Leaves and tubers also as animal fodder (145, 301, 376, 731, 815, 946, 1300, 1411, 1484) |
| *Colocasia esculenta* (L.) Schott | dasheen, taro, 'old' cocoyam, eddo | S, Tr | w | ++ | (Figs. 14 and 15, Tables 8 and 9) About 1000 cultivars, diploid or triploid, sharp or mild tasting, with a single large tuber (dasheen) or many small daughter tubers (eddoes). Forms for cultivation in water (dasheen) or on dry land. Withstands quite high salt contents in the soil. Growing time 6-15 months (145, 162, 284, 301, 359, 376, 731, 815, 876, 946, 1258, 1300, 1315, 1411, 1412, 1794) |

| | | | | | |
|---|---|---|---|---|---|
| *Cyrtosperma chamissonis* (Schott) Merr. | giant swamp taro, maota (Tahiti), gallan, opeves, palauan | Tr | r: Polynesia | + | Tubers up to 50 kg after several years. Leaves 3-4 m high, thrives well on coral islands, also on swampy saline soils. Main source of food on many Pacific Islands (145, 301, 731, 815, 946, 1300, 1409, 1657) |
| *Xanthosoma sagittifolium* (L.) Schott | yautia, new cocoyam, tannia, tarias, Chinese taro, taye, ocumo, mangarito, taioba | Tr | w | ++ | (Figs. 14 and 15, Table 8) The many cultivated forms have also been described as separate species (e.g. *X. atrovivens* K. Koch et Bouché, *X. nigrum* (Vell.) Stellf.). They differ in size and shape of the leaves, size, colour and taste of the tubers. Young leaves eaten as a vegetable. In S America, a form which builds hardly any tubers is grown mainly as a leaf vegetable (*X. brasiliense* (Desf.) Engler). All forms are used as animal fodder, especially for pigs (24, 301, 359, 731, 815, 946, 1215, 1258, 1300, 1315, 1411, 1657, 1794) |
| **Cannaceae** | | | | | |
| *Canna edulis* Ker-Gawl. | edible canna, gruya, achira | S, Tr | r: W India, S America, SE Asia, Oceania | + | Supplies the "Queensland arrowroot" starch of commerce. Yields up to 50 t/ha. Important animal fodder (571, 731, 946, 1201, 1864) |

Table 12. Further root and tuber crops, cont'd (for explanation of symbols see footnotes page 55)

| Botanical Name | Vernacular names | Climatic region | Distribution | Econ. value | Remarks and literature references |
|---|---|---|---|---|---|
| **Cyperaceae** | | | | | |
| *Eleocharis dulcis* (Burm.f.) Trin. ex Henschel (*E. tuberosa* Schult.) | Chinese water chestnut, waternut, pi-tsi, teker, matei (China) | Tr | r: SE and E Asia W Africa, Madagascar, USA | ++ | Cultivated in water. Yields 20-40 t/ha. Eaten raw or cooked as a vegetable. Exported from China to other Asian countries and to the USA. Leaves used for basket-work (731, 745, 946, 1051, 1162, 1201, 1411, 1670) |
| **Marantaceae** | | | | | |
| *Calathea allouia* (Aubl.)Lindl. | Guinea arrowroot, sweet corm root, llerén, allouya | Tr | r: W Indies, S America | + | Small tuber, similar to potato, eaten cooked. Valued as a delicacy (27, 248, 946, 1139) |
| *Maranta arundinacea* L. | St. Vincent arrowroot, herbe aux flèches, araruta | Tr | w: particularly W Indies, Brazil, SE Asia, India | ++ | (Figs. 14 and 15) Only cultivated for starch production. Provides the arrowroot starches "St. Vincent", "Jamaica", "Natal", and "W Indian", which are particularly easy to digest. Harvest after 11-12 months. Yields up to 30 t/ha. Starch content 8-16% (27, 376, 484, 485, 731, 1201, 1411) |

| | | | | | |
|---|---|---|---|---|---|
| **Taccaceae** | | | | | |
| *Tacca leontopetaloides* (L.) O. Kuntze (*T. pinnatifida* J.R. et G.Forst.) | Tahiti arrowroot, East Indian arrowroot, boure, pia (Tahiti), yabia (Hawaii) | Tr | r: SE Asia, Polynesia, Africa | + | Taste of the tubers is burning-bitter. Provides "Tahiti" arrowroot starch. Has lost importance (731, 815, 946, 1201) |
| **Zingiberaceae** | | | | | |
| *Curcuma angustifolia* Roxb. | East Indian arrowroot, tikhur, tavakhira | Tr | l: India (northern highlands) | + | Provides "East Indian" arrowroot starch. Also other *Curcuma* species occasionally used for starch production (376, 1446) |

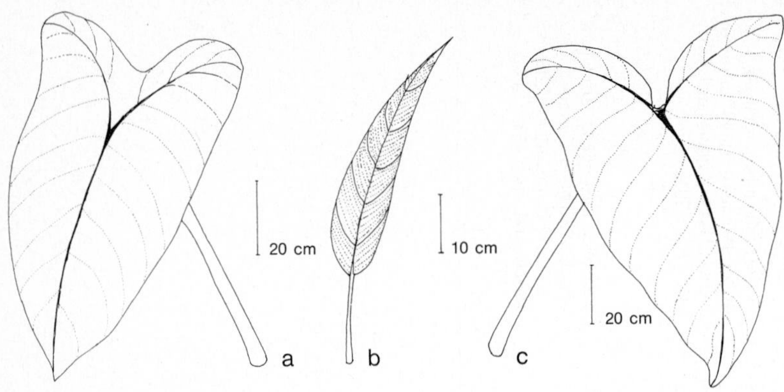


Fig. 14. Leaf forms of (a) *Colocasia esculenta* (peltate), (b) *Maranta arundinacea*, (c) *Xanthosoma sagittifolium* (sagittate)

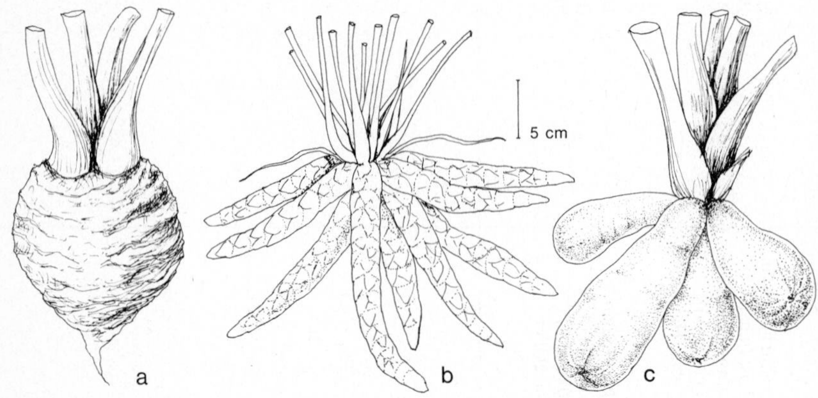


Fig. 15. Roots of (a) *Colocasia esculenta*, (b) *Maranta arundinacea*, (c) *Xanthosoma sagittifolium*

The other species are grown only in limited areas, but they can still be important locally as sources of food (*Plectranthus* and *Solenostemon* varieties in Africa, yam beans in Southeast Asia), or they can be important commercially (e.g. *Eleocharis dulcis* in Southeast Asia). Apart from the species listed in Table 12, there is another range of plants which have starch-rich, edible roots or tubers, but they are included among the vegetables listed in Table 25 because:

- they possess a distinctive taste (e.g. parsnips, *Pastinaca sativa*),

- other parts of the plant are equally or more important food sources (e.g. *Sechium edule*),
- or starch is not the most valuable substance contained in it (e.g. tiger nuts, *Cyperus esculentus*).

Others in this category are *Arracacia xanthorrhiza*, *Coccinia abyssinica*, *Psophocarpus tetragonolobus* (see Table 25).

Apart from the *Pachyrhizus* species, all tuber plants are generally propagated vegetatively, and most have been cultivated for thousands of years. Consequently of most species exist many varieties, which have different shapes, colours, tastes, and cooking characteristics. For *Plectranthus*, *Solenostemon*, *Pachyrhizus* and *Xanthosoma*, the classification of the species is uncertain; some authors separate more strains into species, apart from those named in Table 12. A consequence of vegetative propagation over such a long time is that some species produce few or no seeds.

# FURTHER STARCH PLANTS

Apart from the starch plants which come under the headings "cereals", "pseudocereals", and "roots and tubers", there are other food plants that do not fit under these 3 headings which are also utilized mostly because of their starch contents:

a) numerous seeds belong in this category, such as the European chestnut (*Castanea sativa* Mill. and other spp., see Table 33), the water chestnut (*Trapa natans* L. var. *bicornis* Makino, and var. *bispinosa* Makino, Table 33), the breadnut (seed-bearing varieties of the breadfruit tree, *Artocarpus altilis* (see p. 65), and also other *Artocarpus* species Table 30). Some of the grain legumes contain up to 60% starch, such as the bambara nut and the cowpea (see pp. 120-122, Tables 21 and 25, Fig. 31);
b) some bulbs (e.g. *Lilium lancifolium* Thunb., Table 25);
c) rhizomes (e.g. *Nelumbo nucifera* Gaertn., Table 25);
d) stems (*Ensete ventricosum* (Welw.) Cheesm.) (179, 1650), *Cycas circinalis* L. and other Cycadaceae, and sago palms;
e) and fruits (peach palm (Table 17), plantains, and breadfruit).

Of these, the sago palm, the plantain and the breadfruit tree have great importance as the main sources of food for the population of quite large areas.

Mainly in Malaysia, Indonesia and the Pacific region sago is extracted from a type of palm tree which flowers only once (145, 543, 731, 1098, 1732, 1778, 2006). In some of these areas, it is the staple nourishment instead of rice. The main species is *Metroxylon sagu* Rottb., followed by *M. rumphii* Mart. (which, according to some authors belongs to *M. sagu*), and there are six further species, which are mostly limited to single island groupings in the Pacific Ocean (145). Occasionally, palms of other genera are used for starch production (*Corypha*, *Caryota*, *Phoenix*, and others). The species of the genus *Metroxylon* are natives of the freshwater swamps of the tropical lowlands. The trunks are cut down shortly before the appearance of the inflorescence, the hard outer parts are discarded, the pith is crushed, and the starch is separated from the fibres by washing. In the sago gardens, up to 130 trunks per hectare can be harvested each year, and the starch yield is 15-25 tons per hectare per year (1732). Apart from planned plantations, stands of trees growing in the wild are often harvested. For some islands, sago is an economically important export product. As a food source, it is an almost pure supplier of energy, containing only 0.6% protein.

The plantains (*Musa* x *paradisiaca* L., see p. 181, Table 29, Figs. 39 and 40) are grown everywhere in Southeast and South Asia, and also in tropical Africa and America (7, 376, 544, 843, 1020, 1385, 1751, 1791). They are an important source of food, and in some regions form the staple diet of the people. In 1988, the total world production was 24 million tons, of which 17 million tons were produced in Africa; 7 million tons were grown in Uganda alone. The cultivation of plantains is identical to that of bananas (see p. 184). Most cultivars grow quite tall (3-6 metres), and there are great differences between their fruits with regard to shape, colour, size, and taste (dry/floury, sweet, or sweet/sour). In less well-cared-for plantations in Uganda, the yields are 15-20 t/ha, but with good tending, 38-50 t/ha and per year are achieved. The plantains have a low protein content (see Table 9), and are also low in minerals. Vitamins A and C are found in a satisfactory amount only in a few cultivars. Where plantains form the staple diet in an unbalanced way, numerous cases of nutritional disorders appear; apart from the insufficient supplies of proteins, vitamins and minerals, the high contents of serotonin (40-100 ppm), norepinephrine (2-122 ppm) and dopamine (8-700 ppm) can cause high blood pressure, heart disorders, and psychotic reactions (1058, 1257, 1905). Plantains are mostly cooked, roasted or fried when fresh. They can also be used dried (banana figs), or for flour production (382). In East Africa and some other areas, they are also used for beer-brewing. They are an important product in the local markets.

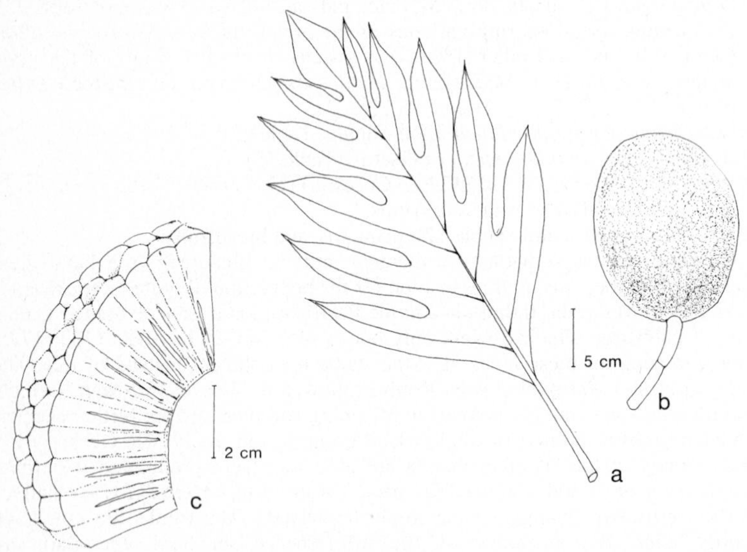


Fig. 16. *Artocarpus altilis*. (a) leaf, (b) aggregate fruit, and (c) part of fruit, in cross section (seedless type)

The breadfruit tree, *Artocarpus altilis* (Parkins.) Fosb. (*A. communis* J.R. et G. Forst, *A. incisa* (Thunb.) L.f.), Moraceae (Fig. 16), originates from the Sunda Islands of Indonesia and from Polynesia. The seedless cultivars are preferred for food purposes. Nowadays it is grown in all tropical countries, often only as an ornamental tree, on account of its attractive foliage. It is an important food plant in its native areas, and also in small districts of South and Southeast Asiatic countries, and in the Caribbean region (145, 376, 382, 584, 1297, 1657). It is propagated from root-shoots and root-cuttings. One tree produces about 40 kg fresh fruit each year. Artificial pollination encourages the growth of fruit (even with seedless forms). 100 g fruit flesh contains 66-75 g $H_2O$, 20-28 g starch, 1.1-1.5 g protein, 0.4 g fat, some vitamin A, a moderate amount of B vitamins, and 0-50 mg vitamin C. The flesh of the fruit can be prepared by cooking or frying. The fresh fruit does not keep; the storage technique is to cut it into slices 1-1.5 cm thick, and to dry these, either in air or in a copra oven. The production is almost entirely used for the peoples' own requirements.

# Sugar Plants

Sugar is a fine food with numerous uses such as sweetening drinks (tea, coffee, lemonade, cola, etc.), in baked goods, for enhancing the flavour of many food preparations, in the canning industry, and in the production of sweets of all types. Its character as a luxury food is expressed by the fact that its use increases as prosperity increases, and that taxes are put on it in many countries. Its nutritive value is equivalent to that of starch, however it plays a subordinate role in the amount used. In only a few countries does it provide more than 10% of the energy needs of the population. In terms of Joules, the world's sugar production only provides 7% of the energy provided by cereals.

A variety of sweet tasting mono- and oligo-saccharides are found in the higher plants, however, agriculturally, only the extraction of sucrose (disaccharide of glucose and fructose) plays a major role. The most important sources are sugar cane and sugar beet. In addition there are some other plants in the tropics from which the sugar-containing juice is obtained, but only in very few cases are they used for the production of crystallized sugar (see p. 74). The statistics quote amounts of centrifuged and un-centrifuged sugar. With centrifuging, the sucrose crystals are separated from the mother liquid (molasses). In uncentrifuged sugar, the whole liquid solidifies after it has been thickened by boiling ("gur" and "jaggery" in India, "muscovado" in Peru, and many other types).

The development of the world's sugar production over the last 10 years is set out in Fig. 17. The increase for sugar cane is 27%, for beet only 5%. In the previous decades, the production of sugar beet increased greatly, particularly because of intensified cultivation in the developing countries of the subtropics. Chile, Greece, Iran, Morocco, Turkey, and many other countries are today nearly self-sufficient or even exporters of sugar, whereas they had to import almost all sugar before the Second World War. The production of beet-sugar in the subtropics has trebled since 1945. Private initiatives and state aid have contributed equally to this success. At the moment, sugar beet provides 40% of the centrifuged sugar, of which about a fifth is produced in the subtropics. The production figures for the main producer countries are set out in Table 13. Brazil harvests by far the most sugar cane, but a large proportion is used for the production of alcohol for use as fuel for vehicles, and therefore Brazil only takes third place in the production of sugar.

With sugar, the farmers do not produce the finished product: they produce a plant material containing a high proportion of water, which contains the eventual product in a concentration of 10-20% sugar. A factory process is needed to extract the finished product. This involves high capital investment in the factory, and the use of a lot of energy to evaporate the water. To function efficiently, a modern sugar factory must process at least 2,000-3,000 tons of cane or beet each day. That requires a well-functioning organization for cultivation, harvesting and transport. To achieve the daily production volume quoted, an area of 4,000-

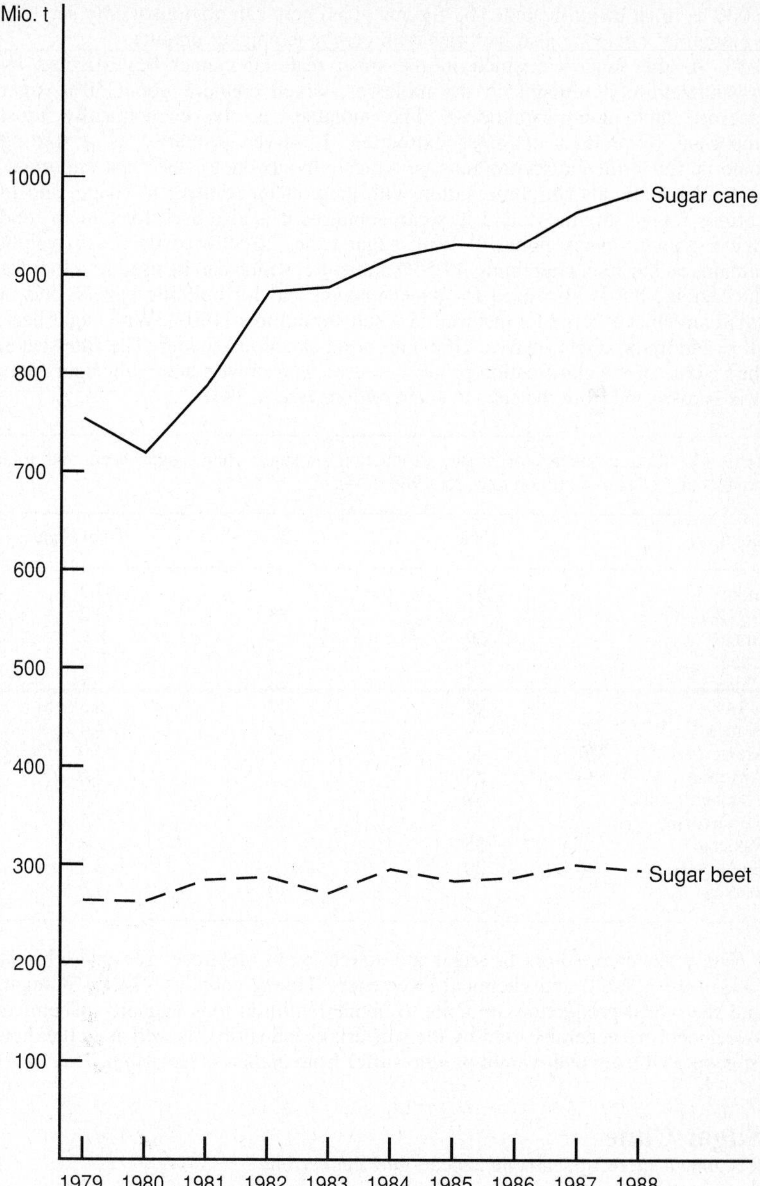


Fig. 17. World production of sugar cane and sugar beet, 1979-1988

6,000 ha must be cultivated. The figures given here can obviously only serve as a guideline. There are also factories with very much higher outputs.
7-8% of the sugar contained in the fresh material cannot be extracted by crystallization. It remains in the molasses, which contains about 50% sugar (sucrose and monosaccharides). The molasses is the economically most important by-product of sugar extraction. It serves primarily as a starting material for fermentation products, of which ethyl alcohol is the most important, as fodder materials (in combination with urea, as an additive to silage, and in various forage mixtures), and in some countries it is also used for human food (the Egyptian "black honey"). With sugar cane, 20-30% of the fresh weight remains as bagasse, containing 49-55% moisture, which can be used as a fuel for the factory, but is also used for paper-making and for building panels, and in small amounts as a fodder material or a soil conditioner (1909). With sugar beet, all of the by-products (leaves, chips) are used as animal fodder. The filter cake, the residue of the clarification process, is used as fertilizer, and with sugar cane, wax is extracted from the cake in some regions (see p. 394).

Table 13. Main producers of sugar. Production of sugar cane, sugar beet, and total production of sugar in million tons, for 1988 (514).

| Country | Cane | Beet | Total sugar |
|---|---|---|---|
| India | 197 | - | 17.2 |
| USSR | - | 88 | 9.2 |
| Brazil | 259 | - | 8.7 |
| Cuba | 74 | - | 7.5 |
| China | 55 | 13 | 6.4 |
| USA | 28 | 22 | 6.3 |
| France | - | 28 | 4.4 |
| Mexico | 42 | - | 3.9 |
| Australia | 28 | - | 3.6 |
| Thailand | 37 | - | 3.3 |
| FR Germany | - | 20 | 3.1 |
| Pakistan | 35 | - | 2.8 |
| South Africa | 20 | - | 2.3 |
| Turkey | - | 11 | 1.6 |

The main competitors of sugar are starch syrup, dextrose (see p. 7), honey (383) (see p. 402), and chemical sweeteners. Honey contains 17-21.5% sugar, and the world production amounts to about 1 million tons annually. Chemical sweeteners are generally used by the soft-drinks industries, as well as in the diets of people who are overweight or who suffer from diabetes (see also p. 75).

## Sugar Cane

fr. canne à sucre; sp. cana de azúcar; ger. Zuckerrohr

All of today's commercially grown cultivars of sugar cane, *Saccharum officinarum* L., Gramineae, are crosses of the noble cane *S. officinarum* (in the original sense), which comes from Melanesia, with three other *Saccharum* species. The noble cane has contributed the most important quality characteristics - high sugar content, good purity of juice, and low fibre content. From the old Indo-Chinese species *S. sinense* Roxb. (which includes *S. barberi* Jeswiet.) comes the ecological adaptability, and the disease resistance was introduced from two wild species, *S. spontaneum* L. and *S. robustum* Brandis et Jeswiet ex Grassl (376, 521, 771, 1657).

**Production**. Sugar cane is one of the world's economically most important cultivated plants. The world production amounts to 69 million tons of sugar, which is produced from 968 million tons of cane (Fig. 17 and Table 13). The export market for sugar is dominated by the sugar cane-growing countries, which altogether export 29% of their production. In 1988, the major exporters were (figures are million tons of raw sugar): Cuba 6.5, Australia 2.7, Thailand 1.7, Brazil 0.9, South Africa 0.9, Mexico 0.9, Mauritius 0.7, Dominican Republic 0.5.

**Breeding**. Sugar cane is cultivated in the tropics and subtropics between 37°N (Spain) and 31°S (South Africa). Thus it is grown under very different climatic conditions, which not only require adaptation to either continuous or limited growing seasons, but also to different diseases and pests. Therefore, sugar cane breeding is carried out at more than 30 breeding stations in all types of climatic zones. The cultivars are designated by breeding station, and also by a reference number (for example, C.P. is Canal Point, Florida; POJ is Proefstation Oost Java; full lists are available (187, 1737)). In noble cane, *S. officinarum* L.s.a., the number of chromosomes in the egg cells is not reduced, but retains the somatic number of 80. By pollination with the haploid pollen of the other species, hybrids with higher chromosome numbers are produced. The $F_1$ hybrids are useless, and usable cultivars usually need repeated back-crossing with noble cane (nobilization) (1436). Breeding goals are to improve:

- weight of canes, sugar content, juice purity, and climatic adaptation (short and long vegetative times, cold and drought tolerance),
- disease resistance, especially against viruses,
- adaptation to the photoperiod (non-flowering),
- good tillering, and good ability to re-grow (ratooning),
- upright growth, and leaves that drop when old (these qualities are important to facilitate harvesting) (1).

**Ecophysiology**. Under favourable climatic conditions, sugar cane is one of the most efficient of cultivated plants, with regard to the total plant substance produced, as well as the value of the end product (28, 495, 1153). The optimum average temperature is 25-26°C. Tropical cultivars grow noticeably slower at 21°C, and at 13°C do not grow at all. Temperatures under 5°C lead to damage (chlorosis). The subtropical cultivars are more tolerant of cold, but their optimum temperature is about the same. Short days induce unwanted flowering, but there are many cultivars which hardly flower at all, even under short-day conditions. The water requirement is high; in most growing areas, the cane needs 1500-1800 mm rain. In hot dry irrigation areas, the annual water need can be 2500 mm or more. When possible, the cane is harvested after a dry period of about 2 months, which causes a cessation of growth and an increased sugar content. Sugar cane is cultivated on soils of very different textures and chemical compositions. In general, heavy, fertile soils with a high water holding capacity are preferred. Prolonged waterlogging is not endured, and the water table should lie at least 1 metre under the surface (44). Because of the large plant mass produced, the need for nutrients is high. Each ton of fresh cane contains 0.5-1.2 kg N, 0.2-0.3 kg P, 1.0-2.5 kg K, 0.3-0.6 kg Ca, and 0.2-0.4 kg Mg (779). The nutrient content of the leaves and tops is many times higher. None of these nutrients are contained in the finished product of pure sugar; they can mostly be returned to the soil using the residues from processing, when this is economically and technically possible. With favourable soil conditions, an endomycorrhiza is always present which increases P uptake (1187), and the supply of N is supported by free-living $N_2$-fixing bacteria (1556, 1758).

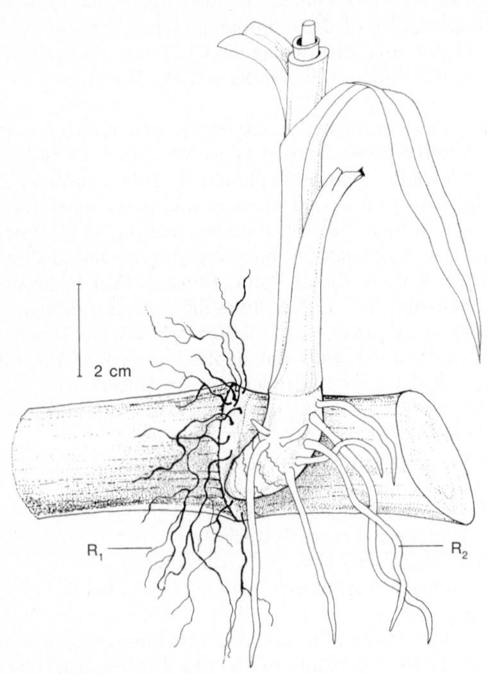


Fig. 18. *Saccharum officinarum*, cutting after sprouting. $R_1$ = short-lived roots from the node, $R_2$ = roots from the new shoot

**Cultivation**. The cane is vegetatively propagated using stem cuttings, usually with two nodes. The best material for cuttings comes from 8 to 9-months old, unripened canes, preferably from one's own propagation nurseries, where strict phytosanitary measures can be enforced. One hectare gives enough cuttings for 5 or 6 ha (1-1.5 tons of cane will plant 1 ha). Before planting, the cuttings should be treated with fungicides and insecticides. Frequently a hot water treatment (1.5 hours at 52°C) is also used to kill viruses and insects. The row-spacing depends on the cultivar, and also on the width of the implements which will be used; it is normally 1.2-1.5 m. The pieces of cane are laid end-to-end in furrows, either by hand or mechanically, and then covered with 3-5 cm soil. Roots are quickly formed at the nodes, then the eyes sprout, and after only a few weeks, a new permanent root system has been formed (Fig. 18).

Until the closure of the leaf canopy, the weeds must be kept under control, either mechanically or chemically (2,4-D, prometon, atrazin, etc.). It is most efficient to use herbicides only within the plant row, and to use mechanical methods to keep the rest of the space in between the rows free of weeds (1372).

Even on good soils, top-dressings of nitrogenous fertilizer, mostly urea, are necessary on oxi- and ultisols, as well as an application of KCl. Sugar cane is better suited then any other plant for being fed with exact dosages of fertilizer; analysis of samples from the leaves or internodes allows a calculated control of the fertilizer requirement (338, 779).

Sugar cane is self-compatible, and is sometimes grown in monoculture for many years without a break. In cooler subtropical regions, the vegetative period ends with the onset of the cool season. There, the cane is freshly planted every year. In the tropics and warmer parts of the subtropics, growth begins again after harvesting; under favourable conditions, one field can be harvested up to 8 times. The yield from ratoon crops can be higher than that from the plant crop, because the vegetation time is often shorter. Diseases, pests, gaps in the standing crop caused by mechanized harvesting, and other causes usually lead to only 1 or 2 ratoon crops being taken.

Uncontrolled fires can cause major losses in sugar cane. Obvious precautions are prevention of fires, and cultivation measures to prevent their spread.

**Diseases and Pests**. Until the start of the programs for resistance breeding, especially crossing with *S. robustum* (see p. 69), the most severe losses to the harvest were caused by virus diseases (chlorotic streak, mosaic, sereh) (1285). Today they play a subordinate role, especially when the necessary precautions are taken (the cultivation of resistant cultivars, virus-free planting materials, sanitary measures). Through resistance breeding, correct choice and disinfection of planting material, and also sanitary measures, most of the fungus and bacterial diseases can also be controlled without too much expenditure and effort. There is a great number of these diseases (1507, 1682). The mostly noteworthy ones are above all eye-spot diseases, such as *Drechslera sacchari* (Butl.) Subram. et Jain.; pineapple disease, *Ceratocystis paradoxa* (Dade) Mor., and gumming disease, *Xanthomonas vasculorum* (Cobb.) Dows. Soil-borne diseases (such as *Pythium* spp., *Rhizoctonia* spp., *Fusarium* spp., etc.) can be controlled by crop rotation, or by special disinfection measures.

Biological pest control of insects plays a bigger part with sugar cane than with many other crops. Leafhoppers, aphids and mealybugs have numerous enemies which keep their populations at low levels. The Cuban fly (*Lixophaga diatraeae* Towns.) and other parasites have been successfully introduced to control most of the stem-borers (*Diatraea* spp. and others) and other insects, especially in tropical regions where helpful insects can become permanently established. Sanitation of the fields, the hot water treatment and disinfection of cuttings are further important aids. In spite of this, situations occur when the only way to prevent the damage caused by stem-borers is to use insecticides at the correct time (1072, 1951). Rats and other rodents often cause more damage than insects. Though modern rodenticides have much improved their control, they are still troublesome in some regions (771).

**Harvest and Processing**. Depending on the climatic and growing conditions, the cane is harvested between 9 months and 2 years after planting (and in some cases even longer). The best time for cane-cutting depends on the sugar content (high total sugar content, small amounts of reducing sugar). With canes that have already been growing for a year, new canes grow in the second year, and the average sugar content must be calculated from all the stems of different ages (779, 780). In humid tropical areas, where ripening is not determined by cool or dry weather, the entire crop can be made to ripen at a chosen time by the use of a chemical treatment ("sugarcane ripeners", glyphosphate, glyphosine, ethephon, etc.) (28, 1282, 1283, 1542). The determining factor in the choice of cutting time is in practice not only the sugar content, but also the schedule of the sugar factory. In the tropics, the factories operate up to 10 months of the year, and must be supplied with a regular flow of cane. The cane can have a sugar content of more than 15%, but mostly it is harvested with 10-11%. Locally, even this figure is often not achieved, but in some countries (e.g. Cuba, Australia) it can be exceeded by 2-4%. With all statements of yields, the length of growing time must also be considered. Under favourable conditions, a harvest of 120 tons of cane per year and per hectare can be achieved, which corresponds to a sugar yield of 13-14 tons. In the countries with advanced production systems, the average yield is 40-80 tons of cane/year/ha, but in many areas of cultivation it is under 30 tons. The global average of sugar yield per hectare per year is no higher for sugar cane than for sugar beet. Cane is still often harvested by hand: the stems are cut off near to the ground with a heavy knife (panga, machete), the green tops are cut off, and the leaf sheaths stripped off. The work is made easier by burning the fields before harvesting. In large plantations, heavy machines speed up the harvest; they lop off the tops, cut the canes near the ground or uproot them (push-rakes), and sometimes also chop the cane into smaller pieces (choppers) (558, 764). Suitable harvesting machines have also been developed for smaller plots of land and also for sloping land (1402).

Transport to the factory once used small-gauge railways, but now it is mostly done with tractor-drawn lorries and trucks. Processing should take place as soon as possible after cutting - in general within 48 hours - otherwise heavy losses can be suffered, particularly with burned cane which has been chopped into pieces. Mechanically harvested cane must be cleansed before sugar extraction (blowers, washing stations). In the traditional method, after fine chopping, the cane passes through 4-6 roller-mills, in order to separate the juice from the fibre (bagasse). For some years now, the diffusion techniques which were developed for sugar beet have also been used for cane, preferably in combination with a roller or screw press. The crystallized and centrifuged sugar is then produced (as for beet-sugar) by clarifying, evaporating, fractionally crystallizing, and centrifuging off the mother liquid (molasses) (767, 881).

Fresh sugar cane is sold in all countries which grow it, so that people can enjoy the sweet juice. That was the only purpose for which noble cane was grown by the Papuans in New Guinea. Cane is sold for chewing in local markets everywhere in the tropics. In many countries, the juice is extracted from the cane in small presses and sold as a fresh drink. For the production of alcohol for motor fuel, the mechanically extracted cane juice is immediately fermented (772, 1575). The use of the by-products - molasses, bagasse, and filter residues - has already been indicated on p. 68. The fermented molasses can be distilled to make rum, especially in Central and South America (1362). The green matter which remains after the harvest is good animal fodder. Also the cane can be processed to make fodder (1434, 1435).

# Sugar Beet

fr. betterave sucrière; sp. remolacha de azúcar; ger. Zuckerrübe

The sugar beet, *Beta vulgaris* L., ssp. *vulgaris* var. *altissima* Döll, Chenopodiaceae, is a very newly cultivated plant, whose original wild forms come from the Mediterranean region (1657, 1971). Sugar beet is well suited for cultivation in the parts of the subtropics with winter rain (861, 1601, 1603, 1725).

**Production**. The total world production is 38 million tons of beet-sugar, which is extracted from 295 million tons of beet (Fig. 17). The most important subtropical producers were (in million tons of beet), in 1988: Italy 13.4, China 13.3, Turkey 11.0, Spain 9.1, Romania 6.5, and Yugoslavia 4.6. Of these, only Italy and Spain exported noteworthy amounts (224,130 and 121,550 tons of sugar respectively).

Recommendation of cultivars, and often even prescription of cultivars, is exercised generally by the local sugar factory. Especially important qualities are bolting and Cercospora resistance. A number of countries import the seed, as seed-setting is often not satisfactory in the warmer countries.

**Ecophysiology**. The temperature requirements for sugar beet do not have narrow limits. The optimal germination temperature lies between 20°C and 25°C, the minimum is around 4-5°C, and the maximum around 35°C. The sugar beet withstands light frosts, and warm sunny days encourage the formation of sugar. The respiration is diminished on cool nights, which favourably influence the sugar content of the beet. In hot seasons, the beets have a low sugar content, a high proportion of invert sugar, reduced purity of juice, and easily become woody. Rainfall of 500 mm, evenly spread throughout the winter months (the growing season in the subtropics) is sufficient for a normal harvest. Where less rain falls, and where the cultivation extends into the summer months, the crop must be irrigated (539, 1602).

A calcareous deep loam, with a neutral to weakly alkaline reaction, is optimal for sugar beet cultivation. Their salt-tolerance is especially important for cultivation in irrigation regions (401, 896, 932). Heavy soils, which are usually used for their cultivation, are not without problems. For example, when the soil has not been worked deeply enough, unwanted forking can occur. The muddiness and crust formation after rainfall or irrigation can cause damage, especially to younger plants. Therefore, in irrigation areas, it is preferable to cultivate the sugar beet on ridges, and double rows make mechanized working easier (495).

**Cultivation and Harvest**. For cultivation without irrigation in areas with a Mediterranean climate, the seed should be sown as soon as possible in autumn. Sowing can be carried out at the end of September, even when the soil is dry. The recommended cultivation system in Morocco is as follows: the first third of the cultivated area should be sown by the middle of October, the second third by the middle of November, and the remainder by the middle of December. Under irrigation, the seeding can even stretch out over a longer time (1601). About 30 kg/ha of seed are needed, with monogerm seed only about 15 kg/ha.

Sugar beet is an intensively cultivated crop, and as such needs sufficient and well-balanced mineral fertilization. 1 ton of beet, plus 0.5 tons of leaves, remove from the soil about 4.5 kg N, 0.9 kg P, 4-7 kg K, and 1.5 kg Ca (921). The nitrogen top-dressing should not be given too late, otherwise the plant forms too many amides, which reduce the purity of the juice (1226). Controlling the growth of weeds takes a lot of manual labour, however the introduction of selective herbicides makes care of the crop easier. For phytosanitary reasons (e.g. *Heterodera schachtii* Schmidt), beet can only be grown every fourth year in any particular field, so the area needed to supply a sugar factory is very large (at least 24,000 hectares).

The harvest, which begins for example in Morocco in the middle of April, should be completed before the start of the hottest months (July), and is generally carried out by hand. Because the beets spoil easily after harvest, transport to the factory and processing should be carried out as quickly as possible.

Yields in the subtropics range from 10-100 tons/ha, with an average of 30-40 tons/ha, but yields over 100 tons/ha have been achieved under especially favourable conditions.

## Further Sugar and Sweetening Plants

Palms are the oldest sources of crystallized sugar. They exude a liquid when the sieve-tubes are tapped, and this contains 12-17.5% sucrose (370). The methods for tapping the palm, and syrup and sugar production by cooking, were developed in India. There, and in other parts of Southeast Asia, this source of sugar still plays an important role in spite of the laborious method of collection. With most palms, the juice is collected from a cut made through the inflorescence or through the stalk just below the inflorescence, and stronger secretion is stimulated by blows to stalk or inflorescence (sugar palm, *Arenga pinnata* (Wurmb) Merr.; palmyra palm, *Borassus flabellifer* L.; fish-tail palm, *Caryota urens* L.; coconut palm, *Cocos nucifera* (see p. 86); nipa palm, *Nypa fruticans* Wurmb; and others). The flow of liquid amounts to 2-4 litre per day and per inflorescence, and 7-14 litres for *Caryota*. Annually, 3-4 (to 10) tons of sugar per hectare are produced. The trunk is tapped in the case of *Phoenix sylvestris* (L.) Roxb., *P. dactylifera* L. (see p. 199), and *Jubaea chilensis* (Mol.) Baill. A major proportion of palm juice is nowadays used industrially for the production of palm wine (toddy) (370, 376, 565, 688, 1085, 1219, 1853). From toddy, arrack is distilled or vinegar produced.

Numerous cultivars exist of *Sorghum bicolor* (L.) Moench (see p. 24), which have sugar-rich juicy stems (mostly with sugar contents of 8-10%, but there are specially bred cultivars with up to 18% sugar). They originate from Africa, where they are cultivated for chewing and are often sold in the markets. The productivity of sugar sorghum is as high as that of sugar cane, when the short growing time is borne in mind (3.5-5 months; production is 30-45 tons fresh weight, 2-5.5 tons of sugar per hectare). The cultivation of "sorgo" for syrup production was widespread in the USA, but it is now much reduced. There is now fresh interest in its cultivation, in order to use the capacity of sugar cane processing factories more fully, and also for sugar production in warmer countries where there is not enough rainfall or irrigation water for sugar cane (38, 235, 434, 533, 736, 815, 1690, 1890). *Echinochloa stagnina* too is grown in small quantities in the Niger basin for sugar and alcohol production (Table 51, it has in its stems 10% sucrose, and 7% reducing sugar).

The extraction of sugar-containing juice from agaves is limited to Mexico. The top of the young pole is cut off and the stem is hollowed out. The juice which comes out is drunk fresh (aguamiel), or fermented to form pulque (production of more than 3 million hectolitres pulque per year). For this purpose, large areas of *Agave atrovirens* Karw. and *A. lehmannii* Jacobi are cultivated (200, 1039, 1445). The distillates "mezcal" and "tequila" come from the fermentation of the crushed sugar-containing stems of *A. tequilana* Weber, and other *A.* species (125).

Occasionally, the sweet pith of some cacti (e.g. *Ferocactus wislizeni* (Engelm.) Britt. et Rose, *Stenocereus thurberi* (Engelm.) Buxb.) is used in the southwest of the USA and in Mexico for the production of syrup and a type of sugar. To the plant products whose most important component is sugar, we must finally add some types of fruit, especially the dates (see p. 199).

The interest in natural sweeteners which do not contain carbohydrates has been reawakened because of the suspicion that synthetic sweeteners, especially cyclamates, are carcinogenic (449, 453, 813). Of particular importance are:

- the extremely sweet-tasting protein monellin, found in the berries of *Dioscoreophyllum cumminsii* (Stapf.) Diels, Menispermaceae (1764);
- thaumatin, from the aril of *Thaumatococcus daniellii* Benth., Marantaceae (77, 1223, 1316);
- also, the glycoprotein miraculin, from the fruits of *Synsepalum dulcificum* (Schum. et Thonn.) Baillon, Sapotaceae, which changes "sour" tastes into "sweet" (117, 1223);
- *Stevia rebaudiana* (Bertoni) Hemsl., Compositae, containing in its leaves the glycoside stevioside, which tastes 300 times sweeter than sucrose. It is the only one of the four species which is cultivated. Commercial production is established in Paraguay, Brazil, Indonesia, Japan, China and Thailand. The export of dried stevia leaves from Thailand reaches 50-80 tons per month (75). Cultivation trials are in progress in several other countries, e.g. Spain and the USA (California) (242, 649, 943, 971, 1646, 1647, 1701).

# Oil Plants

The production of oil plants (609, 807, 1138, 1519, 1856, 1915) takes third place in world production in terms of value, after starch plants and fruit, and ahead of beverages and stimulants (tea, coffee, tobacco), and sugar. More than 90% of the oil plants are produced in the tropics and subtropics, a far higher percentage than with the other groups of food plants (about three-quarters of the cereals and sugar are produced in the hot countries). About a third of the crops are exported; the value of oil and fat exports from the hot countries is only exceeded by that of the non-food goods. Therefore, the cultivation of oil plants plays a major role in development politics (508, 1848).

Fig. 19 shows the production of oil plants over the last 10 years. The diagram of the amount of oil produced (Fig. 20) shows a different ordering of the plants from that in Fig. 19. This is explained by the fact that not all of the seed produced is used for oil extraction (e.g. cottonseed, groundnut), and that the oil contents of the raw materials exhibit considerable differences (soya 18%, peanut 46%). Soya oil has strengthened its dominant role. Notable increases are seen in the production of sunflower (+89%), rape (+167%) and palm (+136%) oils. Most of the other oils show little change, apart from seasonal fluctuations. Further important oil plants have the following production figures, in 1,000 tons of oil: cottonseed 3192, linseed 720, maize 1165, palm kernel 1040, safflower 235, nigerseed 120, tung 95, babassu 110, cocoa butter 100, tea seed 100. All other plant oils are produced in substantially lower amounts; their production figures are given in Tables 17 and 18, as far as they are known (1308).

Over 90% of the plant oils are used as edible oil, the remainder being used for technical purposes. As foodstuffs, oils have an important place in the energy supply; 1 g of oil gives on average 38 kJ energy. In cooking, they serve as frying fat in the creation of numerous tasty dishes, improve the quality of the meal as salad oil or mayonnaise, add their own taste, or enhance the taste of other foodstuffs, modify the taste and fragrance of seasonings - in short, they are the basis of all the finest cuisines. The linoleic acid, which is found in many plant oils (Tables 14 and 15), is the essential fatty acid ("vitamin F"), the shortage of which leads to deficiency symptoms, especially in growing children and animals (513, 1950). The polyunsaturated fatty acids (particularly linoleic and linolenic acids) lower the fat and cholesterol contents of the blood, and are therefore important for the prevention of coronary sclerosis. Plant oils contain vitamin A (e.g., carotenoids, which serve as pro-vitamin A), vitamin E (tocopherol), and ergosterol (pro-vitamin $D_2$), also phospholipids and sterols which are important for health. Through refining and hydrogenation of the vegetable oil, the majority of these useful components are generally removed or destroyed.

From ancient times, oils have also been used technically for lighting and as burning materials, in soap-making, and as painting and lubricating materials. In more recent times, their use has developed in synthetic materials, detergents,

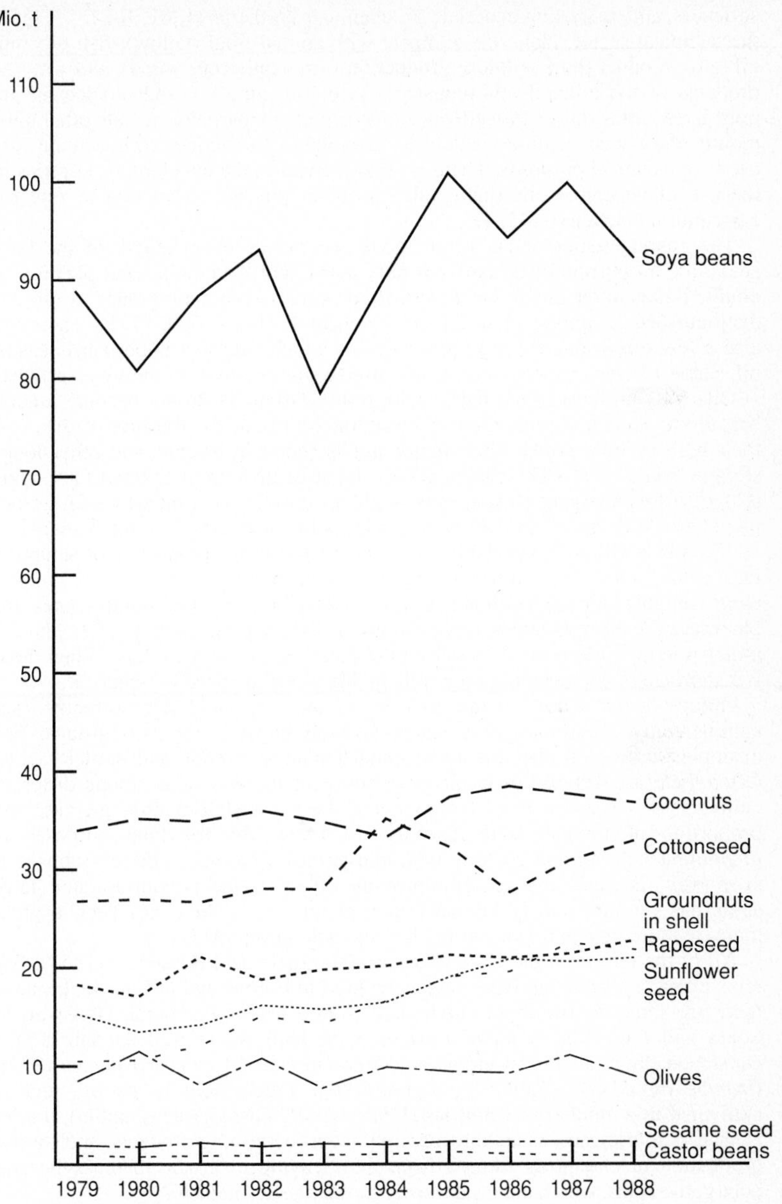


Fig. 19. World production of oilseeds, 1979-1988

softeners, and as starting materials for chemical syntheses (1361, 1614, 1771). In the technical sector, plant oils compete with animal offal (tallow), fish oils, tall oil (a by-product from cellulose production from coniferous wood), and with the products of the mineral oils industry. Castor oil, tung oil, and oiticica oil are only used industrially, linseed oil and coconut oil partially so; all other oils, mainly the cheap qualities which are unsuitable for human consumption, are used for technical purposes. There is great interest in the developing countries in the use of vegetable oils (palm oil, sunflower oil) for conversion to fuel for agricultural machinery (43, 49).

The characteristics of an oil are governed by the chain length of the fatty acids, the proportion of unsaturated fatty acids, and the number and position of double bonds in the chain. For practical purposes, seven main groups of oils are distinguished, examples of which are given in Table 14 (609, 1771). There are also a few oils with oxidized fatty acids, of which the most important is castor oil. Table 15 gives the most commonly used representatives of the seven groups.

Oils (fats, in fact) in the lauric-acid group (Group 1) do not become rancid, because of their low proportion of unsaturated fatty acids. Because of this, and their high melting points, their major use is found in biscuits and other long-storage baked goods. The chemical industry uses their short-chained fatty acids ($C_{10}$-$C_{14}$) for detergents; these fatty acids are also the reason why soaps from palm kernel oils have particularly good cleansing properties (best toilet soaps).

The oils in Group 2, vegetable butter, also have a high proportion of saturated fatty acids, and so seldom become rancid. Their melting point (32-35°C) makes them suitable for confectioneries that will "melt in the mouth", and for pharmaceutical preparations (suppositories). Cocoa butter (see p. 262) plays a major role in world trade; it is one of the most expensive plant fats. When there is a shortage of cocoa butter, other oils in this group are used to replace it.

Groups 3 and 4 contain the majority of the oils which are normally used commercially. There is not a sharp division between the two groups. For example, sesame oil contains about equal amounts of oleic and linoleic acids. Often there are differences in the proportions of the two acids among different cultivars of a species. Cool temperatures during seed formation increase the proportion of linoleic acid. The use of these oils for food purposes is predominant (salad and cooking oils, margarine). The oils in these groups still keep well. Because the health-improving effect of the polyunsaturated fatty acids has become widely known (since about 1950), there has been a great increase in interest in the oils of the linoleic-acid group (513).

All of the oils from the Cruciferae contain erucic acid (Group 5) (1857). For rape, erucic acid-free cultivars have been bred in Europe and in Canada, because there are some doubts about erucic acid from a health viewpoint. However, in South and East Asia, a lot of rape oil with high erucic acid content is still consumed; there also exist erucic acid-free cultivars of *Brassica juncea* (973). *Crambe abyssinica* (Table 18) contains 60% erucic acid in its oil, and is cultivated in a number of countries (USA, USSR, East Germany, India); the oil is used as a lubricant, as an anti-foaming agent in washing products, and for the production of chemicals formed from the cleaving of erucic to brassylic and pelargonic acids, which are used mostly in the plastics industry.

The oils in Group 6 are much less stable than those in Groups 1-4. Linseed oil and soya oil are the most important economically. Soya oil plays a major role as a foodstuff (see p. 94). In Eastern Europe, linseed oil is also used as a food oil;

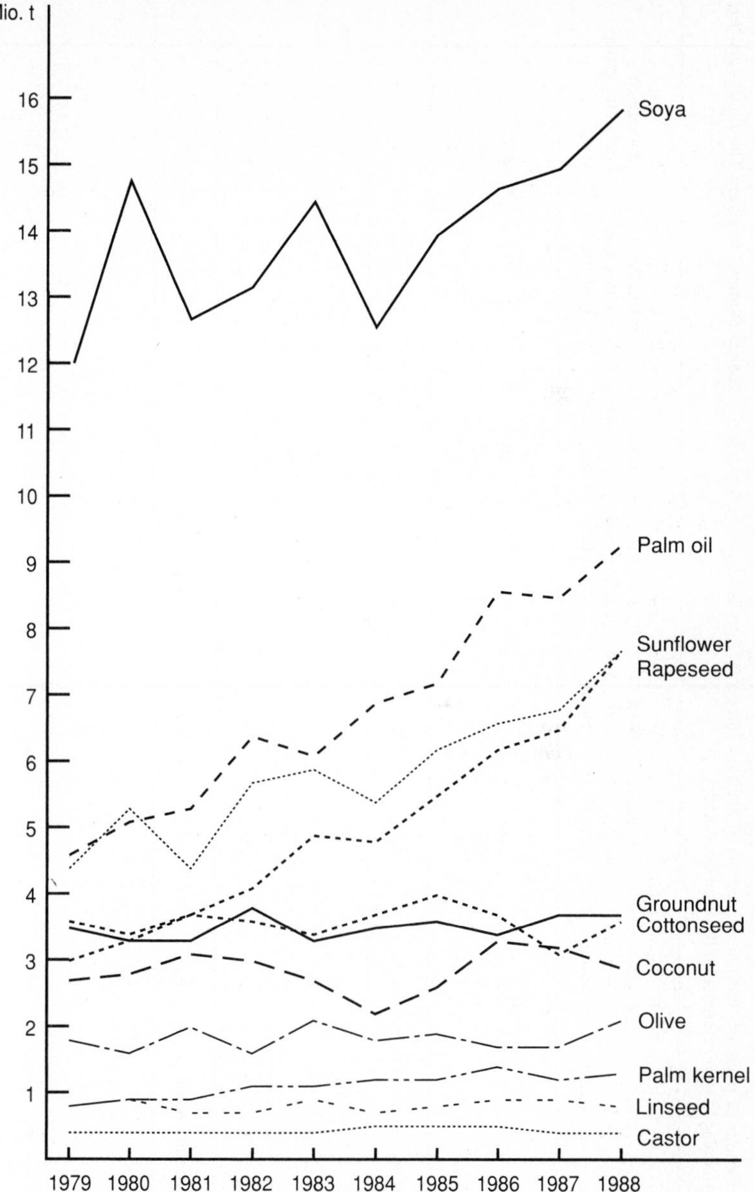


Fig. 20. World production of oils from the most important oil crops, 1979-1988 (517)

Table 14. Groups of plant oils with their contents of the most important fatty acids. For each group, a typical example is given. The figures are the average percentages with reference to the total fatty acids

| Group / Fatty acid | Example | 1 Lauric acid group Coconut | 2 Vegetable butter Cocoa | 3 Oleic acid group Peanut | 4 Linoleic acid group Safflower | 5 Erucic acid group Rapeseed | 6 Linolenic acid group Linseed | 7 Conjugated fatty acids group Tung |
|---|---|---|---|---|---|---|---|---|
| Lauric | | 50 | 0 | 0 | 0 | 0 | 0 | 0 |
| Myristic | | 15 | 0 | 1 | 0 | 0 | 0 | 0 |
| Palmitic | | 9 | 24 | 8 | 4 | 2 | 6 | 4 |
| Stearic | | 2 | 35 | 4 | 2 | 0.5 | 4 | 1 |
| Saturated fatty acids | | 91.5 | 60 | 20 | 7.5 | 6 | 12 | 5 |
| Oleic | | 7 | 38 | 60 | 15 | 14 | 20 | 8 |
| Erucic | | 0 | 0 | 0 | 0 | 45 | 0 | 0 |
| Linoleic | | 1 | 2 | 20 | 75 | 14 | 20 | 4 |
| Linolenic | | 0 | 0 | 0 | 0 | 10 | 45 | 3 |
| Eleostearic | | 0 | 0 | 0 | 0 | 0 | 0 | 80 |
| Unsaturated fatty acids | | 8.5 | 40 | 80 | 92.5 | 94 | 88 | 95 |

Table 15. Plant oil groups, and the oils belonging to them

1. **Lauric acid group**
   babassu
   coconut
   palm kernel
   tucum
   and other palm kernel oils

2. **Vegetable butter**
   Borneo tallow
   Chinese tallow
   cocoa butter
   mowra fat
   sheabutter
   ucahuba
   and some others

3. **Oleic acid group**
   avocado
   date kernel
   peanut
   almond (peach, apricot)
   olive
   palm oil (and other palm fruit oils)
   pecan nut
   pistachio nut
   rice bran
   sesame
   teaseed

4. **Linoleic acid group**
   cotton seed, and other Malvaceae
   kapok seed
   pumpkins and other cucurbits
   maize germ
   poppy seed
   nigerseed
   safflower
   sunflower
   sorghum germ
   tobacco seed
   tomato seed
   grape seed

5. **Erucic acid group**
   crambe
   rapeseed
   mustard
   and other Cruciferae

6. **Linolenic acid group**
   hemp seed
   linseed
   candle nut
   perilla
   soya
   tallow berry
   walnut
   wheat germ
   citrus seed

7. **Conjugated fatty acids group**
   oiticica
   tung

the oil, which dries well, is mostly used in the West in paints and varnishes. Perilla oil has the highest linolenic acid content of the oils in this group (65%).

Group 7 contains only technical oils. Because of the conjugated position of the three double bonds in eleostearic acid, they dry very quickly, and provide hard, water-resistant coats of paint.

Oils are produced from seeds (556, 1771) by pressing (hydraulic presses, screw presses) or by extraction with cyclohexane or other solvents. The separation of the oil from the solid parts of the flesh or seeds is usually followed by a purification (refining) process, in order to remove any unwanted cloudiness, colour, taste and odour factors. By-products of the refining of some oils are phosphatides (soya, maize, cotton), phytosterols (soya) and tocopherols (soya, cotton, rice bran, maize germs, etc.). The most important by-products of oil production from seeds are the protein-rich cake and meal from the leftovers of extraction, which are used for high-protein animal feed, and also as human food (see p. 121, Table 20). Soya has importance as a supplier of protein (30-40% protein in the seeds) at least as much as an oil source (18% oil), but also with most of the other oil sources, the economics of their cultivation depends on the value of the by-products from pressing and extraction. During earlier times, almost all seeds were exported, but now it is generally accepted that the processing to oil and cake and meal should take place in the country of origin, and to export the finished products, as the remains from pressing are often needed in these countries as sources of food and fodder. For cotton, peanut, linseed, rape and sunflower, the exports of the finished products, oil and cake, now exceed the exports of unprocessed seeds.

Crucifer seed cakes contain glucosinolates, which limit their usage as fodder materials. For cotton cakes, the gossypol content prohibits their use as fodder to monogastric animals (those with single stomachs, e.g. pigs), or else as a foodstuff for humans; plant breeding has produced gossypol-free cultivars, and their cultivation began several years ago (see p. 341). Even soya flakes contain compounds which must be destroyed before they can be used as a major source of protein (see p. 97). The extraction remnants from castor and tung oils are so poisonous for both men and animals that they can only be applied to the fields as fertilizers.

## Oil Palm

fr. palmier à huile; sp. palmera de aceite; ger. Ölpalme

The African oil palm, *Elaeis guineensis* Jacq., belongs to the subfamily Cocoideae of the Palmae, which contains numerous other oil-providing palms, apart from the coconut (Table 17). The American oil palm, *E. oleifera* (H.B.K.) Cortes (*Corozo oleifera* Bail.) is also assigned to the genus *Elaeis* (132, 287, 550, 1182, 1387, 1657). Both of these can easily be crossed and produce fertile hybrids (664, 696, 697, 1558).

**Production**. Apart from tropical West Africa, from Guinea to Angola, the oil palm is cultivated especially in Malaysia and Indonesia, and recently in increased quantities in South and Central America (particularly Colombia and Ecuador) (711, 1198). In 1988, the largest producers were (figures in 1000 tons of oil): Malaysia 5033, Indonesia 1370, Nigeria 750, Côte d'Ivoire 235, Zaire 170.

The Southeast Asian countries dominate the export trade (given in 1000 tons of oil): Malaysia exported 4342, Indonesia 886. The exports from Africa are small, since there, the largest part of the oil is used domestically; only the Côte d'Ivoire exports a significant amount: 110,000 tons.

Palm kernel oil is another export product. The market is dominated by Malaysia (535,665 tons). The total export from Africa is 43,130 tons. Palm kernels themselves, which previously were exported in considerable quantities, no longer play a role in the world markets.

**Morphology and Anatomy**. Palms in good growing conditions develop a unisexual inflorescence in each leaf axil. On each tree, there are periodic changes between the production of male and female inflorescences. With young plantations, it can occur that there are insufficient male inflorescences present to pollinate all of the female ones; then the yield can be increased by artificial pollination. The dry pollen can be stored for a month without any special treatment (89). Through importation of *Elaeidobius* beetles, which improve the pollination in West Africa, the yields in Malaysia have been notably raised (1773).

The inflorescences begin to develop 33-34 months before flowering, and their sex is determined 24 months before flowering (1087). The development of the fruit takes 5-9 months from pollination to ripening. The spherical fruit bunches, which weigh 15-25 kg, are formed from the aggregation of 1000-4000 egg-shaped, 3-5 cm-long individual fruits (Fig. 21). The largest proportion of oil is found in the fruit pulp (mesocarp); about an eighth is in the endosperm. The fruit oil belongs to the oleic acid group, the kernel oil to the lauric acid group (Tables 14 and 15). Three types are classified, depending on the thickness of the endocarp: dura (DD), shell 2-8 mm thick and 35-55% of the fruit's weight; tenera (Dd), shell 0.5-3 mm thick and 1-32% of the fruit's weight; pisifera (dd), no shell, mostly sterile (Fig. 21).

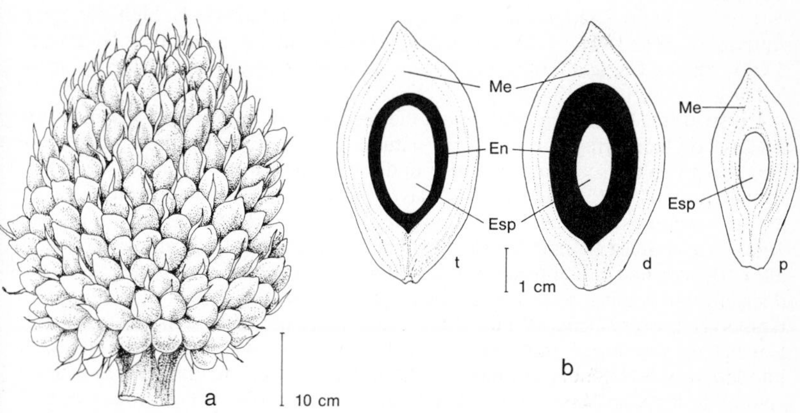


Fig. 21. *Elaeis guineensis*. (a) fruit bunch, (b) fruits of tenera (t), dura (d), and pisifera (p) in longitudinal section. Me = oil-containing mesocarp, En = hard endocarp, Esp = endosperm

**Breeding**. The selection of high-yielding strains, and the production of hybrid seed (tenera, from the cross between dura and pisifera) have led to an extraordinary increase in the potential yield of oil palms. The African wild palms ("bush palms") gave 0.6-1 tons of oil per hectare; currently yields of more than 7 tons mesocarp oil and 1.25 tons kernel oil per hectare are achieved in Malaysia. Thus, the oil palm has become by far the highest yielding oil plant. The hybrid seeds for tenera plantations are produced in special breeding nurseries (90, 535). Apart from the major aims of a thin endocarp and a thick mesocarp, the breeders strive to achieve earlier fruiting, slower stem elongation, higher fruit weight in the fruit bunch (currently about 65%), and resistance against fusarium wilt and dry basal rot (1519). The crosses with *E. oleifera* have not yet resulted in a commercially usable cultivar (1, 364, 711). The productivity of oil palms can be further increased through fertile pisifera forms (325, 855). The propagation by tissue culture technique, which is already practised on a fairly large scale, can speed up the breeding of improved oil palm cultivars considerably, but there are problems because of somaclonal variations (365, 456, 985, 1309, 1448, 1704).

**Ecophysiology**. The oil palm needs as even a temperature as possible in the range 24-28°C. Its cultivation is therefore essentially limited to regions between 10° North and 10° South, and to altitudes below 500 m. The yields are lower in cooler conditions. For the best results, oil palms need a lot of sunshine; an average of 5-6 hours per day is optimal. The annual rainfall should be between 1500 and 3000 mm. The plants will endure a dry period of about three months without an appreciable decrease in the yield (44, 711, 1087).

The soil should be deep and well drained. Brief and occasional flooding does not cause any damage. The best pH range for the soil is 5.5-7.0, although the oil palms thrive even at pH 4, if they are provided with the correct fertilizers. The majority of the nutrient-absorbing roots of fully grown palms lie 0-30 cm deep, in a circle with a radius of 5 metres. The nutrient intake is high for good cultivars (1281, 1841). One ton of fruit bunch contains about 6 kg N, 1 kg P, 8 kg K, 0.8 kg Ca, and 0.6 kg Mg. Of the trace elements, B and Cl are the most important (399, 1775, 1844).

**Cultivation**. In modern cultivation, the seeds are put into special germinators, because of their irregular and slow natural germination. The best results are obtained when the seeds are first held in dry heat (moisture content of seed 18%, temperature 38-40°C, time length 40 days). Then the seeds are allowed to absorb as much water as they can (water content of tenera seed 28-30%, length of time 7 days). The moist kernels are then held at 28-30°C, and the germination begins after 10 days, and is complete after 30-40 days. The newly germinated seeds are dispatched from the breeding station, and transplanted at the plantation into temporary nursery beds, or into small polythene bags or baskets. When 4 or 5 leaves have developed, they are transplanted into a growing bed in the fields, or into larger polythene bags or baskets. After about 10-12 months, they are strong enough to be planted out in the plantation (115, 353, 1841). The propagation of oil palms by cloning (tissue culture) can play an increasing role in the future (see p. 83).

In the plantation, the distance between palms should be 8-9 metres (130-150 palms/hectare). Between the palms, it is best to sow legumes as a ground-cover crop and to provide nitrogen (*Pueraria phaseoloides*, *Centrosema pubescens*, etc., see Table 53) (364, 1319).

In the first years, intercropping is possible (groundnuts, millet, etc.) (711). An area around each palm is kept free from vegetation (with fully grown palms, 2.5 m diameter), and also the connecting paths, so that harvesting is possible. For this, herbicides are usually used in the larger plantations (MSMA, gramoxone, etc.) (1787).

**Diseases and Pests**. With oil palms, one can generally avoid most diseases and pests by taking the following measures:

- raising healthy plants (treating the nuts with TMTD and streptomycin, weeding out any weak plants in the nursery, applying pesticides in the nursery beds) (320),
- choosing an appropriate site for the plantation,
- breeding cultivars resistant to the main diseases (spear and bud rot, *Fusarium oxysporum* Schlecht. f. *elaeidis* Toovey (1496) and other micro-organisms;

leaf-spot disease, *Cercospora elaeidis* Stey.; dry basal rot *Ceratocystis paradoxa* (de Sey.) Moor). (982, 1497).

With good care of the plants, disease control measures in the stand are not usually necessary (1839, 1840). Also, the insect pests, such as *Oryctes* spp. and *Rhynchophorus* spp., are kept within satisfactory limits by sanitary measures in the plantation; the use of insecticides is necessary only occasionally (1310, 1981). The development of biological pest control methods is in the forefront of research (591, 1310, 1982). In some places, the control of rats can be important, as they gnaw the stems of young plants, and will also eat the fruit bunches.

**Harvest and Processing**. The palms begin to bear fruit after 4 or 5 years. The fruit bunches are cut off or knocked down when the fruit is ripe. The correct time for harvesting is indicated when the fruit changes colour (colour change from black to orange for the most commonly grown cultivars). In practice, the fall of individual fruits is used as the sign of ripeness. In order to cut off the fruit bunches on old, tall palms, curved knives fastened to bamboo poles are used; alternatively, someone has to climb up the palm. In some of the large plantations, motorized lifting platforms are in use. After 30 years, the palms are so tall that harvesting is too difficult; then, replanting is necessary. The fruit bunches which have been cut down must be taken to the factory as soon as possible (by narrow gauge railway, trucks, lorries), because they should be processed within 24 hours. The reason for this is that some fruits are always damaged during harvesting; free fatty acids build up in these, reducing the value of the oil. In the factory, the fruit bunches are first treated in an autoclave (sterilizer), so that the lipases are destroyed by heat, and to facilitate the threshing (separation of the fruit from the fruit bunch). The separated fruits are next treated in a digester, in which they are stirred to a pulp at a temperature of 95-100°C (for a time of 20-75 minutes, depending on the method of oil separation). The oil is separated from the fibres with a screw press. Before dispatch, the oil is clarified, in order to remove any dirt, fibres, or gums. In recent years, palm-oil mills for small operations have been developed (1075).

The palm nuts are cleansed (by removing any pulp), dried, and broken in a cracker. The kernels are separated from the shells in a mixture of water and clay, washed, and then dried. Previously, the oil was extracted in the consumer country, but now it is extracted in many of the producer countries themselves (by pressing, or by solvent extraction) (see p. 82).

The best tenera plantations yield 30 tons of fruit bunches per hectare and per year at the peak of their productivity. From that, about 7 tons of palm oil and 0.8 tons of kernel oil are extracted. These figures are for recent plantations in Malaysia and Colombia. In older plantations, especially in Africa, the yields are substantially lower, and seldom amount to more than 3.5 tons oil per ha per year.

After refining, the palm oil is mostly used for the production of margarine and cooking fat (Table 15, Group 3). Palm kernel oil is very similar to coconut oil, and is used for the same applications (see p. 78). The press cakes from palm kernel oil contain 15-16% protein, and are used for animal fodder (664, 1449).

## Coconut Palm

fr. cocotier; sp. cocotero; ger. Kokospalme

**Production**. The coconut palm, *Cocos nucifera* L., originates from the Melanesian region (560, 705, 1274, 1303, 1657, 1792, 1987). This is still the main area of cultivation, even though it has now been grown in all parts of the tropics for a long time. Asia produces 29.9 million tons of coconuts, Oceania 2.3, Central America 1.6, Africa 1.7, and South America 1.0. The countries with the largest production figures are Indonesia 11.5, Philippines 8.6, India 4.6, Sri Lanka 1.4, and Malaysia 1.2; outside the Southeast Asian region, the noteworthy producers are Mexico 1.0, Brazil 0.7, Côte d'Ivoire 0.5, Mozambique 0.4, and Tanzania 0.4 (all figures in million tons of nuts).

The export products which are produced from the coconut palm are coconut oil, copra, coconut flakes (651, 1395) and fresh coconuts. The most important exporters, and the total world export figures, are set out in Table 16.

Table 16. Exports of coconut products 1988. Main exporters and total world exports. Figures in 1,000 tons (518).

| Area | Coconut oil | Copra | Desiccated coconut | Coconuts |
|---|---|---|---|---|
| Philippines | 793 | 80 | 88 | 2 |
| Malaysia | 56 | 18 | 22 | 46 |
| Papua New Guinea | 36 | 74 | - | - |
| Sri Lanka | 5 | 6 | 22 | 12 |
| Pacific Islands (apart from Papua New Guinea) | 25 | 81 | - | 1 |
| World | 1335 | 286 | 170 | 125 |

**Morphology and Anatomy**. An inflorescence can develop in each leaf axil. Male and female flowers are formed in the same inflorescence, the female on the lower part of the panicle branches, the male on the upper. The male flowers open 10-20 days before the female (protandry), so that cross-pollination is the rule (pollination is by insects, especially bees, and by wind). The fruit then takes 12-14 months to develop to full ripeness. It reaches its eventual size after the first 6 months, thereafter first the embryo develops, then the hard endocarp, and lastly the firm endosperm (Fig. 22). These figures are for the high-growing, large-fruited types which are normally cultivated. The dwarf forms mostly have smaller fruit, which ripen 3 months earlier than the larger types.

**Breeding**. Planned breeding has only been carried out for a few decades (535). Earlier yielding and resistance against "lethal yellowing" are in the foreground of efforts, and both are being attained by crossing with dwarf types (Malayan Dwarfs). $F_1$ hybrids between selected tall and dwarf growing parents have better productivity than the original strains; they are already cultivated in substantial numbers (90, 1044, 1090, 1519, 1538, 1586, 1589).

**Ecophysiology**. The primary area of cultivation of the coconut palm ranges from the equator to 15° North and South. For good yields, an annual average temperature of 26-27°C is needed, with small temperature differences between day and night; because of this it is grown no higher than 750 m above sea-level, even at the equator. When sufficient groundwater is available (beaches beside the sea), it grows very well even in dry regions. Where it depends on rainfall, 1250-1550 mm per year is optimal. It needs a lot of sunshine.

The ground should be well drained and aerated, otherwise the coconut palm makes very few demands on the soil. In dry areas, groundwater should be available at a depth of 1.0-2.5 metres. It endures up to 1% salt in the groundwater. As with the oil palm, it needs adequate Cl for optimal growth (1102, 1844). The combination of ecological conditions referred to is provided most completely near the sea, above all in low-rainfall areas. But coconut palms can also be found growing quite distant from the sea; if the soil and the climate are right, they can also be successfully cultivated in these regions (44).

A thousand fruits contain about 10 kg N, 2 kg P, and 10 kg K. The use of K fertilizers may be unnecessary if the fibrous husks are returned to the soil. Chlorosis is prevented in areas far from the sea by giving each plant 2 kg of cooking salt each year (1116, 1775).

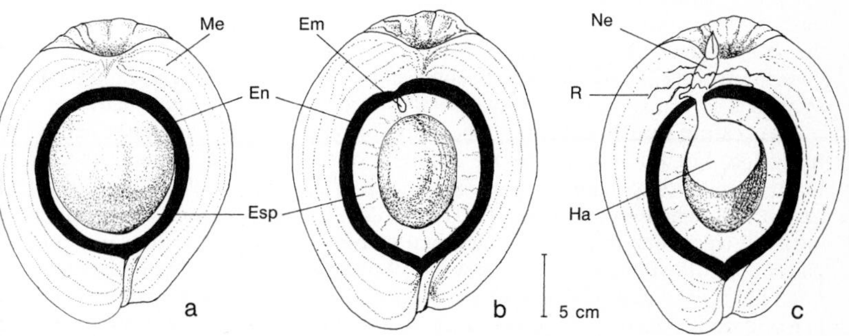


Fig. 22. *Cocos nucifera*. Fruits in longitudinal section. (a) unripe fruit, the formation of the firm endosperm has begun; (b) ripe fruit; (c) seedling after 3 months. Me = fibrous mesocarp, Em = embryo, En = endocarp, Esp = endosperm, Ne = neck of germinating embryo, R = first roots within the mesocarp, Ha = haustorium ("apple")

**Cultivation**. For propagation, the whole fruits are put into a seed-bed, at best horizontally, with the narrowest side downwards, and covered with earth so that the top edge is above the surface. Germination is also carried out in plastic containers (1998, 1999). On germinating, the embryo forms a large haustorium, which digests the endosperm and absorbs the reserve material. The first roots are formed inside the fibrous hull, and only after 4 or 5 months do more robust roots develop which penetrate into the soil (Fig. 22). Propagation by callus culture is possible, but has not yet reached the stage of practical application (227, 1341). After 6-12 months, the seedlings can be transferred to their final position. At this time, the reserve material in the endosperm is not yet all used up. Germination and seedling development are quicker for dwarf than for large types. Only those seedlings which have germinated and developed strongly after 4½ months should be planted out. Planting distances should be 9 m for tall types, and 6-7 m for dwarf varieties and the hybrids. A circle around young palms should be kept free of weeds. Where there is enough rainfall, cropping between the palms can be carried out, especially in the first years of newly planted palms, and then also later with fully grown palms (1246). The highest yields will be achieved when legumes (Table 53) are planted as a ground-covering crop. Economically, a double usage with the grazing of animals can be the most profitable (1410, 1505, 1694).

**Diseases and Pests**. The greatest damage to fully grown palms is engendered by diseases for which the causal factors are not yet sufficiently clear; these diseases are lethal yellowing in Jamaica, Cuba and Haiti, cadang-cadang in the Philippines, and Kerala disease (root (wilt) disease) in India (251, 358, 877, 1285). Lethal yellowing is a mycoplasm (436, 820), and Malayan Dwarfs are resistant to it (159). Of the fungus diseases, bud rot (*Phytophthora palmivora* Butl.) is particularly dangerous in regions with high rainfall.

The insect pests to be careful of are chiefly the rhinoceros beetle, *Oryctes rhinoceros* L., and the palm weevil (*Rhynchophorus* spp.). There is also a range of leaf-eating beetles and moth caterpillars (982, 1047, 1310). Sanitation of the plantation is the most important action to be taken for their control. The use of insecticides is only necessary for severe attacks. Much to be guarded against are the rats, shrews, and perhaps also bats and squirrels, which enjoy eating the growing fruit (324, 1047). The rodents can be prevented from climbing the trees by means of "rat slopes", i.e. metal bands shaped similarly to lampshades.

**Harvesting and Processing**. Coconuts for oil production are harvested when fully ripe. For this, the palms are climbed, or the coconuts are cut off using a knife attached to a bamboo pole. In Indonesia, trained monkeys are used, and one of these can harvest up to 1,000 nuts a day. In many regions, the people wait until the ripe fruits have fallen, and then gather them from the floor in a regular pattern of collection, about once a week (if there is no fear of them being stolen). Each palm produces between 30 and 50 nuts a year, and 8,000 nuts per year and per hectare is considered a good yield. With hybrids, 200-600 nuts/palm/year are achieved. High yields can be expected from a coconut palm up to its 40th year, after which production distinctly decreases. In order to separate the oil from the flesh, the endosperm must be dried to become copra. For this, the nuts (which have often been stored for several weeks) are first separated from the fibrous husks, then the nuts are split with a blow from a heavy knife or hatchet, and left to lie in the sun for a while to dry, so that the endosperm is more easily removed from the shell. Drying takes place quite quickly in the sun or in an oven (651). Copra contains 65-70% oil, and most of the oil is won by pressing (1797). Maximum yields of the new cultivars are 9 tons copra/ha, from which 6 tons oil and 3 tons cake are produced. Therefore, the oil yield is not significantly less than that for oil palms, and the production of coconut oil involves less capital investment in the processing plant.

There is increasing significance in the production of oil in the country where the nuts are grown, and copra is hardly exported nowadays (Table 16). The by-product is the cake, which contains about 20% protein, and is a good animal fodder. Attempts have been made to produce the oil directly from the fresh

Table 17. Further oil-producing palms (242, 370, 550, 1039, 1518)

| Sub-family and botanical name | Vernacular names | Distribution | Used part for oil extraction | Remarks and literature references |
|---|---|---|---|---|
| **Cocoideae** | | | | |
| *Acrocomia aculeata* (Jacq.) Lodd. ex Mart. | Paraguay palm, macaya, micauba, gru-gru | Tropical S America, W Indian Islands | pulp, seeds | Industrially utilized in the State of Minas Gerais. Fruit oil for soap, edible kernel oil, oil content of kernels 50-55% (1065) |
| *A. mexicana* Karw. | coyol, babosa, Mexican gru-gru palm, wine palm | Mexico to Guatemala | seeds | Seeds contain 40% fat, sold in local markets (1960) |
| *A. totai* Mart. | mbocayá palm, totai palm, gru-gru palm | Bolivia, Paraguay, Argentina | pulp, seeds | In Paraguay the fruit and kernel oils are industrially extracted, and to a small extent exported (1129, 1145) |
| *Astrocaryum jauary* Mart. | awarra palm, jauary palm | Brazil | pulp, seeds | Edible fat |
| *A. murumuru* Mart. | murumuru | Amazonas region | seeds | Kernels contain 43% fat with high melting point (33-36°C), used for margarine |
| *A. tucuma* Mart. | tucuma palm, tucum | Tropical America | pulp, seeds | Oil from the pulp for soap. Kernels contain 30-50% fat, used for margarine, also exported (1425) |

Table 17. Further oil-producing palms, cont'd

| Sub-family and botanical name | Vernacular names | Distribution | Used part for oil extraction | Remarks and literature references |
|---|---|---|---|---|
| *A. vulgare* Mart. | awarra palm, cumara palm, aoura palm | S America | pulp, seeds | Grows well on sandy ground. Oils used as for the other *A.* species (1425) |
| *Attalea funifera* Mart. ex Spreng. | Bahia piassava, coquilla nut | Brazil | seeds | Fat utilized in the margarine and chocolate industries. Main product of the palms is piassava fibre |
| *Bactris gasipaës* H.B.K. (*Guilielma gasipaës* (H.B.K.) L.H.Bailey) | pejibaye, peach palm, parépon, piritu, gachipaes, masato, uvito, chonta ruru, pupunha, amana | Central and S America | seeds | Important food plant in Costa Rica, Panama, Colombia, Venezuela and Ecuador. Old crop plant of the Indios, mostly cultivated because of its high-starch fruit flesh (see p. 63); also seedless forms which are vegetatively propagated using basal side-shoots. Red pigment of the fruit flesh used for colouring food. Seeds provide macanilla fat (37, 185, 186, 336, 1258, 2024) |
| *Maximiliana maripa* Drude | maripa, naxáribo | Guyana | seeds | Fruit flesh tastes of apricots. Seed oil known under the name maripa fat (133, 1425) |

| | | | | |
|---|---|---|---|---|
| *M. regia* Mart. | cocorite palm, cucurita palm, inajá palm, jaguá palm | Brazil | seeds | Similar to *M. maripa* (609) |
| *Orbignya cohune* (Mart.) Dahlgr. ex Standl. | cohune palm, manaca | Mexico, Central America | seeds | Fruit flesh edible, also as animal fodder. In Mexico, 20,000 tons nuts harvested annually. Oil used as salad oil, and for technical purposes |
| *O. phalerata* Mart. (*O. barbosiana* Burret, *O. martiana* Barb.-Rodr., *O. speciosa* (Mart.) Barb.-Rodr.) | babassu palm, babaçú | Brazil | seeds | Large wild stands in Brazil. Annual production 250,000 tons kernels, giving 75,000 tons oil. 4,000 tons exported. Press cakes a valuable fodder. Shells to a large extent used for charcoal production (56, 575, 1130, 1145, 1159, 1258, 1425) |
| *Scheelea martiana* Burret (*Attalea excelsa* Mart. ex Spreng.) | ouricoury, urucuri | Central America, Northern S America | seeds | Kernels eaten fresh or pressed for oil extraction. Wax of the underside of the leaf used as with carnauba wax (see p. 394). Also with other *S.* species, the seed fat is used, sometimes they also have edible fruit flesh (1960) |
| *Syagrus coronata* (Mart.) Becc. | ouricury palm, licuri | Brazil | seeds | Fruit flesh edible. Kernels contain 70% of a valued oil. Leaves provide ouricury wax (Table 50) |
| **Coryphoideae** | | | | |
| *Erythea salvadorensis* (H.Wendl. ex Becc.) H.E.Moore | palma | Central America | seeds | Fruit flesh sweet and edible, seeds provide an oil used locally |

Table 17. Further oil-producing palms, cont'd

| Sub-family and botanical name | Vernacular names | Distribution | Used part for oil extraction | Remarks and literature references |
|---|---|---|---|---|
| **Arecoideae** | | | | |
| *Euterpe edulis* Mart. | juçara, jiçara, assai | Brazil | pulp | The oil-containing fruit flesh is used to make a creamy drink. Tips of shoots preferred source of palm cabbage (palmito) in Brazil (see p. 127). *E. oleracea* Mart., from the equatorial regions of Brazil, is used in the same way (1158) |
| *Jessenia bataua* (Mart.) Burret | ungurahui, seje, pataúá, kumbu | Brazil | pulp | Good tasting oil (22-42% of the mesocarp), similar to olive oil. Planted near the dwellings of natives (133, 134, 135, 136, 1145) |
| *J. polycarpa* Karst. | seje grande, coroba, milpesos, jagua, | Colombia | pulp | Cooking oil of the Chocó Indios |
| *J. repanda* Engel | aricaguá, aricacúa | Venezuela | pulp | Oil content of mesocarp and taste of oil similar to seje |
| *Manicaria saccifera* Gaertn. | sleeve palm, monkey cap palm, temiche, ubussu, guágara, bassu | Northern S America | seeds | Oil used locally as food (ubussu oil) |

| | | | | |
|---|---|---|---|---|
| *Oenocarpus bacaba* Mart. | turu palm, bacaba | Northern S America | pulp | Good-tasting oil, used as cooking fat and for the preparation of a drink (bacaba branca). Planted by Indio settlements. Other *O.* species also provide fruit oil (133, 134, 135, 1145, 1425) |
| **Lepidocaryoideae** | | | | |
| *Mauritia flexuosa* L.f. | burity do brejo, aguaje, bâche, miriti, ita, moriche | Northern S America | seeds | Fruit flesh floury, used for making an acidulous drink. Seeds eaten raw, or used for oil extraction (133, 1145) |

endosperm, but this process has not yet replaced the longer process of drying the endosperm to give copra (676, 1395).

For the preparation of desiccated coconut (grated coconut) (Table 16), the brown testa is removed, and the flesh is washed, sterilized, grated and dried (905).

The hard shells are used as a fuel for the copra ovens, and for the production of activated charcoal (exported particularly by Sri Lanka). When finely milled, they are used as a filler material for plastics, and locally they are made into buttons, containers and craft objects. The fibrous husks of the ripe fruit provide coconut fibre (coir), which is produced especially in India and Sri Lanka (see p. 352).

Fruits for fresh consumption should not be quite ripe, and can only be harvested by being cut off. Six to seven month old fruits contain the most coconut water, which is a refreshing drink. Fresh endosperm is used in many dishes. Coconut milk is produced by crushing grated endosperm. In Polynesia, the "apple" of the germinated coconut is a much-valued, vitamin C-rich foodstuff (815).

Of considerable importance in many regions is the utilization of the sap (containing 15% sugar) collected from the inflorescence, especially for the preparation of palm wine (see p. 74) (1390). Trunks and woven leaves are used locally as building materials for housing, and for stabling animals (1171).

## Further Oil-Producing Palms

Apart from oil palms and coconut palms, the fruits of many other palms are used for oil production in South and Central America. Table 17 names most of the genera of which the fruit flesh or seed provide oil; only the best known of these species are presented. Most of the genera belong to the sub-family Cocoideae, as do *Elaeis* and *Cocos*. These palm oils are an important part of the nutrition of the native Indians, and only a small number are made use of commercially. Some of the species have been brought under cultivation. Without exception, the kernel oils belong to the lauric acid group, and the fruit oils to the oleic acid group. The only ones with a definite importance on the world market are babassú and tucum kernel oils. Further general statements can be found in (88, 132, 200, 281, 370, 442, 550, 609, 1039, 1378, 1518, 1519, 1856, 1960). Literature references are given for the individual species in Table 17.

## Soya Bean (American: Soybean)

fr. soya; sp. soya; ger. Sojabohne

The soya bean, *Glycine max* (L.) Merr., Leguminosae, subfamily Papilionoideae, originates in China, where it is a long-established cultivated plant, as in the other East Asian countries (791, 1101, 1519, 1657, 1762, 1942). It has become the most important oil and protein plant only since 1945. The largest producers in 1988 were the USA 41.9 million tons, Brazil 18.1, China 10.9, and Argentina 9.8; these four countries grow 90% of the world's production total. The large-scale cultivation of soya began in Brazil in 1970, and in Argentina in 1977. Although the soya bean possesses excellent qualities, its cultivation in all other countries remains relatively slight, in spite of great efforts to encourage its production. In the whole of Asia apart from China, 4.5 million tons were grown in 1988, and in the whole of Africa, 472,000 tons. More than half of the world's production of soya is exported, as beans or as oil and cake. About 50% of the export is supplied by the USA, 17% by Brazil and 10 % by Argentina. The latter two countries mainly export oil, the exports from the USA are mainly beans (18 million tons of beans, 0.9 million tons of oil).

**Morphology**. The plant is an annual and resembles the bush bean in its habit; all parts of the plant are distinctly hairy. The fruit is formed in the leaf axils on short hanging racemose inflorescences, having few flowers (Fig. 23). In the cultivars with determinate growth, an inflorescence also develops from the terminal bud of the shoot. The pods contain 1-5 seeds, mostly 2-3.

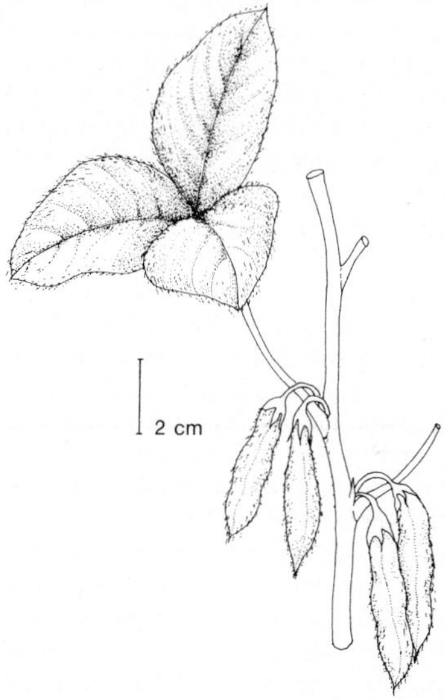

2 cm

Fig. 23. *Glycine max*. Part of a shoot with fruits

**Ecophysiology**. The parts of the subtropics which are always humid provide the best climate for the soya bean. The optimal temperature is 24-25°C. Soya has a pronounced sensitivity to photoperiod, and most cultivars only bloom when the day-length is less than 14 hours. Very short days (12 hours and less) lead to premature flowering, so that the plants remain small and have reduced yields. The cultivars react very differently to day-length, and they are usually adapted to a narrow geographical latitude (495, 1146, 1291, 1292, 1617).

In the main areas of cultivation, the growing period is 4-5 months. In warm countries, 500-750 mm rainfall are necessary for good yields, short dry spells are endured, and it should not rain too much during the ripening period. Growth and yield are definitely influenced by symbiosis with rhizobium (see p. 96).

The demands made on the soil are slight, with a pH value of 6-6.5 being optimal. There are, however, also cultivars which thrive well on acid (pH 5.5) or alkaline (pH 7.5) soils. Ca is important for nodule formation, and Mo fertilizers can be necessary on acid soils. The production of 1 ton of seeds removes from the soil about 15 kg P and 50 kg K (853, 921).

**Cultivation**. The cultivation of soya bean offers few problems. The inoculation of the seeds with the correct strain of *Bradyrhizobium* is essential. *Glycine max* is not only symbiotic with one particular species (*B. japonicum*, (Table 22)), but cultivar-specific strains have also been isolated. In practice, it is recommended to inoculate with a mixture of different strains (composite strain cultures). The nitrogen fixing is very active (about 100 kg N per season), so that on most soils no additional nitrogen fertilizers are necessary, especially if the preceeding crop has been well fertilized (704, 1073, 1146, 1291, 1708, 1765). High soil temperatures reduce the $N_2$ binding potential of the nodules. For this reason, the seed is sown densely in hot sunny areas, to quickly provide a closed canopy for shading the ground. A spacing between the rows of 20-30 cm, with a spacing within the row of 10-15 cm (seed usage about 100 kg/ha) gives better yields in the tropics than the usually recommended row-spacing of 60-90 cm or more (which gives a seed usage of about 65 kg/ha) (321, 373, 495, 1702). Good control of the weeds is important in the first weeks. They are mostly mechanically removed, but there is also a range of herbicides which are available (182, 1679, 1761, 1945).

**Diseases and Pests**. Major losses due to the various diseases (e.g. *Diaporthe phaseolorum* (Cke. et Ell.) Sacc. var. *sojae* Lehm., *Pseudomonas glycinea* Coerper.) are so far controlled by the choice of resistant cultivars, seed selection (and possibly disinfection), sanitary measures, and crop rotation that major losses only occur occasionally. Of the animal pests, the leaf-eating beetle (*Plagiodera inclusa* Stål.), and in East and Southeast Asia, the soya bean moth (*Laspeyresia glycinivorella*) can cause considerable losses, and make the use of insecticides necessary (1291, 1660, 1661, 1915).

**Harvesting and Processing**. Harvesting can be carried out with combine harvesters. In the USA, 2.5 tons/ha is a good yield, and under extremely favourable conditions, over 4 tons/ha can be achieved; the average is 2.2 tons/ha. In the tropics, the yield is mostly about 1 ton/ha, but here also, yields of up to 5 tons/ha have been achieved in experiments. The deciding factor here, apart from the agricultural techniques, is the choice or breeding of locally adapted cultivars (609, 1291, 1942).

More than any other oil seed, soya is a plant with a dual purpose: the seeds of the high-yielding commercial cultivars contain 18% oil and 38% protein (with individual cultivars, the oil content can reach 25% and the protein content 43%, and some small-seeded Asiatic cultivars contain 50% protein); the meal and flakes from the oil extraction represent more than 40% of the value of the crop. Without this dual purpose, the cultivation of a crop with such a low oil content would not be economic (756).

Because of the low oil content of the seeds, the oil is preferably won by solvent extraction. It belongs to the linolenic acid group (Tables 14 and 15) (3-11% linolenic acid), and because of that, it becomes rancid more easily than other vegetable oils. Its main uses worldwide are for the production of margarine; in the USA, after refining and partial hydrogenation, it is also used as a vegetable oil. As an oil which dries well, it is also used for a variety of technical purposes. Of all oils, soya has the highest content of lecithins (1.1-3.2%), which are centrifuged off after the steam treatment of the "miscella" (the mixture of solvent and oil during extraction). Soya lecithin is a surface-active compound and is used as an emulsifier in the food industry, in pharmacy, and in a variety of different technical branches of industry (decorating materials, printing inks, pesticides). The protein in the extraction meal (Table 20) is used not only for human and animal food, but also for technical purposes (synthetic fibres, glues, foam-forming agents) (913, 1406, 1689, 1942).

The unripe seeds are eaten in East Asia and the USA as a vegetable, like peas and beans. The ripe seeds are difficult to digest, and in the raw state, they contain toxic compounds (saponines, goitrogens, protease inhibitors), and they also have an unpleasant taste. It is therefore difficult to introduce them as a foodstuff in new regions. They must be soaked and cooked for a long time before they are edible. In East Asia, a range of other preparation techniques are common which produce valuable foods such as tempe, shoyu (soy sauce) and other products obtained by microbial processes, as well as soya milk, from milled beans, and tofu (soya bean curd, which is soya precipitated using acid and salt) (815, 1074). Finally, soya bean sprouts are an important vegetable in East Asia. The green plants can be used for grazing, or given to animals as green forage. It is well suited to silaging.

## Groundnut, Peanut

fr. arachide; sp. cacahuete, maní; ger. Erdnuß

The groundnut, *Arachis hypogaea* L., Leguminosae, subfamily Papilionoideae, is an old cultivated plant of the South American Indians, and originates in Bolivia (1657). The genus *Arachis* (30 species, all geocarpic), is found only in South America. *A. hypogaea* is not a wild species; it is tetraploid (2n = 40), and probably amphidiploid, originating from two diploid wild species.

The groundnut cultivars are usually classified into the groups Virginia, Valencia, and Spanish. A botanical subdivision of the species differentiates the following subspecies and varieties (1657, 1762):

ssp. *hypogaea*. Mostly low-growing. Seeds dormant. Vegetation period 5-10 months.

var. *hypogaea*. Virginia type. 1-2 large seeds in the pod, strongly branching plant. Preferred for salted peanuts.

var. *hirsuta* Kohler. Peru type. Also cultivated in China. Of little importance in modern agriculture.

ssp. *fastigiata* Waldron. Mostly upright, seeds without dormancy, vegetation period 3-5 months.

var. *fastigiata*. Valencia type. Mostly 4 seeds in the pod. Early ripening. Brittle carpophors.

var. *vulgaris* Harz. Spanish type. The high-yielding and relatively low-demanding cultivar 'Natal Common' belongs to this group. Mostly 2 (1-3) seeds in the pod. Vegetation period 4.5 months.

The botanical varieties are by no means uniform in respect to the most important agricultural characteristics. Thus both upright and creeping types are found in all of the subdivisions (bunch or runner types). This classification is useful for breeding, because crosses between the members of the *hypogaea* and *fastigiata* subspecies are only partially fertile. For further classification proposals, and for assigning the individual cultivars to the groups, see (7, 602, 802, 1101, 1729, 1988).

**Production**. Because of the unusually high nutritional value of the seeds and their pleasant flavour, groundnuts are one of the most important food crops in the tropics and subtropics (13, 1322). They are cultivated on all continents (Asia 15.4, Africa 4.6, America 2.7 million tons). The largest individual producers are India (average annual production of unshelled nuts is 6.2 million tons), China (6.0) and the USA (1.7).

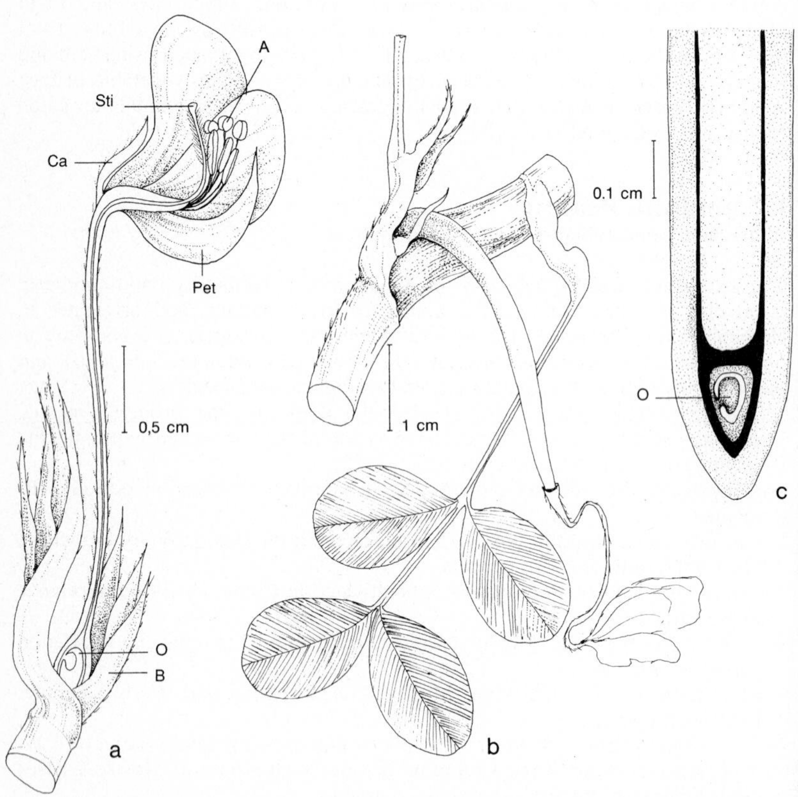


Fig. 24. *Arachis hypogaea*. (a) flower in longitudinal section, (b) carpophore, (c) tip of carpophore in longitudinal section. B = bracts, O = ovary, Ca = calyx, Pet = petals, A = anthers, Sti = stigma

The largest part of the production is consumed in the producer countries. However, for some countries, groundnut products (nuts in shell, shelled nuts, cake and meal) are important export items (69, 1972). The largest exporters of nuts in 1988 were China (227,600 tons) and the USA (158,516 tons); India is still the major exporter of cake and meal (250,000 tons), although these exports are now much reduced. The groundnut exports from Africa have declined. Only Senegal exported notable amounts of groundnut oil and meal in 1988 (140,000 and 251,000 tons respectively); even there, more than 60% of the groundnut production is absorbed by the home market.

**Morphology**. The flowers are borne on a short-stalked inflorescence in the leaf axils. After fertilization (predominantly self-pollination), a meristem develops at the base of the ovary, from which the positively geotropic carpophore develops. The fruit only begins to grow when the tip of the carpophore has reached its final depth in the soil (5-10 cm under the surface); it then grows horizontally (Figs. 24 and 25). When the fruit-bearer does not succeed in entering the soil, the fruit does not develop.

**Breeding**. In the foreground is the resistance against rosette virus (428) and other viruses, *Pseudomonas solanacearum* E.F.Smith and *Mycosphaerella* spp. Recently, resistance against *Aspergillus flavus* Lk. (aflatoxin formation) is receiving much attention (826, 1166, 1173, 1519). After these factors, growth habit, drought resistance, vegetation period, nutrient content, taste, dormancy of the seeds, and firmness of the carpophores have important roles in the breeding programmes (588, 1762, 2002).

**Ecophysiology**. Development and vegetation period for the groundnut are chiefly dependent on the temperature (1762). For germination, 30°C is optimal, and for vegetative growth, 27°C. Sensitive cultivars are damaged (chlorosis) at a temperature of 15°C. Because of this, its cultivation in the northern hemisphere reaches only to the 40° latitude, with a few exceptions, such as southern Russia and East Asia. At sub-optimal temperatures, the vegetation period is extended by 1-2 months. The need for sunlight is not very high, so in Africa, groundnuts are often grown in mixed cropping with maize, and cultivated under oil palms.

Groundnuts are fairly drought tolerant, because they very quickly develop a deep-reaching root system. For early cultivars, 250-300 mm rainfall during the vegetation period are enough (Sahel region). The limit for their cultivation is generally 500 mm annual rainfall. In very rainy regions (over 1,000 mm annually), the soil must be permeable, or else they must be grown on ridges so that sufficient drainage is provided. Moist weather is favourable during the flowering time: the carpophores enter more easily into damp soil which is not too hot, rather than into dry, over-heated soil (1489). At harvest time, the soil should be dry, so that less dirt is attached to the shells.

The soil should be light, well-drained and aerated. Heavy soils, or those with a tendency to form crusts, are unsuitable (entering of the carpophores, harvesting procedures). The pH should lie between 6.2 and 7.5. With acid soils, the nutrient intake is unbalanced, and the nitrogen fixation by the rhizobia is impeded. The nitrogen fixation in the bacterial nodules is very active under favourable conditions (Table 22), so that nitrogen fertilization is not necessary (1758). In respect to the fertilizer needs of the following crop, it should be kept in mind that the entire plant mass is taken away from the field at harvesting. A ton of unshelled peanuts, plus two tons of hay, contain about 60-70 kg N, 5-6 kg P, 40-50 kg K, 20-30 kg Ca, and 8-15 kg Mg. Calcium is important for the symbiosis with rhizobia, and for the fruit development. It is also taken up by the carpophore (top-dressing with gypsum (1016)). Generally, the roots are infected with VA mycorrhiza (1091) which is also found on the carpophores, and improves the supply of P and other mineral nutrients to the growing fruits (634). Larger amounts of K and Na in the soil hinder the calcium intake, and can influence the harvest unfavourably. Deficiencies of Mn, Cu and B have been reported rather frequently (1237).

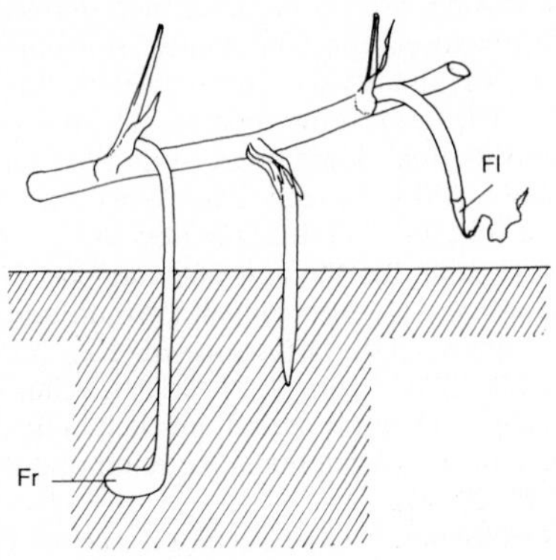


Fig. 25. *Arachis hypogaea*. Carpophores entering the soil (schematically). Fl = dried flower, Fr = fruit

**Cultivation**. Groundnuts should never be grown after another groundnut crop, but only in at least a three-year crop rotation, for example with cotton and maize. In general, with good care of the preceeding crop, no fertilizer needs to be given to the peanuts. The soil preparation is mostly focussed on weed control: on a light soil which is typical for groundnuts, tillage need be no deeper than 20 cm. In primitive agriculture, the whole pod is often sown. In mechanized production, shelled seeds which have been dressed with fungicides are used.

Inoculation with rhizobia is mostly unnecessary, because the symbionts belong to the widely spread *Bradyrhizobium* type (Table 22). The sowing distances depend on the growth habit of the cultivar. Spanish types are more closely sown than Virginia or Valencia types, bush types more closely than creeping. In hot countries, they should be sown as closely as possible, so that the ground is quickly covered. A normal spacing is 30-45 cm between the rows, and 7.5-10 cm within the rows. The quantity of seed required is 65-100 kg/ha. Cultivation on ridges, which can be necessary in moist places, implies substantially larger spacing between the rows, and consequently the yield per unit area is lower, especially with early bush varieties. The seed depth is about 5 cm; deeper sowing weakens the young plants. The control of weeds is important in the first weeks. A range of pre-emergence herbicides have been recommended. Mechanical working of the soil after the beginning of the flowering period mostly results in damage to the yield. A top-dressing with gypsum is given at the beginning of flowering.

**Diseases and Pests**. The root and wilt diseases (*Sclerotium* spp., *Rhizoctonia* spp.) are controlled by crop rotation and seed dressing. The losses from the widespread pathogens causing the leaf-spot diseases (*Mycosphaerella* or *Cercospora* spp.) can be limited by the use of fungicides, though their residues can make the hay unsuitable for animal fodder. Worldwide, the rosette virus is the greatest danger. It is spread by aphids (*Aphis craccivora* Koch), and is practically impossible to control. Resistant cultivars have been found for the first time in recent years (428, 1285, 1762). A range of other viruses are known (1762).

Insect pests (beetles, moth caterpillars) generally play a subordinate role. Insecticides must be used cautiously, because of residue problems in the hay (525, 1762).

**Harvest and Processing**. Manual harvesting of groundnuts involves a lot of work, particularly with the creeping types, i.e. cutting off the main root, pulling up the plant, leaving to dry, stacking for the full ripening of the young pods and the loosening of the pods from the fruit stalks. The dried pods are separated from the hay either by hand or with a threshing machine. In the USA, the harvest is fully mechanized, which however requires several operations. After mowing the tops, the roots are loosened using a cutting bar at a depth of about 15 cm, the plants are lifted, freed from earth using a vibrating sieve, and then left in windrows until the pods are dry enough for threshing (about 8 days). A further drying can be necessary to give a water content of around 5%, so as to protect the nuts from decay (aflatoxin formation) (351, 826, 1148, 1166).

Under favourable conditions, high-yielding cultivars can produce up to 5 tons/ha of nuts in shells. However, the yields are usually very much lower and the global average is about 1.1 ton/ha. In India, the major producer, it is 1 t/ha, in China 2 t/ha. The weight loss by shelling is around 30%.

The oil content of confectionery nuts is 38-47%, of oil nuts 47-55%, the protein content is 24-35%, and the sugar and starch content only 3-8%. The nuts are rich in B vitamins and tocopherol (vitamin E). The oil belongs to the oleic acid group (Tables 14 and 15) (13).

Only about two thirds of the total world production is used to produce oil. Groundnuts provide a good salad and cooking oil, which is readily used in the margarine production industry (vanaspati production in India (717)). In the USA, more than half of the groundnut crop is processed into peanut butter. For this, the nuts are roasted, coarsely crushed, the cotyledons separated from the seed coat, radicle and plumule, and homogenized with the addition of emulsifiers (mono- and di-glycerides) (1988). Peanut butter has been introduced to other countries, especially in preparations with additional aromatic substances (e.g. cocoa). A considerable proportion of the nuts are consumed in the roasted and salted state, a further proportion are used in bakery in the same way as almonds, or used in confectionery. Nuts which are suitable for these purposes bring a higher price than nuts for oil or peanut butter (1967, 1972).

The oil cake contains 40-50% of easily digestible protein, with a high cystine content. Cake meal is suitable for human nutrition, and is contained in numerous protein-rich food preparations (1255, 1322, 1406). Good groundnut hay has about the same nutritional value as lucerne hay. The shells are utilized as roughage, fuel, for particle board (1132), and as a fertilizer material (1988).

## Sunflower

fr. tournesol, (grand) soleil; sp. girasol, mirasol; ger. Sonnenblume

The sunflower, *Helianthus annuus* L., Compositae, is the only crop plant originating in what is now the USA which is cultivated all around the world. It is the third most important oil plant, after soya and oil palm (Fig. 20), and is altogether one of the most important salad oil plants. Through Russian breeding, not only has the oil content of the seeds increased by 50%, but also the growth habit has been greatly changed (dwarf and semi-dwarf types, only 90-150 cm tall), so that mechanical harvesting is now possible (276, 609, 724).

**Production**. The USSR is by far the largest producer (6.2 million tons of seed), followed by France (2.5 million t, 10 times the amount produced in 1980), Argentina (2.9 million t), and China (1.2 million t). The production in the USA is, after a peak in the early eighties (2.3 million t), conspicuously declining (1988: 0.7 million t), whereas production is increasing continuously in the Mediterranean countries (1988: Turkey 1.2, Spain 1.1 million t). The main exporter today is Argentina (938,739 t of oil); the USA still occupies the second place (260,111 t of oil exported in 1988).

**Breeding**. The breeding of Russian short-stemmed, single-headed cultivars with a high oil content has been complemented by the discovery of the male sterility and restorer genes, through which the commercial production of hybrid seed has been made possible. In all the countries in which the production has risen steeply in the last years, the hybrid seed has been the basis of the cultivation (276, 522, 1025, 1519).

**Ecophysiology**. The development of sunflowers is controlled by temperature. The typical regions of cultivation have short, hot summers. Where the summer is warm enough, and the autumn is fairly dry, their cultivation even extends beyond 50 degrees of latitude. They can be cultivated everywhere in the tropics and subtropics, wherever it does not rain too much, especially during flowering and seed formation (276, 471, 807, 883). The Russian dwarf cultivars ripen 2.5-3 months after sowing in hot countries, and need about 250 mm rainfall. Thus it is possible to grow the sunflower in some regions as a second crop; in regions with short rainy seasons, it is often the only saleable crop which can be grown successfully when, due to the late onset of the rains, the cultivation of maize or other field crops is no longer possible. The sunflower is shallow-rooted, with a good absorption power for nutrients. It thrives on a variety of soil types, also on those with little fertility. The pH should not be under 6. The nutrient intake needed for 1 ton of seed is approximately 30 kg N, 2.5 kg P, and 60 kg K (607, 609).

**Cultivation**. Sunflower cultivation offers no particular problems. Due to the unusually quick development of the young plants, the control of weeds is easy. Seeding depth and spacing depend on the cultivar. The small-seeded oil types are sown 3 cm deep, the row-spacing is 30-45 cm, the spacing within the row is 15-20 cm. Large-growing cultivars are planted further apart. For one hectare, 10-15 kg seed are necessary. In countries with non-mechanized agriculture, all of the work from sowing to threshing the heads can be done using simple hand tools (609, 1623, 1829, 1830). In tropical and subtropical countries, seed-eating birds can endanger the harvest (276, 1353). Sunflowers are an outstanding crop for bees (see p. 402), and the bees ensure their pollination (1472).

**Harvest and Processing**. The yield from Russian varieties lies between 1.5 and 2 tons/ha, but it can reach 4 tons/ha under favourable conditions. High-growing cultivars with 4-4.5 months vegetation time can produce even more. The loss in weight from the shells is between 26 and 50 %. The oil belongs to the linoleic acid group (an average of 58-67% linoleic acid) (Table 14 and 15). Its composition is strongly influenced by the temperature during fruit development: for maximum linoleic acid (up to 77%), cool weather is needed; with very high temperatures, the linoleic acid content can fall to 20%. Cold-pressed oil is a valuable salad oil, the lower grades are used for technical purposes (coating materials).

In Eastern Europe, the seeds are eaten - they are enjoyed raw, or roasted and salted (large-seeded types). Cultivars with low oil content find a good market as bird-seed. The cake is valuable as feed concentrate for all types of animals (40% protein, low lysine content) (1406). The shells are used for animal litter, as fuel, as a filler in animal fodders, and as a polishing material. Sunflower provides good green forage, and it can easily be silaged; also the heads, after threshing, are eagerly eaten by cattle, and can also be used for pectin production. The stems can be used in the production of paper pulp (276).

## Sesame

fr. sésame; sp. sésamo, ajonjoli; ger. Sesam

Sesame, *Sesamum orientale* L. (*S. indicum* L.) (1621), Pedaliaceae, originates from the summer-rainfall areas of tropical Africa, from where it spread very early to the Middle East, India, and China. Locally, it is known under a lot of different names: gingelly and til in India, sim-sim in the Arabic countries and East Africa, benniseed in West Africa. The cultivars differ in leaf shape, growing height (0.3-2 metres high), seed size and colour; for mechanical harvesting, the non-shattering cultivars are important (470, 807, 1275, 1519, 1620).

**Production**. The world production amounts to 2.2 million tons of seed (Figs. 19 and 20). The major producers are (given in 1,000 tons): China (531), India (500), Sudan (278), and Burma (190). A fifth of the production comes on to the export market, and the only export product of any quantity are the seeds. The major exporters are China (134,000 tons, plus 28,000 tons via Hong Kong), Sudan (99,000 tons) and Mexico (36,000 tons).

**Morphology**. In the axils of the upper leaves one or several flowers are borne (Fig. 26), the capsules contain numerous small seeds, with colours from white/yellow, to brown, to black (length 1.5-4.5 mm, mostly 3 mm). The plants are annual, with a vegetation period of 3-4 months.

**Cultivation**. With reference to water supply and soil fertility, sesame is less demanding than groundnut (1006). Most cultivars are day-neutral, but there are also definite short-day cultivars (e.g. winter types in India). Because of the small seeds (2.5-3.2 g per 1000 seeds), they require a well-prepared seedbed. The seed must be shallowly drilled (2-5 cm), the row-spacing is 30-60 cm, and spacing within the row 5-15 cm. Seed usage is 2-5 kg per ha. In the early stages of development, the control of weeds takes a lot of work. The vegetation period lasts 70-90 days. The harvest of the seed-shattering cultivars (which are currently most grown) is complicated, particularly because the capsules do not all ripen at the same time. The plants must therefore be cut down before the ripening of the first capsules and be stacked to dry; afterwards, the seeds are gathered by shaking the plants over a cloth (376, 470, 1275, 1914, 1915). Diseases and pests do not generally play a major role (1109, 1883).

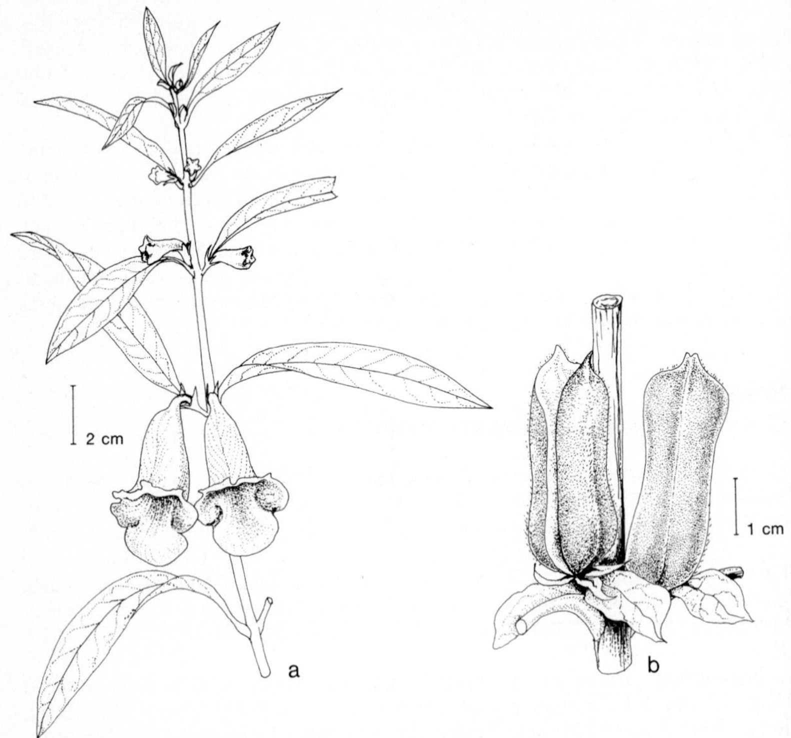


Fig. 26. *Sesamum orientale*. (a) shoot tip with flowers, (b) ripe capsules

**Harvest and Processing**. The yields are small, usually 350-500 kg/ha, because the cultivation is mostly on poor soils, in dry regions, and without fertilizers. In places where the plants are supplied with all they need (Venezuela, Mexico, USA), 1-2 tons/ha are achieved, and at experimental stations with irrigation, over 2.5.tons/ha (609).

The oil (oil content of the seeds is 50%), which is extracted by careful pressing, is the highest-priced of all edible oils; its main applications are found in the pharmaceutical and cosmetic industries. It has a good taste, and keeps well due to the antioxidants sesamin and sesamolin. It contains 35-47% linoleic acid, and is on the boundary between the oleic acid and linoleic acid groups (Tables 14 and 15). Sesamin and sesamolin are easily detected by the Baudoin test. Therefore, in some countries (India, Denmark, Italy), there is a rule that margarine must contain a proportion of sesame oil, so that the plant product can be differentiated from butter.

Only a small part of the production is used for oil extraction. The shelled seeds are used in the producer countries for the preparation of delicacies ("tahini" - a paste which can be spread on bread, "halva" - a sweet-cake), and they are used in baking, as a condiment, and in medicines (376). In the Middle East and Egypt, sesame is used more as a cereal than as an oil plant. The cake contains 35% protein, with a suitable composition (high methionine content), and so it is an animal fodder of great value (895, 1406). Because the oil is extracted from unshelled seeds, the cake contains too much fibre for human consumption. In West Africa, shoot tips and young leaves are used as a vegetable, as with other *Sesamum* species, from which the oil-rich seeds are also used at times (*S. alatum* Thonn., *S. radiatum* Schum.) (480), as also with the African Pedaliacean *Cerathotheca sesamoides* Endl.

## Safflower

fr. carthame; sp. cártamo, alazor; ger. Saflor

Safflower, *Carthamus tinctorius* L., Compositae (Fig. 27, a and b), is a very old cultivated plant. It comes from the Middle East, and was originally cultivated for the yellow pigment from its petals, which was used for the colouring of food and fabrics (1657). It is still used for the colouring of food in the Mediterranean region and India (wild saffron) (see p. 386). In comparison, its oil was used in relatively small quantities in paints and for food. Because of American breeding work, safflower first became of worldwide economic importance after 1948. The knowledge of the importance of linoleic acid (see p. 78) for health especially caused attention to be directed towards the oil, which contains 73-79% of this fatty acid (Table 14 and 15). The new cultivars have 36-48% oil in the fruit, and 18-30% of the fruit consists of shell (979, 1519, 1995).

**Production**. Worldwide, the production of safflower seed has decreased since the beginning of the 1980s; in 1988 it amounted to 873,000 tons. The main producers are India (451,000 t) and Mexico (252,000 t). The USA produced 80,000 t, Ethiopia 34,000 t, Australia 26,000 t, and Spain 13,000 t.

**Ecophysiology**. The best yields are achieved in a Mediterranean climate with irrigation (sowing in the early part of the year). However, the plant is to some extent drought tolerant due to its deep root system, and makes do with only 300 mm rain between sowing and flowering. It is salt tolerant, which is a very useful quality when grown under irrigation. The total vegetation period is 4-5 months, but is longer in cooler regions (376, 609, 1914, 1915).

**Cultivation**. The seed should be shallowly sown (2-3 cm deep). The planting distances are 30-60 cm between, and about 10 cm within the rows, and the weight of seed needed is 20-30 kg/ha (100, 317, 609, 807, 1006, 1287, 1620, 1676, 2013, 2014). The world average yield is only 0.5 tons/ha, in India it is 0.4 tons/ha and in the USA without irrigation 0.4-1.7 tons/ha. With irrigation and good fertilization, yields of 2.8-4.5 tons/ha can be achieved.

**Processing**. The oil is a good salad oil, which can be used for all purposes. As a quickly drying oil which does not become darker, it is valued for lacquers and light-coloured paints. The cake from the shelled fruit contains 36% protein; it is also suitable for human consumption (1406). The roasted seeds are sometimes chewed like nuts, young shoots can be used as a vegetable, and the whole plant can be used for animal fodder, especially the spineless cultivars.

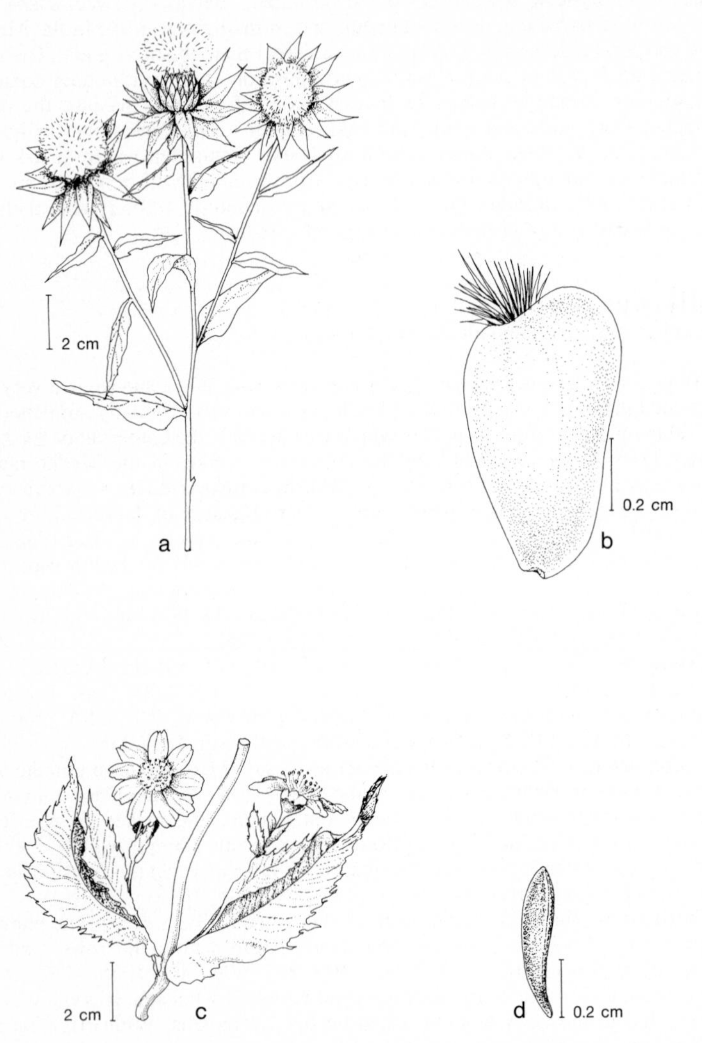


Fig. 27. *Carthamus tinctorius*. (a) flowering branch, (b) fruit. *Guizotia abyssinica*. (c) flowering branch, (d) fruit

# Olive Tree

fr. olivier; sp. olivo; ger. Ölbaum

The olive, *Olea europaea* L., Oleaceae (Fig. 28), originates from the Mediterranean countries (1657). Despite many attempts to develop profitable cultivation in other countries, this has not succeeded anywhere else on a large scale, although it grows well in all tropical and subtropical regions which are not too wet. It is in the Mediterranean area and the Middle East where 98% of the olives are produced. Even there, the olive production is often uneconomic, and must be supported by state subsidies (46). However, olive oil still takes eighth place in the world oil production figures, with an annual production of nearly 2 million tons (Fig. 20). About a sixth of the production is exported; the main importers are (values for 1988 in 1,000 tons): Italy 258, USA 81, Libya 43, and France 30. This concentration of production and use in one region is founded on the cooking habits of the population, who hold on strongly to their traditions in spite of the high price: olive oil is more than twice as expensive as groundnut oil, and almost three times as expensive as sunflower oil. The economy of its production is adversely affected by:

- low oil yields (in the Mediterranean region they average 400 kg of oil per ha per year),
- alternate bearing (full yield only every second year) (1415),
- high labour costs during harvesting,
- and the difficulty of controlling pests (olive fruit fly, *Dacus oleae* Gmel.; scale insects; olive moth, *Prays oleae* Bern., and others (84, 982, 2001)).

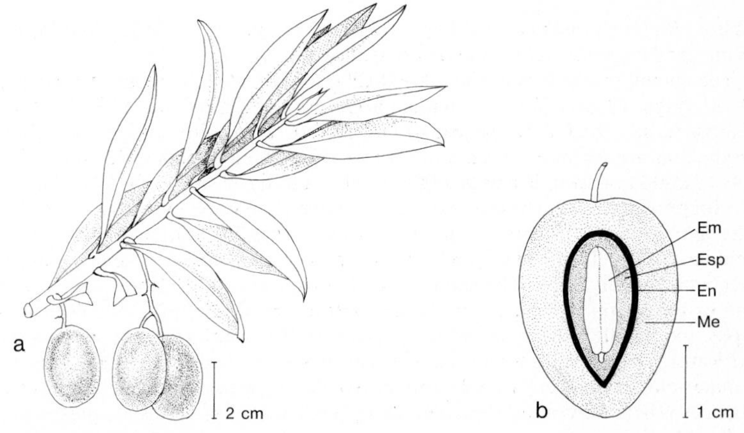


Fig. 28. *Olea europaea.* (a) branch with fruits, (b) fruit in longitudinal section. Me = oil-containing mesocarp, En = hard endocarp, Esp = endosperm, Em = embryo

New plantings of olive trees for oil production are economically sound only in dry regions (e.g. Sfax Province in Tunisia) or where the soil is too stony for any other form of cultivation. It is more profitable in general to cultivate large-fruited, thick-fleshed cultivars with a lower oil content in the flesh, which are used for the production of table olives. Even for these, the costs of picking can be a limiting factor. On the best sites in California, yields of 12.5 tons/ha of table olives are achieved. The total world production of these is about 700,000 tons, with about 200,000 tons being exported.

**Cultivation and Processing**. The olive tree is reproduced only vegetatively (grafting on seedlings, cuttings, wood of the trunk (151, 1087, 1209)). With sufficient rainfall (400-600 mm), 100 trees per hectare are planted, in dry regions (200 mm rainfall is the lower limit), only 17-20 trees/ha. Due to its extended root system (up to 12 m away from the trunk, up to 6 m deep), the tree can take in water from a large volume of soil (3,000 $m^3$). The harvest takes place in the months from December to February. Oil olives can be stripped off using machines, shaken down, or (this is not recommended) knocked down and then collected from the ground. The fruit flesh (mesocarp, Fig. 28) contains 23-60% oil (40-60% with oil cultivars), and the seed (endosperm and embryo) contains 12-15% oil. The oils from the flesh and the seed both belong to the oleic acid group (Tables 14 and 15), and have a similar composition (65-86% oleic acid, 4-15% linoleic acid). The whole fruit is crushed into a paste, out of which most of the oil is separated in two pressings (with a hydraulic press). The remnants from pressing yield an oil after solvent extraction, which is valued for skin preserving toilet soaps (baby soap, Castille soap). The extraction remains are used as fuel or as fertilizer (1078, 1209, 1519).

# Castor

fr. ricin; sp. rícino, higuerilla, mamoeiro; ger. Rizinus

Castor, *Ricinus communis* L., Euphorbiaceae, originates in Africa, but is now found growing wild and in cultivation in all warm regions (1657).

The annual production is 800,000-900,000 tons of seed, or about 350,000 tons of oil (Figs. 19 and 20). The major producers are China (in 1988, 310,000 t beans), India (185,000 t) and Brazil (146,000 t). Two thirds of the total world production are exported (seed and oil); of this, China exported 32,000 t of oil and 127,000 t of seed, Brazil 60,000 t of oil and India 31,000 t of oil in 1988.

**Morphology**. Castor is originally a tree, which can grow over 10 metres high. The agricultural cultivars are usually grown annually, particularly the dwarf cultivars (60-120 cm high), which are best suited for dry areas and for mechanized cultivation. The shoot formation is sympodial (inflorescences at the end of the shoot, and branching under each inflorescence, Fig. 29). With most types, the female flowers are in the upper part of the inflorescence, the male in the lower. The capsule, composed of three carpels, splits open in wild forms, but remains closed in recently bred cultivars, so that the whole crop can be harvested at once. Where harvesting is carried out by hand, spineless forms are preferred.

**Ecophysiology**. Castor comes from the tropical summer rainfall areas, and the best conditions for its cultivation are still found there. But it grows well from the wet tropics to the subtropical dry regions, wherever its temperature requirements are satisfied (growing time is 4.5-6 months, optimum temperature 20-25°C; over

38°C the seed-setting is poor). In summer rainfall areas, dwarf cultivars can be grown with only 500 mm rainfall, although 750-1000 mm is optimal (1006). Due to its deep-reaching root system, castor is quite drought tolerant. Short days delay flowering, however, castor has little sensitivity to day length. It prefers deep, sandy loam with a pH of 6, but it can be cultivated on a wide variety of soils with a pH range of 5-8. It does not tolerate salt. The nutrient removal for 1 ton of seed (with capsules) is about 30 kg N, 5 kg P, 12 kg K, 4 kg Ca, and 3 kg Mg (595). Unbalanced nitrogen fertilization encourages the growth of foliage at the expense of flower and seed formation.

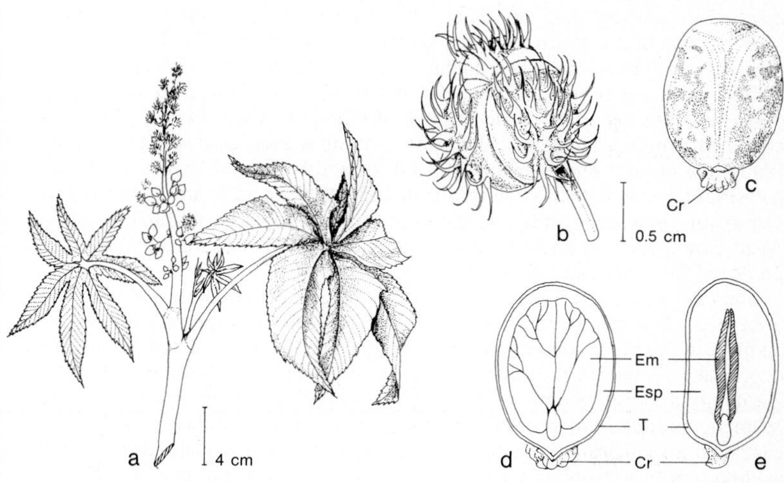


Fig. 29. *Ricinus communis*. (a) inflorescence, (b) capsule, (c) seed, (d and e) seed in longitudinal section - (d) parallel to, (e) across the cotyledons. Cr = caruncle, T = testa, Esp = endosperm, Em = embryo

**Cultivation**. In mechanized cultivation, seeds are sown with a spacing of about 90 cm between the rows and 50 cm within the rows (seed requirement 12-18 kg/ha, depending on the seed size). The seed is sown 5 cm deep. Weed control is especially important in the first weeks, as the plants initially develop quite slowly. Mechanized hoeing should be done with care, as the young plants are very sensitive, due to their brittleness. Good soil preparation and the eradication of persistent weeds before sowing are a substantial help. In the USA, generally herbicides are used, mostly soil herbicides between the rows before germination. In the main areas of cultivation, which have low atmospheric humidity, diseases do not play a major role. Insect pests are generally held in check by sanitary measures and crop rotation. Because of its deep root system and resistance against nematodes (*Meloidogyne* spp.) and *Striga* spp., castor is an excellent crop to be grown in rotation with tobacco, cotton, maize, millet, etc. (376, 520, 1620, 1914, 1915).

**Harvest and Processing**. Indehiscent cultivars are harvested fully ripe by picking the capsules by hand, possibly using a cup-shaped device, or mechanically; with cultivars that burst open, it is recommended that the infructescences are cut off before full ripeness, and then left to dry. The non-shattering capsules are broken to separate the seeds from the fruit walls; in large-scale cultivation, this is done with machines.

The global average yield per hectare is only 0.6 tons; under favourable conditions, 1-1.5 tons should always be achieved, and in the USA, yields of 3 tons/ha are not exceptional under irrigation. The seed contains 42-56% oil. Before pressing or extraction, the seed is usually shelled; the seed coats form 17-20% of the weight. The oil contains 91-95% ricinoleic acid, which is chemically much more reactive than any other fatty acid, due to a double bond on the 9th carbon atom, and a hydroxyl group on the 12th. It can be dehydrated to form a fatty acid with two conjugated double bonds (on the 9th and 11th carbon atoms) which can be easily polymerized. Because of this it finds applications in quick-drying coating materials, as do the oils of Group 7 (Table 14). At 275°C in an alkaline medium, ricinoleic acid is split, forming sebacic acid (used for synthetic fibres and resins) and octanol (utilized as a solvent and in the synthesis of perfumes); vacuum distillation results in undecenylic acid (starting material for a variety of chemical syntheses) and heptaldehyde (used in perfume synthesis). Apart from these important products, there are a number of others which can be produced from ricinoleic acid (128, 1135, 1519).

The castor oil itself retains its viscosity at high temperatures, and is therefore used as a lubricant (1092). Because it does not corrode rubber, it is used as an additive in the production of crumb rubber (see p. 367), and in hydraulic systems (e.g. brake fluids). In the plastics industry, it is used as a plasticizer. Due to its ability to dissolve certain colours, it finds applications in lipsticks, in other cosmetic products, in printing inks, and in the textile industry. Polymerized castor oil gives permanent adhesion and plasticity to paint materials (1914). It is seldom used any more as a purgative. The press cake is poisonous, and is only used as fertilizer (1406). Amazingly, fermented castor seeds are used as a condiment in eastern Nigeria (78); unripe, and also roasted ripe seeds are eaten in Indonesia. Shells and stems are used as fuels and for mulching. Young leaves can be consumed as a vegetable (1788). When animals are accustomed to the leaves, they can be fed these as fodder, in spite of their ricin content (a toxic protein). Castor leaves are the main food of the eri silkworm (see p. 402) (209, 376, 1803).

## Tung Trees

fr. aleuritès, abrasins; sp. tungas; ger. Tungbäume

Tung oil (also called wood oil) is extracted from the seed of three species of the Euphorbiaceae genus *Aleurites*: the Japanese wood oil tree, *A. cordata* (Thunb.) R.Br.; the tung oil tree, *A. fordii* Hemsl.; and the China wood oil tree, *A. montana* (Lour.) Wils. The oils belong to Group 7 in Tables 14 and 15. The seed oil of *A. moluccana* (L.) Willd., the candle nut tree, has a different composition (linolenic acid group), but is economically of little importance (1657). Only *A. fordii* and *A. montana* are cultivated outside their homeland,

China (609, 684). *A. fordii* is by far the most important species; it supplies 90% of the tung oil traded.

**Production**. The world production in 1988 amounted to 85,000 tons. The main producer is China (67,000 tons); the only other major sources are Paraguay (9,000 t) and Argentina (9,000 t). Malawi produces about 1,000 tons/year of *A. montana*.

**Morphology**. The leaves fall in the cool season of the year. With *A. fordii*, the inflorescence appears early in the year before the leaves unfold, on the previous year's growth, and the fruits are smooth and roundish. With *A. montana*, the inflorescence appears on the current year's shoots, and the fruits are strongly ribbed (Fig. 30). The leaf form is very variable with both species.

**Ecophysiology**. *A. fordii* remains smaller (3-6 m high), and is better suited to cooler climates than *A. montana* (10-12 m high). *A. fordii*, and to a lesser extent also *A. montana*, need several cool months to break the dormancy of the buds. In a leafless state, the *Aleurites* can withstand 4 or 5 dry months without damage. In the summer rainy season, 750-1000 mm rain should fall. On well-drained soil, higher rainfalls than this do not cause a problem. Slightly acid, light soils are preferred (pH 6); on very acid soils, difficulties in nutrient uptake can be expected (especially Zn, Cu, Mn) (1719, 1720).

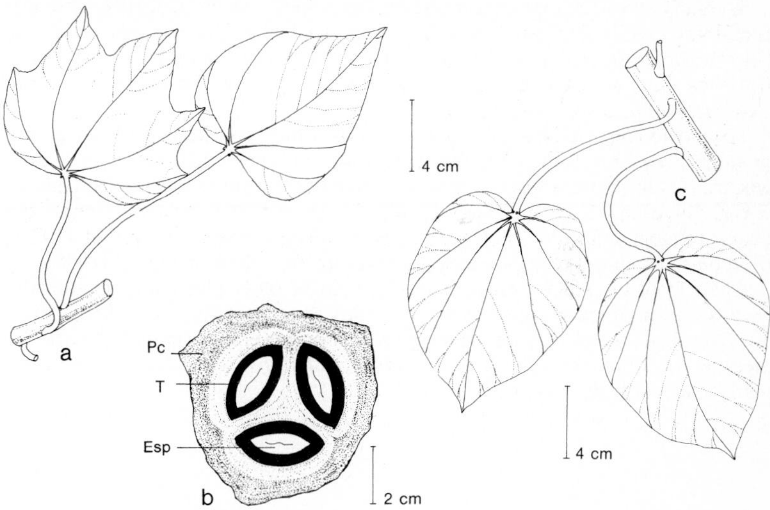


Fig. 30. (a) leaves, (b) fruit in cross section of *Aleurites montana*; (c) leaves of *A. fordii*. Pc = pericarp, T = testa, Esp = endosperm

**Cultivation and Processing**. With both species, the number of female flowers in the inflorescence is genetically controlled (535). To make certain of high yields, the trees must be vegetatively propagated. For this, the buds from selected mother trees are grafted on to seedlings. Good trees begin to bear fruit in their third year, and the full yield can be expected after the tenth year. The normal useful life of the trees is 30 years. Full-grown trees yield 2-3 tons of seed per hectare per year. The oil content of the seeds is 40-58%. The oil (Table 14) is won using a screw press and extraction, after mechanical shelling of the seeds. It is valuable for quick-drying, hard, water-resistant coating materials, floor covers, boat varnishes, and similar uses (613). Bodied tung oil is obtained by heat polymerization and becomes hard particularly quickly. Serious competitors of the tung oils are soya oil, linseed oil, tall oil, castor oil and synthetic products. The cake contains 20-25% protein, but because it is poisonous, it can be used only as fertilizer. The fruit shells are useful for mulching.

## Further Oil Plants

Table 18 describes a range of other plants which are mainly used for oil production. The most important of the families which are included in the table are the Cruciferae. Apart from the species which are mostly grown for oil production, this family includes a number of plants which are primarily grown as vegetables or spices, from which oil can also be extracted from the seed (see Table 25 for: chinese cabbage, rocket cress, garden cress, oil radish; see p. 277 for black and white mustard).

Secondly, the Cucurbitaceae must be mentioned, which generally have oil-containing seed. In part, they are cultivated for these seeds, such as *Cucumeropsis mannii* Naud. (W Africa (496)), types of *Cucurbita* spp., types of *Citrullus lanatus* (Africa, Asia) (Table 30), types of *Lagenaria siceraria* (Africa), *Telfairia* spp., and others (851).

To add to this, there is a long list of oils which can be considered as by-products (see also Table 15): oils from grain embryos (rice, maize, sorghum, wheat), oils from the seeds of vegetables (tomatoes), fruit (apricot, peach, grape, citrus, tamarind), beverages and stimulants (cocoa butter, tobacco seed), fibre plants (cotton and other Malvaceae, kapok, hemp), from hevea seed (1927), from nuts (almond, pecan, walnut), and from the fruit flesh of avocado (958).

Apart from the plants named here, the seed of other plants are occasionally used for oil production; some of these are described in (609, 1519, 1856). There are also plants which are not yet used for this purpose, but a potential exists for them as oil suppliers in future (e.g. *Lupinus mutabilis*, see Table 25, *Cucurbita foetidissima* H.B.K. (894, 1258)).

Table 18. Furhter oil plants (explanation of symbols see footnotes page 55)

| Botanical name | Vernacular name of plant or oil | Climatic region | Main production region | Econ. value | Oil group (Tab. 14 and 15) | Remarks and literature references |
|---|---|---|---|---|---|---|
| **Caryocaraceae** | | | | | | |
| *Caryocar villosum* (Aubl.) Pers. | piquiá, pequiá | Tr | Brazil, Guiana | + | 2 | Mesocarp and seeds contain 70% fat. Used in its home region. Cultivation attempted in Malaysia. Seeds taste good (42, 281, 1011, 1258, 1425, 1518, 1609, 1611, 1854) |
| **Chrysobalanaceae** | | | | | | |
| *Licania rigida* Benth. | oiticica | Tr | NE Brazil | ++ | 7 | Production from wild growing trees is about 50,000 t seeds (15,000 t oil). The oil contains 73-83% licanic acid, and is used in a similar way to tung oil (609, 1518, 1771, 1856) |
| **Compositae** | | | | | | |
| *Guizotia abyssinica* (L.f.) Cass. | nigerseed, niger, tilangi, ram-till, nook, nug (Ethiopia), guja (India) | S, Tr, TrH | Ethiopia, India | +++ | 4 | (Fig. 27 b and c) Production of seed is about 200,000 t in Ethiopia, 135,000 t in India. Exported from Ethiopia and India to USA, Japan and Italy. Valuable vegetable oil (122, 317, 322, 609, 807, 1519, 1620, 1650, 1915) |

Table 18. Further oil plants, cont'd (explanation of symbols see footnotes page 55)

| Botanical name | Vernacular name of plant or oil | Climatic region | Main production region | Econ. value | Oil group (Tab. 14 and 15) | Remarks and literature references |
|---|---|---|---|---|---|---|
| *Madia sativa* Mol. | Chilean tarweed, coast tarweed, pitchweed, madi, melosa | S | Chile | + | 4 | Ancient oil plant of the Indios, low-demanding. Fruit contains 35-40% oil (242, 1856) |
| **Cruciferae** | | | | | | |
| *Brassica carinata* A. Braun | Abyssinian mustard, Ethiopian cabbage | TrH | Ethiopia | + | 5 | Oil and vegetable plant (939, 1519, 1620, 1650) |
| *B. juncea* Czern. et Coss. | Chinese mustard, Indian brown mustard, raya, rai (India) | S, Tr | S and E Asia | ++ | 5 | Many varieties. Strongly-growing plant. Seeds used for oil extraction or as seasoning (see p. 277), leaves used as vegetable (376, 731, 739, 782, 807, 973, 1006, 1519) |
| *B. napus* L. emend. Metzger var. *napus* | rapeseed, colza, colsat, navette | Te, S | Canada, W Europa | +++ | 5 | (Figs. 19 and 20) Edible oil and for technical purposes. Yields up to 4 t/ha (82, 1483, 1519) |
| *B. rapa* L. emend.Metzger (*B. campestris* L.) | | | | | | The oil of all varieties is used as cooking oil and for technical purposes. Press cakes as animal fodder, young leaves as a vegetable (376, 694, 739, 782, 1519, 1834, 2026) |

| | | | | | | |
|---|---|---|---|---|---|---|
| var. *dichotoma* Watt | brown sarson | Te, S, Tr | India | + | 5 | Grown mainly in the Punjab |
| var. *sarson* Prain | Indian colza, yellow sarson, sarisa, | Te, S, Tr | India | ++ | 5 | Numerous forms described, cultivated in several Indian states |
| var. *silvestris* (Lam.) Briggs (var. *oleifera* DC.) | turnip rape, Chinese colza | Te, S | Europe | + | 5 | Cultivation limited to Europe |
| var. *toria* Duthie et Fuller | Indian rape, toria | Te, S, Tr | India | ++ | 5 | Most important variety in India, mostly grown with irrigation |
| *Crambe abyssinica* Hochst. | Abyssinian kale, crambe, colewort | Te, S, TrH | USA, USSR | ++ | 5 | (see p. 78, Tables 14 and 15) The oil is only utilized technically (369, 609, 740, 1519, 1620) |
| **Dipterocarpaceae** | | | | | | |
| *Shorea stenoptera* Burck | Borneo tallow, Pontianak kernels, illipe, engkabang | Tr | Indonesia, Malaysia | ++ | 2 | Seed fat is extracted from several *S.* species. Because only wild trees are used, the production is very variable. Up to 40,000 tons each year exported (Indonesia, Sarawak). Fat used like cocoa butter, 62% saturated, 38% unsaturated fatty acids, 43% stearic acid, 37% oleic acid (259, 609) |

Table 18. Further oil plants, cont'd (explanation of symbols see footnotes page 55)

| Botanical name | Vernacular name of plant or oil | Climatic region | Main production region | Econ. value | Oil group (Tab. 14 and 15) | Remarks and literature references |
|---|---|---|---|---|---|---|
| **Euphorbiaceae** | | | | | | |
| *Sapium sebiferum* (L.) Roxb. | Chinese tallow tree, tallow berry, páu de sebo, chii an shu | S, TrH | China, N India | + | Wax 2<br>Oil 6 | The mesocarp provides Chinese tallow (62% palmitic acid, 27% oleic acid), for cooking fat, soap and candles. The seeds provide stillingia oil, a quick-drying oil for paints and lacquers (53% linoleic acid, 30% linolenic acid). Of commercial importance only in China (376, 609, 1600, 2032) |
| **Labiatae** | | | | | | |
| *Hyptis spicigera* Lam. | moño, nino, kindi, andoka, hard simsim | Tr | Africa | + | 6 | Seeds eaten (Zande) and used for oil production; they contain 24-37% of a drying oil. Cultivation similar to sesame. Leaves eaten as a vegetable (1306, 1788) |
| *Perilla frutescens* (L.) Britt. (*P. ocymoides* L.) | perilla, tzu-su, hsiang-sui | S, TrH | China | + | 6 | Seeds contain 30-51% oil. Cooking oil in China. Technically used for paints, lacquers and plastics. Oil contains 63-70% linolenic acid (259, 376, 581, 609) |

| | | | | | | |
|---|---|---|---|---|---|---|
| **Linaceae** | | | | | | |
| *Linum usitatissimum* L. | linseed, flax, lin, tétard, lino | Te, S, TrH | India, Argentina, Ethiopia | +++ | 6 | (Fig. 20, Tables 14 and 15) Main uses for painting materials, for linoleum, but also for nutritional purposes, also the whole seeds being eaten. About half of the production is in the tropics and subtropics. The press cakes are a valuable fodder. For flax fibres, see Table 44 (322, 376, 609, 807, 1519, 1620, 2012) |
| **Meliaceae** | | | | | | |
| *Antelaea azadirachta* (L.) Adelbert (*Azadirachta indica* Juss.) | neem, nim, sabah-bah, azad-daracht, margosa | Tr | India, SE Asia, Africa | + | 3 | Seeds contain 45% oil, which is used for soap production, and also medicinally utilized (margosa oil). Production is 300,000 tons seeds. Press cakes insecticidal (see p. 391) (16, 234, 259, 376, 609, 807, 1152, 1193) |
| **Myristicaceae** | | | | | | |
| *Virola surinamensis* Warb. | ucuhuba, ucuiba | Tr | Northern S America | + | 2 | Collected from wild trees. Seeds contain 65% fat, of which 63-72% is myristic acid. Other *V.* species also provide seed-fat (242, 1518, 1856) |
| **Papaveraceae** | | | | | | |
| *Papaver somniferum* L. | opium poppy, pavot somnifère, dormideira | Te, S, TrH | Mediterranean countries, E Asia | ++ | 4 | (Table 40) In Asia, the seeds are a by-product of opium. The main producers of seeds and oil are Asia and the Balkans. Good vegetable oil, also used technically. Seed production about 50,000 tons/year, the majority for confectionery (376, 609, 1519, 1856) |

Table 18. Further oil plants, cont'd (explanation of symbols see footnotes page 55)

| Botanical name | Vernacular name of plant or oil | Climatic region | Main production region | Econ. value | Oil group (Tab. 14 and 15) | Remarks and literature references |
|---|---|---|---|---|---|---|
| **Polygalaceae** | | | | | | |
| *Polygala butyracea* Heckel | black beniseed, malukang, ankalaki, cheyi | Tr | Sierra Leone, Togo | + | 2 | Seeds contain 30% fat (malukang butter, 55-60% palmitic acid, 30% oleic acid). Oil and seeds used for cooking. The plants also supply fibres (1306, 1477, 1856) |
| **Sapotaceae** | | | | | | |
| *Madhuca longifolia* (J.G. Koenig) Macbr. var. *longifolia* var. *latifolia* (Roxb.) Cheval. (*M. indica* J.F.Gmel.) | illipé, mahua, moa tree, butter tree, mahwa, mowara | Tr | India, Sri Lanka | ++ | 2 | Seed production more than 100,000 tons. Seeds contain 55-60% fat, with 23% palmitic acid, 19% stearic acid, 43% oleic acid, and 13% linoleic acid. Mostly utilized technically (soap), but also for edible purposes (114, 234, 376, 609, 807) |
| *Vitellaria paradoxa* Gaertn.f. (*Butyrospermum parkii* (G.Don) Kotschy) | sheabutter, karité, bambuk | Tr | Central and W Africa | ++ | 2 | Production potential estimated at 500,000 tons seeds/year. Annual exports from Nigeria, Ghana, Côte d'Ivoire, and other W African countries about 30,000 tons. Oil content of whole seed is 34-44%, with 41% stearic acid and 49% oleic acid. Mostly used for culinary purposes (609, 933, 1319, 1856) |

| | | | | | | |
|---|---|---|---|---|---|---|
| **Theaceae** | | | | | | |
| *Camellia sasanqua* Thunb. (*C. oleifera* Abel) | teaseed, sasanqua camellia | Tr | China, Assam, Japan | ++ | 3 | Seeds utilized from wild and cultivated plants. Oil production about 100,000 tons/year. Seeds and oil scarcely exported. Seeds contain 40% oil with 72-87% oleic acid (similar to olive oil). Predominantly used for culinary purposes (609, 1856, 1859) |

# Protein Providing Plants

## GENERAL

About a third of the world's population is inadequately supplied with protein. The protein deficiency is particularly marked in the countries of the humid tropics, because there the staple diets (plantains - see p. 64, and cassava - see p. 41) contain little protein, and also, supplies of protein from animals are not availabe in sufficient quantities. Acute protein deficiency is limited even in these countries to children and to mothers (in pregnancy and during nursing) (1322).

Table 19. World Production of crude protein

| Source | Amount (million tons/year) | Remarks |
|---|---|---|
| Natural grasslands | 600 | Calculated from: surface area = 3100 million ha. Dry matter = 5 tons/ha/year. Average protein content = 4% |
| Produce from arable agriculture | 259 | Table 20 |
| Animal protein from the oceans | 400 | (1022) |
| Total | 1259 | |

Globally, there is no shortage of protein, as Table 19 demonstrates. The annual crude protein production from land plants is 859 million tons, whereas the total annual protein need of the entire world population adds up to 91 million tons (5 billion people, each needing 50 g protein per day) (1293). The protein from grasslands is only available for animal consumption, by which 80-95% of the protein in the fodder is lost. However, the arable crops alone provide more protein than mankind needs (Table 20) (71, 1978), even if only 50% of the crude protein is considered to be digestible.

As with other foods, the problem is not production, but distribution and use (1422). Within this global picture, attempts to tap other sources of protein (such as single cell proteins from bacteria, fungi and algae) seem of little importance, at least for human nutrition (753, 1354, 1406, 1422), particularly because most of these techniques require large investments and technological expertise.

However, for feeding animals, the production of yeast from agricultural by-products (e.g. molasses), and the extraction of protein from leaves are of interest in the humid tropics (1406). The extraction of algal protein from sewage treatment plants, and the culture of microbes on petroleum substrates already have a definite role.

Table 20. Production of crude protein by the most important food plants

| Plant group | World production[1] (million tons) | Total crude protein content[2] (%) | Quantity of crude protein (million tons) | Share of total crude protein (%) |
|---|---|---|---|---|
| Cereals[3] | 1743 | 10.4 | 181 | 69.9 |
| Root crops[4] | 571 | 1.7 | 10 | 3.9 |
| Soya bean | 92 | 38.0 | 35 | 13.5 |
| Other oilseeds[5] | 104 | 19.7 | 20 | 7.7 |
| Pulses[6] | 55 | 23.0 | 13 | 5.0 |
| Total crude protein | | | 259 | 100 |

1 after (514)
2 weighted average of all species
3 wheat, rice, maize, barley, millet, oats, rye
4 potato, sweet potato, yam, cassava, taro, and others
5 cottonseed, shelled groundnuts, sunflower, copra, rape, linseed, sesame, palm kernel, hempseed
6 beans, peas, chickpeas, faba bean, vetch, pigeonpea, lentils, cowpea, lupins

Also the true protein (pure protein) is of variable value for nutrition. Animal proteins are superior to plant proteins, as the latter mostly contain insufficient quantities of the essential amino acids (lysine, methionine, tryptophane, threonine, valine, leucine, isoleucine, phenylalanine) (507, 1022, 1359, 1422). Examples of this are the low lysine content of most cultivars of maize, and the low methionine content of beans. These deficiencies can be balanced by the combination of different foodstuffs: a diet with 1/3 beans and 2/3 maize gives a biological value of 100; in this system of evaluation, the biological value of a hen's egg is used as the standard (= 100) (920).

The agronomist is interested chiefly in the contribution which the various groups of plants can make to the protein supply. Though calculations such as those in Table 20 can only give a rough picture with many uncertainties, and do not take into account the quality of the protein, they show that the cereals provide by far the largest proportion of the protein supply (1978), that the oilseeds, and especially soya (which is also particularly rich in all the essential amino acids) (920), play an important role in the provision of protein (57), and that the "true" protein plants, the pulses, contribute only a small proportion. Worldwide, improvements in the protein supply can be achieved most effectively by breeding protein-rich grain cultivars, especially those that can thrive in the humid tropics, and by adequate nitrogen fertilization of cereal crops (898, 1888). In this connection, there is special interest in the agricultural use of nitrogen-fixing bacteria in cereal cultivation (1758). Also, the use of oil seeds for protein supply is very important (57, 1322).

Table 21. World production and main producers of grain legumes (in million tons) (514, 517, 947, 1762)

| Species | World production | Main producers | |
|---|---|---|---|
| | | Country | Production |
| *Pisum sativum* | 15.5 | USSR | 7.5 |
| | | France | 2.4 |
| | | China | 1.7 |
| *Phaseolus* spp. and *Vigna* spp. partly[1] | 15.5 | India | 3.5 |
| | | Brazil | 2.9 |
| | | China | 1.6 |
| | | Mexico | 1.1 |
| *Cicer arietinum* | 5.8 | India | 3.6 |
| *Vicia faba* | 4.7 | China | 2.4 |
| *Lens culinaris* | 2.5 | Turkey | 1.0 |
| | | India | 0.7 |
| *Cajanus cajan* | 2.1 | India | 2.0 |
| *Vicia sativa* | 2.0 | USSR | 1.6 |
| *Lupinus* spp. | 1.5 | Australia | 0.8 |
| | | USSR | 0.5 |
| *Vigna unguiculata* | 1.4 | Nigeria | 0.9 |
| *Lathyrus sativus* | 0.8 | India | 0.8 |
| Others[2], and not specified | 2.9 | | |
| Total | 54.7 | | |

[1] *P. vulgaris, P. lunatus, V. radiata, V. mungo, V. angularis*

[2] *Lablab purpureus, Vigna subterranea, Macrotyloma uniflorum, Trigonella foenum-graecum*, and others

While the pulses take a poor fourth place in the list of protein providers, they still make a definite contribution in many countries to filling the protein gap in human and animal nutrition. In Table 21, the production figures are given for the species listed in (514), plus some others, together with the most important regions of cultivation. In the export trade, the pulses play a comparatively minor role. The total exports from the developing countries amounted to 2.2 million tons in 1988. Of this, Turkey alone was responsible for 758,000 tons, mostly lentils and chickpeas. All of the important pulses are listed in Table 25. Descriptions and illustrations of the species cultivated in Africa are given in (1731), Indian legumes are treated in (131, 806), and legumes throughout the world are covered in (31, 444, 947, 1261, 1459, 1685, 1762, 1765).

# CULTIVATION OF PULSES

Most species of pulses are adapted to a wide range of ecological conditions. The majority love heat. Only *Lens culinaris*, *Pisum sativum* and *Vicia faba* do not thrive at high temperatures (Table 24), and so are grown in the cool season of the year, or high above sea-level. Also, *Cicer arietinum*, *Lathyrus sativus*, *Lupinus* spp., and *Trigonella foenum-graecum* prefer lower temperatures, although they can stand more heat than the first three species mentioned. For a high yield of seed, most pulses need full sunshine. Where *Canavalia*, *Lablab*, *Vigna* and others are grown as fodder crops, as ground cover, or for green manuring, half shade, or cloudy weather do not cause any problem. Some pulses were originally definitely short-day plants, but even for these there are now day-neutral cultivars available (*Lablab purpureus*, *Vigna unguiculata*, *Glycine max*) (825).

Many of the pulses do not have high requirements in terms of water supply. Because of their deep-reaching root systems, they have a large volume of soil at their disposal, and can survive long dry periods. The following are particularly drought tolerant: *Cajanus cajan*, *Macrotyloma uniflorum*, *Lathyrus sativus*, *Phaseolus acutifolius*, *Vigna aconitifolia*, *V. subterranea* and *V. unguiculata* ssp. *unguiculata*. On the other hand, there are species which are of interest because they can stand moist conditions, and there is an increased cultivation of these pulses in the humid tropics. They are *Canavalia* spp., *Glycine max*, *Lablab purpureus*, *Lathyrus sativus*, *Psophocarpus tetragonolobus*, *Vigna radiata*, and forms of *V. unguiculata*. Water-logging damages all of them, and in regions with high rainfall, deep, well-drained soil is the precondition for good growth (1889).

There is a range of legumes which make very few demands on soil fertility. *Cajanus cajan*, *Macrotyloma uniflorum*, *Lablab purpureus*, *Lathyrus sativus* and others grow well even on very poor soils. Species such as *Glycine max*, *Phaseolus vulgaris*, and *Pisum sativum* have a greater need for fertile soil, and for high yields, they need a soil with a good structure and high nutrient content. Many tropical legumes also thrive on acid soils (cultivars of *Glycine max* and *Vigna unguiculata*, also *Canavalia ensiformis*, *Vigna angularis*, *V. subterranea*, and others) (58, 444, 825).

As a result of their symbiosis with rhizobia, legumes have the ability to fix molecular nitrogen (116, 237, 432, 1758). With most species, the use of nitrogen fertilizers is not necessary to achieve maximum yields (see p. 96); however, with high-yielding cultivars of peas and beans, nitrogen supplies of up to 120 tons/ha can be economically justifiable (1014).

For efficient $N_2$ fixation, it is frequently necessary to inoculate the seed with rhizobia specific for the genus or the species (Table 22), or even with a bacterial strain selected for a particular cultivar. However, from numerous genera, more than one species of rhizobia have been isolated (particularly *Rhizobium loti* and *Bradyrhizobium* sp., see Table 22), and in tropical soils rhizobia are often so abundant that inoculation is not necessary. Further conditions for efficient $N_2$ fixation include adequate supplies of Ca, Mo, Co, Fe and Cu (1426), a high rate of photosynthesis, and soil temperatures below 32°C.

The first weeks are often decisive for the good development of the plants. If the soil is very low in nitrogen, it is recommended to give a light dressing of nitrogen fertilizer at sowing (about 10 kg N/ha). If the elements named above are deficient, then it is usually sufficient to pelletize the seeds with Ca, Mo, etc. in order to ensure early nodulation (1638). In hot, sunshine-rich regions, close planting is helpful, in order to prevent overheated soil temperatures.

Table 22. Rhizobia groups of important legumes (26, 906).

| Species | Host plants |
|---|---|
| *Azorhizobium caulinodans* | *Sesbania rostrata* |
| *Bradyrhizobium japonicum* | *Glycine*, *Macroptilium* |
| *Bradyrhizobium* sp. | *Acacia*, *Arachis*, *Cajanus*, *Canavalia*, *Centrosema*, *Cicer*, *Crotalaria*, *Cyamopsis*, *Desmodium*, *Indigofera*, *Lablab*, *Lotononis*, *Leucaena*, *Lotus*, *Lupinus*, *Macroptilium*, *Mimosa*, *Mucuna*, *Ornithopus*, *Pueraria*, *Sesbania*, *Vigna* |
| *Rhizobium leguminosarum* | *Lathyrus*, *Lens*, *Macroptilum*, *Phaseolus*, *Pisum*, *Trifolium*, *Vicia* |
| *R. loti* | *Acacia*, *Cicer*, *Leucaena*, *Lotus*, *Lupinus*, *Macroptilium*, *Medicago*, *Mimosa*, *Ornithopus*, *Phaseolus* |
| *R. meliloti* | *Medicago*, *Melilotus*, *Trigonella* |

In dry areas, phytopathological problems are limited in many legumes to a few insect pests (285, 1677). However, in hot humid climates, and especially with insufficient drainage, many legumes are susceptible to root rot and wilt diseases (*Sclerotium rolfsii* Sacc., *Rhizoctonia* spp., *Fusarium* spp., etc.) (29, 183). The choice of suitable sites for cultivation and crop rotation are the most important preventive measures. There can be catastrophic losses of the harvest through bacterial blight (*Xanthomonas phaseoli* (C.F.Smith) Dowson, and others) if the weather conditions are unfavourable, especially with *Phaseolus vulgaris*; this species is one of the legumes which is most susceptible to diseases. In all climates, very severe losses can be caused in storage by weevils (*Acanthoscelides obtectus* Say., the bean weevil; *Callosobruchus chinensis* L., the cowpea weevil, and other species of the family Bruchidae). Infestation occurs in the fields but also in storage, where the weevils can multiply very quickly. If leguminous seeds are to be stored for a long time, they should always be safeguarded by heat treatment, or by fumigation with methyl bromide. On the farm, the treatment of legume seed with small amounts of vegetable oil can suffice to protect the seeds against bruchids.

The progress of the breeding programs for various species show great differences, depending on the economic importance and the genetic variability. The following display a great variety of forms: *Cajanus cajan*, *Lablab purpureus*, *Phaseolus vulgaris*, and *Vigna unguiculata*. The breeding work focusses above all on ecological adaptation, resistance to diseases, and quality (1155, 1762, 1765). Many tropical legumes contain protein which is hard to digest, and only suitable as food after intricate preparation processes, such as fermentation (see p. 97). Many species contain poisonous compounds, which, before consumption, must be removed or destroyed by washing, cooking or heating (e.g. the cyanogenic glycoside linamarin in *Phaseolus lunatus* hemagglutinins in *Canavalia*, *Glycine*, *Lablab*, *Phaseolus*; neuro-toxic amino acids in *Lathyrus*; saponins in soya; alkaloids in lupins) (1058, 1697). There is still great potential for increasing the nutritional value of the pulses through breeding (667, 1190).

# Vegetables

## GENERAL

**Significance**. Vegetables are of great importance in the tropics and subtropics in a twofold sense - as food and as commercial products. In nutrition, vegetables can supply an important part of the minerals, vitamins and proteins (967). Of course, the individual species differ greatly in their nutritional value (Table 23); the leaf vegetables are especially valuable (containing vitamin K, as well as the vitamins listed in the table), whereas the fruit and tuber types contain fewer nutrients, but stimulate the appetite by their colour (tomato, beetroot), smell and taste (cucumbers, radish). Especially in the humid tropics, the leaf vegetables are important sources of protein, which is frequently in short supply there (see p. 120).

Vegetables are of interest economically from several points of view. Where there are good opportunities for sale, vegetables are among the most profitable agricultural products. Market gardening is carried out in all parts of the world using traditional tools and machinery. It is labour intensive, and it requires a lot of experience with regard to suitable cultivars, planting times, cultivation methods and fertilizers, irrigation, and disease and pest control. The expanding cities of the developing world offer a large local market for fresh vegetables. The industrial nations have an increasing need for vegetable imports from the developing countries, because of rising labour costs for their own producers, and the increasing expenses for heating greenhouses (1811). Because of transport costs, the mostly seasonal nature of production, and the limited suitability of transporting most types of vegetables, dehydrated (969) and also canned vegetables are of greater interest for export than the fresh crops. In this way, market gardening can form the basis of secondary industries, which create employment and bring in foreign exchange.

**Species**. The classification of species into the category of "vegetables" is optional, especially with roots, tubers and seasonings, but also with fruit. The onion is mostly used for seasoning, however, as is usual, it is included here among the vegetables, along with all other *Allium* species (except for *A. sativum*, Table 39). Muskmelons and rhubarb are utilized as fruit, and are included there, along with the strawberry (Table 30). The reverse also occurs, and many fruits are used like vegetables in the tropics when they are unripe (e.g. mango), and, in the same way, unripe maize cobs are one of the best-loved of vegetables. Some species, which are traditionally included among the fruits, are more often used as vegetables (e.g. avocado, bilimbi); in these cases, we have followed the normal classification. The legumes are all included in this chapter, as a differentiation is not possible between those which are consumed green as vegetables, and those where the seeds serve as a source of starch and protein (see p. 121-125).

The number of species which are consumed as vegetables in the tropics and subtropics is so large (480, 659, 1143, 1296, 1317, 1788, 1812, 2007, 2031), that Table 25 can only include the most important. To them can be added the leaves of many plants which are grown for other purposes (grapevine, cassava, sweet potato, *Colocasia*, *Xanthosoma*, muskmelon, pumpkins, and others). The number of wild plants of which the leaves or young shoots are regularly gathered and used as vegetables amounts to many hundreds; among them are found many important foodstuffs, such as palm cabbage and bamboo shoots. Finally, a multitude of weeds are eaten as vegetables. The large number of vegetable species is a relic of the days when plant foods were predominantly gathered from the wild. Many species do not taste good when raw, and may even be poisonous, and must therefore be soaked or cooked to make them edible (1058). Some of them may disappear from the "menu", while others will be improved by breeding, so that the abundance of vegetables in the hot countries which are available the whole of the year will be maintained. This refers particularly to leaf vegetables (Table 23), which should not be replaced by tomato, cucumber and beetroot for health reasons (1406). For tropical vegetable cultivation, there are special textbooks available (1211, 1788, 1812). French (316) and American (1806) works include some of the cultivation problems which are of interest in the tropics and subtropics. In addition, there are books devoted to practical market gardening in the individual countries (312, 675, 731, 978, 1296, 1411, 1974, 2011).

**Production**. For all countries of the world, the production of vegetables cannot be accurately described in the statistics, because everywhere a considerable proportion is produced in people's gardens for their own consumption. The quantities sold can only be given in the figures where there are controlled markets, and that is generally only the case in large cities. Even for the most important vegetables traded, the figures given are incomplete (Fig. 31). The largest proportion of marketed vegetables are produced in the temperate zone. Of the two species which are most suitable for transportation, i.e. tomato and onion, about half of the quantity exported nowadays comes from the subtropical regions with a Mediterranean climate. Also cabbage, cauliflower, green beans, and asparagus are produced in increasing quantities in the warm countries, wherever the distance to the importing countries is not too great. To these should be added a range of subtropical and tropical vegetables which do not grow well in the temperate zone, and which have found regular markets in the industrial countries, such as bell peppers, aubergine, artichoke, zucchini, and okra. For countries in the warm zone which are not conveniently situated for transporting their products, those vegetables are of particular interest which can be easily canned and are in demand in the industrial countries. These are primarily tomatoes, green beans, green peas, asparagus, carrots and mushrooms. For dehydrated vegetables, there is a good market for onions, beans, carrots, mushrooms, tomatoes and leeks (969). Some other vegetables from the warm zones may someday find markets in the industrial countries. Possible examples include small-fruited, tasty cultivars of *Cucurbita moschata* and *C. pepo*, the padi straw mushroom, and types of Chinese cabbage.

**Ecophysiology**. The temperature requirements of the most important vegetable species are compiled in Table 24. With many species, there is a great variation in the temperature tolerance of the cultivars. This applies particularly to the species which have been placed in the second column. Also, the choice of

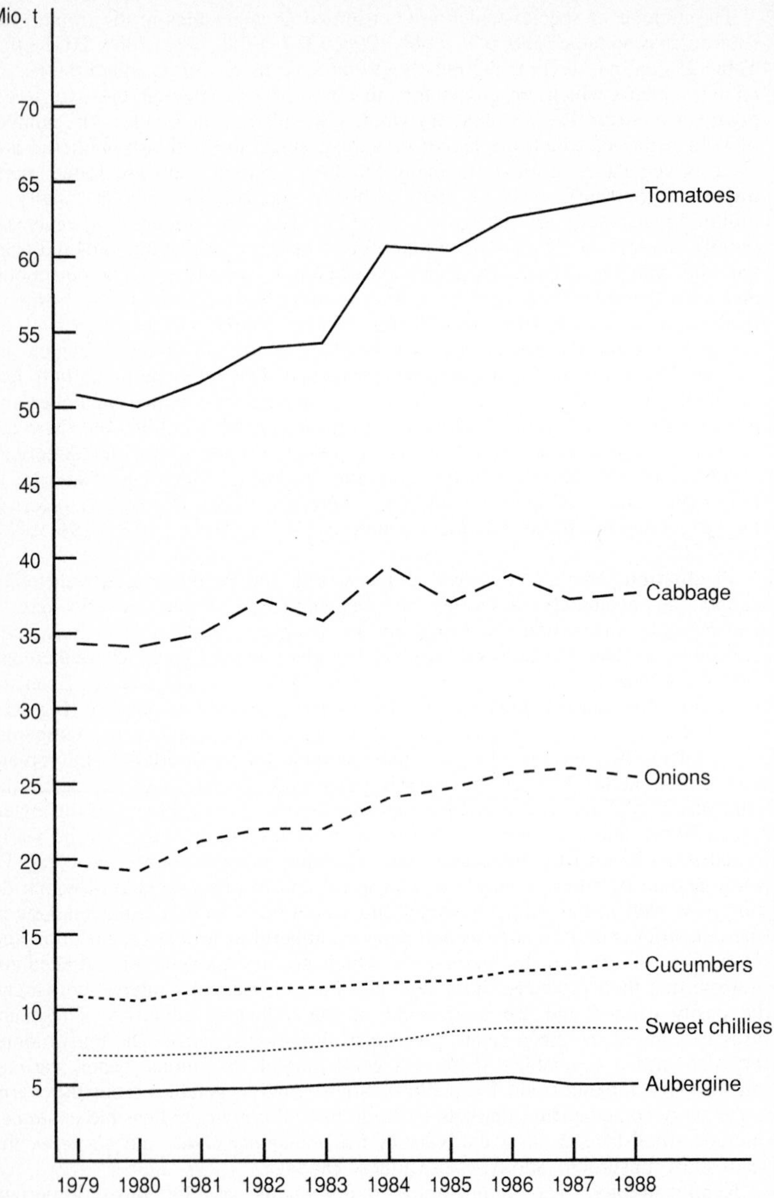


Fig. 31. World production of vegetables, 1979-1988

Table 23. Nutrient contents of some types of vegetables (in 100 g fresh weight) (376, 1812)

| Species | Energy kJ | Crude protein g | Ca mg | P mg | Fe mg | Vit. A I.U. | Thiamine mg | Riboflavin mg | Niacin mg | Vit. C mg |
|---|---|---|---|---|---|---|---|---|---|---|
| Young cassava leaves | 220 | 7.0 | 124 | 82 | 5.6 | 10000 | 0.2 | 0.3 | 0.2 | 256 |
| Young sweet potato shoots | 130 | 3.5 | 73 | 67 | 8.2 | 3300 | 0.1 | 0.2 | 0.9 | 25 |
| *Corchorus olitorius* (leaves) | 200 | 5.0 | 250 | - | 4.0 | 3000 | 0.1 | 0.3 | 1.5 | 100 |
| Green beans | 180 | 2.2 | 44 | 47 | 1.0 | 200 | 0.8 | 0.1 | 0.6 | 13 |
| Tomato | 90 | 1.0 | 10 | 24 | 0.4 | 1000 | 0.05 | 0.04 | 0.7 | 25 |
| Brinjal | 90 | 0.9 | 11 | 27 | 0.5 | 20 | 0.05 | 0.03 | 0.8 | 5 |
| Cucumber | 60 | 0.9 | 12 | 24 | 0.4 | 200 | 0.04 | 0.05 | 0.2 | 10 |
| Garden beet | 170 | 1.9 | 24 | 41 | 0.8 | traces | 0.02 | 0.03 | 0.2 | 5 |

seeding time can be decisive for successful cultivation, especially in regions with seasonal rainfall. Beetroot, Swiss chard, onions, carrots, the subspecies of *Brassica oleracea*, and some other species need cool temperatures for the induction of flowering. These species cannot be propagated in the tropics. The best regions for the production of vegetable seed generally lie in the subtropics with seasonal rainfall.

The day-length has a relatively slight influence on the yield of most vegetable species (1407) (exceptions are Chinese cabbage and onions). Many species require a lot of sunshine for good development, others will tolerate shade so well that they can be grown as intercrops, e.g. under rows of fruit trees (lettuce, spinach, beans, cucumbers, radish).

Most species of vegetables are quick-growing and shallow-rooted. Therefore, they need a good supply of easily available nutrients, and regular irrigation. Market crops must be irrigated in nearly all regions. The best soils are sandy loams or loamy sands, with a pH of about 6; however, vegetables are grown on almost all types of soil. Adverse soil characteristics can mostly be balanced by organic and mineral fertilizers, soil preparation, and irrigation. Deciding factors in the choice of a site are almost always the market conditions (city markets, canning factories).

**Cultivation**. *Sowing and nursing*. In hot countries, it is recommended that the sowing of all small-seeded species be done in seed-beds, pots, boxes, or soil blocks rather than direct sowing, because the top 1-2 cm of soil in the field is very strongly heated, and dries out quickly, so that germination is generally bad. With species that must be directly sown, such as carrot or beetroot, a thin layer of grass mulch is helpful, so that the soil surface is kept sufficiently cool and moist for several days. If irrigation equipment with fine sprinklers is available, then good germination can be achieved outdoors with a few minutes of daily watering. Large-seeded species, such as peas, beans and pumpkins, do not usually have any difficulties with direct sowing. In hot, dry periods, it is recommended that the ground be thoroughly irrigated before sowing, and that sowing be deeper than usual after the soil surface has become dry. The seeds then find enough water to be able to germinate, and the seedlings have no difficulty in breaking through the porous surface, whereas with shallower sowing, an irrigation before germination may be necessary, by which the soil surface may silt up and later form a hard crust. This can be hard to break through, especially for epigeal seedlings.

By sowing in seedbeds, good germination can easily be achieved by watering daily and shading the seedlings. This holds also for seeds which must be sown more shallowly. However, the seedlings should be put out in full sunlight as soon as possible, in order to harden them off for transplanting. In hot climates, transplanting is always a severe shock. In the first days, the plants can be assisted by covering the seedlings with paper caps, or by covering the earth around the plants with grass, coconut husks, or a similar material, until the plants are firmly rooted. Cultivation in soil blocks, small germination pots or boxes is highly recommended, because, in this way the roots remain in earth during transplanting, and growth can continue without interruption. Some species can hardly be transplanted at all, but they can be easily handled when taken from a container, e.g. pumpkins.

*Soil preparation*. The preparation of the soil depends on the climate and the irrigation system. In humid climates, planting is always done on mounds, ridges, or raised beds, as hardly any vegetable species can endure waterlogging. In dry climates, on the contrary, it is better to plant in furrows or sunken beds, so that the water supply is more favourable, especially during the first weeks after planting.

Table 24. Temperature requirements of important vegetable species

| Only in the cool season or at high altitudes | Better in the cool season, but also at higher temperatures | Better, or exclusively in the warm season and at low altitudes |
|---|---|---|
| celery | carrots | beans |
| peas | cabbage (all forms of *Brassica oleracea*) | Chinese cabbage |
| lentils | lettuce | aubergine |
| faba beans | leek | green maize |
| spinach | Swiss chard | cucumbers and gherkins |
| turnips | parsnips | pumpkins and squashes |
| | radishes | New Zealand spinach |
| | beetroot | okra |
| | asparagus | bell peppers |
| | onion | tomato |

*Planting distances*. In hot climates, the heating of the open soil in between the plants has an unfavourable effect on the growth, especially with shallow-rooted species. Therefore it is recommended to plant or sow the plants as densely as possible, so that a complete leaf canopy is quickly formed, under which the soil remains cool and moist. Furrow irrigation, and the necessity of hoeing the weeds, often compels a row-spacing that is inappropriately wide. Thus, the optimal plant density is mostly only practicable when the crop is grown using herbicides and sprinkling systems. Even when these resources are not available, everything should be done that will lead to the formation of a closed leaf canopy as soon as possible.

*Plastic sheets*. Because of the particularly good market conditions for early vegetables, the introduction of plastic sheets for mulching, or as tunnels, has found acceptance in many countries, especially in subtropical regions with low winter temperatures (316).

**Diseases and Pests**. The protection of the plants is one of the most difficult aspects of vegetable cultivation in the warm countries (331, 1211). In warm and humid regions, the fungal leaf diseases (e.g. for tomatoes, *Alternaria solani* (Ell. et Mart.) Sor., *Septoria lycopersici* Speg., *Cladosporium fulvum* Cooke, *Phytophthora infestans* (Mont.) de Bary) cause severe damage, and require regular spraying with fungicides. In dry areas, insects can cause catastrophic damage (e.g. to cabbage, diamondback moth, *Plutella xylostella* L.; bagrada bug, *Bagrada hilaris* Burm., and others). Everywhere, there is a menace from the soil-borne wilt diseases (*Pseudomonas solanacearum* E.F.Smith, *Fusarium* spp., and others), and also from nematodes (*Meloidogyne* spp.). On small farms, an effective control of the various pests and diseases is difficult to carry out because of the large number of vegetable species, and it needs a lot of experience and vigilance. The breeding of resistant cultivars, and the correct choice of cultivars are of exceptional importance in tropical vegetable cultivation. With tomatoes, great successes have already been achieved (full resistance against nematodes, good resistance against *Fusarium*, and others), and with many other species, at least partially resistant cultivars are available. Moreover, slow-releasing systemic insecticides, which are applied to the soil in granular form (e.g. phorate, disulfoton), offer good possibilities of control, even in small-scale cultivation under primitive conditions.

Fig. 32. *Tetragonia tetragonioides*. Top of shoot

Fig. 33. *Amaranthus tricolor*. Top of shoot

**Harvesting and Marketing**. The poor storage quality of most vegetable species is an even greater problem in the warm countries than in the temperate zone (1344). Damages due to high temperatures are of three types: development of fungus moulds; rapid loss of quality because of respiration and biochemical breakdown processes; and wilting. The leaf crops, which are the most valuable vegetables from a dietary viewpoint, are in general more susceptible to these three problems than root and fruit vegetables. To reduce the damage without going so far as major technological measures such as refrigeration, the following precautions can be taken:

- decomposition and quality losses can be reduced if only sound vegetables are harvested;
- injuries through handling (in picking and packing) must be avoided;
- the harvested goods should be protected from overheating (harvesting in the morning or early evening, sorting and packing in the shade, transport by night);
- adequate ventilation has to be provided during storage and transport.

Transpiration losses can be reduced by sprinkling with water, or in the case of root vegetables (carrot and beetroot) by cutting off the leaves at harvest, as well as by using packaging materials which prevent evaporation (banana leaves, plastic sheet). The choice of cultivars can also be decisive with regard to marketing problems. For supplies to a canning factory, the seed materials will usually be supplied by the firm or organization. For supplies to a market for fresh consumption, however, inferior quality cultivars which keep better are often more suitable than high-quality types: thick-skinned tomatoes are more suitable for transporting than thin-skinned ones; beans with strings have a harder subepidermal layer than stringless beans and keep in good condition for substantially longer. The breeders also need to be aware of these factors. Vegetables, like fruit, with a good appearance sell more readily than with high quality.

# SPECIAL SECTION

Apart from the information in Table 25, there are three vegetable crops of great economic importance, for which more detailed indications are necessary for their cultivation in hot countries.

## Tomato

(Fig 31, Tables 23, 24, 25, see p. 386)

There is a difference between canning and market tomatoes. Canning tomatoes are primarily grown for factories which use the tomato pulp. The best known old cultivars are 'San Marzano' and 'Roma'. They have small oval or pear-shaped fruits, with thick, firm flesh. They are usually allowed to grow in a bushy manner without stakes. In some markets they also find easy sales as fresh tomatoes. New breeds are available with resistance against verticillium and fusarium wilt (e.g. 'Roma VF'), and against nematodes and fusarium wilt (e.g. 'Ronita'). Market tomatoes are primarily produced for fresh consumption, but some are also utilized for canning (tomato juice, whole tomatoes, puree).

There is a very large number of cultivars. In humid, hot regions, determinate forms are preferred, since the potential of indeterminate types cannot be exploited because of the incidence of diseases.

The planting distances depend on the cultivars, and on the nutrients available. For indeterminate market tomatoes, a row-spacing of 1.20-1.80 m is the rule, and a spacing of 30 cm within the row. For late types of canning tomatoes on fertile soil, the planting distances can be even greater (2 x 0.5 m). Determinate cultivars will be planted with smaller spacings.

Tomatoes are sensitive to sunburn. In hot regions with a lot of sunshine, the tomato leaves are not pruned, because the protection of the fruit by the leaves is necessary. At temperatures over 32°C, fruit setting is bad, and some cultivars have a tendency to produce mis-shaped fruit ("cat-face"), particularly those with multilocular fruit. At temperatures over 35°C, the fruit flesh can be damaged (brown discolouration), even without the effects of direct sunlight. At high temperatures, the risk of blossom-end rot is also increased; this is conditioned especially by irregular water supplies. Excessive temperatures in the stand occur especially with trellised plants. The temperatures can be lowered if the rows of plants are parallel to the prevailing wind direction (so increasing air movement).

Market tomatoes are ready for picking when the green colour begins to lighten. They are then still firm and well transportable. The full red colour develops just as well in storage as it does in the open, but not at temperatures over 32°C. In the vicinity of the market, it is better to pick in the pink or red condition so that the fruit appears more attractive. For the organization of work, it is often better to pick the fruit only once every 8-10 days. For selling, the tomatoes are then sorted into light green, pink, and red fruit, and offered for sale separately. Canning tomatoes are always harvested when fully red, but before becoming soft (107, 242, 286, 376, 480, 627, 659, 978, 1168, 1211, 1411, 1440, 1867, 1899).

## Onion

(Fig. 31, Tables 24, 25)

Most cultivars need long-day conditions for bulb formation and for ripening. Near the equator, only day-neutral cultivars can be grown, such as 'Texas Grano', 'Yellow Bermuda' and 'Pusa Red'. Another important difference between the cultivars is their ability to be stored and transported. The cultivars which are brown, firm, and medium-sized, like 'Australian Brown', 'Cape Flat', and 'Bon Accord' are best for storage. When they are harvested in hot summer, they can be kept for 4-5 months without any special storage facilities. Except for places where there is a danger of scalding due to intense solar radiation, they can remain in the soil for up to 3 months without any damage (temperatures of 45-50°C, which can be endured without injury, induce dormancy, which lasts for several months). For dehydrating, cultivars with a dry matter content of 14-21% are preferred, such as 'Seco' or 'Southport White Globe' (normal cultivars contain 8-12% dry matter).

In hot countries, the onions are usually started off in seed beds, and then transplanted when they are pencil-thick. Direct sowing is possible when sprinkler irrigation and herbicides are used. The planting distances are usually governed by the tillage and irrigation systems; a normal row-spacing is 30 cm, with 5 cm within the row. Through narrower row spacing, which is possible with sprinkler irrigation, much higher yields per surface area can be achieved.

All onion cultivars require a period of cold weather before they form flowers. Therefore, seed production is only possible in regions with a cool season (316, 376, 469, 659, 903, 1317, 1331, 1411, 1681, 1812, 1821).

## Asparagus

Asparagus thrives very well in warm countries with plenty of sunshine. The only precondition is a cool period of 2 months, during which the plants rest. The preferred cultivars are the American rust-resistant types, such as 'Mary Washington' and 'California 500'. The yields are high because of the long vegetative period, and the cutting season can last for up to 3 months. The lifetime of the plants is shortened by badly drained soil, by planting too deep, and by the supply of too much organic fertilizer (dying-back through fungus attack on the root-stock).

The cut asparagus is very sensitive to high temperatures. The shoots continue growing after cutting, and harden due to sclerenchyma formation. No more than 3 hours should pass between cutting and arrival at the factory. On large farms, cold-storage rooms are necessary, to which the asparagus is brought immediately after harvesting. Because of the quick growth of the spears, it can be necessary to harvest twice a day (316, 713, 1339, 1806, 1812).

Fig. 34. *Basella alba*. Top of shoot

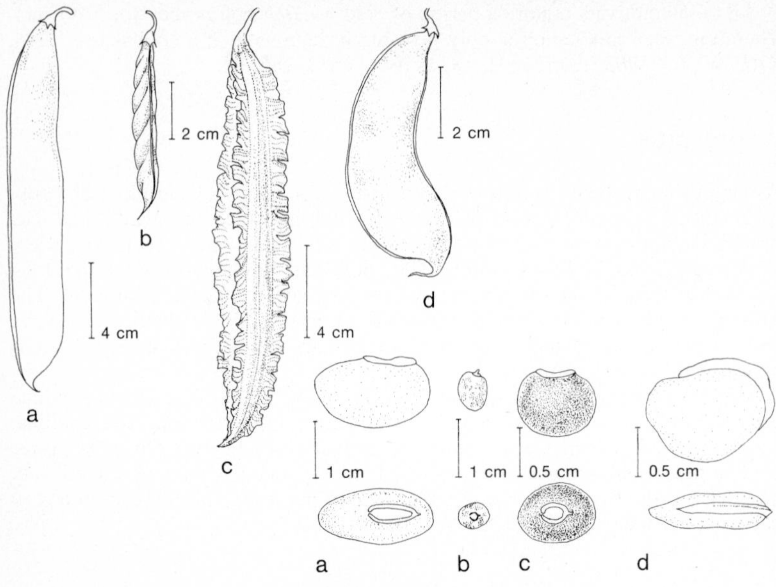


Fig. 35. Legumes, pods and seed. (a) *Canavalia ensiformis*, (b) *Cajanus cajan*, (c) *Psophocarpus tetragonolobus*, (d) *Lablab purpureus*

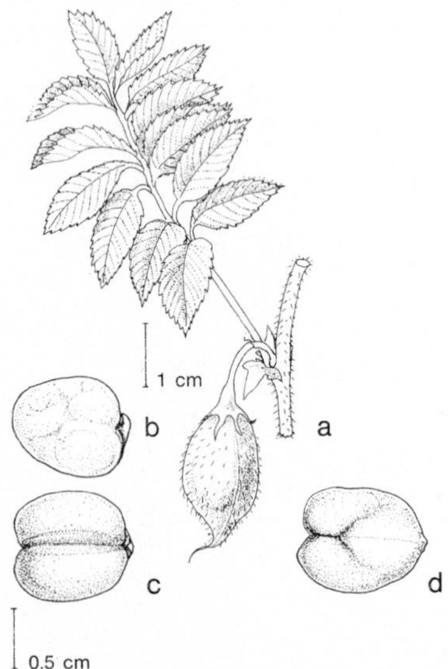


Fig. 36. *Cicer arietinum*. (a) leaf and fruit, (b, c and d) seed from side, from below, and from above

Fig. 37. *Abelmoschus esculentus*. (a) fruiting plant, (b) fruit in cross section

Table 25. Vegetables (explanation of symbols, see footnote and footnotes page 55)

| Botanical name | Vernacular names | Used part[1] | Climatic region | Distribution | Econ. value | Remarks and literature references |
|---|---|---|---|---|---|---|
| | | | **Basidiomycetes** | | | |
| **Agaricaceae** | | | | | | |
| *Agaricus bisporus* (Lange) Sing. | champignon, champignon de couche, fongo comestible | fruit-body | Te, S, TrH | w | +++ | Grown on manure or artificial media (72, 304, 1877, 1996) |
| **Amanitaceae** | | | | | | |
| *Volvariella volvacea* (Fr.) Sing. | padi straw mushroom | fruit-body | Tr | r: SE Asia | ++ | Grown on straw (304, 330, 731, 1060, 1411, 1568, 1996) |
| **Tricholomataceae** | | | | | | |
| *Lentinus edodes* (Berk.) Sing. | pasania mushroom, shiitake (Japan) | fruit-body | S | r: E Asia | ++ | Grown on wood (304, 1576, 1663, 1877, 1996) |
| | | | **Dicotyledonae** | | | |
| **Acanthaceae** | | | | | | |
| *Justicia insularis* T.Anders. | tettu, ondógó, takankola | L | Tr | l: W Africa | + | Also other *J.* species used as leaf vegetables (480, 2031) |

| | | | | | | |
|---|---|---|---|---|---|---|
| **Aizoaceae** | | | | | | |
| *Tetragonia tetragonioides* (Pall.) O.Kuntze (*T. expansa* Murr.) | New Zealand spinach, tetragone cornue | L | S, Tr | w | + | (Fig. 32) Low-demanding and drought tolerant (731, 949, 1143, 1296, 1317) |
| **Amaranthaceae** | | | | | | |
| *Amaranthus tricolor* L. | Chinese spinach, Chinese amaranth, aupa, tampala | L | S, Tr | w | ++ | (Fig. 33) Species with many forms, including var. *gangeticus* (L.) Fiori et Paoletti. Also other *A.* species used as vegetables, as well as some as pseudocereals (Table 7), and as fodder plants (376, 480, 658, 659, 731, 1144, 1211, 1317, 1411, 1513, 1838, 2031) |
| *Celosia argentea* L. | cockscomb, mirabel, soko (Nigeria) | L | S, Tr | w | + | Several varieties cultivated in W Africa. Worldwide as an ornamental plant (480, 658, 659, 1144, 1296, 1317, 2031) |
| **Basellaceae** | | | | | | |
| *Basella alba* L. | Ceylon spinach, Malabar spinach, vine spinach | L, yg. Sh | S, Tr | w | + | (Fig. 34) Also thrives well in humid tropical climates. "*B. rubra* L." is only a red-coloured variety (271, 480, 659, 1296, 1317, 1332, 2031) |

[1] L = leaf, Fl = flower, Fr = fruit, Se = seed, Sh = shoot, R = root, B = bulb

Table 25. Vegetables, cont'd (explanation of symbols, see footnote page 139 and footnotes page 55)

| Botanical name | Vernacular names | Used part[1] | Climatic region | Distri- bution | Econ. value | Remarks and literature references |
|---|---|---|---|---|---|---|
| **Capparidaceae** | | | | | | |
| *Gynandropsis gynandra* (L.) Briq. | African spider-flower, bastard mustard, massarubee | L, Se | S, Tr | w | + | Favourite vegetable with a bitter-piquant taste. Ground seeds used like mustard. Frequent as a weed (376, 978, 1296, 2031) |
| **Chenopodiaceae** | | | | | | |
| *Beta vulgaris* L. ssp. *vulgaris* var. *vulgaris* | red beet, garden beet, remolacha colorada | R | Te, S, Tr | w | +++ | (Table 24) Known everywhere. Also a source of food colouring (376, 978, 1211, 1296, 1611, 1812) |
| var. *cicla* L. | Swiss chard, spinach beet, acelga cardo | L | Te, S, TrH | w | ++ | (Table 24) Best leaf vegetable for the subtropics (659, 978, 1211, 1296, 1317, 1611, 1812) |
| *Spinacia oleracea* L. | spinach, épinard, espinaca | L | Te, S, TrH | w | ++ | Of little importance in the tropics and subtropics. Thrives only in cool weather (Table 24) (376, 731, 1317, 1411, 2007) |

| | | | | | | | |
|---|---|---|---|---|---|---|---|
| **Compositae** | | | | | | | |
| *Chrysanthemum coronarium* L. | garland chrysanthemum crown daisy, tong ho | L | S, Tr | r: | E and SE Asia | + | Weed in the Mediterranean region. Mostly used as a vegetable in E and SE Asia (259, 731, 1029, 1317, 1411, 2007) |
| *Cichorium endivia* L. | endive, escarole, chicorée endive, escarola | L | Te, S, TrH | w | | ++ | Better suited to warm climates than lettuce (731, 1211, 1296, 1317, 1411, 1812) |
| *Crassocephalum biafrae* (Oliv. et Hiern) S.Moore | bologi, morowo | L | Tr | l: | W Africa | + | Propagated by cuttings. Rich in methionine. *C. crepidioides* (Benth.) S. Moore is also cultivated as a vegetable (480, 1306, 1812, 2031) |
| *Cynara cardunculus* L. | cardoon, cardon d'Espagne, cardo de comer | blea-ched peti-oles | S, TrH | r: | Mediter-ranean countries | ++ | Is hardly cultivated outside the Mediterranean region. Some exports to Central Europe (200, 2030) |
| *C. scolymus* L. | globe artichocke, alcachofa | Fl-heads | S, TrH | w | | +++ | Many cultivars. Important export crop. Dried leaves provide bitter substances for alcoholic drinks (376, 731, 2030) |
| *Lactuca indica* L. | Indian lettuce, sawi rana, foo mak ts'oi | L | Tr | r: | SE Asia | + | Mostly eaten cooked. Also a fodder for geese and silkworms (259, 659, 731, 1296, 2007) |

Table 25. Vegetables, cont'd (explanation of symbols, see footnote page 139 and footnotes page 55)

| Botanical name | Vernacular names | Used part[1] | Climatic region | Distri-bution | Econ. value | Remarks and literature references |
|---|---|---|---|---|---|---|
| *L. sativa* L. | lettuce, laitue, lechuga, alface | L | Te, S, Tr | w | +++ | Has an extraordinary range of varieties, some are suitable for the tropics (259, 376, 659, 731, 978, 1296, 1317, 1411, 1812, 2007) |
| *Launaea taraxacifolia* (Willd.) Amin ex C. Jeffrey | wild lettuce, yanrin | L | Tr | l: Nigeria | + | Propagated by root cuttings (258, 262, 480, 2031) |
| *Polymnia sonchifolia* Poepp. et Endl. | yacon strawberry, poire de terre cochet, yacón, llamon, aricoma, jíquima | R | TrH | r: Andes, S America | + | Tuberous root eaten raw or cooked, little nutritional value (inulin as reserve material). Stems used as vegetable, leaves as animal fodder (242, 946, 1039, 1201, 1864) |
| *Scorzonera hispanica* L. | Spanish salsify, black salsify, salsifis noir | R | Te, S | w | ++ | Main area of cultivation is the Mediterranean (316, 376) |
| *Spilanthes oleracea* L. | Para cress, Brazilian cress, cresson de Para, yambo, agrião do Pará | yg. Sh | Tr | w | + | Salad and vegetable with a sharp taste, also a medicinal plant. Other *S.* species used similarly (200, 259, 376, 868, 1296, 1317, 1518) |

| | | | | | | | |
|---|---|---|---|---|---|---|---|
| *Vernonia amygdalina* Del. | bitterleaf, ewuro | L, yg. Sh | Tr | l: | W Africa | + | Propagated by cuttings. Very bitter taste (480, 1023, 1317, 1788, 2031) |
| **Convolvulaceae** | | | | | | | |
| *Ipomoea aquatica* Forssk.(*I. reptans* Poir.) | swamp cabbage, water spinach, kang kong | L | Tr | r: | SE Asia | ++ | Aquatic type cultivated in ditches and ponds, propagated by cuttings; dryland type on moist soil, usually propagated by seed (366, 376, 659, 731, 978, 1411, 1812, 2031) |
| **Cruciferae** | | | | | | | |
| *Brassica oleracea* L. | | | | | | | (259, 376, 1211, 1296, 1317, 1411, 1657, 1806, 1812, 1834, 2007) |
| var. *botrytis* L. | cauliflower, chou-fleur | head | Te, S, TrH | w | | +++ | Good heads only with cool weather (Table 24) (1770) |
| var. *capitata* L. | common cabbage, chou cabus | head | Te, S, Tr | w | | +++ | Cultivars suitable for the tropics are also available |
| var. *gemmifera* DC. | Brussels sprouts, chou de Bruxelles | small heads | Te, S, TrH | w | | +++ | Japanese hybrid cultivars also suitable for warmer countries |
| var. *gongylodes* L. | Hungarian turnip, kohlrabi, chou-rave | tuber | Te, S, Tr | w | | +++ | Thrives also in the tropics, quickly becomes hard |
| var. *italica* Plenck | broccoli | fl. Sh | Te, S, Tr | w | | ++ | Thrives better than cauliflower in warm countries (635) |

Table 25. Vegetables, cont'd (explanation of symbols, see footnote page 139 and footnotes page 55)

| Botanical name | Vernacular names | Used part[1] | Climatic region | Distri-bution | Econ. value | Remarks and literature references |
|---|---|---|---|---|---|---|
| var. *sabauda* L. | Savoy cabbage, chou frisé | head | Te, S, TrH | w | ++ | Unsuitable for the warm tropics |
| var. *viridis* L. | collard, kale, borecole, chou vert | L | Te, S, Tr | w | ++ | This, and other leaf-cabbage varieties have cultivars suitable for warm climates |
| *B. rapa* L. emend. Metzger (*B. campestris* L.) | | | | | | |
| ssp. *chinensis* (L.) Makino (*B. chinensis* L.) | Chinese cabbage, chou de Chine, pak-choi, paak ts'oi | L | S, Tr | r: E Asia | +++ | Annual, warmth-loving, firm heads are not formed. Very many varieties. Seed for oil production (259, 480, 731, 1296, 1411, 1657, 1812, 2007) |
| ssp. *pekinensis* (Lour.) Olsson (*B. pekinensis* (Lour.) Rupr.) | Shantung cabbage, Peking cabbage, chou de Shanton, pe-tsai, pechay | L | Te, S, TrH | w | +++ | Suitable for cooler climate than *chinensis*, biennial, forms compact head. The names pe-tsai (Mandarin) and paak tsoi (Cantonese) are incorrectly used to differentiate the ssp. *chinensis* and *pekinensis* (259, 731, 1320, 1411, 1657, 1812, 2007) |

| | | | | | | |
|---|---|---|---|---|---|---|
| ssp. *rapa* | turnip, chou-navet, nabo | R, yg. L | Te, S, TrH | w | + | Due to its short vegetation time, suitable as catch crop (978, 1211, 1788) |
| *Eruca vesicaria* (L.) Cav. ssp. *sativa* (Mill.) Thell. (*E. sativa* Mill.) | garden rocket, ruce, taramira (Hindi) | L | S, TrH | w | + | Leaves sharp tasting, used as salad and seasoning (see p. 277). Seeds used like mustard. In Asia cultivated for oil production (376, 807, 2007) |
| *Lepidium sativum* L. | garden cress, cresson alénois | L | Te, S, Tr | w | + | Generally used as salad, also cultivated for seed and oil production. Important medicinal plant in Ethiopia (766, 1620, 1650) |
| *Nasturtium officinale* R.Br. | watercress, cresson de fontaine, berro | L | Te, S, Tr | w | + | Cultivated in flowing water in many warm countries. Eaten as a salad or cooked (259, 376, 731, 1296, 1317, 1411) |
| *Raphanus sativus* L. | radish, rave, rábano | R, yg. L | Te, S, Tr | w | ++ | Many cultivars. The large Chinese and Japanese radishes (var. *longipinnatus* Bailey) are cultivated outside SE Asia also as a succulent feed (see p. 400) (259, 376, 659, 731, 946, 978, 1211, 1296, 1411, 1611, 1812, 2007) |
| **Cucurbitaceae** | | | | | | |
| *Benincasa hispida* (Thunb.) Cogn. | waxgourd, Chinese winter melon, calabaza China | yg. Fr | S, Tr | w | + | Cultivated especially in E Asia (376, 659, 1214, 1411, 1812, 1932, 2007) |

Table 25. Vegetables, cont'd (explanation of symbols, see footnote page 139 and footnotes page 55)

| Botanical name | Vernacular names | Used part[1] | Climatic region | Distri- bution | Econ. value | Remarks and literature references |
|---|---|---|---|---|---|---|
| *Citrullus fistulosus* Stocks (*Praecitrullus fistulosus* (Stocks) Pang.) | Indian squash melon, round melon, tinda (India) | Fr | Tr | l: India | + | One of the favourite summer vegetables in some regions of India. Not related to *C. lanatus* (Table 30) (376, 941, 962, 1271) |
| *Coccinia abyssinica* (Lam.) Cogn. | anchoté | R | TrH | l: Ethiopia | + | Root vegetable, also suitable for dry regions (1811, 2026) |
| *C. grandis* (L.) Voigt (*C. cordifolia* (L.) Cogn., *C. indica* Wight et Arn.) | ivy gourd kunduri | yg. L, Sh, Fr | S, Tr | r: SE Asia | + | In India, forms cultivated with slight bitterness in the unripe fruits (376, 731, 1296, 1788) |
| *Cucumis anguria* L. | West Indian gherkin, bur gherkin, cackrey | unripe Fr | S, Tr | r: Central and S America | + | Eaten raw and cooked (659, 1932, 1960) |
| *C. sativus* L. | cucumber, concombre, pepino | Fr, yg. L, Sh | Te, S, Tr | w | +++ | Many cultivars also suitable for the humid tropics (659, 978, 1296, 1411, 1812) |

| *Cucurbita ficifolia* Bouché | Malabar gourd, fig-leaf gourd, lacayote, chilacayote | Fr | S, TrH | w | + | Quality of the fruit flesh lower than for other *C.* species (723, 731, 1039, 1931, 1932, 2007) |
|---|---|---|---|---|---|---|
| *C. maxima* Duch. | winter squash, mammoth squash, pumpkin, calabaza | Fr, yg. L | Te, S, Tr | w | +++ | Species with many types. Especially suitable for the tropics. Fruits keep well (242, 659, 723, 731, 978, 1039, 1411, 2007) |
| *C. mixta* Pang. | ayote | Fr, yg. L | S, Tr | w | ++ | Numerous forms in Central America. Little cultivated in other regions (723, 731, 978, 1039, 1931, 1932) |
| *C. moschata* (Lam.) Duch. ex Poir. | Seminole pumpkin, courge musquée | Fr, yg. L | Te, S, TrH | w | +++ | Most important species of Central America. Many cultivars with good-tasting flesh (indicated by its name) and excellent keeping properties (376, 659, 723, 731, 914, 1039, 1296, 1411, 1932) |
| *C. pepo* L. | gourd, summer squash, field pumpkin, vegetable marrow | Fr, yg. L | Te, S, Tr | w | +++ | Extremely many forms. *C.* species with the widest distribution (316, 659, 723, 731, 1039, 1411, 1932) |
| *Cyclanthera pedata* (L.) Schrad. | wild cucumber, pepino de comer, cahiua | yg. Fr, Sh | Tr | r: Central and S America | + | Mostly cultivated because of its fruit flesh which is similar to cucumber (242, 1039, 1960, 2007) |

Table 25. Vegetables, cont'd (explanation of symbols, see footnote page 139 and footnotes page 55)

| Botanical name | Vernacular names | Used part[1] | Climatic region | Distri-bution | Econ. value | Remarks and literature references |
|---|---|---|---|---|---|---|
| *Lagenaria siceraria* (Mol.) Standl. | bottle gourd, calabash marrow, calabaza, dudhi | yg. Fr, yg. Sh | S, Tr | w | ++ | Cultivars with thick exocarp cultivated for vessels. Non-bitter, mostly soft-shelled cultivars used as a vegetable, especially in India and Africa. Seeds also for oil extraction (376, 659, 723, 731, 1136, 1296, 1494, 1932, 1960) |
| *Luffa acutangula* (L.) Roxb. | ridged gourd, angled loofah, papangaye | yg. Fr, yg. Sh | S, Tr | w | ++ | Only young fruits are edible. Important vegetable in SE Asia. Non-bitter forms of *L. cylindrica* (Table 44) are used similarly as a vegetable (259, 376, 659, 723, 731, 1136, 1296, 1494, 2007) |
| *Momordica charantia* L. | balsam pear, bitter gourd, bitter melon, balsamino | yg. Fr, L, Sh | S, Tr | w | ++ | Particularly in Asia a popular vegetable. Also cultivated are *M. balsamina* L. and *M. cochinchinensis* (Lour.) Spreng. (376, 659, 731, 978, 1296, 2007) |
| *Sechium edule* (Jacq.) Sw. | chayote, chow-chow, chuchu | Fr, yg. Sh, R, L | S, Tr | w | ++ | The fruit flesh is the most important part. The large tubers contain 20% starch and are eaten. Stem fibres are used locally (242, 376, 659, 731, 1039, 1201, 1960, 2007) |

| | | | | | | |
|---|---|---|---|---|---|---|
| *S. tacaco* (Pitt.) C.Jeffrey (*Polakowskia tacaco* Pitt.) | tacaco, chilacayote | Fr | Tr | l: Costa Rica | + | Fruits having only one seed as with *S. edule*, but much smaller (200, 876, 1039, 1960) |
| *Sicana odorifera* (Vell.) Naud. | melocotón, casabanana, cohombro de olor | Fr | Tr | r: Central and S America | + | Aroma of fruit flesh reminiscent of peach. Used as a vegetable and in jams. Also an ornamental plant (1039, 1960) |
| *Trichosanthes cucumerina* L. var. *anguina*; (L.) Haines (*T. anguina* L.) | snake gourd, cohombre vibora, patola, padwal | Fr, yg. L, Sh, | Tr | r: SE Asia, W Africa | ++ | Flesh of the ripe fruit is sweetish, eaten cooked as a vegetable, especially popular in Asia. The similar *T. dioica* Roxb. is also cultivated (376, 480, 723, 731, 978, 1296) |
| **Euphorbiaceae** | | | | | | |
| *Sauropus androgynus* (L.) Merr. | katuk, chekor manis | yg. Sh | Tr | r: SE Asia | + | Popular vegetable in India, Malaysia and Indonesia. Fruit in Java for sweet dishes (259, 376, 1296, 1317) |
| **Leguminosae** | | | | | | |
| MIMOSOIDEAE | | | | | | |
| *Neptunia oleracea* Lour. | lajalu, pag-ka-ched (Thailand) | yg. Se, yg. L | Tr | r: SE Asia | + | Water plant. Especially in Indochina generally cultivated (259, 376) |

Table 25. Vegetables, cont'd (explanation of symbols, see footnote page 139 and footnotes page 55)

| Botanical name | Vernacular names | Used part[1] | Climatic region | Distribution | Econ. value | Remarks and literature references |
|---|---|---|---|---|---|---|
| PAPILIONOIDEAE | | | | | | |
| *Cajanus cajan* (L.) Huth | pigeonpea, red gram, guando, arhar | yg. L, yg. pod, Se | S, Tr | w | ++ | (Fig. 35) (Table 21) (see p. 123, 124) One of the most important pulses in India. Perennial, decidedly short-day plant. Fodder plant, also for lac insects (see p. 403) and silkworms. Green manuring and shade plant (21, 131, 165, 314, 376, 444, 823, 947, 1006, 1023, 1100, 1101, 1220, 1296, 1491, 1657, 1765) |
| *Canavalia ensiformis* (L.) DC. | jackbean, swordbean, horsebean, haricot-sabre, haba de burro | yg. pod, Se | S, Tr | w | + | (Fig. 35) (see p. 123, 124) Only to a limited extent for human nutrition, mostly for green manuring, fodder, and erosion control. Other *C.* species used similarly (e.g. *C. gladiata* (Jacq.) DC.) (376, 444, 947, 1023, 1261, 1296, 1459) |

| | | | | | | |
|---|---|---|---|---|---|---|
| *Cicer arietinum* L. | chickpea, Bengal gram, cicerolle, garvance, garbanzo | Se | S, TrH | w | +++ | (Fig. 36) (Tables 21, 22, 31) (see p. 123, 124, 236) Eaten green and dry like peas. Straw is a good animal fodder. The leaves secrete malic acid, which is absorbed with cloths, used as a vinegar substitute and also medicinally (131, 376, 444, 947, 1006, 1099, 1101, 1292, 1491, 1595, 1596, 1631, 1657, 1696, 1762) |
| *Lablab purpureus* (L.) Sweet (*L. niger* Medik., *Dolichos lablab* L.) | hyacinth bean, bonavist bean, dolique d'Egypte, frijol dólico, lubia | yg. pod, Se, L, Sh | S, Tr | w | ++ | (Fig. 35) (see p. 123, 124) Especially in India used for human nutrition, otherwise cultivated for fodder and as ground cover (creeping forms) (131, 376, 444, 659, 947, 978, 1101, 1261, 1411, 1459, 1731, 1762) |
| *Lathyrus sativus* L. | chickling pea, chickling vetch, gesse, almorta, khesari | pod, Se, L | Te, S, TrH | w | + | (Table 22) Annual production in India 0.8 million tons. Great ecological adaptability (see p. 123). Seeds only edible when well cooked (see p. 124), neurotoxin-free cultivars have been selected. Also a fodder plant (131, 376, 444, 668, 947, 1006, 1058, 1278) |
| *Lens culinaris* Medik. (*L. esculenta* Moench) | lentil, red dhal, lenteja, lentilha | Se | Te, S, Tr | w | +++ | (Tables 21, 22, 24) (see p. 124) One of the most nutritious pulses. High export value. Straw and chaff valuable fodder (131, 376, 444, 947, 1006, 1101, 1596, 1762, 1906) |

Table 25. Vegetables, cont'd (explanation of symbols, see footnote page 139 and footnotes page 55)

| Botanical name | Vernacular names | Used part[1] | Climatic region | Distri-bution | Econ. value | Remarks and literature references |
|---|---|---|---|---|---|---|
| *Lupinus albus* L. | white lupin, lupino blanco, tremoço branco | Se | Te, S, TrH | w | + | (Tables 21, 22) (see p. 123) The Egyptian lupin (*L. albus* var. *termis* (Forssk.) Alef.) is also treated as a separate species. After soaking to remove bitter compounds, the seeds are an important food for the people of Egypt and the other Arabian countries. Alkaloid-free cultivars have been bred. Outside the Mediterranean region, a fodder plant (165, 376, 444, 657, 828, 947, 1659) |
| *L. mutabilis* Sweet | pearl lupin, tarhui, chocho, tauri, tarwi | Se | TrH | r: S America | + | An important foodstuff in the Andes. The seeds are only edible after long soaking. Nowadays, there are selected forms which are alkaloid-free, protein-rich, and oil-rich (up to 25.8%) (see p. 112) (242, 444, 657, 828, 947, 1261) |
| *Macrotyloma geocarpum* (Harms) Maréchal et Baudet (*Kerstingiella geocarpa* Harms) | Kersting's groundnut, bindi, kandela | Se | Tr | l: W Africa | + | Cultivated in a limited region for its seeds (51, 444, 947, 1126, 1261, 1306, 1762) |

| | | | | | | |
|---|---|---|---|---|---|---|
| *M. uniflorum* (Lam.) Verdc. (*Dolichos uniflorus* Lam.) | horse gram, grain de cheval, kulthi (India) | Se | Tr | r: Asia, Africa | + | In dry regions of SE Asia, especially in S India, for human food, otherwise as fodder and for green manuring (131, 376, 444, 947, 1101, 1721, 1722, 1757) |
| *Phaseolus acutifolius* A.Gray var. *latifolius* Freem. | tepary bean, haricot à feuilles aiguës, escomite | pod, Se | S, Tr | w | + | Drought and heat tolerant, quick-growing. Also as fodder plant in arid regions (444, 592, 947, 1039, 1101, 1261, 1762, 1765, 1814, 1885) |
| *P. coccineus* L. (*P. multiflorus* Willd.) | scarlet runner bean, haricot d'Espagne | Se, pod | Te, S, TrH | w | + | In its homeland Mexico and Guatemala, the ripe and unripe seeds and the tuberous roots are eaten, in other countries especially the green pods. Ornamental plant (376, 444, 529, 947, 1101, 1762, 1765) |
| *P. lunatus* L. | Lima bean, butter bean, harricot du Lima | Se, pod | S, Tr | w | ++ | Withstands more humidity than *P. vulgaris*. Seeds eaten ripe and unripe. Some cultivars with a high linamarin content, which must be destroyed by cooking (376, 444, 592, 659, 947, 1023, 1039, 1101, 1411, 1731, 1762, 1765) |

Table 25. Vegetables, cont'd (explanation of symbols, see footnote page 139 and footnotes page 55)

| Botanical name | Vernacular names | Used part[1] | Climatic region | Distri-bution | Econ. value | Remarks and literature references |
|---|---|---|---|---|---|---|
| *P. vulgaris* L. | common bean, kidney bean, navy bean, garden bean, judía común, frijol, feijão | Se, pod | Te, S, Tr | w | +++ | (Tables 22, 23) Bush and runner beans, very many forms. In area of origin Central and S America, the dry beans are the staple diet, together with maize. Green beans and seeds are important export products (131, 242, 376, 444, 592, 630, 659, 947, 1039, 1101, 1612, 1762, 1765, 1889) |
| *Pisum sativum* L. | garden pea, field pea, pois des jardins, guisante | Se, pod | Te, S, TrH | w | +++ | (Tables 21, 22, 24) Many cultivated forms, including the fodder cultivars, which are sometimes separated as *P. arvense* L.. Most important pulse. The canning industry processes unripe seeds on a large scale. Also cultivars with edible pods (sugar peas) (376, 444, 495, 947, 1101, 1411, 1762, 1765) |

| | | | | | | |
|---|---|---|---|---|---|---|
| *Psophocarpus tetragonolobus* DC. | Goa bean, winged bean, pois carré, sesquidilla, pallang | L, yg. pod, Se, R | Tr | w | + | (Fig. 35) Perennial. All parts are edible, the most important product is green pods. Mainly cultivated in SE Asia. For high yields the plants have to be staked. Also planted for fodder, green manuring and ground cover (139, 376, 444, 731, 790, 947, 1140, 1261, 1263, 1296, 1428, 1459, 1513, 1762) |
| *Trigonella foenum-graecum* L. | fenugreek, senegré, alholva, hilbah, methi (Hindi) | L, yg. Fr, Se | Te, S, Tr | w | + | (see p. 123) Cultivated as leaf vegetable in India, for the high-starch seeds in the Middle East, Morocco, and Egypt. Also as a seasoning (p. 277), a medicinal plant (p. 308), and for gum (p. 373). Salt tolerant (376, 409, 444, 1032, 1317, 1538, 1687) |

Table 25. Vegetables, cont'd (explanation of symbols, see footnote page 139 and footnotes page 55)

| Botanical name | Vernacular names | Used part[1] | Climatic region | Distri-bution | Econ. value | Remarks and literature references |
|---|---|---|---|---|---|---|
| *Vicia faba* L. | broad bean, field bean, faba bean, féverole, ful | Se | Te, S, TrH | w | +++ | (Tables 21, 22, 24) The small seeded varieties are most cultivated (var. *equina* Pers., horse bean; var. *minor* Peterm., tick pea, pigeon bean). The dry seeds (protein content 20-35%) are ground, and as gruel (Arabian ful), they are one of the most important foods from the Mediterranean region to China. The yields can be very high (over 6 t/ha). The cultivation of the large-seeded var. *faba*, the broad bean, where the half ripe seeds are used as a vegetable, is limited mainly to Europe. Everywhere also used as animal fodder (205, 206, 376, 444, 719, 722, 947, 1101, 1762, 1765, 1808, 1977) |
| *Vigna aconitifolia* (Jacq.) Maréchal | mat bean, moth bean, haricot mat | pod, Se | S, Tr | w: esp. India, S America | + | Cultivated as a pulse in the dry regions of India, outside India mainly as a fodder plant (131, 376, 444, 731, 1101, 1261, 1762) |

| | | | | | | |
|---|---|---|---|---|---|---|
| *V. angularis* (Willd.) Ohwi et Ohashi | adzuki bean, frijol adzuki | pod, Se | Te, S, Tr | w: esp. E Asia | ++ | Most important pulse in Japan and China, otherwise mainly as a fodder and green manure plant (376, 444, 732, 1101, 1562, 1762) |
| *V. mungo* (L.) Hepper | black gram, urd, ambérique, frijol mungo, moong | Se, pod | S, Tr | w: esp. India | ++ | Valued pulse in India. Otherwise as a fodder and green manure plant (131, 376, 444, 1005, 1006, 1101, 1762, 1765) |
| *V. radiata* (L.) Wilczek | mung bean, green gram, judía de mungo | Se, pod | S, Tr | w: esp. SE Asia | +++ | One of the most popular pulses in India, also in Malaysia and China. As with soya, the bean-sprouts are a vegetable. Green manuring and fodder plant (131, 376, 444, 659, 731, 947, 1006, 1101, 1411, 1731, 1762, 1765) |
| *V. subterranea* (L.) Verdc. (*Voandzeia subterranea* (L.) Thou. ex DC.) | bambara groundnut, groundbean, voandzou | Se, yg. pod | S, Tr | w: esp. Africa | + | (see p. 63) Many forms with different fruit size and colour. Unripe seeds eaten fresh, ripe seeds cooked or milled to make flour (435, 444, 659, 730, 731, 936, 947, 1023, 1101, 1261) |
| *V. trilobata* (L.) Verdc. | pillipesara, mungani, mugan | Se | S, Tr | w: esp. India | + | Seeds only eaten by the poorer people, generally a fodder and green manure plant (131, 376, 1762) |

Table 25. Vegetables, cont'd (explanation of symbols, see footnote page 139 and footnotes page 55)

| Botanical name | Vernacular names | Used part[1] | Climatic region | Distribution | Econ. value | Remarks and literature references |
|---|---|---|---|---|---|---|
| *V. umbellata* (Thunb.) Ohwi et Ohashi (*Phaseolus pubescens* Bl. and *P. calcaratus* Roxb.) | rice bean, haricot riz, frijol de arroz, sutri | Se, pod, L | S, Tr | w: esp. SE Asia | ++ | Widespread as a pulse in SE Asia. Also for fodder, green manuring and ground cover (95, 131, 259, 376, 444, 731, 1101, 1411, 1762) |
| *V. unguiculata* (L.) Walp.(*V. sinensis* (L.) Savi ex Hassk.) | | | | | | Extremely diverse forms in this species, of which the component ssp. have also been classified as separate species (131, 259, 376, 444, 536, 731, 947, 1101, 1411, 1458, 1731, 1762, 1765) |
| ssp. *cylindrica* (L.) van Eseltine | catjang, dolique mongette, agwa ocha, gub-gub, chola | Se, pod, L | S, Tr | w: esp. SE Asia | ++ | Small-seeded, especially cultivated in SE Asia as a vegetable, often also as a fodder crop |
| ssp. *sesquipedalis* (L.) Verdc. | yard-long bean, asparagus bean, dolique asperge, judía espárrago, | Se, pod | S, Tr | w: esp. SE Asia | + | Quick-growing climbing plant, originating in Asia. Also as an ornamental plant for covering fences (659, 731, 978) |

| | | | | | | |
|---|---|---|---|---|---|---|
| ssp. *unguiculata* | cowpea,<br>southern pea,<br>caupi,<br>tua dam,<br>kunde,<br>niébé | Se,<br>pod | S, Tr | w | +++ | (see p. 63) Is closest to the African wild form (ssp. *dekindtiana* (Harms) Verdc.). One of the most important food plants in W Africa ("niébé"). Drought tolerant and low demanding. Much cultivated as fodder plant and ground cover (183, 480, 1023, 1487, 1677, 1678, 1763) |
| **Malvaceae** | | | | | | |
| *Abelmoschus esculentus* (L.) Moench (*Hibiscus esculentus* L.) | okra,<br>lady's finger,<br>gombo,<br>quiabeiro,<br>bamiyahi,<br>habb el mosk,<br>banyah | yg. Fr | S, Tr | w | ++ | (Fig. 37) Cultivated everywhere, young fruits a favourite vegetable, and of considerable trading importance. Mucilage of plants is utilized medicinally and technically (376, 659, 782, 914, 978, 1137, 1211, 1296, 1786, 1812, 2031) |
| *A. manihot* (L.) Medik. (*Hibiscus manihot* L.) | sunset hibiscus,<br>ge'di<br>(Indonesia),<br>aibika | L,<br>yg. Sh | Tr | r: SE Asia | + | Originates in Malaysia-Melanesia. Many cultivated forms (vegetative propagation). In Japan, the root mucilage is used in the paper industry (145, 170, 259, 376, 1296, 1317, 1730, 1788) |

Table 25. Vegetables, cont'd (explanation of symbols, see footnote page 139 and footnotes page 55)

| Botanical name | Vernacular names | Used part[1] | Climatic region | Distri-bution | Econ. value | Remarks and literature references |
|---|---|---|---|---|---|---|
| **Moringaceae** | | | | | | |
| *Moringa oleifera* Lam. | horseradish tree, drumstick tree, ben ailé, murungai | yg. L, yg. Fr, Se | S, Tr | w | + | Drought tolerant tree. The young fruits ("drumsticks") are a favourite vegetable in India and E Africa. The sharp-tasting roots are a seasoning ("horseradish"). Seeds used for oil production. Also, young leaves are used as a vegetable. Many parts of the tree are utilized medicinally. Seed meal recommended for purifying drinking water (259, 376, 1296, 1317, 1468, 1788, 2031) |
| **Nyctaginaceae** | | | | | | |
| *Pisonia alba* Spanoghe | lettuce tree, maluko, kol-banda | L | Tr | r: SE Asia | + | The tree is often planted in India and Malaysia as a hedge and windbreak, especially near the sea (259, 376, 1296, 1788) |

| | | | | | | | |
|---|---|---|---|---|---|---|---|
| **Nymphaeaceae** | | | | | | | |
| *Nelumbo nucifera* Gaertn. | East Indian lotus, lotos, lien, padma, ho-hua | R, Se, L, Fl-bud, nut | S, Tr | r: | S and E Asia | + | Water plant. Main product is the starch-containing rhizome, which is cooked as a vegetable. More seldom, leaves and young shoots are used. The nuts (individual fruits) are a delicacy after removal of the fruit wall and the bitter embryo (259, 376, 731, 946, 1296, 1411) |
| **Polygonaceae** | | | | | | | |
| *Polygonum odoratum* Lour. | nghé, rau ram | L | Tr | l: | S Vietnam | + | Cultivated as a vegetable in S Vietnam. Because of its peppery taste, it is better classed as a seasoning (200, 1788) |
| *Rumex acetosa* L. var. *hortensis* Dierb. | garden sorrel, grand oseille, oseille commun | L | Te, S, TrH | w | | + | Cultivated to a small extent because of its pleasantly sour tasting leaves. Also other *R.* species are cultivated, especially *R. vesicarius* L., and are used as vegetables, medicinal plants, and as seasonings (200, 376, 1296, 1788) |
| **Portulacaceae** | | | | | | | |
| *Portulaca oleracea* L. | purslane, pourpier, verdolaga, perrexi | yg. Sh, L | S, Tr | w | | + | Widespread as a weed; improved forms are used to a small extent as vegetable, salad, and medicinal plants (259, 376, 957, 1143, 1211, 1296, 1317, 1411, 2007, 2031) |

Table 25. Vegetables, cont'd (explanation of symbols, see footnote page 139 and footnotes page 55)

| Botanical name | Vernacular names | Used part[1] | Climatic region | Distri-bution | Econ. value | Remarks and literature references |
|---|---|---|---|---|---|---|
| *Talinum triangulare* (Jacq.) Willd. | Ceylon spinach, Surinam purslane, fameflower, pourpier, Lugos bologi | L | S, Tr | w | + | An important vegetable in W Africa and some other regions. Cultivated, and gathered from the wild (480, 659, 731, 1143, 1296, 1317, 1812, 2031) |
| **Solanaceae** | | | | | | |
| *Capsicum annuum* L. var. *annuum* | bell pepper green pepper, piment, pimiento | Fr | S, Tr | w | +++ | (Fig. 79) (see p. 279, 386) (Table 24) Green and red fruits of the vegetable form are a favourite vegetable because of their high vitamin C content (up to 300 mg/100 g fresh weight) and their good properties for storage and transport. It is exported to a considerable extent (59, 316, 480, 659, 914, 1211, 1411, 1440, 1812, 2007) |
| *Lycium chinense* Mill. | Chinese wolfberry, Chinese boxthorn, gow-kee | L | S, Tr | r: E and SE Asia | + | Esteemed by the Chinese as a vegetable, especially to accompany pork, and as a soup seasoning(731, 1296, 1411) |
| *Lycopersicon esculentum* Mill. | tomato, tomate, jitomate | Fr | Te, S, Tr | w | +++ | (Fig. 31) (Tables 23, 24) (see p. 133) |

| | | | | | | |
|---|---|---|---|---|---|---|
| *Solanum aethiopicum* L. | todo, osun, djackattou, aubergine amère | L, Fr | Tr | r: Central and W Africa | + | Leaves used in soups, ripe fruit cooked (398, 400, 480, 1143, 1306, 1317, 1440, 2031) |
| *S. incanum* L. | garden egg, eggplant, ikan, igba | yg. Fr | Tr | r: W Africa | + | Widespread, also as a weed. In W Africa, bitter and non-bitter forms cultivated, unripe fruits eaten raw or cooked. Near relative of *S. melongena* (480, 659, 1317, 1440) |
| *S. macrocarpon* L. | local garden egg, grosse anghive, djahatu | L, yg. Fr | Tr | r: W Africa | + | Mainly cultivated because of its spinach-like leaves. Also the fruits eaten, cooked (249, 398, 659, 1143, 1317, 1440, 2031) |
| *S. melongena* L. | brinjal, eggfruit, garden egg, aubergine, mélongène, mayenne, berenjena | unripe Fr | S, Tr | w | +++ | (Tables 23, 24) Extremely numerous in forms in S, SE, and E Asia. Outside this region, the black-violet American cultivars are the most important ('Black Beauty', 'Long Purple', etc.); to a limited extent they are also exported (376, 557, 659, 731, 978, 1143, 1411, 1440, 1812, 2007) |
| *S. nigrum* L.s.l. | black nightshade, wonderberry, karakap, morelle noire, | L, yg. Sh | S, Tr | w | + | Widespread weed, in some regions regularly cultivated as a vegetable. Numerous types, which some authors classify as different species (259, 376, 480, 782, 1296, 1317, 1411, 1440, 2031) |

Table 25. Vegetables, cont'd (explanation of symbols, see footnote page 139 and footnotes page 55)

| Botanical name | Vernacular names | Used part[1] | Climatic region | Distribution | Econ. value | Remarks and literature references |
|---|---|---|---|---|---|---|
| *S. torvum* Sw. | turkeyberry, terongan, belangera cimarrona, terong pipit | yg. Fr, L | Tr | r: SE Asia | + | Small-fruited species, cultivation limited to E, SE, and S Asia. Also utilized medicinally (376, 731, 1296, 1440, 1788) |
| **Umbelliferae** | | | | | | |
| *Apium graveolens* L. | celery, celeriac, apio | R, L-stem | Te, S, TrH | w | ++ | Celery is the form with edible petioles, celeriac is the root-celery. Leaves and fruits as seasoning and medicinal uses. An aromatic oil is distilled from the seeds (316, 978, 1317, 1812, 2007) |
| *Arracacia xanthorrhiza* Bancroft | Peruvian parsnip, Peruvian carrot, arracacha, mandioquinha | R | S, TrH | r: Central and S America | ++ | (see p. 63) Important root vegetable. Cultivation is spreading. Annual production in Colombia about 125,000 tons. The tender side roots are eaten; the young leaves are a seasoning, old leaves and roots as animal fodder (242, 643, 946, 1201, 1258, 1864, 2007) |

| | | | | | | |
|---|---|---|---|---|---|---|
| *Daucus carota* L. | carrot, carrotte, carota, zanahoria, cenoura | R | Te, S, Tr | w | +++ | (see p. 133) An important source of vitamin A because of the high carotene content; carotene is also used as a colouring for other foodstuffs. Canned to a large extent exported. Aromatic compounds from the fruits and roots. Leaves are a valuable fodder (316, 376, 659, 978, 1411, 1812, 2007) |
| *Pastinaca sativa* L. | parsnip, panais, pastinaca, chirivia | R | Te, S, TrH | w | + | Good-tasting, high-nutrient vegetable (259, 376, 731, 2007) |
| **Valerianaceae** | | | | | | |
| *Valerianella locusta* (L.) Laterrade (*V. olitoria* (L.) Poll.) | cornsalad, lamb's lettuce, rampon, valerianelle, mache, blanchette | L | Te, S, TrH | w | + | Cultivated in the cool times of the year in the Mediterranean region and some other countries (259, 316, 731) |
| | | | **Monocotyledonae** | | | |
| **Alliaceae** | | | | | | |
| *Allium ampeloprasum* L. var. *porrum* (L.) Gay (*A. porrum* L.) | leek, purret, poireau, porreau, porro, puerro | L | Te, S, Tr | w | ++ | Very adaptable vegetable. Other sub-groups of the species (var. *ampeloprasum* and var. *kurrat* Schweinf.) are especially cultivated in the Near East (659, 903, 1317, 1812, 2007) |

Table 25. Vegetables, cont'd (explanation of symbols, see footnote page 139 and footnotes page 55)

| Botanical name | Vernacular names | Used part[1] | Climatic region | Distribution | Econ. value | Remarks and literature references |
|---|---|---|---|---|---|---|
| *A. cepa* L. var. *cepa* | onion, oignon, cebolla | | | | | (Fig. 31) (see p. 134) |
| var. *aggregatum* G. Don (*A. ascalonicum* L.) | shallot, eschalote, chalotte | B L | Te, S, Tr | w | ++ | An important, easy-to-grow vegetable in many of the tropical countries (480, 659, 731, 1317, 1650, 1680, 2007) |
| *A. chinense* G. Don | rakkyo, ch'iao t'ou | B | Te, S | r: E Asia | ++ | The small onions usually utilized for pickling (731, 903, 1121, 2007) |
| *A. fistulosum* L. | Welsh onion, Japanese bunching onion, cebolla junca, cebolleta | L | Te, S, TrH | w | ++ | Important vegetable in E Asia (annual production in Japan 350,000 tons) (659, 731, 903, 978, 1317, 1411, 1812, 2007) |
| *A. schoenoprasum* L. | chives, civette, ciboulette, cebolleta, cebolinha | L | Te, S, TrH | w | ++ | Mainly grown in house-gardens, in hot countries preferably in the shade (903, 978, 1411, 1812, 2007) |
| *A. tuberosum* Rottl. ex Spreng. | Chinese chives, Oriental garlic, kau ts'oi | L Fl | Te, S, TrH | r: E and SE Asia | + | Similar to chives, generally used as a seasoning (731, 903, 978, 1411, 1812, 2007) |

| | | | | | | |
|---|---|---|---|---|---|---|
| **Cyperaceae** | | | | | | |
| *Cyperus esculentus* L. | earth almond, tiger nut, yellow nutsedge, amande de terre, chufa | R | S, Tr | w | + | A troublesome weed in all hot countries. The cultivation is limited to small regions of Spain and N Africa. The tubers contain 20-28% oil and 20-29% starch. Eaten raw, cooked or roasted, in Spain, the drink 'horchata de chufa' is prepared from them (946) |
| **Gramineae** | | | | | | |
| *Dendrocalamus asper* (Schult.) Backer ex Heyne | bamboo betoong | Sh-tops | Tr | r: SE Asia | ++ | The young shoots of many Bambusoideae are eaten as a vegetable in S, SE, and E Asia. Mostly they are gathered from the wild. *D. asper* is cultivated in Malaysia and Indonesia especially for the production of bamboo shoots. Other uses of the bamboos are for building materials, fibres (Table 44), leaves as animal fodder, seeds as emergency food, windbreaks (Table 55) (259, 376, 731, 762, 1411, 1445, 2007, 2026) |
| **Lemnaceae** | | | | | | |
| *Wolffia arrhiza* (L.) Horkel ex Wimm. | duck weed, water-meal, khai-nam (Thailand) | Sh | S, Tr | w: esp. SE Asia | + | Cultivated in ponds in Indochina. Provides more protein per unit area than soya (180, 1485, 2026) |
| **Liliaceae** | | | | | | |
| *Asparagus officinalis* L. | garden asparagus, asperge, espárrago | Sh | Te, S, TrH | w | +++ | (Table 24) (see p. 135) |

Table 25. Vegetables, cont'd (explanation of symbols, see footnote page 139 and footnotes page 55)

| Botanical name | Vernacular names | Used part[1] | Climatic region | Distri-bution | Econ. value | Remarks and literature references |
|---|---|---|---|---|---|---|
| *Lilium lancifolium* Thunb. (*L. tigrinum* Ker-Gawl.) | tiger lily, oni-yuri | B | S | r: E Asia | + | This, and some other *L.* species are cultivated as vegetables because of their starch-rich bulbs (see p. 63). Fresh bulbs contain 18% starch, and 2.3% protein, when dried, 62% starch (200, 376, 731, 2026) |

# Fruit

We include among fruit all plant products which are eaten mostly because of their refreshing or aromatic taste. Most species have fleshy, juicy fruits. The growth-habit of the plant - whether it is a tree, bush, or herb, an annual or a perennial - is therefore unimportant for this classification.

Fleshy fruits attract men and animals to consume them, and so were among the original foods of the human race, particularly because they are mostly edible raw. Many of the wild fruits were put under cultivation thousands of years ago. The most important species originate from South and East Asia (banana, citrus, mango, *Prunus* species), the Middle East and Mediterranean region (grapes, muskmelons, figs), and South and Central America (pineapple, avocado, papaya).

Apart from the pleasure which fruit gives, it plays a very important role in nutrition and health (768, 1243). First must be mentioned the substances they contain, which regulate or stimulate the digestion. Organic acids (malic acid, tartaric acid, and citric acid) act as mild laxatives or diuretics (the basis of fruit cures). Pectins and phenolic compounds play a part in regulating the pH in the intestines, and in normalizing the intestinal flora, operate bacteriostatically and virostatically, and also detoxify heavy metals; they are also utilized medicinally for these purposes. The flavonoids (e.g. hesperidin and naringin in citrus fruits) act to strengthen the body tissues, and to regulate capillary permeability. Some fruit species are rich in vitamin A (apricot, mango, cape gooseberry, papaya, some types of muskmelon), or vitamin C (Barbados cherry, guava, citrus), while others are decidedly low in vitamins. Some examples are given in Table 26, in which it can also be seen that the majority of fruit species add very little to the energy supply. Most species contain 10-12% digestible carbohydrates, of which, however, 70-100% is supplied as monosaccharides, which are quickly absorbed in the digestive system, and become immediately available as an energy resource. Higher nutritional values are provided by the avocado, banana and sapote, and also by dried fruits (dates, figs, and others). Apart from the avocado, no other fruits have substantial contents of protein or oil. The figures given in the table provide an approximate picture of the nutritional value of the individual species. Not only do cultivar and growing conditions, but also the degree of ripening have a much greater influence on the quality of fruit than with any other foodstuffs.

As with vegetables, the statistics for fruit are very incomplete, and they are generally only trustworthy for countries with controlled markets, and exports. Of the total fruit production, only a small part (about 15%) is exported. In spite of this, fruits take a major place in total plant exports, and for many countries they are the most important export product (in Panama and Honduras, for example, bananas form more than 50% of the total exports).

In Table 27, an allocation of the individual fruits to climatic zones is attempted, and also the production figures for many fruit species are given. The

Table 26. Nutritient contents in 100 g fresh weight of the edible parts of some fruit species (309, 310, 376, 1408)

| Species | Energy | Water | Crude protein | Digestible carbo-hydrates | Crude fat | Crude fibre | Ca | P | Fe | Vit. A | Thia-mine | Ribo-flavin | Niacin | Vit. C | Weight loss before consumption |
|---|---|---|---|---|---|---|---|---|---|---|---|---|---|---|---|
| | kJ | g | g | g | g | g | mg | mg | mg | I.U. | mg | mg | mg | mg | % |
| Apple | 250 | 84 | 0.3 | 14 | 0.4 | 1.0 | 4 | 15 | 0.3 | 20 | 0.04 | 0.02 | 0.2 | 5 | 16 |
| Apricot | 250 | 83 | 1.0 | 14 | 0.4 | 1.1 | 13 | 23 | 0.8 | 3000 | 0.03 | 0.05 | 0.2 | 5 | 9 |
| Avocado | 210-920 | 59-86 | 0.8-4.4 | 1.2-10 | 5.0-32 | 1.5 | 10 | 27 | 1.0 | 200 | 0.07 | 0.15 | 1.0 | 15 | 30 |
| Banana | 460 | 75 | 1.2 | 20 | 0.3 | 0.3 | 7 | 18 | 0.5 | 400 | 0.05 | 0.05 | 0.7 | 10 | 33 |
| Barbados cherry | 125 | 92 | 0.4 | 7 | 0.3 | 0.4 | 12 | - | 0.5 | 20 | 0.03 | 0.05 | 0.4 | 1500-5600 | 17 |
| Cape goose-berry | 290 | 80 | 2.2 | 12 | 1.3 | 2.9 | 14 | 39 | 1.1 | 1000-5000 | 0.01 | 0.04 | 1.1 | 30 | 6 |
| Date (dry) | 1100 | 15 | 2.0 | 64 | traces | 8.7 | 68 | 64 | 1.6 | 50 | 0.07 | 0.05 | 2.0 | 0 | 13 |
| Fig (fresh) | 210 | 85 | 0.8 | 12 | 0.2 | 2.0 | 50 | 20 | 1.0 | 80 | 0.05 | 0.05 | 0.4 | 2 | 0 |
| (dry) | 940 | 17 | 3.6 | 53 | 0.9 | 18 | 200 | 92 | 4.0 | 100 | 0.1 | 0.08 | 1.7 | 0 | 0 |
| Grapes | 310 | 80 | 0.7 | 18 | traces | 1.0 | 20 | 19 | 0.8 | 50 | 0.04 | 0.02 | 0.3 | 5 | 5 |
| Guava | 250 | 80 | 1.0 | 13 | 0.4 | 5.5 | 15 | 27 | 1.0 | 20-1200 | 0.05 | 0.04 | 1.0 | 150-600 | 20 |
| Litchi | 250 | 84 | 0.8 | 14 | 0.2 | 0.5 | 8 | 22 | 0.4 | 0 | 0.04 | 0.04 | 0.3 | 50 | 40 |
| Mango | 210 | 87 | 0.4 | 11 | 0.7 | 0.7 | 14 | 10 | 0.4 | 1000-3000 | 0.03 | 0.04 | 0.3 | 30 | 34 |
| Muskmelon | 80 | 95 | 0.5 | 4 | 0.1 | 0.4 | 12 | 14 | 0.3 | 0-2500 | 0.03 | 0.03 | 0.5 | 30 | 20 |
| Orange | 210 | 86 | 0.8 | 10 | traces | 0.2 | 40 | 24 | 0.3 | 100 | 0.05 | 0.03 | 0.2 | 55 | 25 |
| Papaya | 125 | 90 | 0.4 | 7 | 0.1 | 0.7 | 11 | 9 | 0.4 | 2500 | 0.03 | 0.03 | 0.2 | 50 | 30 |
| Pineapple | 230 | 86 | 0.4 | 13 | 0.1 | 0.5 | 13 | 17 | 0.4 | 20-200 | 0.08 | 0.03 | 0.1 | 30 | 40 |
| Raisins | 1100 | 21 | 1.1 | 64 | traces | 7.0 | 61 | 40 | 1.6 | 20 | 0.1 | 0.06 | 0.6 | 0 | 0 |
| Sapote | 560 | 65 | 1.5 | 31 | 0.5 | 1.5 | 40 | - | 1.0 | 20 | - | 0.03 | 2.0 | 35 | 26 |
| Watermelon | 100 | 93 | 0.5 | 6 | 0.1 | 0.2 | 6 | 9 | 0.2 | 0-200 | 0.02 | 0.02 | 0.05 | 5 | 50 |

Table 27. Main climatic zones of the major fruit species. Production figures in million tons for 1979 and 1988, and percentage increase in this decade

| Climatic zone and species | 1979 | 1988 | % increase |
|---|---|---|---|
| 1. Predominantly tropical species | | | |
| Banana | 38.2 | 41.9 | 10 |
| Mango | 13.0 | 15.0 | 15 |
| Pineapple | 7.5 | 10.6 | 41 |
| Papaya | 1.9 | 3.7 | 95 |
| Avocado | 1.4 | 1.8 | 29 |
| Total | 62.0 | 73.0 | 18 |
| 2. Predominantly subtropical species | | | |
| Citrus | 50.8 | 67.4 | 33 |
| Grape | 69.7 | 59.8 | - 14 |
| Watermelons | 25.6 | 28.5 | 11 |
| Peach | 6.9 | 8.2 | 19 |
| Muskmelon | 6.5 | 9.0 | 38 |
| Date | 2.7 | 3.0 | 11 |
| Apricot | 1.4 | 2.1 | 50 |
| Total | 163.6 | 178.2 | 9 |
| 3. Predominantly temperate species | | | |
| Apple | 36.3 | 40.9 | 13 |
| Pear | 8.6 | 9.9 | 15 |
| Plum and prune | 5.2 | 6.6 | 27 |
| Strawberry | 1.7 | 2.3 | 35 |
| Total | 51.8 | 59.7 | 15 |

leading position of the subtropics in fruit production is clear, though our allocations as well as the production figures must be taken with some reservations; many minor fruits (see Table 30) come from the tropics, and statistical data are lacking for these. The great increase in the production of bananas which took place from the sixties to the middle of the seventies (increase of nearly 50%), has now clearly levelled off. On the other hand, the production of avocados and pineapples has increased considerably, brought about by the potential for exports. Some tropical and subtropical fruit species, though formerly a rarity, are now regularly available as fresh fruit on the European markets (1113, 1823, 1973). Among these species are avocado, mango, passion fruit, cherimoya, guava, kiwi, litchi, watermelon, and muskmelon. Some of these do not keep well, and can therefore only be transported over long distances by air. The market for these fruits will never be large, due to their high price to the consumer (121). The potential is higher for the export of these species as tinned products (e.g. the juice of guava and passion fruit, canned litchi or guava). Work on improving the canning techniques for

tropical fruit is being carried out in many places (509, 768, 860, 1344, 1823). The most important products on the market are:
- tins of whole fruit or segments of fruit for immediate use,
- fruit pulps and pastes for further processing in jam factories, bakeries, etc.,
- jams and jellies,
- fruit juices (unaltered, concentrated, or dried, partially produced as by-products in canning factories),
- fruit syrup, dried fruit, and candied fruit (450).

Products which will keep can also be produced by fermentation into alcohol. The majority of grapes are fermented to give wine. "Wines" are also produced from apples, oranges and other fruits (bananas, cashew apples). A variety of fruit spirits are produced from the distillation of fermented fruits. By-products of the fruit industries are pectin (from citrus and apples), aromatic oils, and aroma concentrates (758). Residues from the processing industries are used as animal fodder.

For many of the lesser known tropical fruits, not only must markets for selling the products be found, and transport and storage methods developed, but also, better production methods need to be evolved. The breeders must improve quality and transport characteristics (535), and modernized propagation and cultivation techniques can save labour, and increase the yields per surface area (584, 1087, 1195, 1572, 1573). Pest control has already made extraordinary progress in the cultivation of fruits for export, but there are still considerable problems with the protection of the minor species (358). There is a wide scope for the introduction of unusual fruit species into other climatic regions, which can be very interesting economically (e.g. the cultivation of apples, peaches and grapes in the tropics (601, 1554, 1594).

## Grapes

fr. raisin; sp. uva; ger. Weintraube

By far the most important species of grape is the Old World *Vitis vinifera* L., Vitidaceae; this supplies 90% of the world production of grapes (1970). Within this species, the European cultivars (var. *occidentalis* Nekr.) are the leading types because of their exceptional qualities. For cultivation in hotter countries, the large-berried oriental cultivars (var. *orientalis* Nekr.) also receive attention (109). For the summer rainfall regions of the subtropics, the American species (*V. aestivalis* Michx., *V. labrusca* L., *V. rotundifolia* Michx., *V. rupestris* Scheele, and *V. vulpina* L.) are of interest because of their disease resistance, strong growth, and their minimal chilling requirement, especially in crosses with *V. vinifera* (better taste, better texture of berries: 'Catawba', 'Delaware', and 'Niagara'), but also as cultivars of the pure American species, or in crosses between the American grapes; the greatest part of these has *V. labrusca* ('Bangalore Blue', 'Concord'), with a minor position taken by *V. rotundifolia* ('Scuppernong') and *V. aestivalis* ('Norton's Virginia') (1311, 1312, 1506, 1535, 1657, 1970).

**Production**. The production of grapes is second only to the citrus species at present (Table 27). The most important regions of production lie in the winter rainfall zones (Mediterranean countries, California, Chile, South Africa, Australia). However, the grape is surprisingly adaptable, and is grown from the temperate zone to the humid tropics. Breeding programmes have been established in many of the subtropical countries in order to develop locally adapted cultivars (1391). In the tropical regions of India, the cultivation of grapes has been carried out for many centuries (296, 376).

**Ecophysiology**. In the temperate zone and the subtropics, the autumnal decrease in day-length and temperature lead to the leaf fall and the onset of the winter rest period (1633). Even for *V. vinifera*, less cold is necessary for breaking the bud dormancy than for the other fruit species of the temperate zone. In the tropics, with their small annual variations in day-length, the natural rest period is absent. Here, special treatments are necessary to achieve good yields (see Cultivation). In summer, the grapes need high temperatures, and some cultivars are decidedly heat resistant, e.g. 'Vadi-Haman-Ali' of the Arabian Peninsula. The grapes develop slowly at moderate temperatures, but develop a better aroma than at high temperatures; the northern wines have a more full bouquet than those from southern countries. High sugar contents are achieved in the tropics also, but only in cultivation regions with plenty of sunshine.

In Mediterranean climates, grapes thrive even with rainfalls of about 400 mm, in spite of the lack of rain in the summer vegetative period. Their extensive root system extracts the water from a large volume of soil. In summer rainfall areas, the grapes must be irrigated in the time before the arrival of the rains, even when there is a much higher total rainfall. This is because the time of first growth, flowering and early fruit development is the time of highest water demand.

Grapes also grow well even on poor, stony soil. The pH should be over 5. Water-logging is not tolerated, however they do well with a relatively high salt concentration in the soil (up to about 0.4%, calculated for dry soil). For high yields, they must have a corresponding amount of fertilizer (595); for a yield per hectare of 30 tons annually, about 240 kg N, 25 kg P, and 100 kg K are necessary.

**Cultivation**. Grapes are vegetatively propagated using 1 year-old wood for cuttings. These cuttings then stay in the nursery for a year. In hot countries, there is no phylloxera, and for that reason the scions there can be directly rooted, or, if special characteristics from the stock are needed (vigourous growth, drought or salt tolerance), then they are grafted. In winter rainfall regions, as in the temperate zones, they are mostly grafted on to phylloxera-resistant stocks (*V. rupestris* and hybrids).

The planting distance depends on the type of training, the growth vigour of the cultivar, the soil fertility, and the water supply. The planting distance is about 2 m for wine grapes, raisins, and for smaller trellises, while for table grapes and for taller trellises it is 3-6 m.

The textbooks on viticulture (1418, 1506, 1970) give details of the various pruning methods, the common aim of which is to obtain every year an amount of fruit-bearing wood (one-year-old growth) which corresponds to the vigour of growth, because only particular eyes on the year-old wood produce inflorescences. In areas with definite seasons, this goal is easy to achieve. The wood ripens in autumn and the pruning is carried out in winter or early in the following year.

In the tropical areas without a cool season (no leaf fall, no dormancy), a variety of techniques are used, depending on the distribution of rainfall (109, 960, 1469, 1633). In the monsoon regions, the vines are cut back to 1 or 2 eyes after harvest, before the onset of the rains. In the following rainy season, growth is allowed to continue unchecked, and, if necessary, the inflorescences which are formed are all removed, as they weaken the growth, and form fruit of inferior quality. At the beginning of the dry period, all the leaves are stripped off, and the bearing wood is cut to the length suitable for the cultivar (4-15 eyes, depending on the cultivar). The new growth begins after three weeks, and the grapes ripen after 3.5-5 months. During the dry season, the vines must be irrigated.

In equatorial lowlands with two rainy seasons, two harvests each year can be achieved. Stripping off the leaves and pruning are preferably carried out so that the grapes ripen in a relatively low rainfall period. Well regulated sprouting after pruning can be achieved by a partial cutting back of the roots (artificial dry period). The time of defoliation and pruning is not essentially dependent on the climatic conditions, and thus it can be organized so that the market is supplied with locally grown grapes throughout the whole year, e.g. as in Bangkok.

The economic lifetime of vines in the tropics is only 10-15 years. European vines can be used near to the equator only up to about 1000 m altitude, because, for them, uniformly high temperatures (23-28°C) are necessary to prevent the onset of the bud dormancy. At greater altitudes, American species (*V. labrusca*) or specially bred hybrids must be planted.

In all grape growing regions, fertilizing with organic materials is advantageous. In winter rainfall regions, either the natural weed growth, or green manure plants which have grown during the rainy season, are worked into the soil in the spring (see p. 429). Where green manuring is not possible, pruned wood and leaves, mulch material, compost and animal manure are utilized as organic fertilizers.

**Diseases and Pests**. The numerous fungus diseases of grapes are much more dangerous in the summer rainfall areas and in the tropics than in the winter rainfall regions. The most important species are *Plasmopara viticola* (Berk. et Curt.) Berl. et de Toni., downy mildew, combated mainly with copper compounds or zineb; *Uncinula necator* (Schw.) Burr., powdery mildew, combated with wettable sulphur or one of the newer fungicides; and, *Elsinoe ampelina* (de Bary) Shear, anthracnose, combated by winter sprayings with lime-sulphur. In summer rainfall areas, less damage is caused to early cultivars, because the harvest is finished before the main rainy season, whereas late cultivars, and those in permanently wet regions, must be regularly sprayed. American grapes are substantially more resistant to all three diseases than *V. vinifera*. Because of this, cross-breeding plays an all-important role in the development of resistant, high quality strains.

There are also numerous insect pests, above all leaf-eating beetles, thrips, red spiders, bugs, mealybugs, and jassids, and these must be sprayed at the correct time, especially at the beginning of the berry growth (problems of pesticide residues). In the tropics, great damage can be caused to the ripening grapes by fruit-eating beetles, birds, and monkeys. For table grapes, it can be necessary to protect the grapes with a paper cover, or plastic wrappers which should be open at the bottom. Otherwise, the beetles must be collected, and the birds and monkeys scared away by watchmen.

**Harvest and Processing**. Grapes do not ripen further after cutting, they must be harvested when they are fully ripe. The harvest is mostly carried out by hand, although mechanical harvesting procedures are available for wine and raisin grapes (897). Special care is needed in the harvesting of table grapes to avoid any damage to the berries. Usually individual berries must be cut out of the bunch using sharp scissors in order to prevent quick decay. Good table grapes keep for about two weeks, even in hot countries. With refrigeration (1-4°C), suitable cultivars remain in a saleable condition for up to two months. Transport by ship from the warmer countries to Europe is therefore possible.

The yields are highest for table grapes, which are irrigated and intensively fertilized and tended in most countries. A good yield is considered to be 20-35 tons/ha, although up to 80 tons/ha are possible. Grapes for wine and raisins often only yield 5-10 tons/ha, even under good conditions.

Of the global harvest of grapes, about 80% is processed to make wine, at least 10% is consumed as fresh fruit (Table 26), 5% is dried (raisins, sultanas, currants), and the rest is made into grape juice, grape syrup, jams and preserves (504). Wine can be made from all grapes. The highest quality standards are set for table grapes, especially for export (apart from the taste, the shape of the bunch, texture of the berries, and ability to be transported are important). For dried grapes, special cultivars are grown: sultanas are seedless, light brown raisins (sometimes artificially bleached); currants are similarly seedless, but smaller, and dark brown. Grape syrup (boiled down grape juice - "grape honey") is a speciality in the Near East. All types can be processed to make juice; hybrids between American and European grapes produce a juice with a special aroma.

The dominant role of wine production is by no means found in all grape-growing countries. In Australia, Greece, Turkey and Iran, the emphasis is on the production of raisins (about half of the grape production is used for these purposes (1087)). Raisins can only be produced in regions where the climate is warm and dry during the harvesting time (sun-drying). Practically the entire production of the countries with tropical climates or from summer rainfall regions is used for table grapes.

The marcs are a by-product of wine production, and can be used as animal fodder. The grape-seed oil is chiefly used in cosmetic products, but also as a salad oil (Table 15). A food-colouring is extracted from the skins of blue grapes ("grape-skin extract") (728) and tartaric acid is another by-product. Grapevine leaves are used as a vegetable, mostly in the orient; they are also eagerly eaten by cattle.

## Citrus Fruit

The genus *Citrus*, Rutaceae, originates from Southeast Asia (India to Japan and Indonesia). Individual species long ago spread to the Mediterranean region (citrons and sour oranges), and to East Africa (sour oranges), chiefly perhaps as medicinal plants. The fruits were first brought to Italy and Spain/Portugal by the Genoese and Portuguese (1430, 1657). Since then they have spread to all parts of the world, particularly the subtropics. Here they have become the most important fruit, at present surpassing even grapes (Table 27). Many forms of citrus emerged in the Mediterranean region and in America, and one group, the grapefruit, has arisen there, which is now considered to be a botanical species.

**Production**. By far the most important species is the sweet orange, *C. sinensis* (L.) Osbeck, (fr. orange; sp. naranja; ger. Apfelsine) (Table 26). In 1988, 46.7 million tons were produced. There exist about 1100 cultivars, which are divided into groups: common oranges ('Valencia', 'Shamouti'); navel oranges

('Wahington', 'Thomson'); blood oranges (Mediterranean region); and sugar oranges (Mediterranean region, oranges with an insipid taste).

The second place in world production is taken by the mandarins, *C. reticulata* Blanco, (fr. mandarine; sp. mandarina; ger. Mandarine) with 8.4 million t in 1988. In East Asia, many more mandarins are eaten than oranges. Approximately 500 types have been described, and they are mostly classified in groups as botanical varieties, e.g. var. *deliciosa* Swingle - tangerine, clementine; var. *unshiu* Swingle - satsuma cultivars.

In third place come the lemons and their relations (world production in 1988 was 6.0 million tons):

- the lemon, *C. limon* (L.) Burm.f. (fr. citron; sp. limon; ger. Zitrone), which grows particularly well in the Mediterranean climate (adapted to relatively low temperatures; the best known cultivars are 'Eureka', 'Lisbon', 'Villafranca');
- the lime, *C. aurantiifolia* (Christm.) Swingle (fr. lime; sp. lima; ger. saure Limette), which prefers warm tropical climates (the main producers are the West Indian Islands and Mexico; 'Mexican' - small fruits with many seeds, 'Tahiti' - large fruited, seedless);
- the sweet lime, *C. limetta* Risso, with the following cultivars developed in the Mediterranean region: 'Millsweet', 'Marakesh-Lime' (more acidic), and 'Palestine Lime' (also called "round lime", and classified as a separate species, *C. limettioides*, by Tanaka).

The last group of importance for cultivation is the pummelo, *C. maxima* (Burm.) Merr. (also called shaddock; fr. pampelmousse; sp. pampelmusa; ger. Pampelmuse) and the grapefruit, *C. paradisi* Macfad., (fr. pomelo; sp. grapefruit, pomelo; ger. Grapefruit), with an annual production in 1988 of 4.9 million tons. Pomelos have the largest fruits of all citrus species, and are spread throughout Southeast Asia in a variety of forms. Grapefruits originated in the West Indies from types of pummelo (1615), and are very popular in the USA. They are, however, cultivated in many countries, also in those with tropical climates (cultivars 'Duncan', 'Marsh', 'Thompson', and others).

In addition to these major species, the following must also be mentioned:

- the sour orange (Seville orange), *C. aurantium* L., (fr. bigarade; sp. naranja agria, or amarga; ger. bittere Orange, Pomeranze). The fruit peel of this is used for making marmelade, for the production of orangeade, and for the liqueurs Curacao and Cointreau; and its ssp. *bergamia* (Risso et Poit.) Engl., the bergamot, from whose peel bergamot oil is produced in southern Italy;
- the citron, *C. medica* L., (fr. cedrat; sp. cidra, poncil; ger. Zedratzitrone), of which the thick peel is processed to make candied peel;
- the small-fruited citrus species calamondin, *C. madurensis* Lour., and the closely-related kumquats, *Fortunella* spp., are used in East Asia for confectionery manufacture, but have become favourite ornamental plants in West Europe and America;
- in Southeast Asia, the fruits and leaves of other citrus species are used as seasonings, e.g. *C. hystrix* DC., the leech lime, djerek purut (1210).

In recent years, hybrids between various species of *Citrus* have acquired commercial importance, especially the "tangelos", which are crosses between the tangerine and grapefruit (cultivars 'Orlando', 'Minneola', and others), and "tangors", crosses between the tangerine and orange (best known cultivar is

'Temple'). These hybrids are the result of spontaneous cross-fertilization, or of systematic breeding.

Lists of cultivars, and accurate information about regions where the various citrus species are cultivated can be found in (204, 332, 393, 624, 765, 822, 1244, 1430, 1486, 1502, 1530).

**Morphology and Anatomy**. The citrus species display a range of morphological and anatomical characteristics (1500). The leaf stalks have large wings, small wings, or no wings, the twigs have thorns, or are smooth (Fig. 38). The degree of thorn formation varies with many species, depending on the stage of development. The thorn-bearing species form long, strong thorns on seedlings and water sprouts, but on fruit-bearing twigs, they have small or even no thorns. The leaves and fruit possess oil-glands (see p. 320). The pericarp of the fruit consists of two layers. The outer is called the flavedo, and is coloured yellow or orange by carotenoids, and the inner, white layer is called the albedo. The juicy fruit flesh consists of segments, which are filled with multi-cellular juice sacs (vesicles). The seeds contain, besides the zygotic embryo, a variable number (1-30, mostly 5-6) of nucellar embryos (489, 915, 1816) (not with pummelos), which are used with some species for cultivar propagation (grafting stocks, Table 28), and for the production of virus-free progeny ("Frost seedlings"). The cotyledons of the dormant seed are white in most species (e.g. oranges, grapefruit), but in others are coloured green by chlorophyll (especially mandarins).

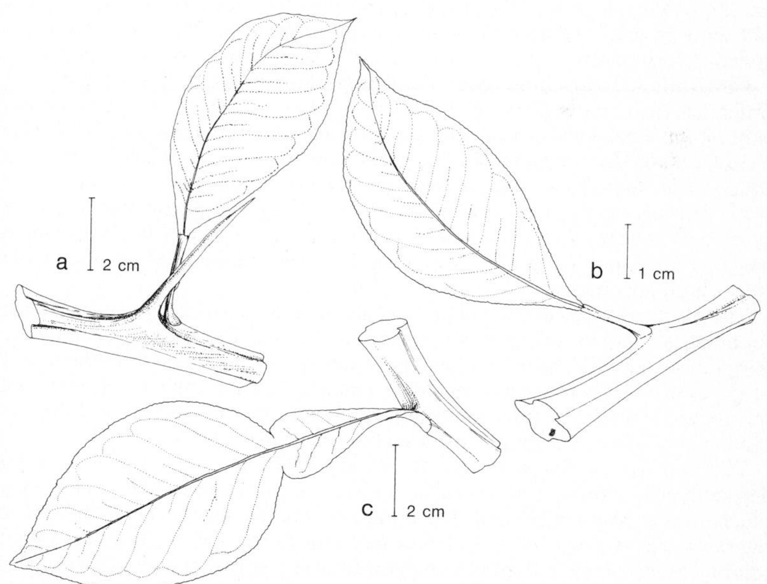


Fig. 38. *Citrus* spp. Vegetative characteristics of (a) *C. limon*, (b) *C. reticulata*, (c) *C. paradisi*

**Ecophysiology**. All citrus species require much sun for healthy growth and for the production of high quality fruit. Most species are not damaged by mild (down to -2°C) frosts of short duration during the winter resting period. The resistance to cold is strongly influenced by the stock (increased by *Poncirus trifoliata* (L.) Raf., and *Citrus reticulata* 'Cleopatra', see Table 28). Cool temperatures during ripening improve the quality of the fruit, hence the best oranges and mandarins are produced in regions with a Mediterranean climate. Cool temperatures (and short days) generally promote the formation of flower buds. Limes and pummelos are suitable for humid, hot climates.

With rainfall which is evenly spread throughout the seasons, 1200 mm is the lower limit for the cultivation of citrus without irrigation, and 2000 mm is optimal. Dry periods of up to 2 months can be endured without damage only during the final stages of fruit ripening, or during the winter dormancy period. Some root stocks improve the drought tolerance ('Rangpur' lime, *Citrus junos* Sieb. ex Tan.). The soil should be deep and well drained. The optimal pH is 5-7. The salt tolerance is generally slight, but is improved by stocks such as 'Cleopatra' mandarin or 'Rangpur' lime. High yields are only possible on fertile soils. Mineral fertilizing with N, P, and K is the rule in commercial citrus cultivation (about 0.6 kg N, 0.3 kg P, and 0.6 kg K per tree per year for oranges) (350, 418, 1499). Unbalanced high N fertilization gives high yields, but the quality of the fruit is impaired (lower sugar/acid ratio). Trace element deficiencies (Zn, Cu, Mn) are generally rectified by foliar sprays.

Citrus trees produce new leaves three times a year. In subtropical climates, the first growth of the year is the strongest and produces most flowers, the second and third growths are irregular, with few or no flowers. The crop from these growths ("off-season crop") can be locally of considerable economic importance.

**Cultivation**. Most citrus species are propagated by grafting on a rootstock which has been grown from seed. The stock often has a definite influence not only on the uptake of nutrients and the growing vigour, but also on the disease resistance and fruit quality (1175, 1572). Some of these characteristics are collected in Table 28 for the most frequently used root stocks. The choice of stock depends on the local conditions: resistance against phytophthora is only necessary on heavy soils, resistance against tristeza is essential in places where no virus-free clone material is available. The breeding and testing of suitable stocks is an important area in citrus research.

The planting distances depend on the cultivar and the stock (growing vigour); for oranges, 7.5-9 m is needed, for mandarins, 4.5-6 m. Citrus trees generally are not pruned, only low-hanging branches, water sprouts and growth in the centre of the crown if it is too dense should be removed (for the entrance of light, and for disease control). In order to make picking easier, or to mechanize it, drastic pruning procedures (hedging) have been developed in the USA.

The soil is usually kept free of weeds. Only in rainy regions with a Mediterranean climate can weeds be allowed to grow during the winter, or a green manure crop can be used. These must be worked into the soil at the end of the rainy season (see p. 429). In large-scale cultivation, soil herbicides are used (diuron, bromacil, etc.). Well-cleaned ground is necessary not only to save water, but also for phytosanitary reasons.

Table 28. Influence of the rootstocks on the qualities of orange trees and fruit. Adapted from (1175, 1430)

| Rootstock | Cold resistance | Vigour | Yield | Fruit quality | Resistance against | | | |
|---|---|---|---|---|---|---|---|---|
| | | | | | Phytophthora | Tristeza | Xyloporosis (cachexia) | Exocortis |
| Rough lemon (*C. jambhiri* Lush.) | - | ++ | ++ | - | - | + | + | + |
| Sour orange | + | + | + | + | + | -- | + | + |
| Cleopatra mandarin (*C. reshni* (Engl.) Tanaka) | ++ | - | + | + | + | + | +- | + |
| Troyer citrange | ++ | + | + | ++ | + | + | + | - |
| Common orange | + | ++ | + | + | - | + | + | + |
| Rangpur lime | - | + | + | + | - | + | - | - |
| *Poncirus trifoliata* | ++ | - | + | ++ | + | + | + | -- |
| Palestine sweet lime | -- | - | + | + | - | - | - | - |

++ = distinctly improved, + = good, - = weakened, -- = very weak

Citrus trees generally give full yields for 20 years (mandarins) to 40 years (oranges). However, there are also hundred-year-old plantations which are still economically viable.

**Diseases and Pests**. Hardly any other group of plants requires as much assistance from phytomedicine as the citrus species (358, 624, 822, 977, 1430, 1486, 1501). Numerous *Phytophthora* species cause foot rot (gummosis), especially on badly aerated soil, and this is controlled primarily by the choice of suitable growing sites and rootstocks (Table 28), but also by the use of fungicides (1019). Fungus leaf diseases are mostly only of local importance, and are kept under control by sanitary measures and by spraying with fungicides. Orchard sanitation, and care in handling during picking and packing control the various fungi, which endanger the fruit during storage (*Penicillium* spp., *Alternaria citri* Ellis et Pearce, *Guignardia citricarpa* Kiely, and many others). Citrus canker (*Xanthomonas citri* (Hasse) Dowson) can be very dangerous; this originated in East Asia, and was eradicated in the USA in 1913-1923, and in South Africa in 1918-1927 by rigorous uprooting and burning of all infected trees. It was first known in 1967 that the diseases "stubborn", "greening", "yellow shoot", and others, which are occurring in many countries, were in fact caused by mycoplasms. These are transmitted by psyllids. Direct control is possible using tetracyclines, and indirect control by eliminating the carriers and by utilizing disease-free grafting materials (1001, 1613).

Disasters in citrus cultivation have been caused by a variety of viruses, especially tristeza, which caused the death of millions of trees in South America in the 1930s. The most important control measure is, besides the utilization of healthy propagation material (from nucellar embryos, see p. 177), the choice of rootstocks which will make the trees resistant to the virus (Table 28).

The number of parasitic insects found on citrus plants amounts to many hundreds. 24 species of the scale insects alone are known, and regular action to control them is required, in order to prevent serious damage to the trees, and to achieve blemish-free fruit for the market. The modern insecticides have led to definite progress, especially the organo-phosphorus compounds. Also, butterfly caterpillars, spider mites, thrips, dwarf cicadas, aphids, and white fly can be troublesome. In almost all regions of cultivation, considerable damage is caused to the fruit by fruit-piercing moths, the caterpillars of false codling moths (several species), and above all the maggots of the various fruit flies. One should also beware of nematodes (*Radopholus* spp., *Tylenchus* spp., *Pratylenchus* spp., etc.), against which some at least partially resistant rootstocks are available.

**Harvest and Processing**. Citrus plants which bear well can yield about 30 tons/ha of oranges or mandarins, and 40 tons/ha of lemons or grapefruit per year. For fresh shipping, oranges are separated from the branch using a knife or shears in order to avoid lesions to the fruit. When the fruit is to be used immediately for processing, it can be plucked with less care, or even mechanically harvested (in Florida (363)). The further handling of fruit, especially for export, is usually carried out by large organizations, which market the fruit under a trade name ('Jaffa', 'Sunkist', 'Outspan'), and which guarantee blemish-free products to their customers. In addition to the sales of fresh fruit, and the production of the various canned goods, an increasing role is played by the processing of juice, wine (1580), concentrates, frozen concentrates, and dried products (utilized mostly in the soft drinks industries (30, 1133, 1244, 1662)). The peels which remain after these processes are used for the production of pectin, or as animal fodder. Further by-products are the aromatic citrus peel oils (orange, mandarin, lemon, grapefruit, lime, sour orange); bergamot is cultivated exclusively for the preparation of its peel oil (southern Italy, see p. 320). Aromatic oil is also produced from the leaves ("petitgrain" oil from lemon, orange, sour orange, bergamot, mandarin). Orange flower oil ("neroli Portugal") is distilled from the flowers of the common orange, and the very high-priced neroli oil ("neroli Bigarade") is produced from the flowers of sour orange (661, 758, 1521, 1523). Further by-products are the citrus kernel oils (linolenic acid group, Table 15) (224, 225, 1181), and the medicinally useful flavonoids hesperidin and naringin (see p. 169).

# Banana

fr. banane; sp. banana; ger. Banane

The cultivated forms of banana belong to two sections of the genus *Musa*, Musaceae, *Australimusa* and *Eumusa*. The cultivars of the *Australimusa* section originate from the pacific region ("Fehi" bananas). They have not been fully investigated botanically, and are only of local importance (145, 297, 1654, 1657). Here, we deal only with the section *Eumusa*.

Table 29. Genetic constitution of the cultivated bananas. After (535, 1751)

| Ploidy | Genome | Cultivars |
|---|---|---|
| 2 n | AA | 'Sucrier', about 60 cultivars, mostly in Southeast Asia |
| | AB | 'Ney poovan' from southern India. These cultivars are inadequately described |
| 3 n | AAA | 'Gros Michel', 'Cavendish', 'Robusta', the main types of bananas; Ugandan beer bananas; distributed worldwide, about 30 cultivars |
| | AAB | 'French Plantain', 'Horn Plantain', in total about 100 cultivars; the main types of plantains |
| | ABB | 'Bluggoe', 'Pisang awak', Southeast Asian, drought resistant, floury plantains, about 25 cultivars |
| | BBB | 'Saba', Philippines |
| 4 n | AAAA | recently bred bananas, as yet without importance |
| | ABBB | 'Klue teparod', sweet plantains from Thailand |

After the clarification of the evolution and cytogenetics of the banana (1654), the classification of the old "species" has become untenable, so that the Linnean name should now be used for the entire complex of bananas and plantains, and the hybrid nature of many of the cultivars should be indicated by a cross, thus: *Musa* x *paradisiaca* L. (1611, 1789). The banana originated in Southeast Asia, in the zone where there is an overlap of the two wild species, *M. acuminata* Colla (genom A) and *M. balbisiana* Colla (genom B). An overview of the various forms is given in Table 29. The plantains have been dealt with on p. 64.

**Production**. The banana was brought to Africa in the first millenium BC by Indo-Malayans or Arabs. Bananas were first introduced to America only after 1500 AD. Nowadays, they are cultivated in all of the tropics, and in many of the subtropical regions. Globally, there has been an overproduction of bananas for many years. In the last decade, their cultivation has only slightly increased (Table 27). The main producers and exporters are the Latin American countries, namely Brazil (production in 1988 5.1 million t), Ecuador (2.2 million t), Colombia (1.3 million t) Mexico, Venezuela and Costa Rica (1.1 million t each), Honduras (1.0 million t), and Panama (0.9 million t). The production in Asian and African countries is predominantly for home consumption (India 4.6, Indonesia 1.9, Thailand 1.6, Vietnam 1.5, Burundi 1.5, Tanzania 1.3 million t); the Philippines are an exception here, with a production of 3.6 million tons, of which 0.9 million t are exported, 85% of this going to Japan.

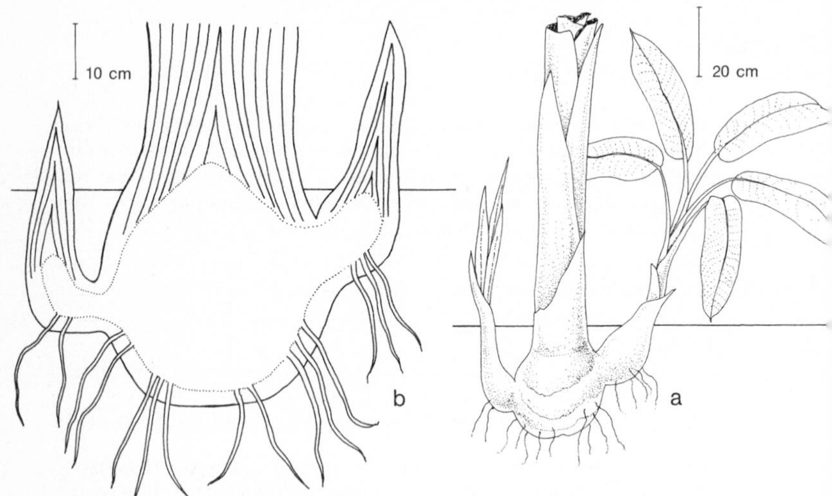


Fig. 39. *Musa paradisiaca.* (a) lower part of pseudostem and corm with maiden sucker (right) and sword sucker (left); (b) corm in longitudinal section, with base of pseudostem, adventitious roots, and young suckers

**Morphology**. The banana (297, 1713, 1751) is a herbaceous perennial with an underground rhizome (corm), from whose sprouts the new fruit-bearing shoots are formed. The sprouts (suckers) are also used for propagation (Fig. 39). The sheaths of the leaves form a pseudostem. The leaf blades are entire, but often are torn by wind. The inflorescence develops 7-9 months after the appearance of the sprout. After the fruiting shoot has grown through the pseudostem, it bends and grows, directed by positive geotropism, more or less vertically downwards. In the axils of the red-violet coloured bracts, groups of flowers are formed, first females, then hermaphrodites, and finally males (Fig. 40). The fruits develop from the female flowers without pollination. The hermaphrodite flowers can be absent; also with cultivars which have them, they do not form fruits.

**Breeding**. In breeding, resistance against Panama and sigatoka diseases are in the foreground (see p. 184); after this in importance are fruit characteristics and tillering capability, which must also be considered (535, 1385, 1545). The unusual technical difficulties (parthenocarpy, polyploidy) which stood in the way of a planned approach to breeding have been overcome since the 1940s; the breeders, however, have not yet succeeded in creating outstanding new cultivars.

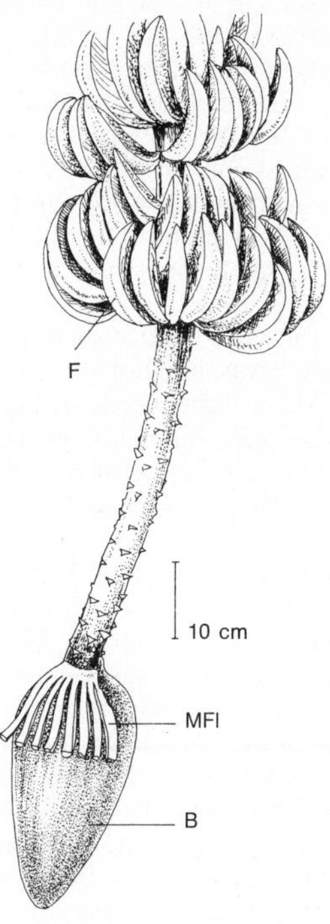


Fig. 40. *Musa paradisiaca*. Tip of inflorescence, F = fruits, MFl = male flowers, B = bracts

**Ecophysiology**. The large, high-yielding bananas ('Gros Michel', 'Giant Cavendish', 'Robusta', etc.) and the plantains require a uniformly warm climate (annual average over 20°C), a lot of sunshine, and evenly spread rainfall of about 2500 mm/year. They are susceptible to wind breakage. The ground should be loose, well drained, and rich in organic material (mulch) in the top 20 cm. The optimum pH range is 5-7, however both lower and higher values are tolerated. For high yields (up to 50 t/ha/year), the uptake of nutrients is high (1 ton of bananas contains about 2 kg N, 0.3 kg P, 5 kg K, 0.4 kg Ca, and 0.5 kg Mg) (853). Because of this, fertile volcanic or alluvial soils are preferred. The dwarf bananas ('Dwarf Cavendish') also thrive in the subtropics (Lebanon, South Africa). Temperatures down to 0°C can be withstood, however they bring growth to a standstill. Because dwarf bananas are planted close together and develop very quickly, they can produce yields similar to those of the tropical cultivars if they are sufficiently irrigated and receive high levels of mineral fertilizers (308).

**Cultivation**. For planting material, there is a preference for using suckers whose leaves have not unfurled ("sword suckers", Fig. 39). Also "maiden suckers" and older sprouts can be used, but then the leaves should be removed, and the terminal growing point is cut out with the aim that a new sprout develops from one of the axillary buds. The sets should be treated against nematodes with hot water (56°C for 5 minutes), and with pesticides against diseases and pests (banana borer) (8). For several years, propagation via meristem culture is also being used commercially. The planting distance depends on the cultivar and climate (about 2 x 2 m for 'Dwarf Cavendish', 3 x 3 m or more for tropical cultivars). Only enough side-shoots are allowed to grow to reap a fruit bunch every 4-6 months (from the appearance of the tip of a new daughter plant to the ripening of the fruit takes 12-14 months for the tropical cultivars). Any other sprouts are cut off (with a spade or scrub cutter), or killed with chemicals (mineral oil).

The working of the soil for weed control should be kept to a minimum, so that the most important feeding roots are not damaged, as these lie close to the surface of the soil (1891). Most of the weeds will be kept down by a 20 cm thick layer of mulch from banana plant remnants, grass, and other organic material, and this will also contribute to the supply of K to the plants. Liberal fertilization with minerals, especially with N compounds (urea, ammonium sulphate) is always necessary.

**Diseases and Pests**. Where the banana wilt (Panama disease), caused by *Fusarium oxysporum* Schlecht var. *cubense* (E.F.Sm.) Snyd. et Hans., is present to a great extent, sensitive cultivars, such as 'Gros Michel' cannot be grown. Other cultivars are resistant, such as 'Giant Cavendish', 'Lacatan', and others. In the tropics, the leaf spot diseases are widely spread (sigatoka disease); they are caused by *Mycosphaerella musicola* Leach (yellow sigatoka) and *M. fijiensis* Morelet (black sigatoka) and are controlled by regular spraying with fungicides (mostly copper compounds in oil, applied from aircraft by the ULV (ultra low volume) technique, or organic preparations, sometimes also mixed with mineral oil). The selection of healthy planting material, disinfection of the sets together with orchard sanitation, and destruction of any infected plants are the most important precautions that can be taken against other fungus diseases, against bacterial diseases (moko disease, caused by *Pseudomonas solanacearum* E.F.Smith), against viruses (bunchy top, transmitted by the aphid, *Pentalonia nigronervosa* Coq.), and against insects, especially the banana root borer (*Cosmopolites sordidus* Germ.) (358, 524, 887, 982, 1285, 1750).

**Harvesting and Processing**. For export, bananas are harvested green-ripe. For this, the trunks are broken down with a machete, the bunches are cut off (they weigh 30-45 kg), and brought to the packing shed with as little mechanical damage as possible. There, they are washed and treated with disinfectants (sodium bisulphite or hypochlorite). The hands are detached from the bunches and packed in standard cartons for shipment (mostly 18 kg, recently also 12 kg). In the cold-storage room of the ship, the bananas are kept at 14-15°C, and in the harbour of their destination, they are kept at a somewhat higher temperature for ripening to yellow, in an atmosphere that contains 0.1% ethylene. Details of the high nutritional value of bananas have been given on p. 169 and in Table 26. Compared to fresh sales, preserved bananas play a subordinate role (pulp, confectionery, dried bananas) (382, 815, 944, 1713, 1968). Starch is produced from unripe bananas and from plantains.

In some countries, the leaves and other waste are used for animal fodder after harvesting (60, 1026). Locally, fibres are extracted from the leaf sheaths of bananas (see also *Musa textilis*, Table 44).

## Mango

fr. mangue; sp. mango; ger. Mango

Mango, *Mangifera indica* L., Anacardiaceae, is the most important tropical fruit after the banana (Table 27), but plays a minor role in world trade, because it needs very careful handling due to the sensitivity of the fruit to bruising. It originates from the Indo-Burmese monsoon region. In Southeast Asia (Java), some other *Mangifera* species are cultivated (*M. caesia* Jack, *M. foetida* Lour., *M. odorata* Griff.) (1012, 1119, 1296, 1297, 1439, 1657, 1675), however they are of only local importance.

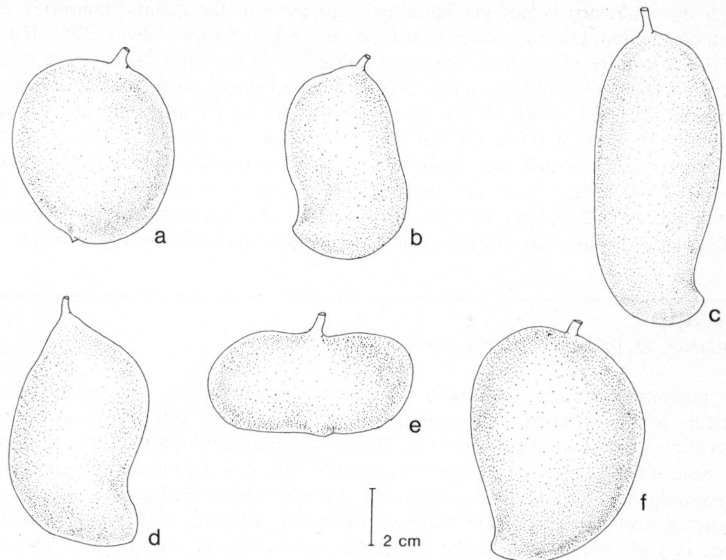


Fig. 41. *Mangifera indica*. Different fruit shapes. (a) 'Singapur', (b) 'Kidney', (c) 'Sabre', (d) 'Long green', (e) 'Extrema', (f) 'Pico'

The mango was brought from Arabia to Africa in the first millenium AD, and is nowadays grown in all warm countries, and far into the subtropics. The cultivars differ in fruit form (Fig. 41), fruit size, texture, taste, and climatic requirements. Some of the best cultivars for marketing have been selected in Florida (1555). The largest number of types (more than 1000) are found in India (376, 535, 839, 1675, 1759), which alone produces 9.5 million tons (two thirds of the world's production). The world production figures show a steady but slight increase (Table 27).

**Ecophysiology**. The mango grows best in tropical summer rainfall regions. In spite of their full foliage, the trees are surprisingly drought tolerant; a dry period of several months encourages the flowering and the fruit setting (1823). Even in the cooler subtropics, their cultivation can be profitable (Egypt, Israel). Some cultivars tolerate light frosts of short duration. The soil requirements are moderate, and even wet conditions are endured for some time (44).

**Cultivation**. Cultivars with polyembryonic seed (nucellar embryos, similar to citrus, see p. 177) can be generatively propagated, but as with other fruits, monoembryonic cultivars are propagated by grafting (by a variety of methods (584, 1555, 1693, 1823)) or by budding on seedling rootstocks. As rootstocks, seedlings of polyembryonic cultivars are preferred. Some cultivars even bear fruit in their third year, but most only from their fifth year onwards. Full yields can first be expected in the 12-15th year. A plantation of full-grown mango trees produces 10-25 tons/ha/year.

**Utilization**. Mangoes find a variety of uses. The young leaves, and fruit in which the endocarp is not yet hardened, are used in the Asiatic countries as vegetables, salad, and in a variety of types of pickles (see p. 126) (1296). Ripe fruits are eaten fresh (nutrient content in Table 26), or used for the preparation of juice (a variety of recipes), candied fruit, confectionery, and canned or dried products (282, 568, 1759, 1823). Overseas exports are by airfreight or by ship (cold-storage room at 10-12°C), but because the fruit is prone to decay, exports are not expected to reach large quantities (1745). On the other hand, the markets for conserves, namely juice and chutney, can certainly be increased (900, 910). All remains from the plant, also the nutritionally rich kernels and any fruit which is unsuitable for sale, are valuable as animal fodder, especially for pigs (1471).

## Pineapple

fr. ananas; sp. piña de América, ananás; ger. Ananas

The pineapple, *Ananas comosus* (L.) Merr., Bromeliaceae, originates from tropical South America, where today many forms are still found in semi-cultivation by the Indians (599, 1024, 1657). The most important cultivar groups are: 'Smooth Cayenne', with large and smooth fruit; it is the most important type for canning, and is cultivated worldwide. 'Spanish', with medium-sized fruit, is flavourful and cultivated especially in Malaysia. 'Queen', with small, prickly fruits, is very aromatic, and the preferred type for fresh consumption (South Africa). 'Abacaxi', with medium sized, elongated fruits, and slightly acidic, aromatic flesh, is generally consumed fresh (Brazil) (352, 1445, 1451). For all cultivar groups, there are local lines, which have arisen from mutations (1912). Cross-breeding is possible (all pineapple cultivars are self-sterile, and therefore they are seedless, cultivar crosses are generally fertile, and thus produce seeds), but cultivars of any commercial importance have not yet been produced.

**Production**. After the discovery of America by Columbus, the pineapple rapidly spread to all parts of the world. Nowadays, it is cultivated everywhere in the tropics between 25°N and S. However, there are individual regions of production at higher latitudes, e.g. near East London in South Africa (33°S), although the yields are lower there than in the tropics due to lower temperatures in the winter months. During the last 10 years (Table 27), the world production did not increase as vigourously as in the preceding decade. The Philippines and

Brazil extended their production considerably, but it receded in Thailand and Mexico. In 1988, the Philippines produced 2.3, Thailand 1.8, Brazil 1.0, India 0.8, and the USA 0.6 million tons. Fresh fruit is exported in large quantities only from the Philippines and the Côte d'Ivoire. The majority comes on the market in canned form; here again, the Philippines and Thailand are the major suppliers.

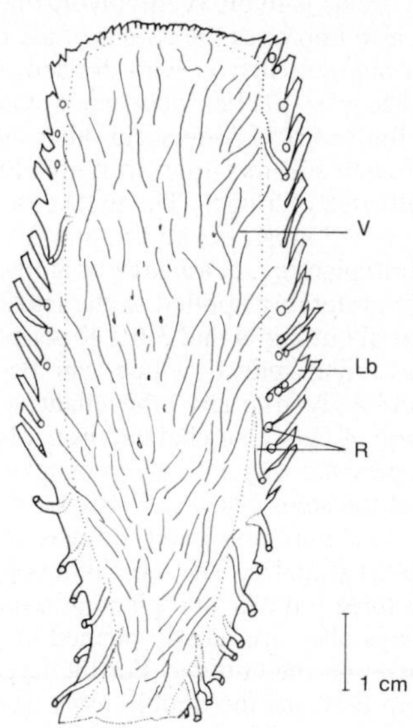


Fig. 42. *Ananas comosus*. Stem of young plant in longitudinal section. Lb = leaf bases, R = roots, V = vascular bundles

**Ecophysiology**. The pineapple is xerophytic by nature, surviving droughts of many months, and its photosynthesis follows during dry periods by the Crassulacean acid metabolism ($CO_2$ fixation in the night, stomata closed during the day), except when the night-time temperature is very high (44, 1276). Rainwater, mist, and dew are collected by the leaves, and are, eventually together with nutrients, absorbed mainly by the roots which emerge from the stem (Fig. 42). For good yields, 1000-1500 mm rain are necessary, and 600 and 2500 mm rainfall can be taken as the extreme limits for cultivation. In dry periods, the yield can be substantially improved by a little additional sprinkler irrigation. The temperatures should be high, and as equal as possible throughout the year; cooler temperatures at night (as low as 20°C) encourage flowering (566). Therefore, pineapples are preferably grown near to the sea, and are seldom grown over 700 m above sea-level. Temperatures under 20°C retard growth, and induce metabolic disturbances (chlorosis). However, mild, short-lasting frosts do not kill the plants ('Queen'). The best yields are obtained when there is full exposure to light, but pineapples can also be found growing as a crop under trees (palms, young rubber trees, etc.) (1246). Intensive direct solar radiation falling on ripening fruits often causes sunburn, especially if the fruits are not standing upright and are not sheltered by their crown of leaves (protection using paper covers, straw, fine wood-shavings, tying together the basal leaves, etc. (671)). Flowering is induced by ethylene. For more than a hundred years in the greenhouses of the Azores, smoky fires have been lit when the flowering should begin; these produce a lot of ethylene (1323). In the open air, one can use carbide, napthyl acetic acid, β-hydroxyethylhydrazine (352), and now mostly ethephon (672), which also improves the colour of the fruit.

Pineapples grow on any soil that is well-drained, whether it be sandy soil, oxysol, or peat soil. The best pH lies between 5 and 6, with the lower limit around 4.5, the upper limit at 6.5. Almost no other cultivated plant reacts with such sensitivity to too much soil moisture. In rainy climates and on heavy soils, pineapples must be cultivated on ridges. The uptake of nutrients for 1 ton of fruit is about 1 kg N, 0.2 kg P, 2.5 kg K, 0.3 kg Ca, and 0.1 kg Mg (595). P is mostly supplied to the soil before planting. N (as urea or ammonium sulphate) and K (as potassium sulphate) are preferably applied to the leaves (solutions of up to 10% are tolerated), each year about 2 g N and 3.6 g K per plant (671, 1785). All trace elements are given as foliar sprays, which can be applied at much higher concentrations than can be given to any other plant, without fear of burning the leaves, e.g. a 2% solution of $FeSO_4$ or chelated iron, also 1% zinc sulphate.

**Cultivation**. Pineapples are vegetatively propagated through lateral shoots. Shoots from the base of the stem ("suckers") are best suited for this, those from underneath the fruit ("slips") are less good; crowns are only exceptionally used as planting material (573) (Fig.43). The side shoots can be stored in the shade without moistening for three months, and planted in dry soil, without any loss of growth vigour. Nowadays, they are mostly planted in double rows, in which the planting distances depend on the cultivar. For the large-growing 'Cayenne', it is 90 x 90 cm, with 1.20 m between the double rows, and for 'Spanish', it is about 60 cm between the twin rows, 30 cm within the rows, and 1 m between the double rows. Mechanical weeding is very laborious, so soil herbicides are used in large plantations (bromacil, diuron, monuron), or mulching is carried out using tarred paper or sheets of black plastic. Also, organic materials (leaves, grass, straw) are often used for mulching.

With good plant material, flowering begins after 15-18 months (with some cultivars, it begins even after 10 months if conditions of temperature and day-length are favourable). The fruits are ripe about 4 months after flowering. After the harvest, the field is ploughed up and replanted, or, one allows a second generation of plants to grow from the suckers; the fruit from these ripens about 12 months after the first harvest (ratoon crops). In Malaysia, pineapples are continuously grown for up to 30 ratoon crops (1912). 'Queen' and 'Spanish' are better suited for multiple harvests than 'Cayenne'.

**Diseases and Pests**. Under favourable growing conditions, pineapples suffer little from diseases and pests. The heart and root rot, which is caused by *Phytophthora cinnamomi* Rands, and other fungi, are only a danger when the soil is too wet (see above). Thrips (carrier of the virus disease "yellow spot") are easily controlled by sanitation, weed control, and if necessary by spraying, and similarly with the mealybug (*Dysmicoccus brevipes* Ckl.), which is the carrier of the pineapple wilt disease, and presents no great problems (control of the ants which spread them, and by sound or decontaminated planting materials). On imported pineapples, fruitlet core rot (black rot) can sometimes be observed; this is caused by mildly parasitic fungi (*Penicillium*, *Fusarium*, *Cladosporium*), which enter the fruit through small injuries to the outer surface (197, 352, 358, 982, 1451, 2001).

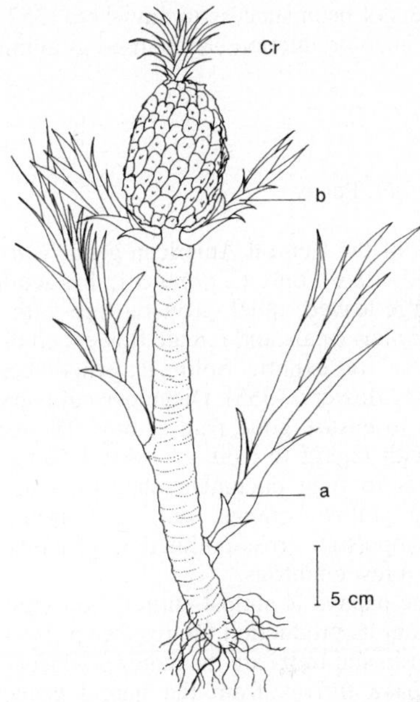


Fig. 43. *Ananas comosus*. Ripe plant without leaves. a = suckers, b = slips, Cr = crown

**Harvest and Processing**. The fruits are harvested fully ripe for the canning industry (for this, the crowns can be removed in the fields). For the local fresh markets, they can be harvested in the almost ripe state, for exports, green-ripe, or preferably half ripe (when the change of colour begins at the base of the fruit). In large-scale cultivation, the harvesting is carried out semi-mechanically. For this, it is necessary to induce simultaneous flowering by the use of chemical agents (see p. 188), so that a field need be harvested only once (352, 1450). Half-ripe fruits can be kept at 6-7°C for 4 weeks, and can thus be transported to the customer countries by ship. For 'Cayenne', the average yields achieved are 35-40 t/ha (the maximum is up to 100 t) for the first harvest, and 37.5 and 25 t/ha for the following ratoon crops. 'Singapore Spanish' yields 30 t/ha in the first harvest, then about 16 t/ha in the following years (1912).

A large part of the production is consumed in the countries of origin as fresh fruit (nutritional contents in Table 26). For exports, canned goods are the most important (slices, cubes, grated). All remnants (skins, heart) are processed to make juice (352, 703, 1270, 1451, 1832). The global exports of fresh fruit amount to about 600,000 tons.

By-products are: bromelain, a mixture of proteolytic enzymes, which is utilized in the food industry, the leather industry, and in medicine (452, 1205, 1815), and the fibres from the leaves, which are one of the finest leaf fibres (Table 44). They are however only extracted in East Asia and Brazil for special purposes; the industrial extraction of fibres from the leaves of pineapples grown on fruit plantations has not been successful anywhere (352, 972). The dried and milled remnants from juice production can be used as animal fodder ("pineapple bran").

## Papaya

fr. papaye; sp. papaya; ger. Papaya

Of the various species of the Central American genus *Carica*, Caricaceae, with edible fruits (550, 1657, 2026), only *C. papaya* L. has worldwide and economic importance. It is a large-leaved, quick-growing tree with soft wood (Fig. 44). Most cultivars are dioecious (male and female flowers on different trees), but the most important cultivar for export, 'Solo', is gynodioecious (the trees have female or hermaphrodite flowers) (535). Dioecious cultivars need a male tree for every 25 female trees to ensure good fruit setting. There are great differences among the cultivars with regard to fruit size (some 100 g to 10 kg and more), fruit shape (round-oval to long cucumber shaped), colour of the fruit flesh (whitish yellow, deep yellow, orange, red), and aroma, (561, 579). As a consequence of the compulsory cross-pollination, genetically pure cultivars are found growing in only a few countries.

**Production**. Because papaya is almost entirely consumed locally, there is no precise information about its production figures (see p. 169). According to (514), Brazil, Mexico, Indonesia and India are the major producers.

**Ecophysiology**. Papaya thrives from the humid tropics to the subtropics, wherever frost does not occcur (1463). Its demand for water is high (about 1500 mm). The rainfall should be distributed as equally as possible throughout the year. In regions with longer dry periods, the use of grass mulch is highly recommended. Because of their sensitivity to damage by high winds, shelterbelts are often necessary.

**Cultivation and Harvesting**. Papaya is propagated by seeds. With dioecious cultivars, 5 seedling trees are planted for each planting site (or, a sufficient number of seeds are directly sown). On the appearance of the first flower buds (after about 5 months), the trees are thinned out so that the correct number of male trees remain.

For 'Solo', one tree per planting site is sufficient. Planting distances vary from 2 x 3 to 3 x 3.5 m, depending on the soil and the cultivar. The trees begin to bear at the end of the first year. In tropical climates, flowering and fruiting are continuous, but in the subtropics, the cool season interrupts the fruit formation. The yields are highest in the second year (60-80 t/ha under favourable conditions), they are lower in the third year (about 50 t/ha), and they then decline rapidly. Apart from this, the harvest is made more difficult by the annually increasing height of the trees. Thus, in commercial plantations, the trees are kept for only 3-5 years (113, 2023). The fruits must be very carefully handled at all stages because of their extreme sensitivity to bruising, and the quick decay of the harvested fruit due to *Colletotrichum gloeosporioides* Penz., which develops from wound sites. The danger of infection is decreased by hot water treatment (49°C for 20 min). For local sale and for export, the fruits are harvested when their tips turn yellow. They can be stored for up to 3 weeks at 10-13°C. Fruits which are picked when green do not develop their full aroma. Fruits picked when fully ripe are extremely sensitive to bruising, and can only be stored for a few days (1028).

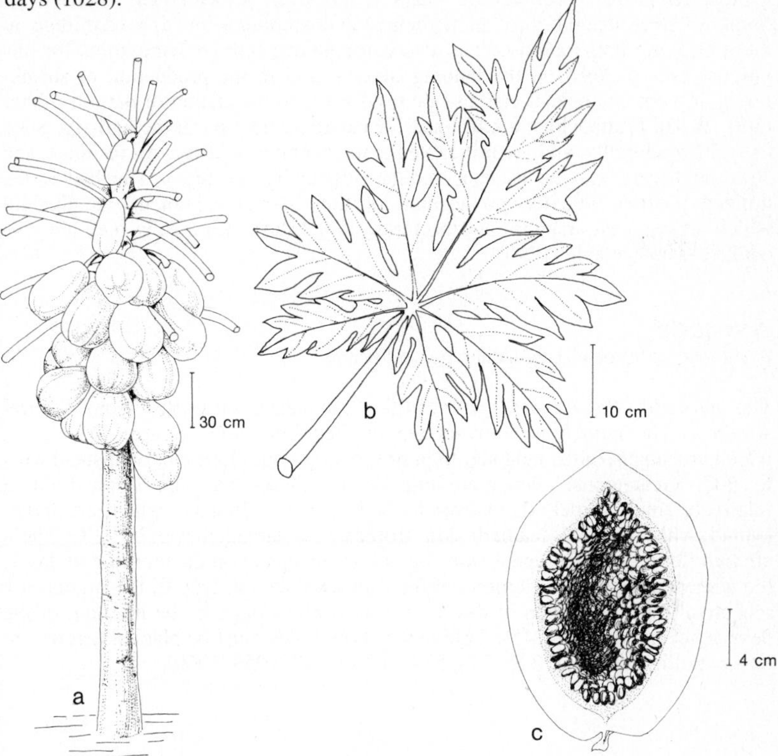


Fig. 44. *Carica papaya*. (a) fruiting tree (leaves cut), (b) leaf, (c) fruit in longitudinal section

**Utilization**. Papayas are eaten fresh with lemon and sugar, or in fruit salads. In the countries which grow them, they are a much-loved, refreshing fruit (nutritional value, see Table 26), but the market for the fresh fruit will always remain limited in Europe, even if better quality fruit is someday made available. Substantially better prospects for exports are found with papaya pulp and nectar (fruit juice containing the flesh of the fruit) (231, 910, 1015, 1823).

In some countries (especially Zaire, Sri Lanka, and Tanzania), papain, a protein-splitting enzyme, is collected after scratching the green fruit (454, 546, 579, 1204, 1427). The papain which is traded is the dried latex. One tree yields up to 500 g per year, but in practice often only 100-150 g. The sale of fresh fruit is more profitable, but if there is no market for it, then the production of papain can be economically viable. Papain is by far the most important of the proteolytic enzymes of plant origin (others are bromelain, see p. 190, and ficin). The USA alone imports about 200 tons annually, that being 70-80% of the world production. It finds numerous uses in the food industries (for stabilizing beer, as a meat tenderizer - for several years it has been intravenously injected 10 minutes before slaughtering), in medicine (against indigestion, as a vermifuge, in tooth cleaning preparations - it is also compounded into chewing gums for this purpose (see p. 369)), in the tanning of furs, and in the production of shrink-resistant wool and silk. It can also be used for the coagulation of natural rubber (360, 2010). Fruits which are utilized for papain extraction in the half-ripe state then ripen normally, and are utilizable for the canning industry. All residues, and also the leaves, can be used as fodder (especially for pigs). The leaves are utilized pharmaceutically, on account of their carpaine content, an alkaloid which acts against amoebas and bacteria (253, 1884), and locally they are also cooked as a vegetable (1296).

## Avocado

fr. avocat; sp. avocado, aguacate; ger. Avocado

The avocado, *Persea americana* Mill., Lauraceae, originates from Central America. The numerous cultivars are divided into three groups: "Mexican", which are small-fruited, adapted to poor growing conditions, and withstand frost to -6°C; "Guatemalan", which are large-fruited, thick- and rough-skinned with a relatively small kernel, stand frost to -4.5°C; "West Indian", which are large-fruited, with a smooth, leathery skin, tropical, are damaged even by -2°C. There are two floral types: in type A, the stigmas are receptive on the morning of day 1, the anthers open in the afternoon of the following day; in type B, the stigmas are receptive in the afternoon of day 1, and the anthers open in the morning of the next or second day (Fig. 45). Cultivars of both types must be planted together to ensure pollination (242, 376, 535, 574, 1039, 1657, 1959, 1960).

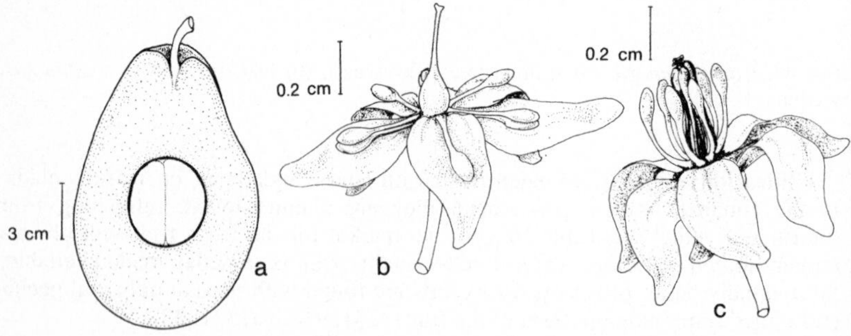


Fig. 45. *Persea americana*. (a) fruit in longitudinal section, (b) flower at female stage, (c) flower at male stage

**Production**. The avocado has been cultivated outside its home regions since the beginning of this century. Only after the Second World War did it become an export fruit. Its cultivation is possible from the tropics to the subtropical winter rainfall regions, and it is now being taken up in more and more countries. At present, 79% of the world's production is grown in the Americas. The main producers are Mexico (324,000 tons), the USA (308,000 t), Brazil (133,000 t), and the Dominican Republic (133,000 t).

**Cultivation**. The soil should be deep and well-drained, but otherwise the tree makes no particular demands, although it prefers a pH of 6-7.5, and will not tolerate salinity. In ecologically unfavourable situations, a grass mulch is highly recommended. Avocados are vegetatively propagated, mostly by grafting. Budding is also possible. The planting distance depends on the cultivar, distances of 6-10 m are usual. Most cultivars bear fruit from the 4th or 5th year onward. Full-grown trees give yields of 5-12 t/ha/year (574, 646, 1112, 1599, 2025).

**Harvest and Processing**. The high nutritional content (see p. 169, Table 26) (815), and the nut-like taste of some cultivars (e.g. 'Fuerte') make the avocado one of the most valuable fruit types. For local sale, the fruits are picked when they are beginning to soften. Exceptions are the cultivars such as 'Hass' and 'Fuerte', where the ripe fruits remain hard as long as they are hanging on the tree (1813). For export, all cultivars are harvested before softening. When the fruits are picked too soon, they do not ripen properly later, the skin becomes wrinkled, and the full aroma of the flesh does not develop. The choice of the correct picking time requires much experience, because the degree of ripeness is indicated by a colour change of the skin in only a few cultivars. Some cultivars can be kept at 5-7°C for storage and transport (however, the "West Indian" cultivars can only stand 12-13°C), and can be transported by refrigerator ship to the importer countries. Their consumption in Europe has increased considerably since the 60s (1746); West Europe receives the majority of its avocados from Israel and South Africa; the only European country which produces notable amounts of avocado is Spain (1988: 28,000 t).

From fruits which are too ripe for sale, avocado oil can be extracted. It belongs to the oleic acid group (Table 15, 77% oleic acid), is very highly priced (prices about 10 times higher than for peanut oil), and is used particularly for cosmetic preparations, but also as a salad oil (574, 758, 845, 1112).

## Annonas

A large number of fruit-bearing small trees and shrubs belong to the Annonaceae. Species of the genera *Annona* and *Rollinia* are cultivated (376, 500, 501, 550, 584, 1039, 1258, 1378, 1960). Many of them have only local importance, such as *R. mucosa* (Jacq.) Baill., the biribá, in Brazil, *A. muricata* L., the soursop (guanábana), and *A. reticulata* L., the bullock's heart (Fig. 46a), the last two being particularly grown in Central America and the West Indian islands. Up till now, the only ones of greater commercial importance are *A. cherimola* Mill. (Fig. 46c), the cherimoya, and *A. squamosa* L., the custard apple or sweetsop, and the crosses between these species, "atemoya".

The cherimoya originates from the highlands of Peru and Ecuador. It thrives only in tropical highlands and in the subtropics, it stands cool temperatures better than the other annonas, and is fairly drought tolerant. The fruits on the European market come mostly from Spain and Israel. They should be harvested before fully ripe, and can only be stored at temperatures above 14°C. In the ripe condition, they are very sensitive to bruising, and spoil quickly. Cherimoyas are eaten fresh (best if chilled), or processed for fruit salads, ice-cream, and fruit drinks.

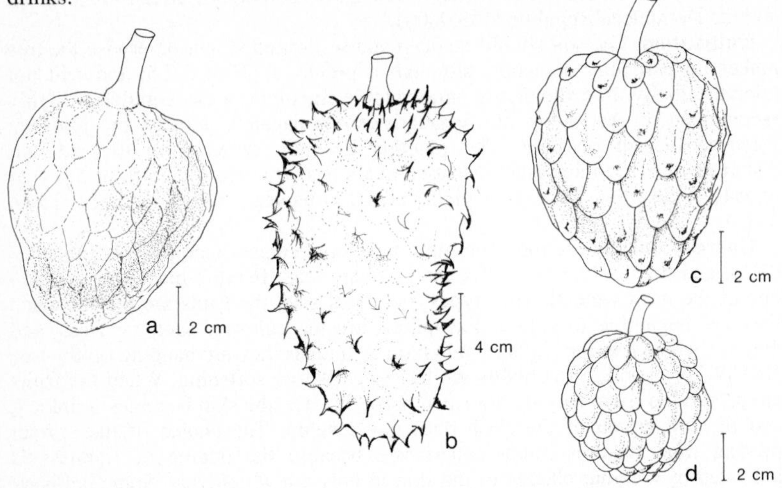


Fig. 46. *Annona* species. (a) *A. reticulata*, (b) *A. muricata*, (c) *A. cherimola*, (d) *A. squamosa*

The custard apple is the most important species commercially. In India, Thailand, the Philippines, China, and some other countries, large quantities are available in the local markets. It makes very low demands on soil and water (in India, the majority of the fruit comes from wild shrubs). It needs a significantly warmer climate than the cherimoya, but is also grown in the subtropical summer rainfall regions. Most of the custard apples are consumed fresh, a small proportion is processed to give nectar. The exports to Europe are insignificant (e.g. from Madeira to England). Transport and storage conditions are similar to those of the cherimoya, however, the ripe fruit is even more sensitive to damage, and spoils quickly (238, 910, 914).

## Guava

fr. goyave; sp. guayaba; ger. Guave

The genus *Psidium*, Myrtaceae, is tropical American. Numerous species have edible fruit. Only *P. guajava* is cultivated worldwide, locally also *P. cattleyanum* Sabine, the strawberry guava, *P. friedrichsthalianum* (Berg) Niedenzu, and *P. guineense* Sw. (550). *P. guajava* is an extraordinarily tough and undemanding small tree (in cultivation, it is kept to shrub shape), which nowadays can also be found to have run wild everywhere in the Old World tropics. A not inconsiderable part of the world production comes from such wild plants. For commercial cultivation, only large-fruited, specially bred cultivars should be utilized. These have thick flesh (as in Fig. 47), few seeds (there are also seedless cultivars), a uniform flesh colour (white, pink, or salmon red), a resiny taste which is not too noticeable, and medium (for canning) or low ("dessert" guava) acid content (472, 914, 1643, 1707). Good cultivars must be vegetatively propagated by grafting or budding on seedlings, layers, or stem cuttings (584, 1643).

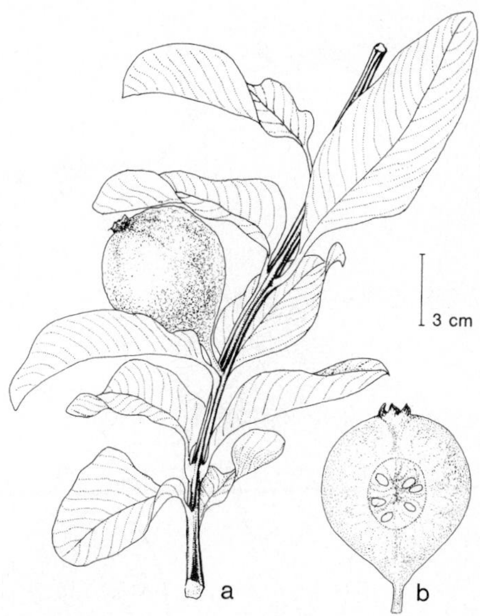


Fig. 47. *Psidium guajava*. (a) fruiting branch, (b) fruit in longitudinal section

Guavas can be grown from the humid tropics to the outermost subtropics with winter or summer rainfall, as long as no severe frosts occur. Despite their undemanding nature, for good yields they need deep soil, rich fertilizers, and sufficient water. In the subtropics, growth and flower formation come to a standstill in the cooler months; in the humid tropics, they bear fruit more or less all year round. The fruit ripens about 5 months after flowering. Yields up to 35 t/ha are reported (376). At harvesting, the fruit should be ripe; for export, they should be picked only a few days earlier than for the local market. They keep up to 3 weeks at 8-10°C, and can be transported in refrigerator ships. However, the export of fresh fruit does not appear to make very much sense, as even in the tropics, guava is seldom eaten raw. It is outstandingly well-suited for export in a number of canned goods, which should find good markets (see p. 171) because of their excellent flavour and high vitamin C content (Table 26). The following products have potential: clear juice, nectar (see p. 191), jams, jelly, thick paste ("guajabade"), and whole or half fruits in syrup (910, 1616, 1643). The leaves (Folia Djamboe) are used medicinally for digestive disturbances, and are exported, e.g. from India (376).

## Passion Fruit, Granadilla

fr. grenadille; sp. granadilla, maracuya; ger. Grenadilla

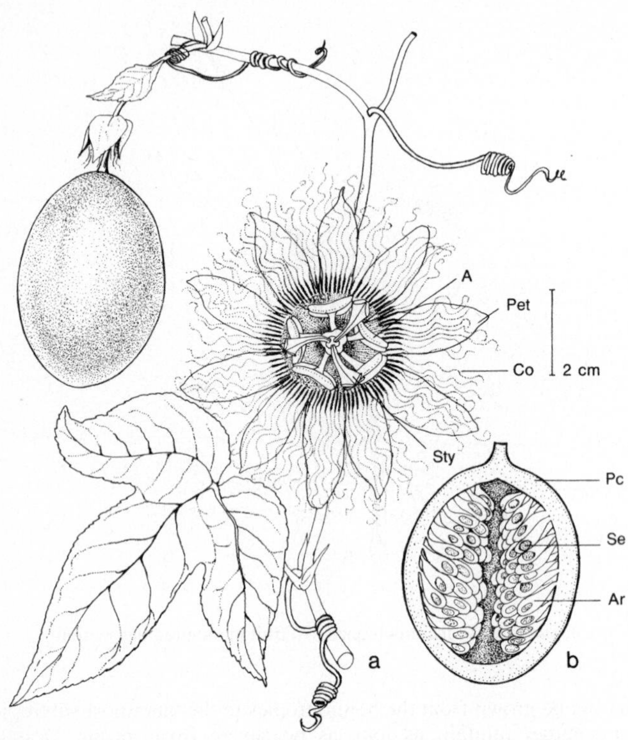


Fig. 48. *Passiflora edulis*. (a) branch with flower and fruit, (b) fruit in longitudinal section. A = anther, Ar = aril, Pet = petal, Co = corona, Pc = pericarp, Sty = style, Se = seeds

Almost all of the 500 species of the genus *Passiflora*, Passifloraceae, originate from tropical and subtropical America. Most are climbing plants, and many are grown as ornamentals on account of the fantastic shape and striking colouring of their flowers (Fig. 48); about 20 of them have edible fruit (550, 1142). Four species are cultivated to a fairly large extent (*P. edulis* Sims, *P. ligularis* Juss., *P. molissima* (H.B.K.) Bailey, *P. quadrangularis* L.), but of these, only *P. edulis* has great commercial value. The edible part of the fruit of this species is the juicy and highly aromatic arillus, which encloses each seed (with *P. quadrangularis*, the pericarp is also eaten). Two varieties of *P. edulis* are cultivated, *P. edulis* Sims var. *edulis*, the purple granadilla, preferred in tropical highlands and in the subtropics (e.g. Australia, South Africa), and *P. edulis* Sims var. *flavicarpa* Degener, the yellow granadilla, which gives higher yields, is resistant against *Phytophthora nicotianae* B. de Haan var. *parasitica* Waterh. (799, 990), and is better suited to warm tropical regions (e.g. Hawaii). Cultivars of these two varieties are not distinguished.

**Cultivation and Harvesting**. Passion fruit is generally grown from seed. The seedlings are planted out in rows of 4 x 4 m, and trained on wire trellises (178, 669, 691, 1063, 1118, 1529). The useful life of a commercial plantation is generally not more than 5-6 years, although the plants can live for decades, and form stems as thick as an arm. The flowers are pollinated by insects. In tropical countries, the pollination can be inadequate if too few bees and other large insects are flying (1288). The harvest begins in the second year after planting. The fruit is picked when fully ripe (a uniform deep violet for purple granadilla, deep yellow for yellow granadilla; when easily separated from the fruit stalk), or, they are gathered from the ground when they have fallen. A normal yield is 15 t/ha/year.

**Processing**. The fruit is consumed fresh locally (scooping out the fruit when it has been cut open), or is used in fruit salads (without separation of the seeds). The bulk of the production goes to factories for the extraction of juice (see p. 171) (280, 910, 1938), which is processed into a variety of drinks because of its special aroma (alone, or mixed with other fruit juices). The export of fresh fruit is insignificant.

Species of *Passiflora* contain alkaloids in all their parts (harman and others), which act as blood pressure reducers, antispasmodics, and sedatives. The leaves of *P. incarnata* L. are the main source of the drug Herba Passiflorae, but, the leaves of *P. edulis* are also utilized medicinally in several countries (376).

## Litchi (Lychee), Longan, and Rambutan

fr. litchi, longan, ramboutan (litchi chevalu);
sp. litchi, longan, rambutan;
ger. Litchi, Longan, Rambutan

The tropical family of the Sapindaceae contains a number of fruit trees, of which only the three named are of significant economic importance.

The litchi, *Litchi chinensis* Sonn. (Fig. 49a), is the most important, and it originates in southern China, where more than 100 cultivars are known (376, 1054, 1297, 1338). It is now cultivated in all subtropical regions, wherever the climate and soil are suitable. The climate should have a definite cool but frost-free season, without too much heat in summer, and high air humidity during fruit development (1177, 1178, 1249). The water requirement is high (about 1500 mm), and in regions with a long dry period, it must be irrigated during the winter. The soil should be fertile, deep, and well-drained. Propagation is carried out vegetatively, in Asia mostly by air layering, but it is also possible using cuttings. The trees grow large, and should be planted 10-15 m from each other. Cultivars which are adapted to the local climate produce about 150 kg fruit per tree annually. The fruits of most cultivars have bright red shells, which become brown after several days. Underneath the shell is the translucent white arillus, which has a sweet/sour, finely aromatic taste; it also keeps its flavour when canned. The fruits can be kept for 5 weeks when they are correctly packed and kept at a temperature of 1-2°C (1817), but they quickly spoil at temperatures over 20°C. They can be shipped from the producer countries to Europe, however, often reaching the customers in very poor condition (914). With the outstanding quality of some of the preserves, these appear much more suitable for export (910, 1823). In China, the fruits are dried (litchi nuts), and can in this way be kept for an indefinite period.

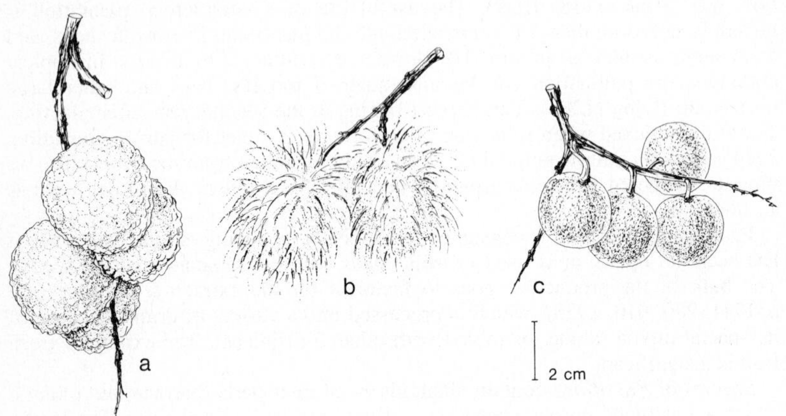


Fig. 49. (a) *Litchi chinensis*. (b) *Nephelium lappaceum*. (c) *Dimocarpus longan*

Longan, *Dimocarpus longan* Lour. (*Euphoria longana* Lam.) (376, 1030, 1297), appearing like a small, almost smooth-skinned litchi (Fig. 49c), originates from eastern India, and is cultivated from there to southern China as a fruit and ornamental tree. It endures much more winter cold than the litchi, even several degrees of frost, and makes slight demands on the soil. It even withstands occasional flooding. In northern Thailand, the longan trees stand in the river basins, and the litchis grow on the lower slopes of the mountains. Propagation is carried out mostly by seed, but also vegetatively. The taste is not quite so fine as the litchi's, the fruits (i.e. the arillus) are eaten fresh, preserved, or dried. Longan is one of the most beautiful foliage trees of the outer tropics, and deserves a wider distribution. Its canned fruit could be of interest for the European market.

Rambutan (Malayan "rambut" = hairy), *Nephelium lappaceum* L., is well described by its French and Malayan names (Fig. 49b) (376, 416, 1297). It originates from the Malayan archipeligo. Outside Southeast Asia, it is seldom grown commercially. In its home regions, there are numerous cultivars, which are maintained by vegetative propagation (some cultivars of the tree are dioecious!). Propagation is carried out mostly by budding on seedlings, or air layering (584). Climatically, it belongs to the hot and humid tropics, it demands 2500-3000 mm rain, and is seldom grown over 300 m above sea-level. The trees are small, and the yields much lower than for litchi. The fruits are consumed locally, either fresh or cooked, and there is not yet a canning industry or export trade. The pulasan, *Nephelium mutabile* Bl., is a close relative whose cultivation is limited to West Java (1297, 1324).

## Date Palm

fr. palmier dattier; sp. datilera; ger. Dattelpalme

Of the palms with sweet fruit pulp, only the date palm, *Phoenix dactylifera* L., has a major agricultural importance. It is an ancient cultivated plant, and its relationship with wild species of the genus *Phoenix* is uncertain (derived from *P. atlantica* A. Chev. in West Africa, or from *P. sylvestris* (L.) Roxb in India (516, 1229, 1230, 1657)). The numerous cultivars (439, 535, 1227, 1421) are divided into soft, semidry, and dry types. For export, those of the semidry type are of most interest.

**Production**. The main areas of cultivation of the date palm are identical to its original regions, that is, the Afro-Asian dry zone from Morocco to Pakistan. Worldwide, the production has changed only slightly during the last 10 years (see Table 27). In some countries it increased considerably (e.g. Oman +130%, Pakistan +45%, Iran +43%), in others it declined (e.g. Morocco -55%, Algeria -13%). In 1988, the largest producers were (in 1,000 t): Egypt 545, Saudi Arabia 495, Iran 430, Iraq 350, and Pakistan 290. Iraq remained the largest exporter (1988: 100,000 t), Pakistan took second place (35,454 t), followed by Saudi Arabia (25,975 t) and Tunisia (15,767 t). One should be aware, that in the FAO statistics (514), under "dates", the dried fruits of *Ziziphus jujuba* (Table 30) have been included; the East Asian exports of "dates" all consist of this species.

**Ecophysiology**. The date thrives only in arid regions with a high summer temperature. Rain during flowering can hinder the pollination. Rain during the fruit development can lead to bursting and decay of the fruit. The water requirement of the palm must be provided for by water-courses, ground water (roots up to 6 m deep), or by irrigation. The soil should be deep, well drained and well aerated. The date is the cultivated plant which is most tolerant of salt, 1-1.5% in the soil solution will be tolerated. A pH of up to 8 does not cause any harm. In the Middle East, animal dung is given to supply its nutritient needs, in other countries, mineral fertilizers are used (about 500 g N, 300 g P, and 250 g K per tree per year) (853, 1227).

**Cultivation**. The date palm is dioecious. To propagate the female trees true to type, shoots formed on the lower part of the trunk with an age of 5-10 years are utilized. If they have already formed roots at the mother plant, they can be directly planted (planting distances 9 x 9 to 10 x 10 m), otherwise they are kept in a nursery for 6 weeks for rooting. Recently, propagation by tissue culture techniques has been successfully applied to the date palm (394, 570).

The female inflorescence is a panicle with long stalks (Fig. 50). From ancient times, pollination has been improved by hanging a twig of a male inflorescence on the female inflorescence. In modern cultivation, pollination is carried out using hand devices or machines to apply the pollen, which can be stored for 2-3 months (1381). To achieve the same yield each year, and to ensure fully developed fruits, the infructescences are thinned out, or some inflorescences removed entirely (535, 1227).

**Harvest and Processing**. For export, fruits are often picked before they are fully ripe (colour change to brown, softening of the flesh, loosening of the skin), and ripened in the sun or artificially. For this, the fruit body is steeped in boiling 1% NaOH solution for 1 minute, then dried in an oven at 50-55°C (440, 1823). The average annual yield is 80 kg per tree, with good care even 100 kg; in the oases of North Africa, usually only 20-30 kg are harvested.

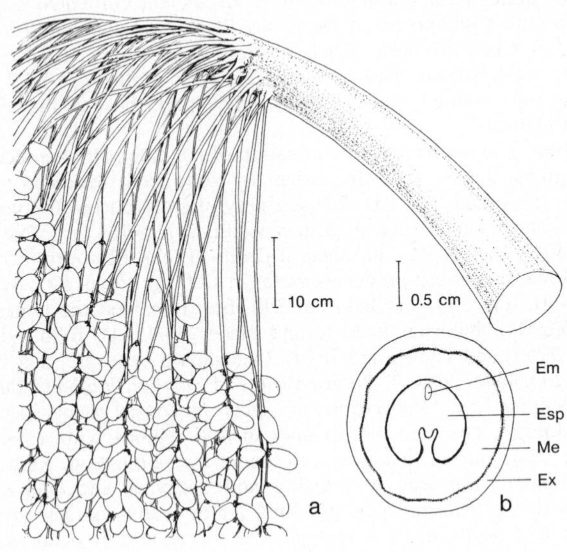


Fig. 50. *Phoenix dactylifera*. (a) part of infructescence, (b) fruit in cross section. Em = embryo, Esp = endosperm, Ex = exocarp, Me = mesocarp

Dates are not only eaten fresh, they are also used for the preparation of jams, and to make a paste which is used in the confectionery industry. To prepare date syrup, the ripe dates are extracted with water, and the extract is then thickened. The kernels (Fig. 50) are used as animal fodder (65% hemicellulose, 7% oil, 5.2% protein). The oil belongs to the oleic acid group (Table 15) (376). Also, whole dates are fed to animals. In the countries which grow dates, there is a shortage of wood, and all parts of the tree are utilized - the trunk for structural timbers, the leaves for basket work and for roofing houses etc. By tapping the upper part of the trunk, a sugary juice is obtained (see p. 74) (376); also, the preparation of crystallized sugar from the dates can be profitable (1569).

# Fig

fr. figue; sp. higo; ger. Feige

In the tropics and subtropics of the Old and New World, more than 1000 species of the genus *Ficus*, Moraceae, are found. Many of them have edible fruit, but only with *F. carica* L., the common fig, are they large and good-tasting. It originated from the Near East, and had already spread to the whole Mediterranean region in antiquity, where nowadays more than 90% of the world's production is grown (about 1.1 million t). It also thrives in tropical and subtropical regions with summer rainfall; however, there they cannot be dried, and can therefore only be consumed fresh or canned.

The fig tree is deciduous (354, 376, 535, 1242, 1657). The edible figs come from cultivars (about 100 named cultivars) in which the inflorescence contains only long-styled female flowers. The pollination is carried out by gall wasps (*Blastophaga psenes* Gravenh.), which have developed in the short-styled female flowers of another plant, the caprifig, of which several dozen cultivars are known. When they leave this inflorescence, they are loaded with pollen (Fig. 51) (157, 158, 576, 593). Edible and caprifigs produce three generations of figs each year, the spring figs, whose primordia were already formed before the autumnal leaf-fall, and summer and autumn figs, which are formed on the new year's growth. The spring figs are the best-tasting, the summer figs provide the greatest quantity (in some cultivars (Smyrna type), and in some climates, these are the only figs produced), and the autumn figs are of the lowest quality.

Where figs are planted only in home gardens, for sale in the local market, then common figs are mostly used, because their fruit ripens without pollination (parthenocarpy). For dried figs, the Smyrna types are preferred, as they contain seeds, and these give the fruit a pleasantly nutty taste. Their fruits develop only after pollination, and therefore they must be planted together with caprifigs. The same applies to the San Pedro types, whose spring figs are parthenocarpic, but whose summer figs only develop after pollination. With some cultivars, parthenocarpic fruit development can be induced by the application of phytohormones (1199).

**Ecophysiology**. Figs grow best in regions with winter rainfall. They survive dry summers well, because they have a sufficiently large volume of soil at their disposal; their roots can extend to a radius of 15 m and a depth of 8 m. The vegetative growth is stronger in the summer rainfall regions of the tropics and subtropics. However there, because the fruits are growing in the humid season, they often crack, and then spoil within a few hours (acid formation by bacteria and yeasts; this can also occur with undamaged fruit, due to the entrance of the microorganisms through the opening of the pseudocarp). The demands on the soil are slight, other than the soil should be sufficiently supplied with lime. The figs are one of the salt tolerant fruit types, and withstand 0.6% salt in the soil solution without any damage.

**Cultivation**. Figs are vegetatively propagated by cuttings (well-ripened one-year-old shoots, or two- or three-year-old), which easily take root, and are then planted out in the following winter. The planting distance mostly depends on whether the ground will be irrigated or not, on the cultivar, and on the pruning type (bush or tree). With irrigation, small-growing cultivars are planted at distances of 3 x 3 to 5 x 5 m, large-growing ones at 5 x 5 to 7 x 7 m. In dryland conditions, the planting distance depends on the rainfall. With planting far apart (10 x 10 m , or even 20 x 20 m), figs can survive with very little rainfall. In irrigated plantations, they are fertilized with about 50 kg N, 40 kg P, and 80 kg P per ha per year. When cultivated without irrigation, in arid and stony regions they are supplied with animal dung. In regions with sufficient rain, an undercrop, such as wheat, is sown in winter, and this receives the fertilizer.

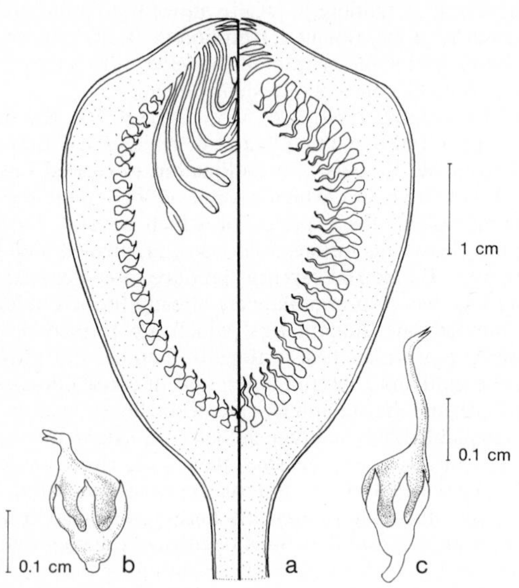


Fig. 51. *Ficus carica*. (a) longitudinal sections through two inflorescences. Left: monoecious variety with short-styled abortive female, and fertile male flowers. Right: purely female variety with long-styled fertile flowers. (b) short-styled flower. (c) long-styled flower

**Harvest and Processing**. In good plantations, one can expect 12 t of fresh figs per ha and year (= 4-5 tons dried figs). For fresh consumption, the figs should be harvested as ripe as possible, but before they have become completely soft. They are the most sensitive of fruits, and will only keep for a few days after picking, even with the most careful handling. This limits their commercial potential. Because of this, figs are cultivated on a large scale only where dried figs can be produced. In Mediterranean climates, the figs are left on the tree until they begin to shrivel. In this condition, they easily drop, and are then gathered from the ground. Drying is performed mostly in the sun, but also using drying installations. Fresh figs are also preserved with sugar, and processed to make jam (best mixed with ginger). Fig paste (for sweets) is produced from dried figs.

The nutritional values of fresh and dried figs are given in Table 26. Roasted dried figs are used as an additive or substitute for coffee (354, 335). A fairly considerable part of the fig harvest is used as animal fodder in the Mediterranean region.

## Stone Fruit

fr. fruits à noyau; sp. fruta de huesco; ger. Steinobst

Stone fruit comprise here those fruits of *Prunus* species, Rosaceae, in which the endocarp forms a hard shell around the seed (481, 1554, 1657). By far the most important is *P. persica* (L.) Batsch, the peach, (fr. pêche; sp. melocotón, durazno; ger. Pfirsich) (Table 27). This species originates from China. There is a very large number of cultivars. The most important characteristics for the cultivars are: soft and juicy or firm flesh; white or yellow flesh; flesh firmly attached to the kernel or loose; and ripening time. Nectarines are peaches with smooth rather than downy skins. Peaches are mostly produced in countries with a Mediterranean climate, however they also grow well in subtropical summer rainfall regions, and in high places in the tropics, as long as the winter temperature is cool enough to break the bud dormancy. In the need for cold in winter, there are great differences among the cultivars. Through defoliation, peaches can be made to flower and bear fruit even in the uniformly warm tropics (see also grapes, p. 173, apple types, p. 205) (554, 601, 1055, 1057, 1248, 1593, 1594). In tropical highlands, defoliation can induce suitable peach cultivars to bear fruit twice a year (1641).

Second place in world production is taken by the plums and prunes (fr. prune; sp. ciruela; ger. Pflaumen) (Table 27), which belong to several botanical species: *P. domestica* L., which contains the damson and many other forms (126, 200), is native to West Asia and southeastern Europe, and is cultivated in countries with a Mediterranean climate (Yugoslavia), and in those with temperate climates. For warmer climates, the East Asian species are suitable, particularly *P. salicina* Lindl., of which many well-known cultivars and hybrids have spread worldwide ('Satsuma', 'Santa Rosa', 'Wickson') (126, 376, 481, 1055, 1554). Their need for cool temperatures in winter is less than that of the peaches.

For the warmer countries, the apricot, *P. armeniaca* L., takes second place after the peach (fr. abricot; sp. albaricoque, damasco; ger. Aprikose) (Tables 26 and 27). It also originates in China. The few cultivars differ with regard to the fibre content of the flesh, fruit size and colour. Its requirement for winter coldness is moderate (about the same as *P. salicina*).

From the temperate zone of Europe and West Asia come the cherries, the sweet cherry, *P. avium* (L.) L., (fr. cerise; sp. cereza; ger. Kirsche), and the sour cherry, *P. cerasus* L. These play a subordinate role in the tropics and subtropics, with the exception of the European Mediterranean region, although cultivars of both species exist, which on suitable sites (those with sufficient winter cold, but with no late frosts) will also grow well in the highlands of the summer rainfall regions.

**Ecophysiology**. The requirements of stone fruit for soil texture are minimal, as long as the soil is deep and well-drained. The water supply must be guaranteed, especially in spring. In summer rainfall areas, therefore, the trees must be irrigated before the beginning of the rains; but, also in winter rainfall regions, quality fruit is only achieved with irrigation, especially with late ripening cultivars. Stone fruits are sensitive to soil salinity, and equally sensitive to high soil pH, at which Fe and Zn particularly are absorbed in insufficient quantities. Suitable applications of N, P, and K are necessary for high yields, depending on the soil type. Unbalanced fertilizing with N encourages excess vegetative growth. For the production of large, well-developed peaches, there must be a balanced relationship between fertilization and pruning, and it is usually necessary to thin out the young fruit. The market value of large fruits is higher than that of many small fruits.

**Cultivation**. Peaches and apricots are self-fertile, cherry and plum cultivars are often self-sterile, so that at least two, better three or four, cross-fertile cultivars must be planted together to ensure good fruit setting. The pollination is carried out by insects, especially bees, which also improve the pollination of the self-fertile species.

All stone fruit species are vegetatively propagated, mostly by budding or grafting on seedling rootstocks:

- peach is the most usual rootstock for peach, apricot and plum (the tree nurseries generally procure seed of suitable cultivars from the canning factories);
- for apricots, also apricot, plum and cherry plum (*P. cerasifera* Ehrh.) can be used as rootstocks;
- for plums, plum can also be used;
- cherries are grafted on cherry, cherry plum or rock cherry (*P. mahaleb* L.).

The trees are grown in bush form, and recently also as hedgerows.

**Utilization**. Most stone fruits are sold in the fresh markets. Only apricots are predominantly, and sour cherries exclusively eaten after processing. Many peach cultivars, most plum cultivars, and all apricots and cherries can only be stored for a limited time, since they do not develop their full aroma when harvested green-ripe. Because of this, they can only be sold in the local markets, or transported for short distances (from the Mediterranean countries to Central, East and North Europe). There are, however, some cultivars of peach and plum which can be stored at about 0°C for up to 6 weeks after harvesting, and which are therefore suitable for overseas export.

Apart from fresh consumption, processing into jams, conserves, and canned products also plays a role. Nectar is also produced from peaches and apricots (see p. 191). A further considerable part of the production is dried, especially peaches and apricots. Dried fruit is almost exclusively produced in the winter rainfall areas, where the drying can be carried out in the sun. In some countries, the fermentation of the fruit, and the distillation of fruit spirits plays an important role.

By-products from the processing industries are the oil-containing kernels; apricot kernels ("Chinese almonds") (1424), and to some extent also peach kernels, are used in the same way as almonds (e.g. "persipan", a marzipan made from peach and apricot kernels). The oil (53% in kernel, oleic acid group, Table 15) is used in the same way as almond oil, or is admixed to almond oil (958). A considerable part of the bitter-almond oil which is traded, in fact originates from apricot kernels (see p. 238).

# Pome Fruit

fr. fruits à pépins; sp. fruta de pepitas; ger. Kernobst

Pome fruit is the designation of the fruits of several genera of the Rosaceae, in which the fruit flesh is formed from the tissue of the floral axis, which fully envelops the ovary.

The apple, *Malus domestica* Borkh., (fr. pomme; sp. manzana; ger. Apfel) is the fourth most important fruit species, and by far the most important in the temperate zone (Table 27). Its hardiness in winter is responsible for this position, and the ability of the fruit of some cultivars to be stored for long periods (also without modern storage techniques). It is the fruit type which reacts in the most sensitive way to lack of winter cold, by delayed foliation (601, 1593). There are great differences in need for cold among the cultivars, so that some are very successful in Mediterranean climates, and can even be cultivated in the subtropical summer rainfall regions ('Granny Smith', 'Rome Beauty', 'Golden Delicious', and others) (376, 601, 1554, 1594). Because the apple is so popular with the Europeans, there has been no lack of attempts to cultivate it in the higher places in the equatorial region. This is possible with some cultivars ('Rome Beauty'), if the leaves are removed before the beginning of bud dormancy (stripping off, or chemical defoliation), and if the twigs are tied down, to encourage the formation of flower buds. As with grapes (see p. 173), two harvests per year can be achieved with these methods (481, 865, 1055, 1056, 1592).

The pear, *Pyrus communis* L. (fr. poir; sp. pera; ger. Birne), also thrives best in temperate climates, however, its demand for winter coldness is less than that of the apple, especially the cultivars which originated from crosses with the Japanese pear, *P. pyrifolia* (Burm. f.) Nakai ('Kieffer'); of course, these are of lower quality, and are almost only suitable for use as cooking pears (1154, 1554, 1593).

The quince, *Cydonia oblonga* Mill., (fr. coing; sp. membrillo; ger. Quitte) originates from Iran, where nowadays it is still found in large wild stands, and is substantially less demanding climatically than apple or pear. It thrives in subtropical summer and winter rainfall regions without any difficulty. It is only eatable after processing, and therefore commercial cultivation is only profitable where supply to a canning factory is possible (1554).

**Cultivation**. All three species are propagated vegetatively by budding or grafting, quinces mostly by layers on their own roots. For apple, the rootstocks are themselves vegetatively propagated from specially bred lines with particular characteristics ("Malling", "Merton", etc.), or the apple is grafted onto seedlings of another species, such as *M. baccata* (L.) Moench. Also for pears, there is a range of possible rootstocks, such as quince, *P. communis*, *P. pyrifolia*, *P. pashia* Buch.-Hamm., and other Asiatic species. The choice of rootstock depends on the type of soil, the preferred growth form, and the climatic conditions (376, 1919).

The pome species make fewer demands on the soil structure than the stone fruits; especially, they endure heavy, badly drained soils, and pears and quinces growing on *P. communis* or quince rootstocks withstand even temporary flooding.

Most pear cultivars are self-sterile, so that several cultivars must be planted together. Also, some apple cultivars are not fully self-fertile. Quinces are self-fertile. For apples and pears, the differentiation into flower-producing short shoots and vegetative long shoots requires more skill in pruning the tress than with stone fruits. Quinces form their flowers on the current year's shoots. The formation of fruit-bearing short shoots can be much increased, even in unfavourable climates, by bending down the twigs. When too much fruit is set, the excess fruit should be removed at the correct time, so that healthy, well-shaped fruits are achieved.

**Processing**. The majority of apples and pears are sold as fresh fruit. Many cultivars can be stored at 0-4°C for up to 6 months. Thus, there is no problem with overseas exports, and this plays a major role with apples, and to a lesser extent also with pears. A large share of the production of pome fruits is processed (canned goods, purée, jelly, dried powder, juice, wine, cider, dried fruit). A by-product from the canning industry is pectin from apples, and the seed of quinces (the seed coat contains 20% gum, which is utilized particularly in cosmetics, see p. 373).

# Further Fruit Species

Apart from the species which have been described, there is a multitude of tropical and subtropical plants which are used as fruit. The most important cultivated species are compiled in Table 30. Some of these are of great economic importance, even for export (e.g. strawberries, watermelon, muskmelon), others are well known food plants in their areas of cultivation (e.g. jackfruit, pomegranate, *Pandanus tectorius*). Besides the cultivated species, there are many wild species in all countries of the tropics and subtropics, whose fruits are gathered regularly, or perhaps only occasionally by children, and are eaten in times of shortage; these cannot be included here. Further information is available in the literature (126, 200, 259, 281, 376, 550, 1095, 1297, 1444, 1445, 1573, 1611, 1846, 2026).

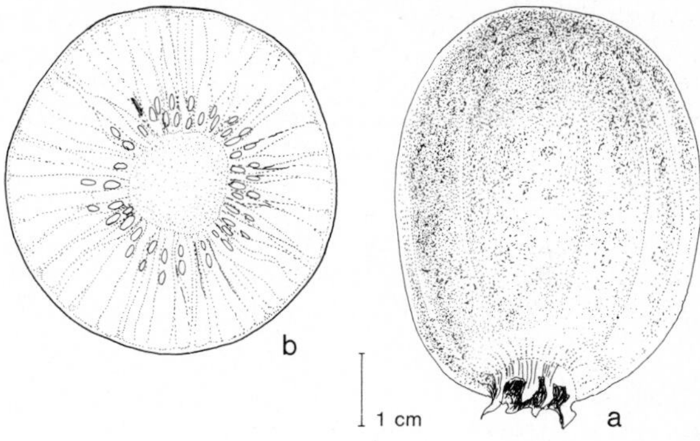


Fig. 52. *Actinidia deliciosa*. (a) fruit, (b) fruit in cross section

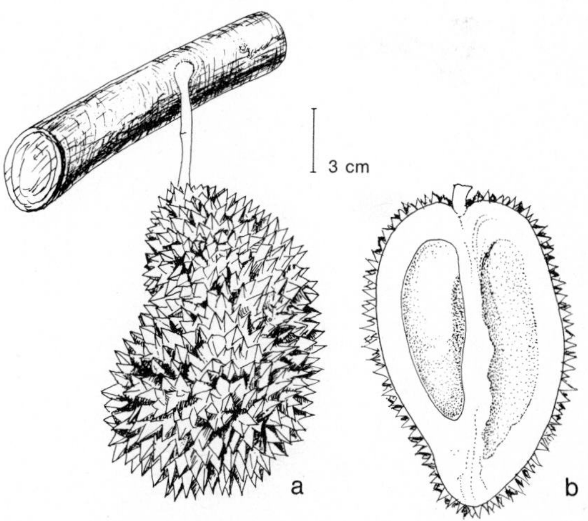


Fig. 53. *Durio zibethinus*. (a) fruit, (b) fruit in longitudinal section

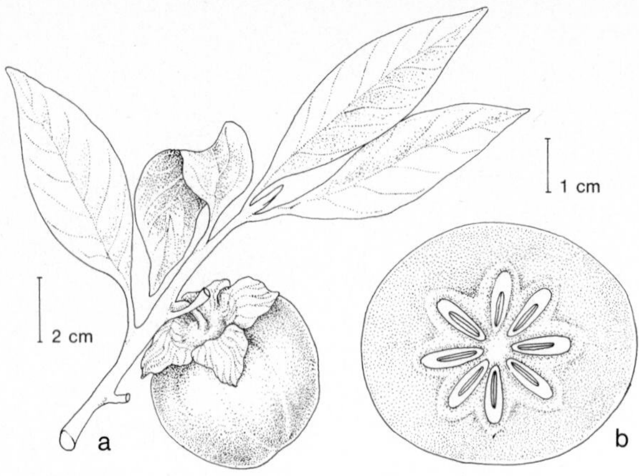


Fig. 54. *Diospyros kaki*. (a) branch with fruit, (b) fruit in cross section

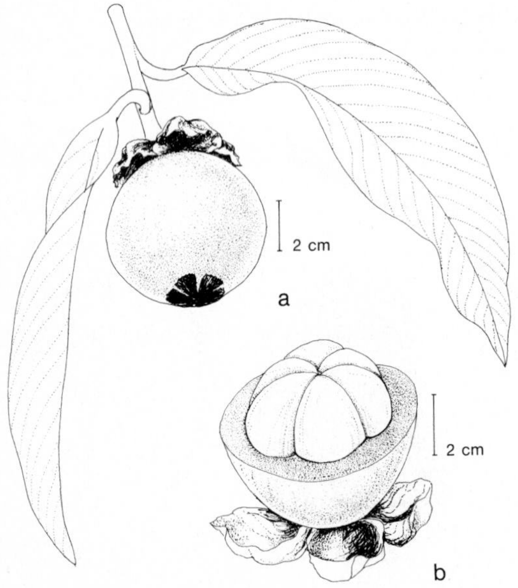


Fig. 55. *Garcinia mangostana*. (a) Tip of branch with fruit, (b) fruit in cross section

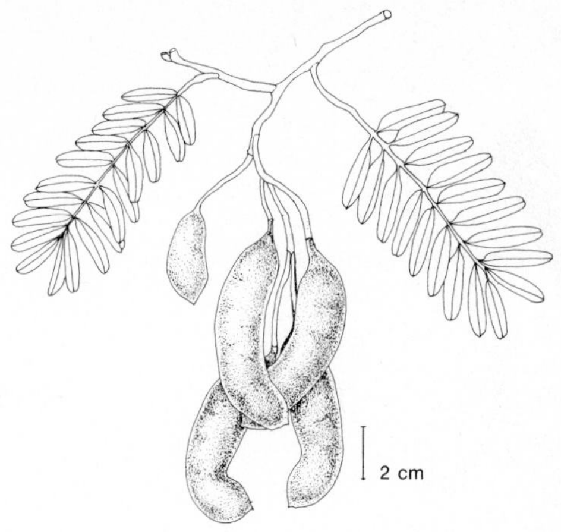


Fig. 56. *Tamarindus indica*. Fruiting branch

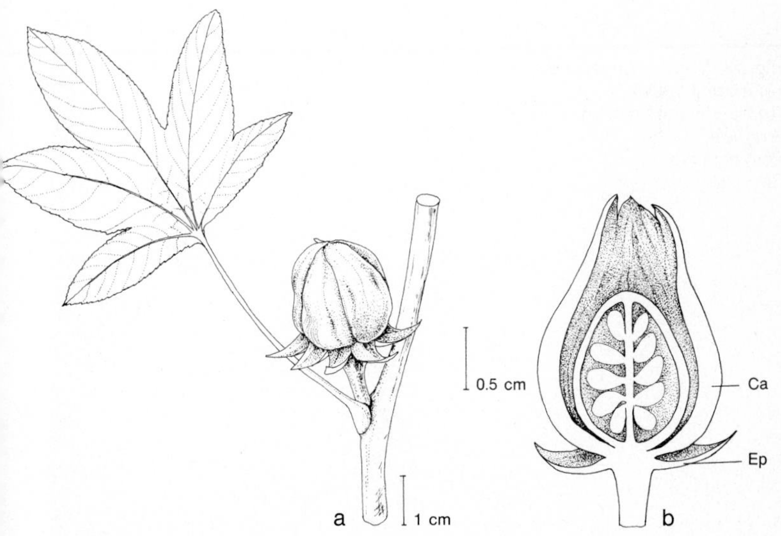


Fig. 57. *Hibiscus sabdariffa*. (a) nearly ripe fruit, (b) fruit in longitudinal section. Ep = epicalyx, Ca = enlarged calyx

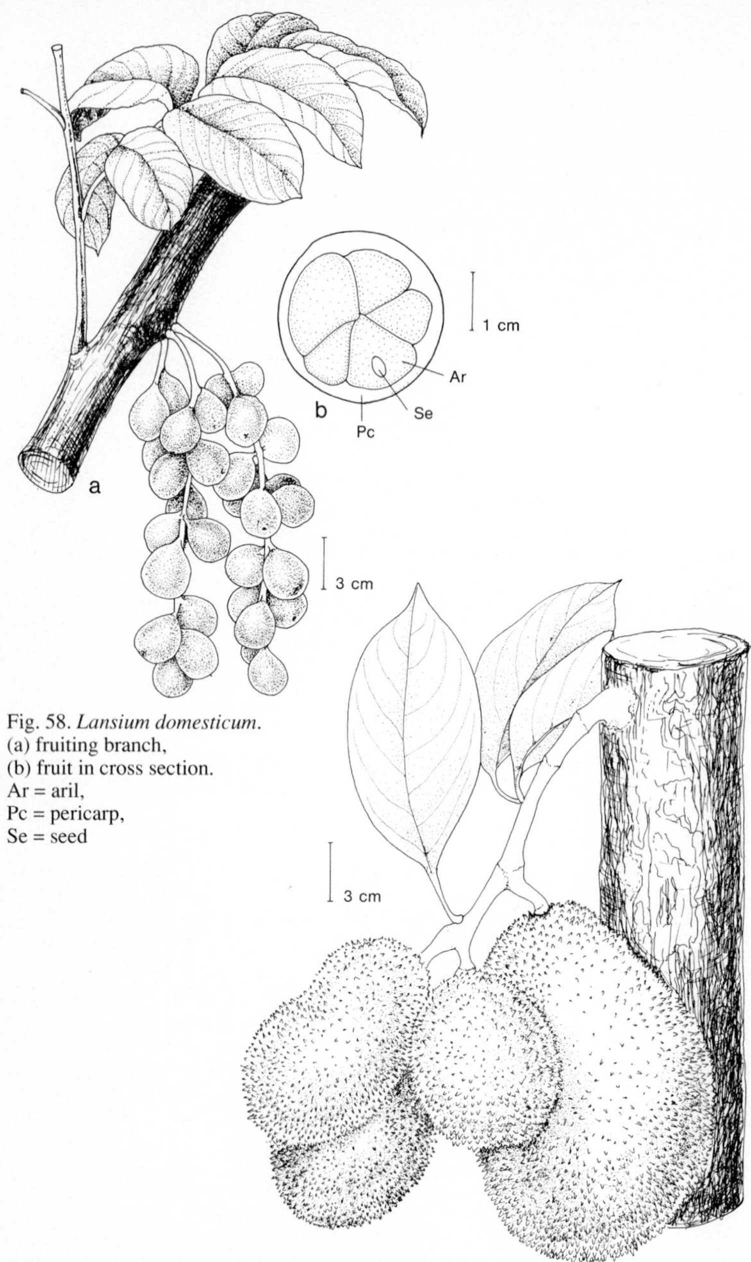


Fig. 58. *Lansium domesticum*.
(a) fruiting branch,
(b) fruit in cross section.
Ar = aril,
Pc = pericarp,
Se = seed

Fig. 59. *Artocarpus heterophyllus*. Fruit-bearing branch

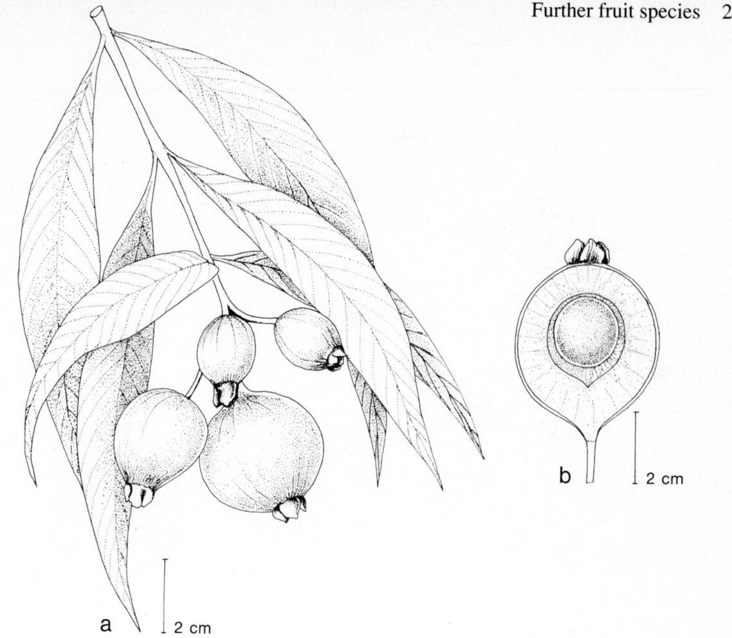


Fig. 60. *Syzygium jambos*. (a) branch with unripe fruits, (b) fruit in longitudinal section

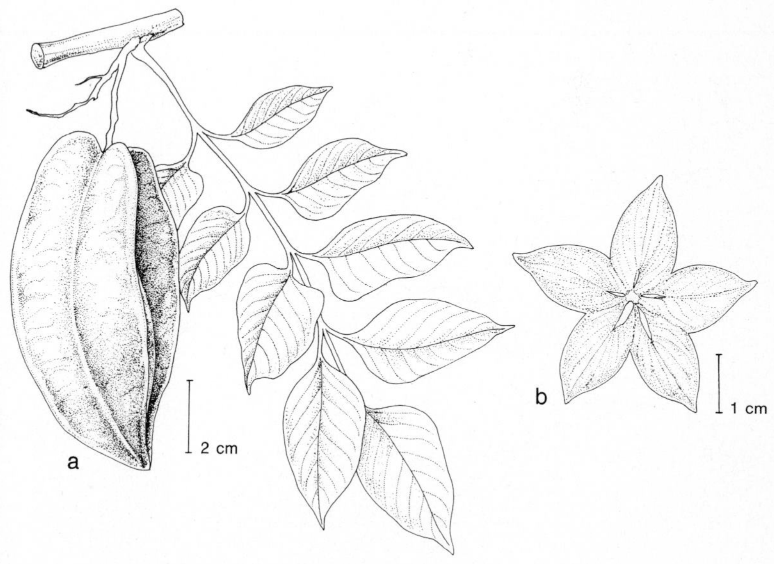


Fig. 61. *Averrhoa carambola*. (a) fruiting branch, (b) fruit in cross section

Fig. 62. *Punica granatum*. (a) flowering branch, (b) fruit in longitudinal section

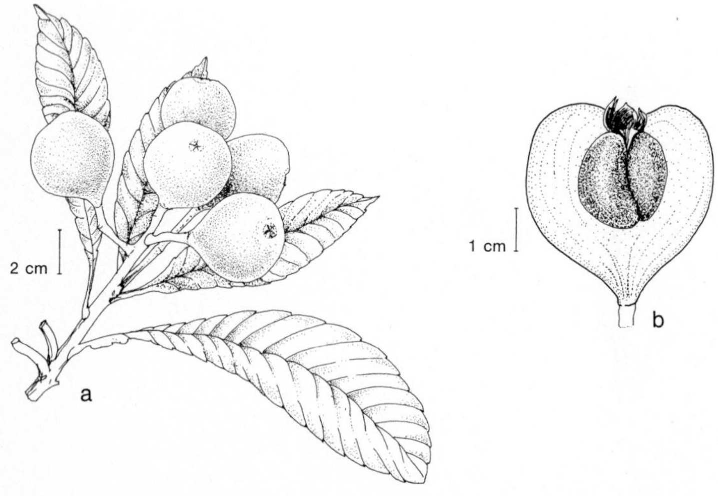


Fig. 63. *Eriobotyra japonica*. (a) fruiting branch, (b) fruit in longitudinal section

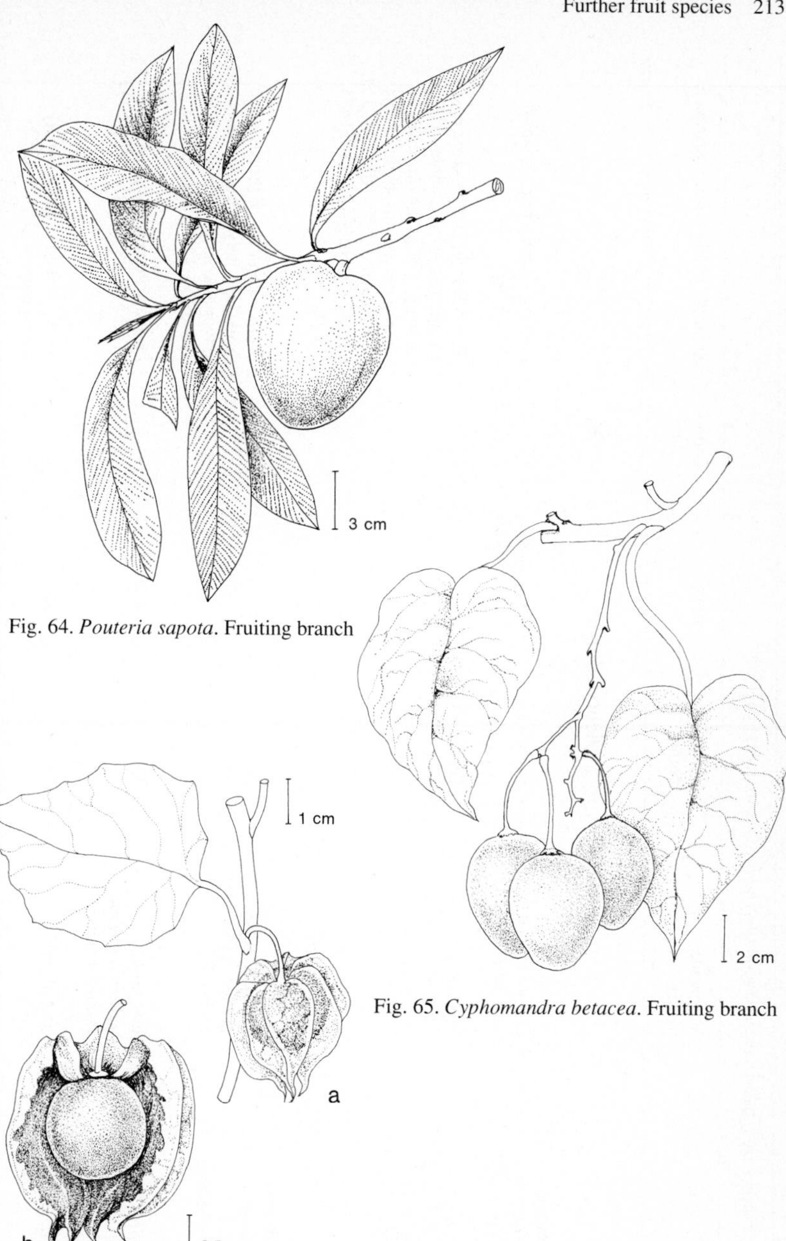


Fig. 64. *Pouteria sapota*. Fruiting branch

Fig. 65. *Cyphomandra betacea*. Fruiting branch

Fig. 66. *Physalis peruviana*. (a) piece of branch with fruit in enlarged calyx, (b) fruit in opened calyx

Table 30. Further fruit species (explanation of symbols, see footnotes page 55)

| Botanical name | Vernacular names | Climatic regions | Distribution | Econ. value | Remarks and literature references |
|---|---|---|---|---|---|
| | | | **Dicotyledonae** | | |
| **Actinidiaceae** | | | | | |
| *Actinidia arguta* (Sieb. et Zucc.) Miq. | tara vine, saru-nashi (Japan) | Te, S | r: China, Japan, Korea, USSR | + | Fruits are cherry to plum sized (188, 1846) |
| *A. deliciosa* (A. Chev.) C.F. Liang et A.R. Ferguson (*A. chinensis* Planch.) | kiwifruit, Chinese gooseberry, mao-yang tao | Te, S | w | ++ | (Fig. 52) Keep well and are easily transported. Rich in vitamin C (300 mg/ 100 g), grown in all subtropical regions, main exporter New Zealand (105, 188, 356, 529, 632, 673, 911, 1013, 1286, 1805, 2033) |
| **Anacardiaceae** | | | | | |
| *Spondias dulcis* Parkins. (*S. cytherea* Sonn.) | Otaheite apple, golden apple, ambarella | Tr | w | + | Fruits eaten raw, or processed to drinks and jelly. Young shoots and leaves also as a vegetable (203, 376, 1296, 1297, 1383, 1420) |
| *S. mombin* L. | hog plum, yellow mombin, mombin jaune | Tr | r: America | + | Fruit astringent, less palatable (376, 550, 888, 1297, 1425, 1774) |
| *S. pinnata* (Koen. ex L.f.) Kurz | hog plum, amara | Tr | r: India, Malaysia | + | Much planted. Leaves supply amara gum (259, 376, 1383) |

| | | | | | |
|---|---|---|---|---|---|
| *S. purpurea* L. | Jamaica plum, red mombin, ciruela roja | Tr | r: America | ++ | Most frequently planted *Spondias* species. Hedge plant, vegetatively propagated (888, 1039, 1297, 1383, 1960) |
| *S. tuberosa* Arr. Cam. ex Koster | hog plum, imbú, umbú | S, Tr | l: NE Brazil | + | Very drought tolerant. Usually semi-cultivated (1039) |
| **Apocynaceae** | | | | | |
| *Carissa carandas* L. | karanda | S, Tr | r: India | + | In India much used as a preserved fruit, mostly planted as hedges (259, 376, 406, 1823) |
| **Bombacaceae** | | | | | |
| *Durio zibethinus* Murr. | durian, civet fruit | Tr | r: SE Asia, E Africa | ++ | (Fig. 53) The arillus is the edible part, the smell of the fruit wall is unpleasant. Roasted seeds eaten locally (203, 259, 584, 914, 1114, 1258, 1296, 1297, 1328) |
| **Burseraceae** | | | | | |
| *Canarium album* (Lour.) Raeusch. | Chinese white olive, paak-laam | Tr | r: Indochina to SE China | ++ | Fruit flesh very rich in oil. Processed like desert olives (259) |
| *C. pimela* C. König | Chinese black olive, moo-laam | Tr | l: S China | + | Wild and planted. Fruits utilized as for *C. album* (259, 1039) |
| *Dacryodes edulis* (G.Don) Lam (*Canarium edule* (G.Don) Hook., *Pachylobus edulis* G.Don) | safou, bush butter tree, eben | Tr | r: W Africa | + | Fruit flesh contains 30-65% fat. Cooked with salt, and despite the resinous taste, very popular. Also other species have edible fruits (200, 214, 1306, 1319, 1477, 1874, 2021) |

Table 30. Further fruit species, cont'd (explanation of symbols, see footnotes page 55)

| Botanical name | Vernacular names | Climatic regions | Distribution | Econ. value | Remarks and literature references |
|---|---|---|---|---|---|
| **Cactaceae** | | | | | |
| *Hylocereus ocamponis* (Salm-Dyck) Britt. et Rose | pitaya roja | Tr, TrH | r: Central and S America | + | Climbing, needs support on stakes or trees. Fruits very tasty. Also other *H*. species, e.g. *H. triangularis* Britt. et Rose, pitaya amarilla, cultivated (526, 1378, 1611) |
| *Opuntia ficus-indica* (L.) Mill. | Indian fig, prickly pear, tuna de Castilla | S, Tr | w | ++ | Especially important in dry areas. Locally important trading product. Spineless cultivars also used as animal fodder (see p. 400). Apart from this species, many other cacti are used as fruits in Central America (392, 490, 550, 1039, 1378, 1557) |
| **Caricaceae** | | | | | |
| *Carica papaya* L. | papaya | S, Tr | w | +++ | see p. 190. Fig. 44 |
| *C. pentagona* Heilborn | babaco | S, TrH | r: S America, Mediter-ranean countries, New Zealand, Australia | ++ | Cold tolerant. Cultivars with seedless fruit are vegetatively propagated. Fruit eaten stewed or preserved (200, 356, 1709, 1827) |

| | | | | | | |
|---|---|---|---|---|---|---|
| **Cucurbitaceae** | | | | | | |
| *Citrullus lanatus* (Thunb.) Matsum. et Nakai | watermelon, pastèque, sandla | S, Tr | w | | +++ | (Table 26) One of the most important subtropical fruits. High value in local trading. Also forms which are used for fodder (see p. 399), and others, which are only cultivated for the oil-rich seeds (Table 15, see p. 112) (376, 659, 731, 1296, 1411, 1932) |
| *Cucumis melo* L. | muskmelon, cantaloupe | S, Tr | w | | +++ | (Table 26) Trading value similar to the watermelon, but greater importance as an export product (376, 659, 731, 1296, 1932) |
| **Dilleniaceae** | | | | | | |
| *Dillenia indica* L. | elephant apple, chalta, karmal, hondapara | Tr | r: | India to Malaysia | + | The edible parts of the "fruit" are the fleshy calyx leaves. Mainly utilized for refreshing drinks and jelly. Also an ornamental tree (200, 259, 376, 406, 1383) |
| **Ebenaceae** | | | | | | |
| *Diospyros kaki* L.f. | kaki plum, Japanese persimmon, caqui, kik-we, shi-tse | S, TrH | w | | +++ | (Fig. 54) Old E Asian cultivated plant with many cultivars. Fairly drought tolerant. Cultivation is also important in the Mediterranean region. Fruits taste good when fully ripe. Dried fruit ("kaki figs") a favourite dried fruit in E Asia, also exported from there (259, 376, 914, 974, 1061, 1297) |

Table 30. Further fruit species, cont'd (explanation of symbols, see footnotes page 55)

| Botanical name | Vernacular names | Climatic regions | Distribution | Econ. value | Remarks and literature references |
|---|---|---|---|---|---|
| *D. lotus* L. | date plum, hei-tsao | S, TrH | r: Mediterranean countries to Japan | + | Fruits cherry-sized. Used as a rootstock for *D. kaki* (376) |
| **Euphorbiaceae** | | | | | |
| *Antidesma bunius* (L.) Spreng. | Chinese laurel, bignai, buni | Tr | r: SE Asia | + | Fruits sour, eaten fresh, or prepared in jellies and sweet dishes (high pectin content). Young leaves eaten as a vegetable (203, 259, 376, 1039, 1296, 1297) |
| *Baccaurea dulcis* (Jacq.) Muell. Arg. | tjoopa, kapoondong | Tr | l: Indonesia | + | Cultivated in Sumatra and Java (200, 203, 259) |
| *B. motleyana* Muell. Arg. | rambai, burung | Tr | l: Malaysia | + | Fruits eaten raw or cooked (259) |
| *B. ramiflora* Lour. (*B. sapida* Muell. Arg.) | mafai, ranbi | Tr | r: SE Asia | + | Cultivated from India to China. Apart from the slightly sour fruits, the flowers are also eaten (200, 259, 376) |
| *Phyllanthus acidus* (L.) Skeels | Otaheite gooseberry, ceresier du Tahiti, grosella, guinda | Tr | w | + | Fruits very acid, mostly processed to jelly or sweet dishes (200, 203, 1039, 1296, 1383) |

| | | | | | |
|---|---|---|---|---|---|
| *P. emblica* L. (*Emblica officinalis* Gaertn.) | emblic, myrobalan, aonla, ambla | Tr | r: S and SE Asia | ++ | Fruits sour, very rich in vitamin C (500-1000 mg/100 g), mostly processed before eating (376, 406, 1213, 1383, 1823) |
| **Guttiferae** | | | | | |
| *Garcinia mangostana* L. | mangosteen, mangis | Tr | r: SE Asia | ++ | (Fig. 55) Fruit very tasty. Cultivation difficult. Trees bear only after 10-15 years (36, 203, 259, 376, 584, 910, 914, 1258, 1297, 1823, 1948) |
| *Mammea americana* L. | mammee apple, mami, abricotier d'Amérique | Tr | r: West Indian Islands | + | Fruit pulp acidic, with some similarity to apricot. Raw in fruit salad with other fruits, and cooked as jams and other preserves. Hardly cultivated outside its homelands (550, 1297) |
| *Platonia esculenta* (Arr.Cam.) Rickett et Stafl. (*P. insignis* Mart.) | bacuri, bacury, parcouri | Tr | r: Brazil | + | In the Amazon region much valued for sweet dishes, preserves and drinks (172, 550, 1425, 1518) |
| *Rheedia acuminata* (Ruiz et Pav.) Planch. et Triana (*R. madruno* Planch. et Triana) | madroño, naranjito, cozoiba, madruno | Tr | r: Colombia, Peru, Venezuela | + | Fruits acidic, eaten raw or prepared. The fruits of several other *Rheedia* species are used in a similar manner (172, 550, 1039, 1378) |

Table 30. Further fruit species, cont'd (explanation of symbols, see footnotes page 55)

| Botanical name | Vernacular names | Climatic regions | Distribution | Econ. value | Remarks and literature references |
|---|---|---|---|---|---|
| **Leguminosae** | | | | | |
| CAESALPINIOIDEAE | | | | | |
| *Cassia fistula* L. | pudding pipe tree, golden shower, Indian laburnum, cañafistula | S, Tr | w | + | Originates in S Asia, found everywhere in the tropics and subtropics as an ornamental bush ("Indian laburnum"). Fruit traded in Europe as "manna". Fruit pulp is a mild laxative, and utilized for flavouring chewing tobacco (259, 376, 1261) |
| *Ceratonia siliqua* L. | St. John's bread, carob, locust bean, algarrobo caroubier | S | w | ++ | Originates in the Middle East, found worldwide as an ornamental tree. Main production of the pods in the Mediterranean area. Especially important as animal fodder. The juice pressed out ("kaftan") is used as a syrup. Kernels provide gum (see p. 373) (376, 406, 408, 444, 1261, 1383) |

| | | | | | |
|---|---|---|---|---|---|
| *Tamarindus indica* L. | tamarind, magyi, imli | S, Tr | w | ++ | (Fig. 56) Originates in the summer rainfall regions of tropical Africa. Mainly cultivated in India. Fruit pulp eaten fresh, and processed, often into drinks and sauces (ingredient of Worcester Sauce). Kernels are a foodstuff and source of gum (see p. 373). Young leaves and pods as a vegetable (167, 168, 273, 376, 406, 444, 914, 1261, 1306, 1774, 1823, 1948) |
| **Malpighiaceae** | | | | | |
| *Byrsonima crassifolia* (L.) H.B.K. | golden spoon, maurissi, nance | Tr | r: Central and S America | + | Fruits sour, eaten raw, or processed into refreshing drinks. Vitamin C content 90-240 mg/100 g. Several other *Byrsonima* species are used similarly (281, 550, 1039, 1425, 1960) |
| *Malpighia glabra* L. and *M. punicifolia* L. | Barbados cherry, acerola, cerise carrée | S, Tr | r: Central and S America | + | Famous because of their high vitamin C content (Table 26). Grows on poor, stony soils. Also introduced to other regions, but never of great commercial value outside its homeland. Because of its high acidity, mostly processed into fruit juices (550, 584, 1039) |

Table 30. Further fruit species, cont'd (explanation of symbols, see footnotes page 55)

| Botanical name | Vernacular names | Climatic regions | Distribution | Econ. value | Remarks and literature references |
|---|---|---|---|---|---|
| **Malvaceae** | | | | | |
| *Hibiscus sabdariffa* L. var. *sabdariffa* | Jamaica sorrel, roselle, karkadeh, oseille de Guinée | S, Tr | w | ++ | (Fig. 57) The plants originate in the summer rainfall regions of tropical Africa. Used are the shining red calyces, which become fleshy, and have a high acid and mucilage content (see also *Dillenia indica*). Processed into a refreshing drink and jellies, also consumed as a hot tea. The dried calyces ("hibiscus flowers") are an important export product. The leaves are used as a vegetable, the seeds used like sesame. High-growing forms are cultivated as fibre plants (see p. 347) (259, 340, 376, 659, 1039, 1948) |
| **Meliaceae** | | | | | |
| *Lansium domesticum* Correa | langsat, langsep, duku | Tr | r: SE Asia | ++ | (Fig. 58) Reputed as one of the best fruits of Malaysia. The juicy arillus is the edible part. Many forms, also seedless (203, 259, 376, 584, 1297, 1948) |

| | | | | | |
|---|---|---|---|---|---|
| *Sandoricum koetjape* (Burm.f.) Merr. (*S. indicum* Cav.) | red santol, kechappi | Tr | l: Malaysia | + | Fruits often of inferior quality. Outside Malaysia planted as a quick growing shade and street tree (200, 203, 259, 376, 1039, 1383, 1936) |
| **Moraceae** | | | | | |
| *Artocarpus heterophyllus* Lam. (*A. integrifolia* L.f.) | jackfruit, jacquier, po-lo-tan | Tr | w | ++ | (Fig. 59) Originates in India; much cultivated in SE Asia, less in Africa and America. One of the largest tropical fruits, consumed fresh, seeds eaten cooked or roasted (see p. 63). The wood provides "basanti", the yellow dye for the clothes of Buddhist monks (203, 259, 376, 406, 584, 910, 1296, 1297, 1425, 1774, 1798, 1823, 1948) |
| *A. integer* (Thunb.) Merr. (*A. champeden* (Lour.) Spreng.) | chempedak, champedak, lemasa | Tr | r: Malay Archipelago | + | Similar to jackfruit. Flesh softer and more aromatic (203, 259, 1297, 1948) |
| *A. odoratissima* Blanco | morang, marang | Tr | l: Philippines, Kalimantan | + | Flesh sweet and juicy. Kernels eaten roasted (259) |
| *Morus nigra* L. | black mulberry, mûrier noir, moro | S, TrH | w | + | Planted in house gardens in all subtropical countries. Fruits used fresh, or for juice, jelly or jam. *M. alba* L. is preferred for silkworm feeding (see p. 402) (200, 1823) |

Table 30. Further fruit species, cont'd (explanation of symbols, see footnotes page 55)

| Botanical name | Vernacular names | Climatic regions | Distribution | Econ. value | Remarks and literature references |
|---|---|---|---|---|---|
| **Myrtaceae** | | | | | |
| *Acca sellowiana* (Berg) Burret (*Feijoa sellowiana* (Berg) Berg) | feijoa, pineapple guave, guajabo del pais | S | w | ++ | Drought tolerant small tree, fruits acidic, tasty, mostly eaten raw. Also introduced into subtropical regions of Africa and Asia; New Zealand is recently supplier for international markets (119, 356, 550, 1039, 1328) |
| *Eugenia dombeyi* (Spr.) Skeels (*E. brasiliensis* Lam.) | Brazil cherry, gruixameira, grumichama | Tr | l: S Brazil | + | Tasty fruit, eaten raw or cooked. Little cultivated outside its homeland (550, 584, 1039, 1297) |
| *E. uniflora* L. | Surinam cherry, pitanga | S, Tr | w | + | Fruit mostly processed into drinks and jams. Apart from the two *Eugenia* species mentioned, there are many others with edible fruit (550, 1039, 1297, 1948) |
| *Myrciaria cauliflora* (Mart.) Berg | Brazilian grape tree, jaboticaba | Tr | l: Brazil | ++ | Important fruit tree in the surroundings of Rio de Janeiro. The acidic fruit is usually processed into drinks and jams. Further Brazilian *Myrciaria* species have edible fruits ("pitangas"), as in (200, 550, 1039, 1297) |

| | | | | | |
|---|---|---|---|---|---|
| *Syzygium aqueum* (Burm.f.) Alston (*Eugenia aquea* Burm.f.) | water rose apple, jamlac, tambis | Tr | r: SE Asia | + | Hardly cultivated outside its homelands (India - Malay Archipelago). Fruits eaten raw and processed to make drinks (203, 259, 376, 584, 1039, 1383, 1948) |
| *S. cumini* (L.) Skeels (*Eugenia cumini* (L.) Druce) | Java plum, Peru naval, jambolan, jaman | Tr | w | ++ | Found in many forms in its homeland, the Malay Archipelago. The good cultivars are mostly eaten raw, acidic cultivars used for jams. Also planted as a shade tree (259, 273, 376, 406, 815, 1039, 1383) |
| *S. guineense* (Willd.) DC. (*S. owariense* Benth.) | waterberry, mukute, sinti, yinti | Tr | l: Togo | + | Over-ripe fruits are very tasty, and much valued by the natives (1477, 1846) |
| *S. jambos* (L.) Alston (*Eugenia jambos* L.) | rose apple, jamun, pomarosa | S, Tr | w | ++ | (Fig. 60) Fruits eaten raw or cooked (259, 376, 541, 584, 1039, 1383, 1425, 1948) |
| *S. malaccense* (L.) Merr. et Perry (*Eugenia malaccensis* L.) | Malay rose apple, Surinam cherry, manzana de agua, pomerac | Tr | w | + | Outside its homeland (Malaysia), it is mostly planted as an ornamental, shade tree and windbreak. Fruits pear-shaped, mostly eaten cooked. In SE Asia and Africa, the fruits of several other *Syzygium* species are eaten; some of them are cultivated locally to a small extent (203, 259, 376, 1039, 1383) |

Table 30. Further fruit species, cont'd (explanation of symbols, see footnotes page 55)

| Botanical name | Vernacular names | Climatic regions | Distribution | Econ. value | Remarks and literature references |
|---|---|---|---|---|---|
| **Oxalidaceae** | | | | | |
| *Averrhoa bilimbi* L. | bilimbi, cucumber tree, groselha China | Tr | w | + | Generally cultivated in its homeland, Malaysia, elsewhere only occasionally in house gardens. Fruits very acid, mostly only eaten cooked with sugar, or pickled (203, 259, 376, 584, 1328, 1948) |
| *A. carambola* L. | carambola, starfruit | Tr | w | ++ | (Fig. 61) Originates in SE Asia, but to a small extent cultivated in all tropical regions. Fruits larger and less sour than bilimbi, good cultivars can be enjoyed fresh, often cut into slices ("starfruit") mixed into fruit salads, otherwise processed into drinks and jams (59, 259, 376, 584, 1297, 1328, 1823, 1948) |
| **Polygonaceae** | | | | | |
| *Rheum rhaponticum* L. | garden rhubarb, rhubarbe, ruibarbo | Te, S, TrH | w | ++ | Cultivated in many tropical and subtropical countries, and available in the local markets (731, 1444, 1657) |

| | | | | | |
|---|---|---|---|---|---|
| **Punicaceae** | | | | | |
| *Punica granatum* L. | pomegranate, grenadier, granada | S, Tr | w | +++ | (Fig. 62) Found wild from India to Iran. Very low demanding, drought and salt tolerant. Fruits keep and transport well. The edible part is the fleshy outer seed coat (sarcotesta). Eaten fresh, but mostly processed into cool drinks (sherbet, sorbet). Pericarp contains 30% tannin. Hedge plant, ornamental bush (376, 406, 1297, 1383, 1823) |
| **Rhamnaceae** | | | | | |
| *Hovenia dulcis* Thunb. | Japanese raisin tree, kemponaschi, sika | S, TrH | r: India to Japan | + | Planted especially in China and Japan as an ornamental and fruit bush. The edible part is the fleshy swollen fruitstalk, which contains 25-30% sugar, and tastes pear-like. It is also utilized medicinally because of its diuretic effect (200, 376) |
| *Ziziphus jujuba* Mill. | common jujube, Chinese date, jujubier, yuyuba | Te, S | w | ++ | Frost and heat tolerant. One of the most important fruit species in China. Candied fruit an important trading item in Asia. Silkworm fodder (see p. 402) (376, 406, 919, 1228, 1383, 1401) |

Table 30. Further fruit species, cont'd (explanation of symbols, see footnotes page 55)

| Botanical name | Vernacular names | Climatic regions | Distribution | Econ. value | Remarks and literature references |
|---|---|---|---|---|---|
| *Z. mauritania* Lam. | Indian jujube, ber, bor | S, Tr | w | ++ | Especially important in India. Very drought tolerant. Fruits candied or dried. Apart from *Z. jujuba* and *Z. mauritania*, a number of other *Ziziphus* species provide edible fruit (e.g. *Z. lotus* (L.) Lam., *Z. spina-christi* (L.) Willd.); they are also cultivated in dry regions of N Africa and the Middle East (127, 273, 376, 584, 1228, 1306, 1348, 1383, 1774) |
| **Rosaceae** | | | | | |
| *Eriobotrya japonica* (Thunb.) Lindl. | loquat, Japanese medlar, bibacier, níspero del Japon | S, TrH | w | ++ | (Fig. 63) Widespread in all subtropical countries. Beautiful shade tree. Good cultivars have large, good-tasting fruit with small kernels, which are mostly eaten fresh. Locally important trading product (102, 1297, 1514) |

| | | | | | |
|---|---|---|---|---|---|
| *Fragaria* x *ananassa* Duch. | strawberry, fraiser ananas, fresa, morango | Te, S, Tr | w | +++ | Cultivars have been bred from the American "Everbearing" types which make cultivation possible in almost all climatic zones. In warm countries, strawberries are often grown for only 1 year. The propagation (runner formation) must sometimes be carried out in other climatic regions of a country than the cultivation. The yields can be very high (20 t/ha or more), the quality is mostly not as good as in the temperate zone (less sugar, weaker aroma). The commercial possibilities are limited due to the poor keeping quality (147, 357, 376, 481, 550, 552, 553) |
| *Rubus* hybrids (*R. loganobaccus* L.H. Bailey and others) | blackberry, boysenberry, loganberry, marionberry, youngberry | Te, S, TrH | w | ++ | The genus *Rubus* is cosmopolitan, and species which bear edible fruit occur in the tropical highlands and subtropics of all continents. Nowadays, mostly only the American hybrid cultivars are worth considering for cultivation, and these can have considerable importance commercially (juice, jams) (200, 481, 550, 1378, 1657) |

Table 30. Further fruit species, cont'd (explanation of symbols, see footnotes page 55)

| Botanical name | Vernacular names | Climatic regions | Distribution | Econ. value | Remarks and literature references |
|---|---|---|---|---|---|
| **Rutaceae** | | | | | |
| *Aegle marmelos* (L.) Corrêa | Bengal quince, bel, bael tree | S, Tr | r: S and SE Asia | ++ | Cultivated mostly in India, the homeland of the tree. Fruits eaten fresh, processed into drinks, or cut into slices and dried (259, 376, 406, 1383, 1547) |
| *Casimiroa edulis* Llave et Lex. | Mexican apple, white sapote, zapote blanco, | S, TrH | r: Mexico, Guatemala, California, West India | + | Important fruit in the highlands of the Andes and Mexico, also cultivated in California and Florida. Fruit pulp eaten fresh (550, 1378, 1629, 1960) |
| *Clausena lansium* (Lour.) Skeels | wampee, vampi, wong pei | Tr | l: S China | + | Little cultivated outside China. Fruits eaten raw or processed into jam (259, 376, 1383) |
| **Sapindaceae** | | | | | |
| *Blighia sapida* C. König | akee, seso vegetal | Tr | r: W Africa, West Indies | + | The fully-ripe arillus is the edible part. Mostly eaten cooked (e.g. with fish). In the unripe state it is poisonous due to the amino acid derivative hypoglycine A (reduces blood sugar) (262, 1306, 1574, 1874, 1948, 1960) |

| | | | | | |
|---|---|---|---|---|---|
| **Sapotaceae** | | | | | |
| *Chrysophyllum cainito* L. | star apple, caimitier, caimito | Tr | r: Central America | + | Also planted outside its home region, principally as an ornamental tree. Fruit tastes good, eaten fresh or processed to jam (203, 550, 584, 1039, 1960) |
| *Manilkara zapota* (L.) von Royen (*Achras zapota* L.) | sapodilla, chicle tree, zapote | Tr | w | ++ | Planted also outside its home region, C America, especially in SE Asia. Is considered the best of the Sapotacean fruits, mostly eaten fresh. Can be stored or transported for 5 weeks, if harvested before fully ripe. The latex of the tree supplies chicle (Table 45) (203, 273, 376, 406, 550, 584, 914, 1039, 1328, 1518, 1823, 1948, 1960) |
| *Pouteria sapota* (Jacq.) H.E.Moore et Stearn (*Calocarpum sapota* (Jacq.) Merr.) | sapote, mamee sapote, mamey, zapote | Tr | r: Central America | ++ | (Fig. 64, Table 26, see p. 169) Cultivated in a wide region. Fruits sweet, but rather tasteless, mostly used for preserves and jams. Seed powder admixed to cocoa. Apart from the species mentioned, many other Sapotaceans have edible fruits, but they are not of noteworthy commercial importance (200, 550, 1039, 1611, 1960) |

Table 30. Further fruit species, cont'd (explanation of symbols, see footnotes page 55)

| Botanical name | Vernacular names | Climatic regions | Distribution | Econ. value | Remarks and literature references |
|---|---|---|---|---|---|
| **Solanaceae** | | | | | |
| *Cyphomandra betacea* (Cav.) Sendtn. | tree tomato, palo de tomate, tomate de árbol | S, TrH | w | ++ | (Fig. 65) Well suited for preserves and jams, otherwise used similarly to tomatoes (199, 242, 550, 1039, 1297, 1378, 1440, 2007) |
| *Physalis peruviana* L. | Cape gooseberry, capulí, alkékénge de Pérou, tiparree | S, TrH | w | ++ | (Fig. 66, Table 26) Cultivated outside its original region (the Andes from Venezuela to Chile) in India, S and E Africa and Australia. Fruits rich in provitamin A. Taste good when fully ripe, sometimes eaten raw, but mostly processed to jams and preserves (242, 376, 550, 914, 1039, 1161, 1440, 2007) |
| *P. philadelphica* Lam. (*P. ixocarpa* Brot. ex Hornem.) | Mexican husk tomato, jamberry, tomatillo, miltomate | S, Tr | w | ++ | Cultivated mainly from Guatemala to S Texas, in other areas only on a small scale. Fruits mainly used for soups and sauces, in India also for jam (242, 376, 400, 550, 731, 1598) |
| *P. pruinosa* L. (*P. pubescens* auct. non L.) | strawberry tomato, hairy groundcherry | S, TrH | r: S USA, Antilles | + | Cultivation limited to its homelands. Usage as for *P. peruviana* (550, 1039, 1440, 2007) |

| | | | | | |
|---|---|---|---|---|---|
| *Solanum muricatum* L'Hérit. ex Ait. (*S. variegatum* Ruiz et Pav.) | melon pear, pepino, pera melón, cachum | S, TrH | w | + | Outside its homeland (Peru), also cultivated in the Mediterranean region, New Zealand, the USSR, and Ethiopia. After removal of the skin, eaten raw, or used for preserves (200, 242, 1039, 1378, 1440, 1650, 2007) |
| *S. quitoense* Lam. | naranjilla, morella de Quito, lulo | TrH | l: Ecuador, Colombia | + | Originating in the Andes of northern S America, nowadays also cultivated in other S American countries. Fruits are mostly used for the preparation of a good-tasting drink, otherwise also for preserves (242, 421, 442, 725, 1039, 1258, 1297, 1378, 1440, 1504, 1529) |
| *S. topiro* Humb. et Bonpl. (*S. hyporhodium* A. Br. et Bouché) | peach tomato, cocona, topiro, lulo jibara, cubiu | TrH | l: Colombia, Bolivia | + | Fruits edible raw, used especially for preparing a refreshing drink, also for jam and preserves (242, 550, 725, 1297, 1330, 1440, 1529) |
| **Tiliaceae** | | | | | |
| *Grewia asiatica* L. (*G. subinaequalis* DC.) | phalsa (India), falsa | S, Tr | r: India, Sri Lanka | + | Low demanding, drought tolerant. Frequently cultivated in India. Fruits sweet/sour, eaten raw or processed into refreshing drinks (376, 406, 1383) |

Table 30. Further fruit species, cont'd (explanation of symbols, see footnotes page 55)

| Botanical name | Vernacular names | Climatic regions | Distribution | Econ. value | Remarks and literature references |
|---|---|---|---|---|---|
| | | **Monocotyledonae** | | | |
| **Araceae** | | | | | |
| *Monstera deliciosa* Liebm. | delicious monstera, ceriman, harpón, piñanona | Tr | w | + | Homeland S Mexico. Found worldwide as an ornamental plant. Fully ripe fruit bodies are edible raw, mostly utilized for flavouring ice-cream, and for preserves (550, 1388) |
| **Palmae** | | | | | |
| *Salacca edulis* Reinw. (*Zalacca edulis* Reinw.) | lizard-skinned salak, roftan-zalak, salak (Java) | Tr | r: Malay Archipelago | + | The fruit flesh is soft and sweet, and is mostly eaten fresh. Many palms have good tasting fruit flesh (Table 17) Outside S and Central America, particularly dates (see p. 199) and *S. edulis* are cultivated as fruits (203, 259, 1296, 1297) |
| **Pandanaceae** | | | | | |
| *Pandanus tectorius* Soland. ex Parkins. (*P. odoratissimus* L.f., *P. odorifer* (Forsk.) O. Kuntze) | screw pine, pandang | Tr | r: Polynesia, S Asia, Australia | + | Numerous forms, sometimes described as separate species. Important food in the Micronesian Islands. Eaten fresh, or processed to preserves. Leaves for basketwork. Hedge plant (145, 259, 406, 1383, 1742) |

# Nuts

Nuts are fruits with dry shells (hazel nuts), parts of fruits (coconuts, walnuts), or seeds (Brazil nuts), which, due to their pleasant flavour, are eaten raw, roasted or cooked, and which find uses in confectionery as aromatic agents, and also in other foods (446, 1176, 1539, 1986).

Nuts have a high nutritional value (Table 31). The same generally applies to the energy content, and mostly also to the protein content. In primitive societies, they made a substantial contribution to nutrition (e.g. the use of seeds from araucarias and the Chile hazel (*Gevuina avellana* Mol., Proteaceae) by the aboriginal inhabitants of Chile).

Table 31. Nutritient content of some nuts (without shells) (1986)

| Species | % protein | % oil | % starch and sugar |
|---|---|---|---|
| Almond | 20 | 55 | 20 |
| Brazil nut | 15 | 66 | 9 |
| Cashew nut | 20 | 45 | 26 |
| Chestnut | 7 | 4 | 79 |
| Chickpea | 20 | 5 | 61 |
| Macadamia | 9 | 76 | 16 |
| Pistachio | 25 | 54 | 7 |
| Walnut | 15 | 65 | 13 |
| Watermelon seeds | 25 | 45 | 19 |

The world production of nuts amounts to at least 3.6 million tons (514). Fairly reliable information is available only for very few nuts (Table 32). Even for these, the figures are certainly incomplete, however an impression is given of the economic importance of these products. There is no means of covering the nuts which are collected from wild trees, or from cultivated forest trees (Gymnosperms: *Araucaria* spp. in America, *Pinus* spp. in Central and North America, Europe and Asia, *Torreya* spp. in East Asia (550, 958); Angiosperms, *Buchanania latifolia* Roxb., chironji, in India, *Carya* spp. in North America, *Juglans* spp. in America and East Asia, *Lecythis* spp. in South America, etc.). Furthermore, great quantities of the fruits and seeds of many other cultivated species are nibbled as nuts, such as the kernels of *Citrullus lanatus* (Table 30) and *Cucurbita pepo* (Table 25) in Asia, peanuts (see p. 96) and chickpeas (Table 25) everywhere in the world, sunflower seeds in the Balkans and Russia, and the fruits of *Nelumbo nucifera* in Southeast Asia (Table 25) (731, 1424, 1856).

Ecologically, the adaptation of the various nut species ranges from pronouncedly arid regions (pistachios) to the humid equatorial zone (Brazil nuts). Almonds and pistachios are of great importance for several countries in the winter rainfall regions, particularly because nuts are highly suitable for export, and fewer quality problems are found than with juicy, easily spoiled products. However in recent years, thousands of tons of Brazil nuts and hazelnuts have been rejected by the importing countries because unacceptable levels of aflatoxin were present in the nuts. Aflatoxin formation can be prevented by proper drying and storage of the harvested nuts.

Table 32. Production of the most important nuts (world, and main countries of cultivation) (514)

| Species | World production (1,000 tons) | Country | Amount produced (1,000 tons) |
|---|---|---|---|
| Almonds | 1170 | USA | 439 |
| | | Spain | 146 |
| Walnuts (several species) | 843 | USA | 181 |
| | | China | 153 |
| | | Turkey | 110 |
| | | USSR | 85 |
| | | Italy | 20 |
| Hazelnuts (several species) | 548 | Turkey | 362 |
| | | Italy | 120 |
| | | Spain | 18 |
| Cashew nuts | 476 | India | 163 |
| | | Brazil | 143 |
| | | Mozambique | 30 |
| | | Tanzania | 25 |
| Chestnuts (several species) | 456 | China | 95 |
| | | Turkey | 92 |
| | | Italy | 50 |
| | | Japan | 48 |
| Pistachios | 149 | Iran | 70 |
| | | USA | 43 |
| | | Turkey | 30 |

From nuts of inferior quality, the various nut oils are derived, which are used as aromatics in the pharmaceutical and cosmetics industries (758).

The cultivated nut species are compiled in Table 33, provided they are not included in other chapters. Diseases of nuts are dealt with in (358). For two important nut species, cashew and almond, and for the newcomer, macadamia, rather more full descriptions are given below.

## Cashew Nut

fr. acajou, anacarde; sp. marañon, cajú, acajú; ger. Kaschunuß

The cashew nut (Fig. 67) (44, 805, 892, 959, 1211, 1245, 1302, 1319, 1404, 1856, 1948, 1966) is indigenous to South and Central America (Brazil to Mexico), where the cashew apple (the pear-shaped swollen fruit stem) was especially utilized as a fruit. The tree was quickly spread to all tropical countries by the Spanish and Portuguese, for use as a provider of fruit. After the development of an industrial roasting procedure for the separation of the shell oil, the nuts became the main product. The separation of the shells and seed coats from the kernels was carried out by hand until a few years ago (there was a monopoly of this in India up to about 1967). Nowadays, there are fully mechanized plants, which make the processing possible in all countries (457, 458).

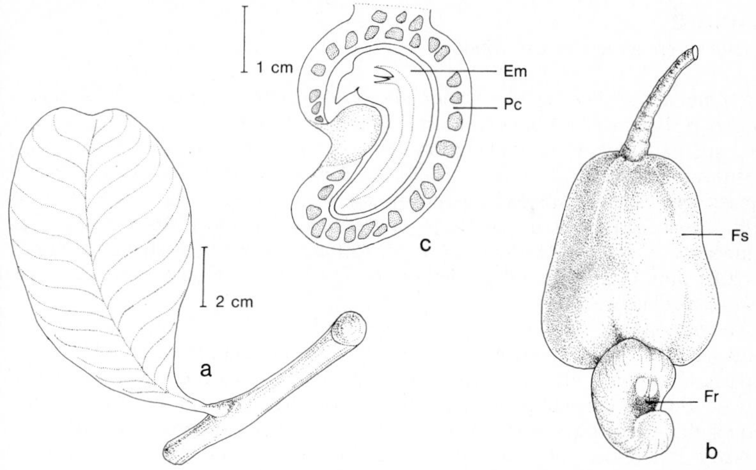


Fig. 67. *Anacardium occidentale.* (a) leaf, (b) fruit (Fr) with swollen fruit stalk (cashew apple, Fs), (c) fruit in longitudinal section. Pc = pericarp with oil-containing cavities, Em = embryo

The cashew tree grows best in tropical summer rainfall regions with 500-3500 mm rainfall (467). It is very drought tolerant, and on dry sloping sites is often the only cultivated plant which produces a cash income, even if the yields are then very low (8-10 kg unshelled nuts per tree per year; with good conditions, it can be up to 70 kg). On such sites, it is also important for reducing erosion, and as a supplier of firewood (625). With high rainfall (over 1500 mm), it only thrives on well-drained soil.

The propagation is mostly carried out using seeds, although these germinate irregularly, and sometimes only at a rate of 20% (small fruits with a high specific gravity germinate better than large ones). For the propagation of good cultivars, vegetative methods are possible (cuttings, grafting, or budding) (349, 584). The kernels contribute 35-47% of the fruit weight, and the shells are 53-65%; the shells contain 15-30% oil. Commercially, 1 ton of nuts provides 200 kg of kernels and 180 kg of oil (cashew nut shell liquid - CNSL; ger. Kaschuschalenöl; fr. baume cajou). CNSL is obtained by a roasting process. It consists of anacardic acid, cardanol and cardol, and is primarily processed to make synthetic resins, which are used especially for brake linings, clutch plates, heat and corrosion resistant paint materials, and in building panels (326, 1239).

The world production of CNSL amounts to about 40,000 t. The shelled kernels are dried in the sun or in hot air, and after peeling, they are best vacuum packed in cans. The cashew apples, which are separated from the fruit immediately after picking, are sold as fresh fruit, or are processed to make drinks (juice, wine), conserves or vinegar. In India, a gum is obtained from the trunk, which can be used as a substitute for gum arabic (see p. 373) (376, 760).

## Almond

fr. amand; sp. almenda; ger. Mandel

The almond was, besides the walnut, always the most important nut, especially for export. It grows best in winter rainfall regions, where it can survive even with very small amounts of rainfall. With intensive cultivation, irrigation is profitable (California) (388, 1539, 1554, 1686). It flowers earlier than other *Prunus* species, and is therefore endangered by late frosts (1593). The many cultivars differ from each other in the shape of the seed, and the texture of the shell (endocarp), which can be thick and hard, soft and porous (soft-shelled almond), or paper thin. Propagation is carried out by budding on almond, peach, or other *Prunus* species (see p. 203).

All cultivars are self-sterile, and in any location, at least 2, preferably 3 or 4, cultivars are planted together for reciprocal pollination. For cultivation, nowadays almost only sweet almonds (var. *dulcis*) are used, because bitter-almond oil is extracted more cheaply and in greater quantities from apricot kernels than from bitter almonds (var. *amara* (DC.) Buchheim). In the cosmetic industry, it is replaced by benzaldehyde. A large part of the almonds are used in confectionery (marzipan, etc.). Almonds which are unsuitable for sale as nuts are used for pressing out the fatty oil, which finds applications in the pharmaceutical and food industries.

## Macadamia

Macadamia (Fig. 68) won its place in the world market only some twenty years ago, after its spectacular economic success in Hawaii (604, 952, 983, 1211, 1539, 1645, 1948, 1986). In a variety of regions (South and East Africa, Australia), large plantations have been established. Successful cultivation is certainly possible in many other countries, probably in all places where arabica coffee or guava thrive.

The botanical separation of species is difficult (983). *Macadamia ternifolia* F.von Muell. is the wild-growing "gympie nut". Because both of the cultivated forms can be crossed with each other, and new cultivars can contain genetic material from both of them, the botanical differentiation is of little practical importance. *M. integrifolia* is preferred in the tropics, *M. tetraphylla* withstands mild frosts (but only when the tree is fully grown), and is less demanding.

For new plantings, only the cultivars which have been bred in Hawaii should be considered. They are propagated by grafting on seedling rootstocks (689). For harvesting, the nuts are allowed to fall, and then collected from the ground. Fully-grown trees (from about 15 years onward) supply about 50 kg unshelled nuts per year. The processing should be undertaken in a central factory (650,

966). The commercial product is the nuts, which are shelled, fried in coconut oil or roasted in an oven, and usually salted. They keep for up to 2 years when vacuum packed.

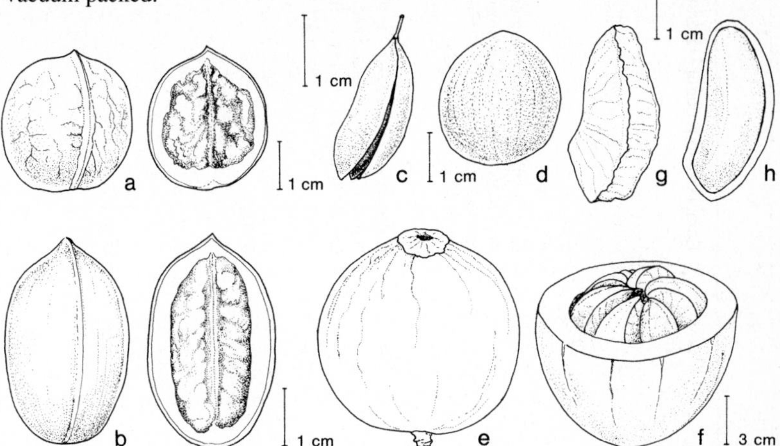


Fig. 68. Nuts. (a) *Juglans regia*, in shell and opened, (b) *Carya illinoensis*, in shell and opened, (c) *Pistacia vera*, fruit with splitting shell, (d) *Macadamia integrifolia*, seed, (e-h) *Bertholletia excelsa*, (e) whole fruit, (f) fruit wall cut across, (g) seed, (f) seed in longitudinal section

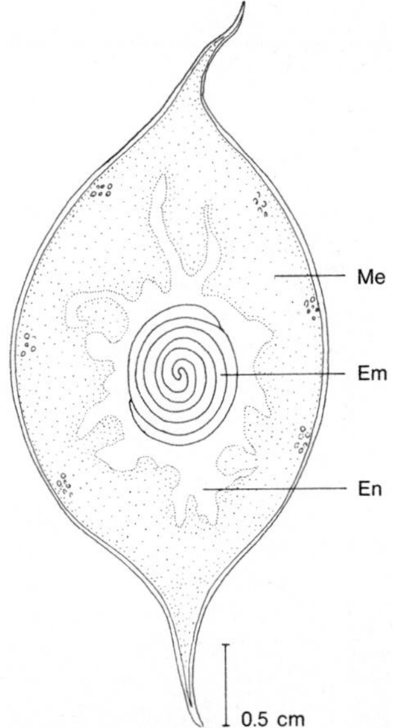


Fig. 69. *Terminalia catappa*. Fruit in cross section. Em = embryo, En = endocarp, Me = mesocarp (air-containing to allow floating)

Table 33. Nuts (explanation of symbols, see footnotes page 55)

| Botanical name | Vernacular names | Climatic regions | Distribution | Econ. value | Remarks and literature references |
|---|---|---|---|---|---|
| | | **Gymnospermae** | | | |
| **Ginkgoaceae** | | | | | |
| *Ginkgo biloba* L. | ginkgo, maidenhair tree, noyer du Japon | Te, S | w | + | Chiefly planted as an ornamental tree. Starch containing seeds eaten roasted, traded in E Asia (376, 1176, 1383, 1424, 1842) |
| **Gnetaceae** | | | | | |
| *Gnetum gnemon* L. | joint fir, buko, malinjo, genemo, bagoe, bulso | Tr | r: SE Asia | + | Cultivated in this region for the starch containing seeds. Seeds eaten roasted. Young leaves and unripe seeds cooked as vegetables. Fibres of inner bark for nets and ropes (259, 376, 1296, 1317, 1383, 1856) |
| | | **Dicotyledonae** | | | |
| **Anacardiaceae** | | | | | |
| *Anacardium occidentale* L. | cashew nut, anacardier, merei, cajú | Tr | w | +++ | (Fig. 67, see p. 236) |

| | | | | | |
|---|---|---|---|---|---|
| *Pistacia vera* L. | pistachio nut, pistachier cultivé | S | w | +++ | (Fig. 68c) Winter rainfall regions. Drought tolerant, mostly budded or grafted on *P. terebinthus* L. Dioecious (378, 388, 858, 961, 1103, 1321, 1539, 1986) |
| **Barringtoniaceae** | | | | | |
| *Barringtonia procera* (Miers) R.Knuth | nua | Tr | l: Solomon Islands | + | Cultivated because of the oil-rich seeds. Also hedge plant (145, 1365, 2015) |
| **Burseraceae** | | | | | |
| *Canarium indicum* L. (*C. commune* L.) | Java almond, kanari nut, | Tr | r: SE Asia | + | Planted as a shade and street tree. Seeds aromatic, utilized for baking, sweets and drinks (259, 273, 371, 815, 1211) |
| *C. ovatum* Engl. | pili nut, nuéz pili | Tr | l: Philippines | + | Regularly exported from the Philippines. Pili nuts of trade are also obtained from other species (e.g. *C. luzonicum* Mig.). Cultivation has begun in Central America. Other *Canarium* species cultivated locally (259, 372, 815, 1211, 1856) |
| **Caryocaraceae** | | | | | |
| *Caryocar nuciferum* L. | souari nut, butter nut | Tr | r: Northern S America | + | The nuts traded are mostly gathered from wild trees. Some cultivation on W Indian islands (337, 1176, 1425, 1854) |

Table 33. Nuts cont'd (explanation of symbols, see footnotes page 55)

| Botanical name | Vernacular names | Climatic regions | Distribution | Econ. value | Remarks and literature references |
|---|---|---|---|---|---|
| **Combretaceae** | | | | | |
| *Terminalia catappa* L. | Indian almond, badam, kottamba | Tr | w | + | (Fig. 69) Originated in S Asia and Pacific. Shadow tree with slight demands on the soil. Nuts only used locally (259, 376, 406, 815, 1218, 1383, 1774, 1856, 2015) |
| **Corylaceae** | | | | | |
| *Corylus avellana* L. | hazel nut, cobnut, filbert, noisettier | Te, S | w | + | Bush type, widely distributed, but of little economic importance, self-sterile (388, 1539, 1986) |
| *C. colurna* L. | Turkish hazel, Turkish filbert, noisetier de Byzanze | S | l: Turkey | +++ | Large tree. Black Sea coasts of Turkey. Most important species in world trade, self-sterile (1539, 1986) |
| *C. maxima* Mill. | giant filbert | Te, S | r: SE Europe | ++ | Bush type, calyx encloses the nut. Main producers Italy and Spain, self-sterile (1539, 1986) |
| **Cucurbitaceae** | | | | | |
| *Acanthosicyos horridus* Welw. ex Benth. et Hook.f. | naras | S | l: Namibia | + | Most important foodstuff of the Hottentots. Seeds exported, used like almonds (1172, 1905) |

| | | | | | |
|---|---|---|---|---|---|
| *Telfairia occidentalis* Hook.f. | fluted gourd, iroko, oroko, krobonko | Tr | l: W Africa | + | Cultivated from Nigeria to Sierra Leone for seeds and oil production. Young leaves and shoots eaten as vegetables (200, 262, 397, 480, 731, 1307, 2031) |
| *T. pedata* (Sm. ex Sims) Hook. | oyster nut, Zanzibar oil vine | Tr | r: E Africa | + | After removal of the bitter shells, kernels eaten and used for baking. Also for oil production (200, 731, 1905) |
| **Fagaceae** | | | | | |
| *Castanea crenata* Sieb. et Zucc. | Japanese chestnut, châtaignier du Japon | S | l: Japan | ++ | Of little importance outside Japan. Nuts eaten cooked (1986) |
| *C. mollissima* Bl. | Chinese chestnut, châtaignier de Chine | Te, S | w | ++ | Great climatic adaptability. Many cultivars. Can be crossed with *C. sativa* (1168, 1366, 1986) |
| *C. sativa* Mill. (*C. vesca* Gaertn.) | European chestnut, châtaignier | Te, S | w | ++ | Originated from Mediterranean region. Considerable exports. Wood supplies tanning material (Table 47) (1539, 1986) |
| *Castanopsis sumatrana* (Oerst.) A. DC. (*C. inermis* Benth. et Hook.f.) | berangan | Tr | r: Malaysia, Kalimantan | ++ | Cooked seeds much eaten in regions of origin and cultivation (259) |
| **Juglandaceae** | | | | | |
| *Carya illinoensis* (Wagenh.) K.Koch | pecan nut | S | w | +++ | (Fig. 68b) Cultivated particularly in the southern USA, S Africa and Australia. Many cultivars (356, 721, 1539, 1554, 1856, 1986) |

Table 33. Nuts cont'd (explanation of symbols, see footnotes page 55)

| Botanical name | Vernacular names | Climatic regions | Distribution | Econ. value | Remarks and literature references |
|---|---|---|---|---|---|
| *Juglans ailantifolia* Carr.(*J. sieboldiana* Maxim.) | Japanese walnut, heartnut, shan-hu-tao | Te, S | r: E Asia | + | Hardly cultivated outside its home region (1986) |
| *J. neotropica* Diels (*J. honorei* Dode) | Ecuador walnut, tocte, nogal | TrH | l: Ecuador | + | Up to now, as with other S and C American species, of little economic importance (200, 1846) |
| *J. nigra* L. | black walnut | Te, S | l: USA | + | Of economic importance only in its homeland (1986) |
| *J. regia* L. | Persian walnut, noyer commun, nogal común | Te, S | w | +++ | (Fig. 68a) Widespread in subtropical winter rainfall regions. In subtropical summer rainfall regions, pecans are mostly superior (306, 388, 406, 1383, 1539, 1554, 1856, 1986) |
| **Lecythidaceae** | | | | | |
| *Bertholletia excelsa* Humb. et Bonpl. | Brazil nut, noix du Brésil, nuez del Brasil, castanha do Pará | Tr | l: Brazil, Bolivia, Peru | +++ | (Fig 68e-h) Important export product of the State of Amazonas (annual exports about 50,000 t). Only a small portion originates from cultivated trees (200, 550, 1211, 1425, 1539, 1856, 1948, 1986) |

| | | | | | | |
|---|---|---|---|---|---|---|
| **Olacaceae** | | | | | | |
| *Anacolosa luzoniensis* Merr. | galo | Tr | l: | Philippines, Malaysia | + | Shrub with good-tasting nuts. To a small extent cultivated (259, 1846) |
| **Proteaceae** | | | | | | |
| *Macadamia integrifolia* Maiden et Betche | macadamia, smooth-shell Queensland nut | S, Tr | w | | +++ | (Fig. 68d) (see p. 238) |
| *M. tetraphylla* L. Johns | macadamia, rough-shell Queensland nut | S, TrH | w | | +++ | (see p. 238) |
| **Rosaceae** | | | | | | |
| *Prunus dulcis* (Mill.) D.A.Webb (*P. amygdalus* Batsch) | sweet almond, amandier, almendro | S | w | | +++ | (see p. 238) |
| **Trapaceae** | | | | | | |
| *Trapa natans* L. var. *bicornis* (Osbeck) Makino (*T. bicornis* Osbeck) | water chestnut, water caltrops, horn nut, ling | Tr | r: | SE Asia | ++ | Fruits large and black. Much eaten in their areas of cultivation. Main cultivation regions in S China to Thailand, some also in Java. Only edible when cooked (200, 731, 1383, 1411) |
| var. *bispinosa* (Roxb.) Makino (*T. bispinosa* Roxb.) | singhara nut, hishi | S, Tr | r: | India, Sri Lanka | ++ | Fruits small and brown. In Kashmir, general food of the people. Eaten raw, cooked, roasted, or used for flour production (200, 263, 376, 731, 1162) |

# Beverages and Stimulants

Since the earliest history of mankind, there has been a knowledge that some plants contain chemical compounds that increase the physical and mental effectiveness, quench thirst and hunger, break down psychic inhibitions, or produce fantastic dreams (200, 243, 1608). There are few people who do not have the daily custom of using one of these stimulants in the form of drinks, chewing, sniffing or smoking materials. One can even be tempted to see the usage of these materials as a differentiation between man and the animals. Some of the plants which provide the sought-for stimulant or sedative effects were brought under cultivation at an early stage of human development (e.g. tobacco, betel nut), others only in historical times (e.g. coffee), and many are nowadays still collected from wild plants (e.g. *Psilocybe*, *Lophophora*). Man discovered very early that the enjoyment from sugar-containing drinks (fruit, agave, palms) could be much increased by alcoholic fermentation. He has also learned to increase the effect of the active principles by technical procedures (the addition of lime to betel and coca), or to improve the taste (drying and fermentation of tobacco, roasting of coffee). A consequence of this second development is that some people use the stimulants only because of their aroma (caffeine-free coffee, coffee substitutes, nicotine-free tobacco). Such plants are included in Table 38, in as far as they are not already included in other chapters (e.g. ginger in spices, herb teas in medicinal plants, fruit juices in fruits). The reverse also applies, in that the active constituents of some stimulants are used in other areas of activity: caffeine, theobromine, morphine, codeine, cocaine and arecoline as medicines, nicotine as an insecticide.

All of the plants named here originate from the tropics and subtropics. With the exception of tobacco, the tropics have retained a total monopoly on these plants. This is of great economic importance for the developing countries, and in fact the export value of these products (about US $ 20 billion) is greater than that of the products of any other plant group. Exports are promoted not only by the very constant demand, but also by good storage properties and by low transport costs relative to the value of the products. Quantitatively (Fig. 70), tobacco and coffee are at the top; coffee leads in terms of monetary value, which is approximately equivalent to the total of tobacco, tea and cocoa (518). The use of several stimulants and beverages remains limited to a locality or region. They play little or no role in world trade, but they can be of considerable economic importance in their homelands. The following examples can be especially given: maté (South America, annual production about 200,000 tons); betelnut (South and Southeast Asia, annual production in India about 600,000 t); kola nut (West Africa and West Indian islands, annually about 175,000 t); coca leaves (South America, annual usage in Peru amounts to about 100 tons cocaine); guaraná (Brazil), opium (Asia) and kat (Yemen and Ethiopia); for the last three, no figures for their consumption are available. For the economy as a whole, illegal trading, especially in opium, cocaine and hashish, is significant. For national economies, the taxes on such luxury goods are an important factor. In this respect, the taxes on tobacco are by far the highest, and the value of the taxation revenue is several times greater than tobacco's basic value in many countries. On a percentage basis, tobacco is usually followed by alcoholic drinks, which often take first place for tax yields ahead of the other luxury goods.

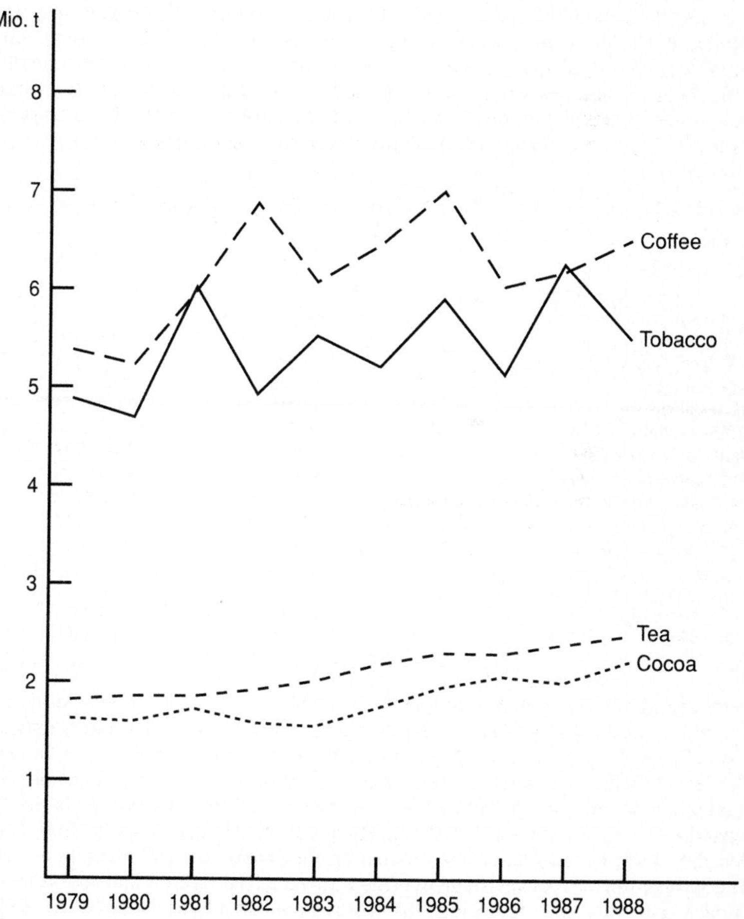


Fig. 70. World production of the most important beverages and stimulants, 1979-1988

# Coffee

fr. café; sp. café; ger. Kaffee

The genus *Coffea* L., Rubiaceae, is found throughout the tropics of the Old World, with about 60 species. All of the cultivated species originate in Africa. The most important are *C. arabica* L., "arabica coffee", which provides 74%, and *C. canephora* Pierre ex Froehner, "robusta coffee", which provides 25% of the world's production. The remaining 1% is supplied by *C. liberica* Bull ex Hiern., "liberica coffee", and *C. stenophylla* G.Don., "highland coffee", and others (335, 339, 375, 1657, 1994). The most important differences between arabica and robusta are set out in Table 34; what is quoted here should only serve as a general guide. The various cultivars of *C. arabica* show considerable differences in their sensitivity to the diseases mentioned. The figures for yields are for well-kept plantations with fully grown bushes. The world average is about 574 kg/ha, and thus is substantially lower than the figures quoted.

Table 34. The most important differences between *Coffea arabica* and *C. canephora*

| Characteristic | *C. arabica* | *C. canephora* |
|---|---|---|
| chromosomes (2n) | 44 | 22 |
| fertilization | self-fertile | self-sterile |
| time from flowering to ripeness | 9 months | 10-11 months |
| flowering time | after rain | irregular |
| berries at ripening | fall off | remain on bush |
| caffeine content (%) | 0.6-1.5 | 2.0-2.7 |
| yield (kg beans/ha) | 1500-2500 | 2300-4000 |
| root system | deep | shallow |
| optimum temperature range (annual mean) | 17-23°C | 18-27°C |
| optimum rainfall | 1500-2000 mm | 2000-3000 mm |
| *Hemileia vastatrix* (rust) | sensitive | resistant |
| koleroga (web blight) | sensitive | tolerant - immune |
| tracheomycosis (*Gibberella xylarioides*) | resistant | sensitive |
| coffee berry disease | sensitive | resistant |
| nematodes | sensitive | resistant |

**Production**. The greatest amounts of arabica coffee are produced in the Americas, while the centre of robusta production lies in Africa and Southeast Asia (Table 35). The total world production of coffee has altered little in the last decade (Fig. 70). However, there are considerable rises and falls in the production figures for the individual countries in this period: Indonesia +58%, Uganda +53%, Kenya +51%, Costa Rica +49%, Mexico +30 %, and Côte d'Ivoire -32%; in Angola, coffee production for export virtually came to an end. The cultivation of rust resistant hybrids (arabusta, see below) and rust-tolerant arabica cultivars are replacing the cultivation of robusta coffee in some countries. The prices of arabica and robusta coffee have come near together, only the highest qualities of arabica (which are hardly ever processed into instant coffee powder) are priced substantially higher than robusta.

**Breeding**. For cross-breeding with *C. arabica*, the chromosome number of all the other species must be doubled; they are mainly used for the introduction of resistance genes (104, 535). After disease resistance, the following have played a role in breeding: adaption to wet soils (*C. congensis* Froehner), adaption to dryness (*C. stenophylla*), and lower caffeine content (*C. bengalensis* Heyne ex Willd., *C. eugenioides* S.Moore, and also Madagascan species) (104, 375). Crosses of *C. arabica* x *C. canephora* (tetraploid) are given the name "arabusta", and are cultivated in Indonesia and the Côte d'Ivoire; they have also been described as a botanical species, *C.* x *arabusta* Capot et Ake Assi (177, 270). By selecting from pure arabica populations, a number of cultivars have also been obtained with sufficient resistance to the various diseases (339).

Table 35. Production of arabica and robusta coffee in the most important producer countries in 1988 (in 1,000 tons) (514)

| Arabica coffee | | Mostly robusta coffee | |
|---|---|---|---|
| Brazil | 1321 | Indonesia | 358 |
| Colombia | 780 | Côte d'Ivoire | 187 |
| Mexico | 283 | Uganda | 184 |
| Ethiopia | 180 | Philippines | 141 |
| Guatemala | 162 | Cameroon | 138 |
| El Salvador | 152 | India | 125 |
| Costa Rica | 145 | | |
| Kenya | 125 | | |

**Morphology and Anatomy**. In its vegetative structure, the coffee bushes differ from other plants in that they form two types of axillary buds (1992). On upright-growing (orthotropic) main shoots, some millimetres above each leaf is formed the "primary" or "legitimate" bud, from which originate the horizontally-growing (plagiotropic), fruit-bearing side shoots. Between the legitimate buds and the base of the leaf stalk are found the "secondary" or "serial" buds, which usually remain dormant on undisturbed bushes. These however are induced to sprout by topping or bending-over the main shoot; they form only orthotropic shoots. Sprouting of the serial buds can also be induced by growth hormones. Cuttings from such orthotropic side-shoots can greatly accelerate the vegetative propagation of hybrids (1736). The bunches of flowers are formed in the axils of the leaves of the horizontal branches. By cutting off their tips, buds can also be induced to form vegetative shoots there. These side shoots are always plagiotropic (1087, 1444).

The fruits (berries) consist of the tough exocarp, the juicy mesocarp (pulp), and the endocarp which surrounds each seed (parchment skin). The seed coat is thin (silver skin), and the bulk of the seed consists of the folded endosperm, which encloses the small embryo (Fig. 71).

**Ecophysiology**. As shown in Table 34, *C. arabica* is adapted to cooler temperatures, and therefore grows best in tropical highlands and in the summer rainfall regions of the subtropics. High temperatures during the differentiation of flower buds induce the formation of many flowers (1918). Because of its deep root system, it is quite drought tolerant. 2-3 months of dryness in the period of bud dormancy are favourable for its cultivation, because then the first rain creates a simultaneous opening of the flowers (41, 1918), and the berries ripen within a short time span.

*C. canephora* needs higher temperatures, and because of its shallow root system, it needs a more evenly spread rainfall. But even in this case, 1-2 months of dry period encourages flowering. Coffee tolerates a lot of shade, however the highest yields are achieved in full sunlight. Without shade, a substantially higher level of fertilizer application is necessary, especially of N; shortages of individual elements are more rapidly noticeable than under shade trees (160, 1911, 1946). In highlands, shade has a favourable effect, because it moderates the temperature fluctuations between day and night. In regions with slight rainfall or long dry periods, coffee cannot be cultivated with shade trees, as these would use too much water. The lack of shade can in part be compensated for by a thick layer of mulch (more uniform and lower soil temperature) (7, 1444).

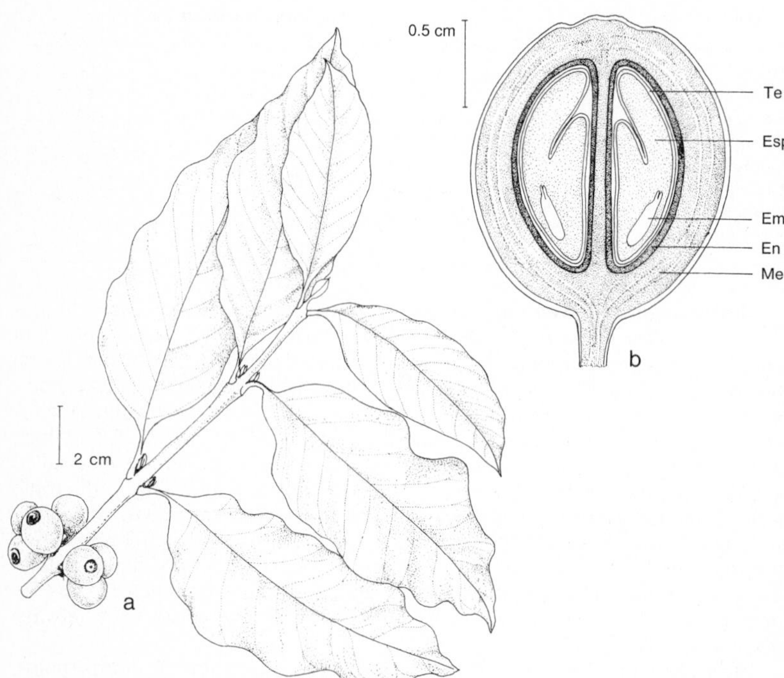


Fig. 71. *Coffea arabica.* (a) branch with fruits on last year's wood, (b) fruit in longitudinal section. Em = embryo, En = endocarp, Esp = endosperm, Me = mesocarp, Te = testa

Both species need deep, well-drained and aerated soil. The optimal pH value is 6-6.5. Coffee has high demands for fertilizer, especially for N and K. For 1 ton of coffee beans, corresponding to 5 tons of berries, the amounts removed from the soil are about 35 kg N, 2.6 kg P, and 42 kg K. The amounts of fertilizer recommended for high-yielding coffee are 150-200 kg N, 25-40 kg P, and 80-160 kg K per ha (44, 277, 339, 595, 921).

**Cultivation**. Coffee is mostly propagated from seeds. The seeds maintain their full germination capacity for only 2 months, and up to 4 months with storage in moistened charcoal powder. Coffee seed is stored in its parchment skin, and mostly also sown like this, although shelled coffee germinates rather more quickly. The germinaton is epigeal: 4(-8) weeks after sowing, the seeds are lifted above the soil surface by their hypocotyls, and after another 4 weeks the cotyledons unfold, which in the meantime have absorbed the remaining reserve materials from the endosperm. After the unfolding of the cotyledons, the seedlings are transplanted from the seed-bed into nursery beds, or better, into plastic bags. It is preferable to plant out the seedlings in the open after the development of 6 pairs of leaves - that is, before the appearance of the first side shoots.

*Coffea canephora* and hybrids are mostly or exclusively propagated vegetatively, usually using single-node cuttings of unripe wood. For this, only orthotropic shoots can be used. The wood is split lengthways so that two cuttings are obtained from each node. The shoot is cut closely above the node to remove the primary axillary bud, as this would give a plagiotropic shoot. With *C. canephora* and its hybrids, because of their self-sterility, two clones which are mutually fertile must be planted together. Grafting on seedlings is also possible. Because of the labour-costs involved, this is only done in places where a rootstock with special characteristics is needed (e.g. resistance against nematodes, tolerance of wet soil). The planting distance depends on the cultivar, pruning system and rainfall. It is between 4 x 3 m in low rainfall areas, and 2 x 1.5 m in humid regions. The root development in young plants is assisted if transpiration losses are limited by shading them with leaf-bearing twigs, or by planting them in aisles made between previously sown, high-growing cover crops (*Tephrosia*, *Crotalaria*, *Cajanus*, etc., see Table 53). For shade trees, legumes with wide-spreading crowns and light foliage are preferred (*Albizia* spp., *Gliricidia sepium* (Jacq.) Steud., *Inga* spp., etc., Table 55). In cultivation without shade, a layer of grass mulch 20 cm thick has proved to be useful (*Pennisetum purpureum* (Table 51), *Cymbopogon* spp., *Hyparrhenia* spp., etc.) (7, 277, 1444).

*C. arabica* was earlier trained with a single main stem. The pruning of the side-shoots required considerable labour and experience. Therefore, nowadays the multiple-stem system is preferred, by which several orthotropic stems are formed by the repeated capping of orthotropic shoots, or by bending the main stem over. Usually, four stems are allowed to grow, one of these being removed each year. At the same time, a new stem is allowed to grow to replace it. It is even more simple, if each year every fourth row is cut back to the base of the stem, from which new orthotropic shoots rapidly sprout. Using this method, only two of the four rows yield a harvest (Hawaii method). *C. canephora* is grown using the multiple-stem system (7, 44, 339, 375, 1087, 1444).

The most important measure for the control of weeds is to clean the land before planting, especially of rhizome-forming grasses (*Digitaria scalarum*

(Schweinf.) Ciov., *Cynodon* spp., etc.). The control of weeds can be carried out mechanically or chemically, and it is facilitated by ground-cover plants between the rows, mulching, and shade trees.

**Diseases and Pests**. Numerous fungus diseases of coffee have been described. Usually they cause significant damage only in unfavourable growing conditions (nutrient deficiencies, wet soil, dense growth, too much shade). Among the serious pests, coffee rust, *Hemileia vastatrix* Berk. et Br., is the worst. It can be so devastating in the humid tropics that the cultivation of *C. arabica* becomes impossible (Ceylon 1869, Indonesia 1876). In dry climates, the tolerance of some cultivars and spraying with Cu compounds are sufficient to keep the damage within tolerable limits. Other diseases, which require control measures in many places are:

- koleroga (web blight), caused by *Corticium koleroga* (Cke.) v.Hoehnel, controlled by the elimination of diseased plants, and spraying with Cu and Zn compounds;
- coffee berry disease, caused by *Colletotrichum coffeanum* Noack, especially dangerous in East Africa, controlled by regular spraying with Cu or other fungicides (540);
- South American leaf spot, *Mycena citricolor* (Berk. et Curt.) Sacc., the most serious disease in South and Central America (spraying with Cu or Zn compounds, as with koleroga);
- brown eye spot, *Cercospora coffeicola* Berk. et Cke., widespread especially in South America (controlled with Cu compounds);
- and, wilt disease (tracheomycosis), which in Africa is caused by *Gibberella xylarioides* (Stey.) Heim et Sacc.; it can only be controlled by the uprooting of infected trees, and disinfection of the soil (982).

Among animal pests (1043), the following should be mentioned: the coffee-berry borer, *Hypothenemus hampei* Ferr. (*Stephanoderes coffeae* Haged.), which originated in Africa but which is now widespread in South America and Indonesia, and further, the numerous leaf-mining moths (*Leucoptera* spp.), and the stem-borers (*Anthores leuconotus* Pasc., and other species). Sanitation and other hygienic measures can help against the berry borers, and also spraying or dusting with insecticides; it is particularly troublesome where coffee is dry-processed. The leaf-miners can nowadays be easily controlled with systemic insecticides.

Stem-borers cause less damage in multiple-stem growing than in single-stem systems; contact insecticides are used against them. Apart from the pests mentioned, considerable damage can also be caused to coffee by aphids, bugs, scale insects, mealybugs, leaf-cutting ants, and spider mites; the usual control measures are used against these insect groups (982).

**Harvest and Processing**. Only fully ripe berries provide top quality coffee. They are obtained by picking several times (even if the flowering was limited to a narrow time interval), or, with arabica coffee, by gathering the fallen berries from the ground, this being made easier by using outspread cloths or nets (269). An experienced worker picks 30-60 kg berries per day, and the cost of labour is thus quite high. The fallen berries decay only slowly. Gathering with a time interval of 6 weeks involves little loss of quality. This requires only a quarter of the labour costs of picking, the plants are not damaged, and bringing in the harvest can be accomplished in periods of dry weather.

For processing, two procedures, the dry and the wet, are usual. Using the dry process, the berries are spread out in a thin layer, and dried in the sun. That takes 15-25 days, but with damp weather, there is the danger that the berry flesh will rot, thus affecting the flavour of the beans. The quality is mostly lower than with the wet process ("harder" taste). The degree of ripeness of the berries is not so important in this procedure, so that the harvest can all be undertaken in a single operation (stripping off all the berries at once). Large surface areas are necessary for drying. In Brazil and Angola, the coffee is mostly prepared dry.

The wet processing was developed for humid climates, in which drying is difficult. Here, the berries are crushed in a pulper, in which the greater part of the fruit flesh is separated from the parchment skin. The mucilaginous remaining parts of the fruit flesh, which are still sticking to the skin, are removed in a short fermentation process (12-24 hours, in cool weather 2-4 days). They can also be removed by treatment with pectinase or alkali, or by mechanical procedures (special machines, passing through centrifugal pumps). After this, the coffee is washed, and dried in the sun or with hot air. The glass-hard dried parchment coffee (or with dry processing, the whole dried berries) is then mechanically hulled and polished to remove the shell and silver-skin. For sale, the beans are sorted according to size and colour. The green coffee can be stored for an almost unlimited time. For the development of aroma, the raw coffee is roasted at 200-250°C. The roasting must take place in the consumer country, because roasted coffee quickly loses its aroma, even with good packaging. Instant coffee powder is to a large extent produced in the country of cultivation (929). For this, the roasted coffee is ground, extracted with hot water, and the powder is formed in a spray-drying procedure (lower qualities), or by freeze-drying (top quality) (339, 1683).

The residues from coffee processing are utilized as fertilizer and mulch; on phytosanitary grounds, the berry-flesh should first be composted if possible. They can also be used as fuel and animal fodder (India). In Ethiopia, drinks are prepared from dried leaves and dried and roasted berries (1650). A tea is made from the leaves in Indonesia and Malaysia (259).

# Tea

fr. thé; sp. té; ger. Tee

Tea, *Camellia sinensis* (L.) O. Kuntze, Theaceae, originates in the mountains of Southeast Asia. The plant is predominantly cross-pollinated, and thus varies strongly in its characteristics. There is no point in giving botanical names to the various forms. In practice, it has proved useful only to name the two extreme varieties, var. *sinensis*, China tea (leaves small, hard, clearly serrated, without tips, strongly aromatic, tolerant of drought and cold, low yields), and var. *assamica* (Mast.) Kitam., Assam tea (leaves large - up to 35 cm long, soft, smooth edged, with tips, mildly aromatic, sensitive to drought and cold, high yields) (460). Because nowadays the tea grown is almost entirely hybrids of these two varieties, even this differentiation has lost importance.

**Production**. The cultivation is still concentrated in East and Southeast Asia: 73% of the world's production is grown there, 11% in Africa, 6% in the USSR, and 2% in America. The main producers are India (690,000 t), China (566,000 t), Sri Lanka (225,000 t), USSR (160,000 t), Indonesia (144,000 t),

Turkey (140,000 t), and Japan (96,000 t). Since the Second World War, the production in Africa rose from 16,000 to 277,000 t. Tea has become an important export product for Kenya and Malawi. The world market is dominated by India, Sri Lanka and China, which together provide 55% of the exports. The main consumers are the UK, the USA, Australia, Canada, and South Africa, which together take 28% of the world imports (1712); after the Anglo-Saxons, the Arabs are the greatest tea consumers.

**Morphology**. Without cutting-back, the tea plants develop into small trees (up to 10 m high). In cultivation, they are kept as bushes, and every few years they are cut back to a height of 1.5 m, so that picking is possible. The root system is strong; individual roots can penetrate several metres deep into the soil. The majority of the roots are found up to 1 m deep, the most important feeding roots lie near the soil surface.

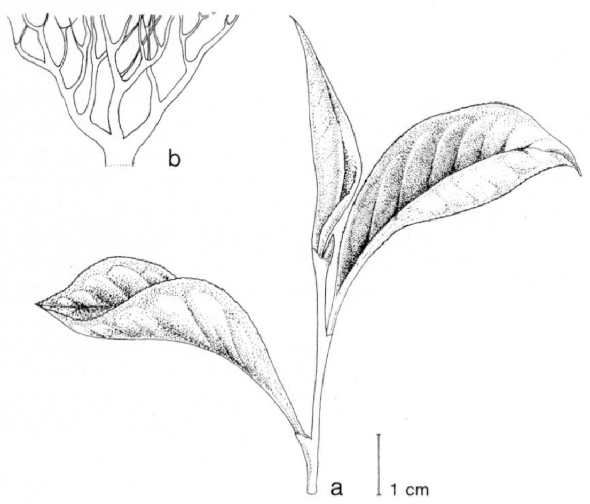


Fig. 72. *Camellia sinensis*. (a) tip of branch, (b) scheme of bush structure

**Breeding**. The breeders seek for climatically adapted, high-yielding cultivars. The greatest progress has been achieved in a few years with vegetatively propagated clones (535, 702). With these, the improved capacity of the cuttings to take root is a special criterion for selection. The quality of the leaves and disease resistance also play a part in the selection of clones. Because of the narrowing of the genetic base of resistance in vegetative propagation, suitable clone combinations (polyclone cultivars) are also sought, with which selected mother trees supply seed which combines a wide spectrum of resistance with high yield; however, the breeding of such cultivars is a very time-consuming project (535).

**Ecophysiology**. Tea grows in a wide range of climates. It finds its best conditions in a uniformly mild temperature (18-20°C), with equally distributed rainfall of 1500-2000 mm, and sufficient sunshine (1087). The production drops where the temperature falls below 13°C or rises above 30°C, when dry periods of several months occur, if the sky is clouded too long or too densely, or where the days in winter are too short (148). The var. *sinensis* withstands frosts of -5°C, whereas the leaves of var. *assamica* are killed even by the mildest frost. The soil should be deep, well-drained, low in Ca, and it should be acidic (the optimal pH value is 4.5-5.5). The nutrient supplies needed by high-yielding clones are very large. One ton of tea removes from the soil 45-60 kg N, 4-7 kg P, and 20-30 kg K. Nitrogen is the most important nutrient for increasing the yield. Even extremely high supplies of N (800 kg/ha/year) do not have a detrimental effect on the quality (44, 595).

**Cultivation**. The seeds are mostly pre-germinated under damp sackcloth, then planted in seed-beds or in plastic bags. When the seedlings are grown in plastic bags, they can be planted in the fields after 6-8 months (about 20 cm high), when they are grown in seed-beds, they are allowed to grow for 2-3 years, then cut back to about 10 cm high before transplanting ("stumps"). With hedge cultivation, the pre-germinated seeds can also be directly sown into the soil. Nowadays, mostly vegetatively propagated tea is planted. For this, single node cuttings are used, with one leaf, and a well developed axillary bud, and these are cut from young shoots (only the three upper internodes are rejected). They are cut directly above the eye, and the stem below the leaf should be 3-4 cm long. Mostly, they are rooted in plastic bags, and planted out when about 20 cm high.

The planting distance depends on climate, soil fertility, and growth form. For individual bushes, distances of 90 x 150 cm are normal, in rainy, cool regions, they can be planted closer (75 x 120 cm). For hedgerows, the distance between the rows is 120-150 cm, and within the rows 20-30 cm. After planting out, light shade is recommended, or planting between rows of a high-growing cover crop (see p. 251). Shade trees are only useful in very warm, sunny places (160, 737, 1946). Well developed tea plants form a closed canopy of leaves after 2-4 years, so that the soil is fully shaded. Until then, mulch is recommended to cover the soil (1739). The young tea bushes are cut back quite early, in order to achieve the broad shape of bush which is desired (Fig. 72b); artificially bending the branches down ("pegging") accelerates this process (7). Cutting-back is repeated many times, in order to achieve many shoots, giving a dense cover of leaves. Picking begins when the tea bushes have the right shape, and are 50-60 cm high. From time to time, the tea plants must be skiffed, so that the bushes are brought back to a convenient height for picking, and also to stimulate the production of new shoots. In regions with a definite winter dormancy (e.g. southern Russia and Japan), the skiffing is done annually during the dormant period, and otherwise about every third year (depending on the strength of growth and the picking system). With each cycle of skiffing, about 5 cm in height must be allowed, so that the bushes always grow taller. After 15-20 years, a drastic pruning to about 40 cm is then necessary, or even much nearer to the ground ("collar pruning") (460). With mechanized harvesting (see p. 256), regular skiffing is also necessary (this can be mechanized too). With correct soil preparation before planting (removal of all perennial weeds), intensive work to control the weeds is only necessary in the first years (by mulching, hoeing, or herbicides such as simazin, dalapon, and paraquat). After this time, the tea generally keeps down the growth of weeds by its own shade. The fertilizer is applied in several instalments during the growing season.

**Diseases and Pests**. With favourable climatic and soil conditions, tea suffers relatively little from parasites (681, 1588). In several regions of Asia, the blister blight (*Exobasidium vexans* Mass.) causes serious losses. It is controlled by Cu compounds, Ni chloride, or organic fungicides; there are also resistant clones. The parasitic alga, *Cephaleuros parasiticus* Karst ("red rust"), may necessitate spraying with Cu compounds in very humid regions. The root fungi, *Armillaria mellea* (Vahl ex Fr.) Kum., *Ganoderma* spp., *Poria* spp., *Fomes* spp., and others (1482), attack the tea especially where wood and root remnants remain in the soil after uprooting bushes, or where the soil is too wet and badly aerated. Remains of infested bushes must be thoroughly removed, and afterwards the soil must be disinfected. Thrips, mites and bugs can make control measures, using a variety of insecticides, necessary. Nematodes (*Meloidogyne* spp.) are alarming in some regions, and can make re-planting necessary after thorough disinfection (e.g. using *Tripsacum fasciculatum*, Table 51); there are also clones which are resistant to them.

Table 36. Catechin and caffeine contents (in % of the dry weight) in the buds and youngest leaves of tea. After (701)

| Part of the plant | Catechin | Caffeine |
|---|---|---|
| Bud | 26.5 | 4.7 |
| 1st leaf | 25.9 | 4.2 |
| 2nd leaf | 20.7 | 3.5 |
| 3rd leaf | 17.1 | 2.9 |
| Upper stem | 11.1 | 2.5 |

**Harvest and Processing**. The young leaves have the highest catechin and caffeine contents (Table 36); also they are soft and can be easily rolled. The usual procedure is that the bud and the top two leaves are picked by hand (Fig. 72a). However, a considerable proportion of the tea is now mechanically harvested (Japan, USSR), as nowadays there is less demand for tea of quality, and there are new processing techniques (see p. 257). With manual picking, one worker can gather 30 kg of fresh leaves per day. With full mechanization, a worker can harvest 100 kg per hour (940). The distinctive picking techniques which were previously used to achieve the highest qualities are not normally used nowadays. In any case, the harvest should be carried out so that the greatest amount of tea of satisfactory quality is produced, while leaving as many leaves as are necessary to guarantee the strong re-growth of the bush. Intervals between picking depend on the growth rate of the tea (1337, 1781). In warm periods, they are picked every 7-10 days, and in cooler weather, every 2 or 3 weeks. Hedgerow tea in cool climates is harvested four times a year. The yields from Assam and hybrid teas in the tropics are between 1.5-2.5 tons of dry tea per ha per year (maximally over 5 t), and in regions with a dormancy period caused by winter or dry conditions, and for China tea, the yield is between 0.8-1.6 t/ha/year (maximally 4 t). The global average lies below 1 t/ha/year because the care and fertilizer supply are inadequate in many regions. About 4.5 kg fresh leaves produce 1 kg of processed tea.

For the production of black tea using the "orthodox" process (701), the fresh leaves (water content 75-80%) are withered in an air stream to give a water content of 58-64%. The rolling takes place in heavy machines in four (up to nine) successive passes, each of which lasts about 30 minutes. Between each pass, the freshly rolled leaves are sieved. By rolling, the cell structure is destroyed to such an extent that the plant's enzymes (polyphenoloxydases) come in contact with the catechins. In the following fermentation process, which should run at a temperature preferably not exceeding 25°C (higher temperatures give lower quality), and which lasts for 3-4 hours, the leaf turns copper-coloured, due to the oxidation of the catechins to theaflavin and thearubigin (201, 1940). Both of these compounds form complexes with caffeine and protein, and define the character of the tea - together with the aromatic compounds, which likewise fully develop during the fermentation process. The fermentation is ended by quick-drying in a drying oven, at 90-95°C. In this way the tea acquires its brown or black colour. After this, the tea is sorted using sieves. Modern continuous processes (701) eliminate the time-consuming rolling process. CTC machines are used ("crushing, tearing, curling"), and rotorvanes, alone or in combination, sometimes also still in combination with a simplified rolling process (690). Also when these machines are used, the leaves are first withered. The damage to the texture of the leaves is more drastic than with rolling, and the fermentation is completed after 1-2 hours; in spite of the shorter time needed, the leaves become darker than with the orthodox process.

With drying, a large part of the fragrant compounds are lost. In East Asia, the tea is often perfumed by flowers (jasmin, gardenia, fresh flowers are put in layers between the dried tea, and later sieved out). Also in Europe, these types of tea are coming more and more on the market.

With the orthodox process, in which the tea is sieved between each rolling and sieved after drying, a range of different types are produced. These are substantially different in their characteristics, because they consist of material of different ages: Orange Pekoe consists mostly of the first rolled tips and youngest leaves, Pekoe chiefly of older leaves, and Broken Pekoe of short pieces of the leaf lamina, without mid-ribs. With the modern processes, these classes are totally or partially absent. Tea which has been broken up too finely (fannings and dust) is mostly utilized in tea bags. Instant tea powder was developed later than for coffee. For its production, the tea does not need to be dried after fermentation. However, instant tea is usually extracted from the lower qualities of dried-tea production. In dissolving and freeze-drying, many of the aromatic compounds are lost. Therefore, aromatics are frequently added to the powder before it is offered for sale.

The wastes from tea processing (stems, leaf stalks, other coarse pieces, dust) can be used for the extraction of caffeine (nowadays, the majority of caffeine and all theophylline are chemically synthesized) (1628).

In East Asia, green tea is the preferred drink. Apart from this it is popular in Arab countries, and is also available in the West. For this, the phenoloxidases in the fresh leaves are first inactivated by steaming or in heating pans, so that the original composition of the catechins remains unchanged. After this follows repeated rolling and drying, often still done by hand. In this way, the colour becomes olive green, and the infusion has a golden colour. It tastes rather more astringent than black tea, but is more refreshing (free caffeine) and more beneficial for the health (tissue strengthening effects of catechins, "vitamin P").

There are intermediate forms - "red", "yellow", and Oolong tea (southern China and Taiwan), whose fermentation is stopped at an earlier stage; they are of little importance in the export trade.

## Cocoa

fr. cacao; sp. cacao; ger. Kakao

The largest quantity of cocoa comes from the variable species *Theobroma cacao* L., Sterculiaceae. In Central America, two other species are also cultivated for cocoa extraction, *T. bicolor* Humb. et Bonpl. ("pataste", fruits on the side branches, with reticulate surface), and *T. angustifolium* Moç. et Sessé ex DC. ("cacao de mono", "cacao de mico", fruits on the side branches, covered with rust-coloured down). In Brazil (Pará and Maranhão) and in the Amazon region of Colombia, *T. grandiflorum* (Spreng.) K. Schum., ("cupuassu", large fruits, ellipsoidal, hard-shelled, on young twigs) is cultivated because of its good-tasting pulp, which is used for making a sweet, aromatic drink, and for desserts; from its seeds, a fat is obtained, similar to cocoa butter (87, 228, 698, 1039, 1319, 1983). The various forms of *T. cacao* have been described as subspecies (and some authors even raised them to the rank of species). Because they all have the same number of chromosomes (2n = 20), and can be crossed, their botanical separation is inappropriate. In practical agriculture one differentiates Criollos (Central America to Colombia), Forasteros (upper Amazon, including the group which originates from the Guianas and the lower Amazon, the "Amelonados"), and the Trinitarios (considered to be hybrids of Criollo and Forastero, indigenous to the region at the mouth of the Orinoco, and in Trinidad). The most important differences between Criollo and Forastero are listed in Table 37. Large-scale cultivation is dominated by the Forasteros (80%), especially by the Amelonados, which are the form almost exclusively cultivated in West Africa.

Table 37. The most important differences between typical Criollo and Forastero cocoas.

| Characteristic | Criollo | Forastero |
|---|---|---|
| Fruit colour | yellow-red, spotted | green-yellow |
| Fruit form | elongated, pointed | oval |
| Fruit surface | uneven, warty, deeply furrowed | smooth, shallow furrows |
| Fruit husk | thin and soft | firm and tough |
| Seed size | large, round | small, flat |
| Number of seeds/fruit | 20-40 | 30-60 |
| Colour of cotyledons | cream-coloured to rose | purple |
| Aroma | strong | weak |
| Yield | low | high |

**Production**. In the years between 1970 and 1980, a considerable shift in the centres of cocoa production took place. In 1976, Ghana was the main producer, but now the Côte d'Ivoire is in the lead (680,000 tons), followed by Brazil

(347,000 t), whereas Ghana now stands a considerable distance behind in only third place (290,000 t). The production in Southeast Asia has gone up due to increasing cultivation in Malaysia, which today produces 220,000 t, thus surpassing Nigeria (140,000 t), formerly one of the leading producers.

**Morphology and Anatomy**. The cocoa tree grows to 8-10 m high, but in cultivation it is mostly kept smaller by cutting it back. Its tap-root reaches about 2 m deep into the soil. The feeding roots arise from the root collar, are 5-6 m long, and are found in the upper 15-20 cm of the soil. The stem grows orthotropically in the first 14-18 months. When a height of 1.20-1.50 m has been reached, it ceases to grow, and the vegetation point divides into five meristems, from which arise the plagiotropic fan branches, which have indeterminate growth (1926). After some time, a dormant bud sprouts on the vertical stem, underneath the branching point. This again forms an orthotropic shoot ("chupon"). Cessation of vertical growth and branch formation follow each other in cycles. Thus, in the course of the years, a number of tiers of plagiotropic branches are formed (637). The flowers develop in the leaf axils on 2-3 years old wood of the orthotropic and plagiotropic shoots. They are borne on extremely short-stalked inflorescences, which over many years form new flower buds over and over again (cauliflory) (1037). In constantly warm and humid tropical climates, the cocoa flowers all the year round. However, strong fruit-setting hinders the development of new flowers and further fruit, so that even under these conditions, the harvest is not evenly distributed throughout the year, but is concentrated in one or two periods. Cooler temperatures and dry periods interrupt the flower formation. In regions with such climatic variations, there is thus a decidedly seasonal yield (40, 535).

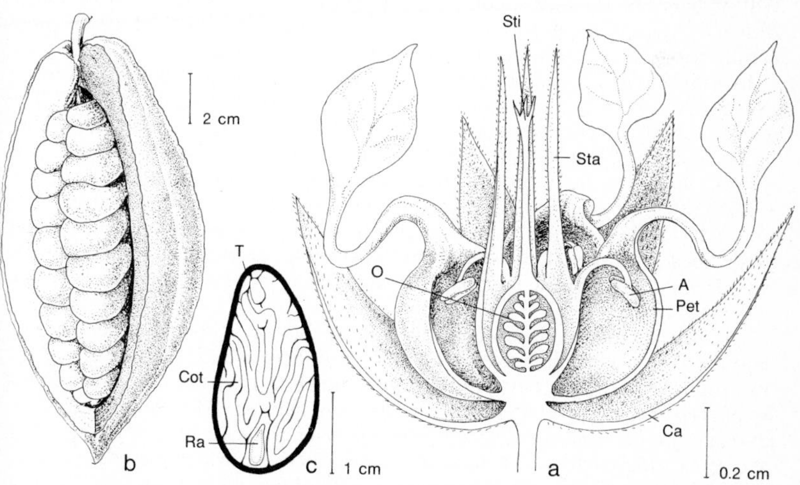


Fig. 73. *Theobroma cacao*. (a) flower in longitudinal section, (b) opened fruit, (c) seed in longitudinal section. A = anther, Pet = petal, Cot = cotyledon, O = ovary, Ca = calyx, Sti = stigma, Ra = radicle, T = testa, Sta = staminode

In the flowers (Fig. 73a), the anthers lie in the pouches of the petals, so that self-pollination is impossible. The most important pollinators are small flies of the genera *Forcipomyia* and *Lasioshelea* (Ceratopogonidae) (1711, 1969). Ants and thrips can also occasionally transport pollen; however because cocoa is often self-sterile, they are of little importance for pollination (479). In some areas of cultivation, the pollination by insects is insufficient; it can then be worthwhile to ensure full fruit-setting by mechanical pollination (1363, 1710). After fertilization, at first only the endosperm develops. The fertilized egg cells begin their development after 40-50 days. As long as the growth of the embryo has not yet begun, numerous young fruits turn yellow and wither, due to competition for nutrients with older fruit, and shortage of growth hormones ("cherelle wilt") (228, 698, 1622). The growing embryos consume the endosperm, so that the ripe seed consists mostly of the large, folded cotyledon (Fig. 73c). The seeds ("beans") are embedded in the sweet pulp (Fig. 73b). The fruits ripen 5-7 months after flowering, depending on the cultivar and climate.

**Breeding**. Breeding seeks especially to develop cultivars which are resistant against swollen shoot (West Africa), witches' broom (Trinidad), and Phytophthora pod rot (everywhere), and some other fungus diseases. Next to disease resistance, important goals of breeding are early bearing (2 years after sowing), regular high yields, and adaptation to local growing conditions. Most of the genes for resistances, strength of growth, and early bearing originate from the Forastero collections from the Upper Amazon region. They have the disadvantage of high self-sterility, which, for vegetative propagation, must be eliminated by back-crossing, or which necessitates the mixed planting of mutually fertile clones. On the other hand, the self-sterility is helpful in the production of hybrid seed from suitable partners. In the future, such hybrid seed will probably be the most important method of propagation (535, 1444, 1983). Vegetative propagation leads most quickly to the cultivation of newly bred combinations, however it can only be recommended for self-fertile clones.

**Ecophysiology**. Cocoa originates in the underwood of the tropical rain forest and grows best in humid tropical climates with temperatures of 25-28°C. The temperature in the coolest month should still lie over 20°C. Such conditions are found in the tropical lowlands up to 15°north and south of the equator. The rainfall should be as evenly distributed as possible throughout the year, and 1500-2000 mm is optimal. In dryer climates additional irrigation is necessary. Cocoa has little need for sunshine, and formerly was generally grown under shade trees. However, it also thrives in full sunlight, as soon as its leaf canopy is closed, assuming that the supplies of water and of nutrients are sufficient (44, 160 , 652, 1946).

The soil should be deep and well-drained, but of sufficient water holding capacity. The pH can lie between 4.0 and 7.5 if there is a rich supply of organic material (1983). This is best provided by a mulch layer. Also, cocoa should not be planted immediately after forest clearing by burning. After felling the trees, natural regrowth is allowed, or tall-growing cover crops are planted (see p. 251), or crop-bearing plants like banana or cassava, between which the cocoa can be planted later. In this way the biological activity of the soil is maintained, and the mycorrhiza of the cocoa can develop immediately. The amount of nutrients removed in the harvest is not very high if the fruit shells are brought back to the plantation for use as mulch. One ton of cocoa beans contains about 20 kg N, 4 kg P, and 10 kg K. In cultivation without shade the amount of nutrients supplied must be at least doubled compared to cultivation under shade; with this, the yields can be trebled.

**Cultivation**. Cocoa can be directly sown if the growing conditions are favourable. However the seedlings are mostly started off in plastic bags, and planted out after about 6 months. At first, they must be sheltered from insolation which is too strong, and from the wind (45). For vegetative propagation, 15-20 cm long tips of shoots from plagiotropic branches are preferred because they retain their direction of growth tilted upward. In this way the trees remain lower, and quickly form a closed leaf canopy. As a rooting medium, compost mixed with coir dust is recommended. The root formation will be accelerated by briefly dipping the cuttings in hormone solution (3 g napthylacetic acid plus 3 g indolbutyric acid dissolved in 100 ml of 50% alcohol) (47, 1241).

The planting distance depends on the cultivar and the soil fertility. The normal distances between the trees are 2.5 x 2.5 to 3.5 x 3.5 m. The care necessary is limited to keeping the soil clean around the trees. The free spaces are used at first for cultivation of crops between the rows, or are planted with ground cover plants. As soon as the leaf canopy is closed, no further efforts are usually necessary for the control of weeds, especially if a good layer of mulch is provided (1640). The pruning is limited to the removal of orthotropic shoots which are not needed, and also of side shoots which are growing too densely or too near to the ground.

In cultivation without shade, and with cultivars which set fruit very strongly (or if the fruit-setting is increased by hand pollination), mineral fertilizers are necessary in amounts of about 100 kg N, 20 kg P, and 70 kg K per ha per year. The fertilizer is given in instalments, at the beginning of the main foliage growth period, at the main flowering time, and at the time of the main growth of the fruit (1444).

**Diseases and Pests**. In all cocoa growing areas, the pod rot (black pod), caused by *Phytophthora palmivora* Butl. and other *Phytophthora* species, is a serious problem (647). The fungus also attacks the bark (bark canker), and the leaves and roots. To control it, the first measure is to thin out the stand, so that through a better flow of air, a quicker drying of the plants can be achieved (1528). Some cultivars are tolerant or resistant (Forasteros). Also Cu compounds can be utilized. The three other severe diseases are limited to particular regions. The witches' broom disease, caused by *Crinipellis perniciosa* (Stahel) Singer, occurs in the American tropics between 10°N and S of the equator. The most important control measures are the cutting out and burning of infected shoots, and especially the cultivation of resistant cultivars. The monilia pod rot, caused by *Moniliophthora roreri* Cif. et Par., is limited to the region from Equador to Nicaragua; so far, it can only be controlled by sanitary measures and spraying with fungicides. The swollen shoot disease, which is caused by viruses, is devastating in W Africa. The viruses are carried by mealybugs, whose spread can be prevented by controlling the ants which tend them. Since 1945 in Ghana, a drastic campaign has been carried out, in which much more than 100 million infected trees have been destroyed, and this has still not exterminated the disease. Modern breeding has created tolerant or almost resistant cultivars (535, 1444, 1809). There are numerous insect pests of cocoa (479): bugs (1019), mealybugs, scale insects, and thrips should especially be mentioned. Their control often makes the application of insecticides necessary (228, 479, 982, 1983).

**Harvest and Processing**. The correct degree of ripening of the fruit is recognized by its colour change, which is characteristic for each cultivar. At this stage, the pulp has separated from the husk. In fruit which is harvested too late, the seeds germinate, and are unusable for processsing. The fruits are cut off with sharp knives, so that any damage to the flower cushions is minimized. About 20 fruits provide 1 kg of dried beans. A good yield is considered to be 1-1.5 t of dried beans per ha and year, but under the best conditions, high-yielding cultivars can produce over 3 t. However, in 1988, the global average was only 399 kg.

The beans are cured after harvesting (288, 1076, 1536, 1983). The fruits are cut open, the beans are loaded into fermenting boxes together with the white, mucilaginous, sweet pulp, where a rapid development of yeasts, acetic acid and lactic acid bacteria immediately takes place. In this way the temperature rises to about 45°C. The alcohol which is formed first is mostly oxidized to acetic acid. In order to obtain an even temperature, and to allow an even distribution of $O_2$, the beans are transferred to other boxes twice. Under favourable conditions, the fermentation is complete after 3 days with Criollo types, and after 6 days with Forastero types. The acetic acid which enters into the seeds kills the tissue of the embryo. By this, phenoloxidases are set free, which oxidize the phenolic compounds (catechins and anthocyanins, particularly leucoanthocyanins) when $O_2$ is absorbed, giving the insoluble chocolate-brown colour. This process continues during the following drying process; in this, the cotyledons lose their astringent, bitter taste, and develop their aroma. With the normal drying process on frames, the beans are only exposed to the sun for a few hours each day; it takes abut 7 days. In regions with little sunshine and high air humidity, the cocoa is dried artificially. This process must also be carried out slowly and at moderate temperatues (60-70°C), so that the enzymatic oxidation is fully completed (597, 856). Typical defects of insufficient or faulty fermentation are cocoa with a slate grey or violet colour of the cotyledons, and an acid taste. The causes are usually a temperature which is too low (below 40°C), and the access of too little $O_2$. The acidity can be reduced by soaking and washing the beans between fermentation and drying. Improvement in the colour is possible through an auxiliary fermentation (re-wetting the dried cocoa, followed by slow drying), and by longer storage. In small operations in some countries, e.g. West Africa (87), cocoa is still fermented in heaps, baskets or barrels.

The dried cocoa beans are graded and exported in bags. In the processing, the beans are roasted for 10-45 minutes, at 90-140°C, depending on the cultivar, then broken, so that the seed coat, remains of the endosperm and radicle can be removed, then ground. The processing to chocolate follows directly. For the production of cocoa powder, most of the cocoa butter is extracted in hydraulic presses. Cocoa butter (Table 15, see p. 78) is an important by-product, the bulk of which is utilized in chocolate and chocolate-cream manufacture, and also in pharmaceutical and cosmetic preparations. Cocoa which is too poor for the production of cocoa powder or chocolate is processed to make cocoa butter. Also theobromine can be extracted, which is also extracted from the residues of normal cocoa processing.

# Tobacco

fr. tabac; sp. tabaco; ger. Tabak

The two economically most important species of tobacco are old American cultivated plants: *Nicotiana tabacum* L., and *N. rustica* L., Solanaceae. Both are amphidiploids, *N. tabacum* from *N. sylvestris* Spegazz. et Comes x *N. tomentosiformis* Goodspeed (636), whose distribution overlaps in northwestern Argentina and Bolivia, and *N. rustica* from *N. paniculata* L. x *N. undulata* Ruiz et Pav., from the west side of the Peruvian Andes (1657). The Indians utilize a range of other *Nicotiana* species for smoking or chewing, and to a small extent these are still cultivated today by them (20, 242, 742, 1444). Only *N. tabacum* and *N. rustica* have a high nicotine content; *N. tabacum* has an average of 3%, *N. rustica* 5%, in the dried leaves. In the other species mostly nornicotine is found. There is a large number of cultivars of *N. tabacum*. However, because drying procedures, soil and climate are more important for character and end-use than morphology, the botanical differentiation into cultivars is of subordinate importance.

**Production**. *N. tabacum* is cultivated in almost all countries of the world, from the equator up to 60°N and 45°S. However, the majority of the production takes place in the tropics and subtropics (over 80%). Here are also produced the best tobaccos from a quality point of view (Sumatra, Cuba). In terms of quantity, the major producers are China (2,353,000 t), the USA (604,000 t), Brazil (439,000 t), the USSR (340,000 t), India (320,000 t), and Turkey (212,000 t). In general, the cultivation of *N. rustica* is limited to Russia ("Machorka") and some countries of the Middle East (Turkey, Persia, water-pipe tobacco, "Tombak", "tonbaco", "Nargileh"). Of the exporting countries, the USA is in the lead (220,000 t), followed by Brazil (199,000 t), Greece (114,000 t), Zimbabwe (104,000 t), and Turkey (78,000 t).

**Breeding**. Breeding for resistance is more in the foreground than for any other plant, because the use of pesticides is limited (except in the seed beds), due to the problem of residues on the leaves. Apart from this, the breeders strive to maintain the typical character for a particular region (colour, texture, aroma, burning characteristics). For the introduction of resistance genes, wild species must frequently be utilized, e.g. *N. repanda* Willd. ex Lehm. for nematode resistance, *N. longiflora* Cav. for wildfire resistance, and *N. glutinosa* L. for mosaic resistance (20, 257, 622, 1734, 1850).

**Ecophysiology**. Tobacco is cultivated under such a variety of ecological conditions that general rules can only be given for some factors. Under no circumstances will it withstand waterlogging or high salt levels in the soil. The leaves are extremely sensitive to hail and wind, but high temperatures will be tolerated with sufficient supplies of water; temperatures below 20°C retard the growth. With regard to water requirements, the range extends from the low-demanding orient tobaccos (annual average about 400 mm) to the cigar tobaccos of the humid tropics (2000 mm and over). For high production, much sunshine is required, however the best cigar tobaccos are produced in the cloudy tropics, or under artificial shade (shade tents in the USA and Cuba). Sandy, or sandy-loam soils with a pH of 5-5.6 are generally the most favourable, but good tobacco can also be produced on vertisols. The need for nutrients shows an equally wide range; certainly, all cultivars need sufficient P and K; with them all, Cl impairs the burning quality. However, with N fertilization, orient tobaccos should receive only 5-15 kg N/ha, light Virginia tobaccos about 20 kg, yet large-leaved pipe and cigar tobaccos 100-150 kg N/ha (328, 595, 621, 1832).

**Cultivation**. The small size of the seeds and their need for light requires that they be sown in seed-beds. Germination follows after 5-7 days. After this, seedlings which are standing too densely are thinned out, and the shading is gradually reduced, so that the young plants harden off. After 40-60 days, the seedlings are large enough to plant out (20).

The planting distance depends on the cultivar: small-leaved orient tobacco is planted at distances of 10 x 30 cm to 20 x 50 cm, whereas for large-leaved, heavy cultivars, a planting distance of 90-120 cm is needed. With good soil preparation, only occasional shallow hoeing is necessary after planting for weed control, and for keeping the soil open, if needed. With the heavy types of tobacco, the inflorescence is broken off at the bud stage, together with the upper leaves (topping), and the side shoots are removed (suckering). In this way the leaves become larger and thicker. This is not done with orient tobacco and other light types, most cigarette tobaccos are first topped after the opening of the flowers (20).

**Diseases and Pests**. Because each injury to a leaf means a reduction in quality and value, phytomedicinal measures are of unusual importance with tobacco. Because pesticide residues must be avoided, the hygiene of the seed-bed, sanitation, crop rotation and resistance breeding (see above) are the measures most commonly used. As a rule, the seed is dressed, the soil in the seed-bed or boxes is sterilized, and the health of the seedlings is regularly inspected, or ensured by preventive spraying. Healthy young plants and clean fields can make expensive or unwanted measures unnecessary (fumigation of the crop area against *Meloidogyne*, spraying at later stages).

Of the numerous fungus and bacterial diseases (567, 1086), the following should be mentioned primarily: blue mould (*Peronospora tabacina* Adam), white mould (*Erysiphe cichoracearum* DC.), frog eyes (*Cercospora nicotianae* Ell. et Ev.), black shank (*Phytophthora parasitica* Dast. var. *nicotianae* (Breda de Haan) Tucker), wildfire (*Pseudomonas tabaci* (Wolf et Foster) Stapp), and angular leaf spot (*Pseudomonas angulata* (Fromme et Murray) Stapp). The numerous viruses require great vigilance; of these, the tobacco mosaic virus is the best known, but there are also all of the potato viruses (beware of potato cultivation in the same region), and some others (1285).

There is a long list of insects which can damage the leaves and plants (flea beetles, moth caterpillars, leafminers, thrips, and others), but they are mostly of lesser importance than the diseases named above. Everywhere, one must beware of the storage pests, the tobacco (cigarette) beetle, *Lasioderma serricorne* Fab., and the tobacco moth, *Ephestia elutella* Hbn. (20).

**Harvest and Processing**. The leaves of a plant are harvested when they have reached the required shade of colour - lighter green for cigars, gold-green for cigarette tobacco. That makes it necessary to reap the leaves 4 or 5 times, at about weekly intervals, whenever they are ripe. Especially with flue curing, care should be taken with the harvest, as every injury to the leaves leads to colour blemishes during curing. If the uniform colour of the leaves does not play a large role (core and binder leaves of cigars), or if the leaves all turn yellow quickly and simultaneously (orient tobacco on poor soils), then all leaves can be harvested in one operation.

The yield (dry leaves) depends very much on the tobacco type. The lowest is for orient tobacco (400-800 kg/ha), for Virginia and related types it is 1000-2000 kg/ha, and for heavy cigar tobaccos it is highest (up to 3000 kg/ha).

The various curing methods influence the character of the end-product to a considerable extent (20, 1444). Light tobaccos are obtained by sun-drying orient tobaccos, by flue-curing Virginia and other cigarette tobaccos, and with some cultivars by shade-drying. With these procedures, the leaves must have largely turned yellow before harvesting. With flue-curing, the remainder of the chlorophyll disappears in 1-2 days at moderate temperatures (38°C) and high air humidity. By then raising the temperature to 50°C, and opening the ventilators (dry air), the leaf tissue is killed, and the deep yellow to orange-yellow or reddish colour is developed (about 20 hours). For full drying, even of the leaf mid-rib, the temperature is raised to 70-77°C; this third phase lasts about 2 days.

Dark tobaccos are achieved by slow drying in the shade, lasting 3-5 weeks, with or without heating. The leaves still contain a lot of chlorophyll, which disappears slowly during the drying, but sometimes not completely. The dark colour results from the oxidation of phenolic compounds. Apart from the development of the brown colour, other enzymatic processes take place (hydrolysis of proteins, respiration reducing carbohydrate content, partial breakdown of nicotine), so that even with the same cultivar, dark tobacco has a substantially different character to quickly dried, light tobacco. If the air is too humid, there is a danger of mould developing; if a particular flavour is wanted, or if a lighter colour must be attained, then the shade-drying is complemented by heating with an open fire (fire-curing). After drying, cigar tobaccos are heaped into large piles for fermentation. Due to the cellular enzymes and bacteria, respiratory processes occur, which raise the temperature to 55-60°C. Uniform fermentation is achieved by repeated re-piling. Through the various biochemical processes, carbohydrates are lost by respiration, pectins, pentosans and hemicellulose are hydrolysed, amino acids and alkaloids are broken down. The main purposes of the fermentation are the change in texture (pectin breakdown) and the development of aroma (1832). Depending on the use of the end-product, the fermentation is ended earlier (wrapper leaves) or later (binder and filler leaves).

The other tobaccos are not strongly fermented, especially flue-cured and orient tobaccos, but in storage they go through an ageing process, in which the temperature does not climb higher than 30°C, and only slight chemical changes take place (e.g. splitting of sucrose into reducing sugars). The full development of aroma takes 1-2 years.

For local types of tobacco, in many countries there are characteristic variations on the harvesting, drying and fermentation procedures given here (800). Many tobaccos, especially pipe and chewing tobaccos, are artificially aromatized before use. Apart from improving the aroma, an increase in moisture retention is often intended (casing). These additions are, among others, sorbitol, sugar, liquorice, coumarin, aromatic oils (20).

Tobacco residues are used for the production of nicotine (insecticide, see p. 391). Nicotine-rich cultivars of *N. rustica* with up to 12 % nicotine are also cultivated solely for this purpose.

Tobacco seed provides a good salad oil (see p. 112, Table 15), which is extracted in Bulgaria, Greece and India.

## Further Beverages and Stimulants

Table 38 includes other plants which are grown primarily as beverages and stimulants. In this group can also be included other species, which are grown in the first place as spices (e.g. nutmeg, with its hallucinogenic, poisonous myristicin, Table 39) or as medicinal plants (*Cinchona* species, *Papaver somniferum*, various Solanaceae, Table 40). The psychotropic drugs have received considerable medicinal interest in recent years (243, 1608). They are mostly extracted from wild plants. Because the most important of these compounds have now been artificially synthesized by the chemical industry, the demand for them has not led to any of these plants being brought under cultivation.

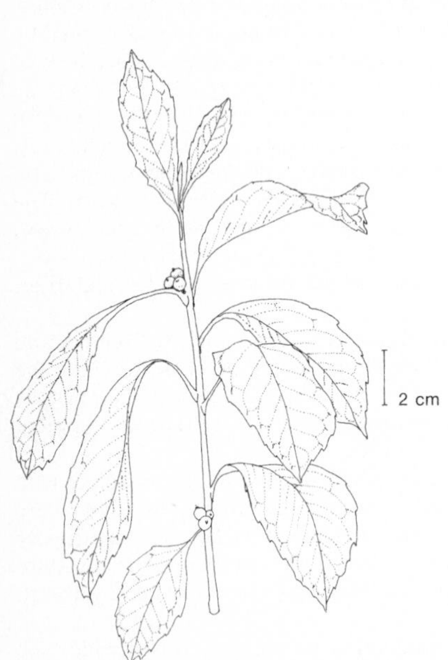


Fig. 74. *Ilex paraguariensis.* Fruiting branch

Fig. 75. *Erythroxylum coca.* Flowering branch

Fig. 76. *Cola nitida*. (a) branch with flowers, (b) fruit

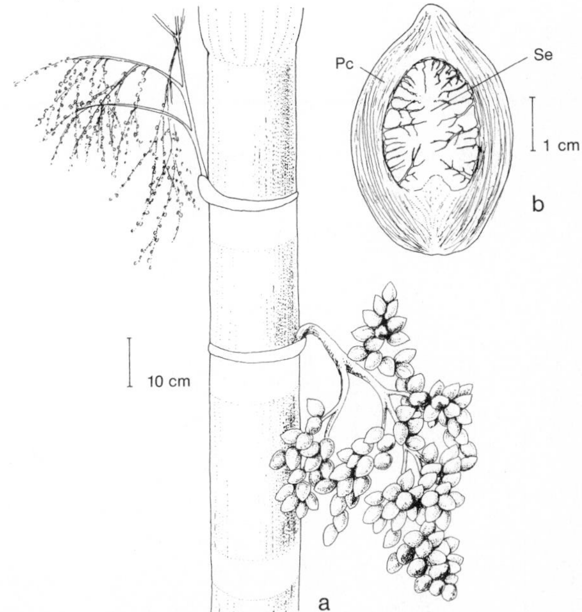


Fig. 77. *Areca catechu*. (a) trunk with inflorescence and fruit bunch, (b) fruit in longitudinal section. Pc = fibrous pericarp, Se = seed with ruminate endosperm

Table 38. Further beverages and stimulants (explanation of symbols, see footnotes page 55)

| Botanical name | Trade name of product | Used part | Climatic region | Distribution | Econ. value | Remarks and literature references |
|---|---|---|---|---|---|---|
| | | | **Dicotyledonae** | | | |
| **Aquifoliaceae** | | | | | | |
| *Ilex paraguariensis* St. Hil. | Paraguay tea, maté, yerba mate, caá, congonha | leaf | S, Tr | r: S America | +++ | (Fig. 74) Active constituent is caffeine. Cultivated in Paraguay, Argentina and Brazil. Caffeine content variable, averages 1.5% (242, 1036, 1423, 1948) |
| **Cannabidaceae** | | | | | | |
| *Cannabis sativa* L. var. *indica* (Lam.) Pers. | Indian hemp, marihuana | female inflorescence | S, Tr | w | ++ | Active constituents cannabidiol, cannabinol. Hashish is the female flowers, marijuana is a mixture of leaves and flowers from the female plant with a substantially lower content of active compounds (376, 476, 909, 1264, 1454) |

| | | | | | | |
|---|---|---|---|---|---|---|
| **Celastraceae**<br>*Catha edulis* (Vahl) Forsk. ex Endl. | kat, qat murungu miraa | leaf | S, Tr | l: Yemen, Ethiopia, Kenya | ++ | Active constituents nor-pseudoephedrine and other ephedrine derivatives. Leaves are chewed when fresh or dried, usage increasing in NE Africa and Yemen. The drug produces apathy (7, 594, 984, 1306, 1503, 1905) |
| **Compositae**<br>*Cichorium intybus* L. var. *sativum* Lam. et DC. | chicory, chicorée à café, achicoria | root | Te, S | w | ++ | No active ingredients. Yields in India 23 t fresh roots/ha, harvested 6 months after sowing. Roots cut in pieces, dried and roasted. Thereby develop from inulin oxy-methylfurfurol and other compounds with an aroma similar to coffee. Up to 30% is added to instant coffee powder (96, 315, 335, 376, 921, 929) |
| **Erythroxylaceae**<br>*Erythroxylum coca* Lam. | coca, hayuelo, ipadú | leaf | TrH | r: S America, Andes | + | (Fig. 77) Active constituent: cocaine. Leaves chewed with ashes or lime. Subdues feelings of hunger and thirst. For medicinal purposes, 2500 kg of cocaine annually are extracted from coca leaves (6, 198, 1150, 1212, 1444, 1948) |

Table 38. Further beverages and stimulants, cont'd (explanation of symbols, see footnotes page 55)

| Botanical name | Trade name of product | Used part | Climatic region | Distribution | Econ. value | Remarks and literature references |
|---|---|---|---|---|---|---|
| *E. novogranatense* (Morris) Hieron. | truxillo coca, hayuelo | leaf | Tr | r: S America, Andes | + | Active constituent: cocaine. Grows at lower heights above sea level than *E. coca*. Used in the same way (1444, 1948) |
| **Labiatae** | | | | | | |
| *Salvia divinorum* Epling et Jativa | hojas de la pastora, yerba de María | leaf | Tr | l: S Mexico | + | Active constituent: unknown. Propagated by the Indios by cuttings, and planted in concealed places (1608, 1849) |
| **Leguminosae** | | | | | | |
| PAPILIONOIDEAE | | | | | | |
| *Aspalathus linearis* (Burm.f.) R. Dahlgr. ssp. *linearis* | rooibos tea | leaf | S | l: S Africa | + | Active constituent: none. Cultivated in S Africa and generally traded. Annual production over 1000 t (319, 395, 1217, 1905) |

| | | | | | | |
|---|---|---|---|---|---|---|
| **Malpighiaceae**<br>*Banisteriopsis caapi* (Spruce ex Griseb.) Morton | ayahuasca, caapi, yajé | stem | Tr | l: NW Amazonas valley | + | Active constituents: banisterine, harmine, harmaline. Little information about cultivation. The decoction of shredded stems is drunk as a stimulant (527, 545, 1495, 1516, 1608, 1846) |
| **Piperaceae**<br>*Piper betle* L. | betel pepper, betelvine, sirih, suruh | leaf | Tr | r: S and SE Asia, E Africa | +++ | Active constituent: none. A variety of cultivars, grown especially in India. Stimulating/aromatic taste, mostly used as a wrapping for betel nut slices (259, 376, 1446) |
| *P. methysticum* G. Forst. | kawa pepper, kava-kava, yangona | root | Tr | l: Polynesia | + | Active constituents: methysticine, kawaine, yangonine, marindinine. After fermentation of the comminuted roots, taken as a narcotic drink (145, 586, 815, 1446) |
| **Rubiaceae**<br>*Borojoa patinoi* Cuatr. | borojó | fruit | Tr | l: Central and S America | + | Active constituent: unknown. Fruits used for the preparation of a refreshing drink. Cultivated on a small scale (442, 1378, 1866) |

Table 38. Further beverages and stimulants, cont'd (explanation of symbols, see footnotes page 55)

| Botanical name | Trade name of product | Used part | Climatic region | Distribution | Econ. value | Remarks and literature references |
|---|---|---|---|---|---|---|
| *Uncaria gambir* (Hunter) Roxb. | gambir, pale catechu, white cutch | juice of leaf and stem | Tr | r: SE Asia | + | Active constituent: none. Utilized as an aromatic for chewing betel. Also a dye and for tanning (see p. 386) (259, 376, 1383, 1948) |
| **Sapindaceae** | | | | | | |
| *Paullinia cupana* H.B.K. | guaraná, cupana | seed | Tr | r: Brazil, Venezeula, Colombia | ++ | Active constituent: caffeine. Very widespread drink in Brazil, also used in the USA for the preparation of drinks. Caffeine content of the seeds 5%. Old cultivated plant of the Indios, grown in the Amazon region. Annual production 250 t (486, 550, 618, 1607) |

| | | | | | | |
|---|---|---|---|---|---|---|
| **Simaroubaceae** | | | | | | |
| *Quassia amara* L. | Surinam quassia, bitter wood, quina de Cayena | wood | Tr | r: Northern S America, S Mexico | + | Stimulant constituents: none. Bitter compounds from the wood used as a substitute for quinine in the drinks industry, also utilized medicinally and as an insecticide. Cultivated in its home region, otherwise used in the tropics as an ornamental bush. The wood of *Picrasma excelsa* (Sw.) Planch., Jamaican quassia, is used similarly (376, 758, 1960) |
| **Sterculiaceae** | | | | | | |
| *Cola acuminata* (P.Beauv.) Schott et Endl. | kola nut, abata kola, noix de kola | seed | Tr | r: W Africa | + | Active constituent: caffeine. To a small extent used for the production of kola nuts (mostly from wild trees, as also with several other *Cola* species). Cultivation limited to Nigeria (465, 1207, 1319, 1327, 1948) |

Table 38. Further beverages and stimulants, cont'd (explanation of symbols, see footnotes page 55)

| Botanical name | Trade name of product | Used part | Climatic region | Distribution | Econ. value | Remarks and literature references |
|---|---|---|---|---|---|---|
| *C. nitida* (Vent.) Schott. et Endl. | kola nut, male kola, gbanja kola | seed | Tr | r: W Africa, West Indian Islands | +++ | (Fig. 76) Active constituent: caffeine. By far the most important species, cultivated to a large degree in Nigeria, but also in Brazil and the W Indian Islands. Annual production about 300,000 t. Caffeine content about 2.5%: with chewing, kola-red is formed from the bitter catechin compound kolanin. The product traded locally is primarily the embryo which consists mostly of the cotyledons of which the seed shell has been rubbed away after fermentation. There is an average yield of 30,000 nuts/ha/year, which can be increased tenfold with the cultivation of selected trees. A considerable export of dried seeds exists from Nigeria and the W Indian Islands to the USA (465, 466, 1207, 1299, 1304, 1313, 1319, 1327, 1774, 1948) |

## Monocotyledonae

| | | | | | | |
|---|---|---|---|---|---|---|
| **Palmae** | | | | | | |
| *Areca catechu* L. | betel nut, areca nut, noix d'arec, aréquier, supari, pinang | seed | Tr | r: S and SE Asia, E Africa | +++ | (Fig. 77) Active constituents: arecoline, arecaidine. The endosperm of the seed is cut into thin slices or small pieces, then dried. For chewing, it is wrapped in betel-pepper leaves together with slaked lime and other additions (see gambir). The effect of the areca alkaloids is stimulant/narcotic. Also utilized medicinally against intestinal worms (154, 155, 181, 376, 406, 808) |

# Spices

The term "spice" is used here in a wide sense (1538, 1825) which includes all culinary herbs, seasonings and condiments of vegetable origin.

The use of spices for improving the taste of foods is a cultural achievement of all races. It has been independently developed in many regions of the earth, and it also led to cultural exchanges very early in history (Egypt - Asia). Spices were the most important products in trading between Asia and Northeast Africa or Europe for thousands of years. The spice trade has had a decisive effect on politics, and was the main reason for the beginning of the European interest in Africa and Asia (1446, 1538, 1825). Even nowadays, it still occupies an important place among foodstuffs, beverages, and technical raw materials. For some developing countries spices are of the greatest economic importance.

Asia is the continent from which most of the tropical spices originate. Of the major spices, America has only contributed paprika and vanilla. Compared to the European spices, which mostly originate from the Mediterranean region (Labiatae and Umbelliferae), the tropical spices are notable for being pungent tasting (chili) or highly aromatic (cloves), or a combination of both factors (ginger).

It is difficult to define which plants should be included as spices. Many vegetables have their own characteristic flavour, not only *Allium* spp., *Eruca vesicaria*, *Moringa oleifera*, but also gherkins and parsnips, mushrooms, fenugreek, and many others (Table 25). In the same way, some fruit species (lemons, candied lemons, candied oranges), as well as nuts (pistachios, almonds) and oil seeds (poppy seed, linseed, sesame) are regularly used for the seasoning of meals. Also plants which are mostly cultivated as medicinal plants (*Cinchona* spp.) or for the production of essential oils (*Cymbopogon citratus*) are used to a large extent as seasonings. They can be found in the relevant chapters. Here, we only include plants which are primarily cultivated for seasoning purposes. Some spices must therefore be neglected, as they are gathered from wild plants, even if they are very widely used (juniper berries). More complete listings are given in (758, 1442, 1740).

The value of spices for human nutrition has often been overvalued in the past. Their health-giving effects are usually indirect. They stimulate the eye (parsley as a decoration of dishes), nose and tongue, and thus stimulate the flow of saliva (starch digestion) and the secretion of enzymes in the stomach-intestine system (protein and fat digestion) (1825). Their effects on the digestibility of food cannot be questioned. However, on the other hand, there is the fact that many spices contain poisons, which even in small quantities cannot be considered harmless (myristicin, isothiocyanate, coumarin, vanillin, safrole). Some spices which were formerly usable are, therefore, no longer allowed in the USA and some other countries (1183).

Total production figures which are accurate to some extent are only available for a few spices (garlic (514)). Better statistics exist for imports and exports

(518, 841). In terms of quantity, mustard, pepper, capsicum peppers, garlic, cinnamon, caraway (including black cumin), and ginger are the major spices. In terms of value, vanilla also joins this group. The spices not only bring foreign exchange to the developing countries, but because their cultivation and processing are very labour intensive, they also substantially improve the labour market. For the same reason, the cultivation of spices severely declined in the industrialized countries. Because it is increasingly common for technical processing to take place in the countries of production (see below), they give a further motivation for the development of local industries. However, it should not be overlooked that the primary producers of the spices often earn very little.

Except for the local markets, most spices are traded in their processed forms (1442). The most important procedure is drying, which is also often the starting point of further processing, supplemented by cleaning, sorting, and packing. Some spices are offered as preparations which are ready for use (mustard paste, capers, green pepper, horseradish). For a long time, the volatile aromatic compounds have also been distilled off as essential oils (1825), which especially find applications in the drinks industries (liqueurs), perfumery and pharmacy. In recent years, the extraction of the total aromatic compounds by organic solvents has taken a leading place (1446, 1636, 1825). Oleoresins are now considered indispensible to the food industries. They are further processed into essences (mostly solutions of the oleoresins in alcohol or other solvents) or emulsions. Finally, decoctions (e.g. spice vinegars), oil solutions, and binding on to easily soluble or almost insoluble carriers (salt, sugar, gum arabic, gelatine) must also be mentioned (1825).

Most spices can only be included here in tabulated form (Table 39). The most important are dealt with somewhat more fully.

## Mustard

fr. moutarde; sp. mostaza; ger. Senf

The mustard which is traded (731, 1442, 1538, 1657, 1857, 2007) comes mostly from three Cruciferae, *Brassica nigra* (L.) W.D.J. Koch, black mustard, *Brassica juncea* (L.) Czern. et Coss., brown mustard (Indian mustard, sarepta mustard, rai (India)), and *Sinapsis alba* L., white mustard. There is also *Eruca vesicaria* (L.) Cav. var. *sativa* (Mill.) Thell., garden rocket, Persian mustard. White mustard is the most used species in the West, and is predominantly cultivated in the temperate zone, but also in the subtropics (the Balkans). The whole corns are used as a spice for gherkins, preserved fish, and sausages. The mustard powder, as produced or freed from oil, is the basis of mustard paste. The sharp taste of the powder develops first with the addition of water (the sinalbin is split by myrosin into p-hydroxybenzylisothiocyanate and sinapine). *Brassica juncea* is especially important in the tropics and subtropics. This species is extraordinarily rich in forms (amphidiploid from *B. rapa* x *B. nigra* (2026)), and includes pure oil plants (Table 18) and various vegetables (731). With some authors, even the various spice-providing forms of *B. juncea* are separated into species. The cultivation of *B. juncea* cultivars has spread in the last decades, while the cultivation of *B. nigra* has decreased. Both species contain sinigrin, which yields isothiocyanate when enzymatically split; this is a mustard oil with a more pungent taste than the p-hydroxybenzylisothiocyanate from white mustard.

## Pepper

fr. poivre; sp. pimienta; ger. Pfeffer

For seasoning use, apart from *Piper nigrum* L., Piperaceae, the black, white, and green pepper, there are also *P. guineense* Schum. et Thonn., West African pepper, West and Central Africa, *P. longum* L., long pepper, India, and *P. retrofractum* Vahl., Java long pepper. *P. cubeba* L.f., cubeb, from Java, provides an oil which to a small extent is used as an aromatic in the drinks industry, and medicinally (1358, 1446, 1461), though as a spice it is scarcely traded any more. In world trade, only *P. nigrum* plays a role (232, 376, 945, 1107, 1446, 1825). World production amounts to about 200,000 tons, and primary exports add up to 178,000 tons. The majority of the exports come from Southeast Asia (140,000 t); the main exporters there are India (47,000 t), Indonesia (42,000 t), and Malaysia (19,000 t). A further large exporter is Brazil (24,000 t). For pepper, there are a large number of cultivars (376, 535, 1825), especially in India; the trade distinguishes primarily according to the country of origin. The fruits contain the pungent piperine (highest concentration in the endocarp, Fig. 78), also resin (especially in the cells of the mesocarp) and essential oil (highest concentration in the inner part of the mesocarp) (1107). Black pepper is the whole, dried fruit, harvested when they are not quite ripe. It contains all the aromatic compounds. White pepper is extracted by a water-retting of ripe berries; in this, the fruit wall is partially destroyed, and the remainder can then be easily rubbed away (1107). It contains fewer aromatics

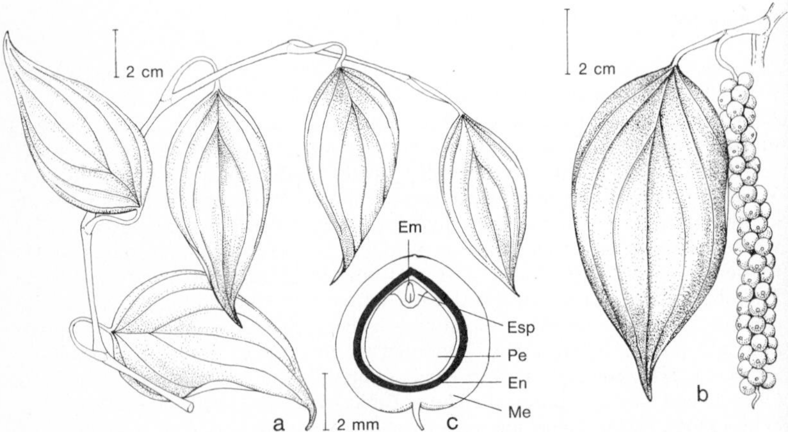


Fig. 78. *Piper nigrum*. (a) branch, (b) fruit spike, (c) fruit in longitudinal section. Em = embryo, En = endocarp (connate with testa). Esp = endosperm, Me = mesocarp, Pe = perisperm

(oil and resin) than black pepper. In the 1970s, a large market developed in West Europe (France and West Germany) for green pepper. For its production, the fully grown berries, which are still green, are submerged in a container full of brine after separation from the fruit-stem. In this way, browning through oxidation (polyphenoloxidases) is prevented. The endocarp is still soft, the taste is highly aromatic, but less pungent than with black pepper. The main producers of green pepper are India, Madagascar, and Brazil (1442, 1446). In India, also dehydrated green pepper is produced (1156).

## Capsicum Peppers

fr. piment, poivre d'Espagne, poivre rouge; sp. ají, chile, pimiento; ger. Paprika

Of all the spices, capsicum peppers are the most widespread in the countries of the tropics and subtropics. The abundance of cultivars (Fig. 79) has led botanists to give names to a range of species. The following are now considered to be true species (726, 1169, 1400, 1446, 1657, 1686):

- *Capsicum annuum* L., chili, red pepper, which is by far the most widespread species,
- *C. baccatum* L., Peruvian pepper, red pepper,
- *C. chinense* Jacq., ají, South America,
- *C. frutescens* L., bird pepper, tabasco, and
- *C. pubescens* Ruiz et Pav., rocoto, Peru.

Within *C. annuum* are included the bell peppers as var. *annuum* (Table 25).

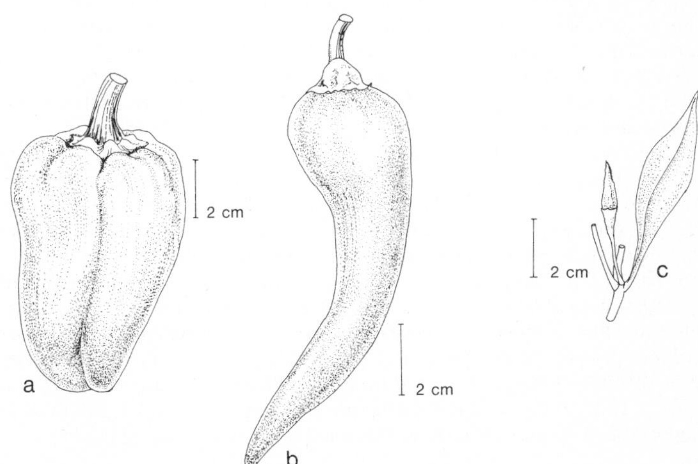


Fig. 79. *Capsicum annuum*. (a) sweet pepper, (b) large-fruited hot pepper, (c) small-fruited hot pepper with upright fruits

The spice peppers differ from each other in size, shape, and position of fruit, thickness of fruit flesh, presence or absence of capsaicin, which causes the hot taste, and the intensity of the red colouring, which is brought about by capsanthin and other carotenoids, which, however, possess no vitamin A activity (β-carotene occurs only in small quantities). Small-fruited types are dried whole, large-fruited types are dried after splitting open and removal of the seeds. The ground fruit flesh is the most important form traded. In the tropical regions of cultivation, fresh whole fruits are often used as seasonings, either green or ripe. Production figures are not available. Probably the world's production is higher than that of *Piper nigrum*. The primary exports amount to 167,000 t, and the most important exporting countries are China, Mexico, Spain, Pakistan, Hungary and Malaysia (518).

## Vanilla

fr. vanille; sp. vainilla; ger. Vanille

Apart from common vanilla, *Vanilla planifolia* Andr. (*V. fragans* (Salisb.) Ames), Orchidaceae, the following are also cultivated to a small extent: *V. pompona* Schiede, West Indian vanilla, fruits traded as "vanillons" (Guadaloupe, Dominican Republic, Martinique), and *V. tahitensis* J.W.Moore, Tahiti vanilla (Tahiti, Hawaii) (217, 1442, 1446, 1538, 1960). The vanillons are less highly valued (mostly used for extracts), but Tahiti vanilla brings substantially higher prices in the USA than common vanilla (1122, 1446). Common vanilla supplies 96% of the world trade. The main producers are Madagascar, the Comoros, and Reunion (since 1964, they have joined in the "Alliance de la Vanille", which supplies 80% of the world's exports; the product bears the trade name "Bourbon-Vanilla"). Apart from these, Mexico, Java and Uganda play a continuous role in the world market (99, 1122, 1446). Through its exceptional fragrance, natural vanilla has again recovered a good market; before, synthetic vanillin (from lignin) had severely endangered production. World production now lies around 1500 t. Vanilla is a large tropical climbing plant. In cultivation, it is trained up living trees, or on frames. Outside its homelands (Mexico, Central America), the bees and humming birds which are necessary for its pollination are absent. The pollination is then carried out by hand, whereby the rostellum is bent upwards using a pin, so that the pollinia can be pressed with a finger on to the stigma (Fig. 80). From fertilization to ripening takes up to 9 months. The vanilla of commerce is produced using an intricate fermentation process, in which the pods are first killed by hot air, or by treatment in hot water at about 70°C. This temperature is not sufficient to inactivate the enzymes, which in the course of several weeks, hydrolyse the glucovanillin to glucose and free vanillin; other enzymes are necessary for the development of the brown-black colour of the fermented pods. The classical fermentation process requires a great deal of manual work. A partial mechanization has been proposed (1442, 1446, 1795). In Uganda, a drastic technique is practised, in which the pods are first chopped up. The fermentation is then completed in 4-7 hours (1122).

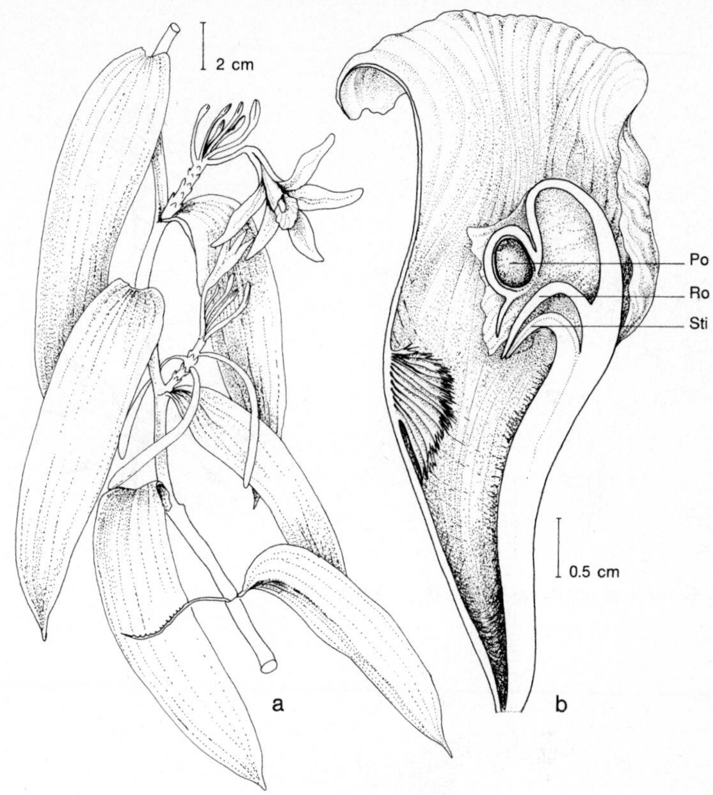


Fig. 80. *Vanilla planifolia*. (a) flowering and fruiting branch, (b) longitudinal section through flower. Po = pollinium, Ro = rostellum, Sti = stigma

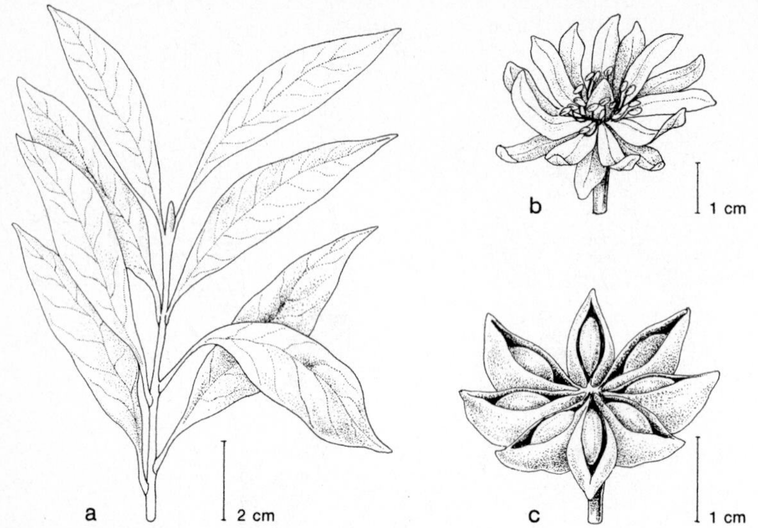


Fig. 81. *Illicium verum*. (a) branch, (b) flower, (c) fruit

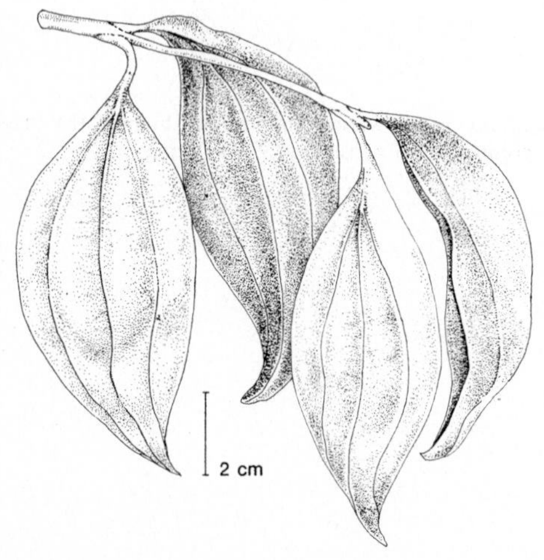


Fig. 82. *Cinnamomum verum*. Tip of branch

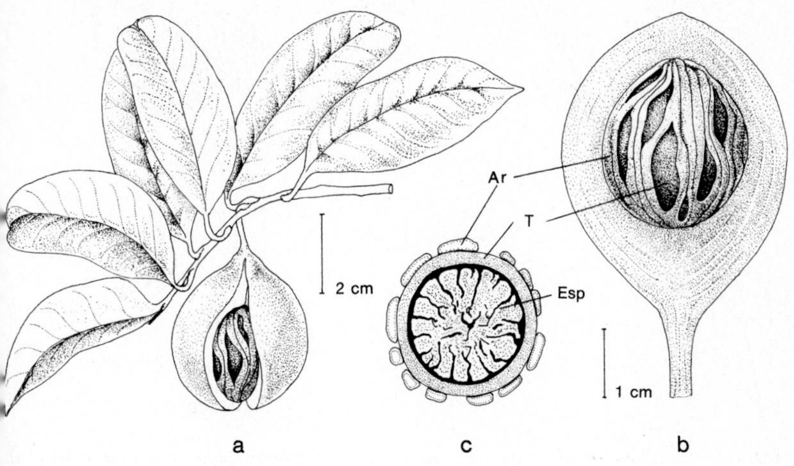


Fig. 83. *Myristica fragrans*. (a) fruiting branch, (b) fruit opened longitudinally, (c) seed in cross section. Ar = aril, Esp = endosperm, T = testa

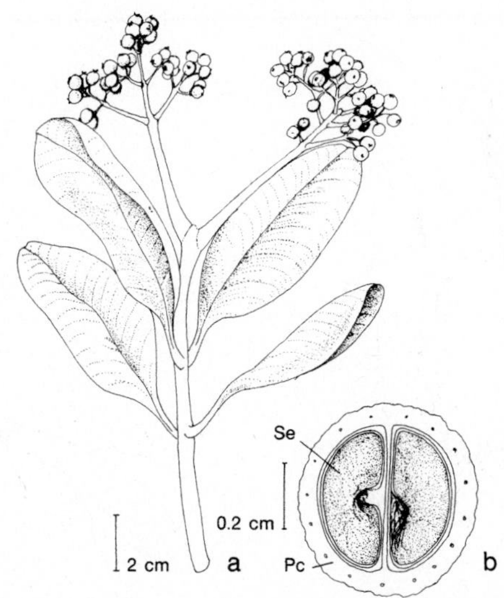


Fig. 84. *Pimenta dioica*. (a) fruiting branch, (b) fruit in cross section. Pc = pericarp, Se = seed

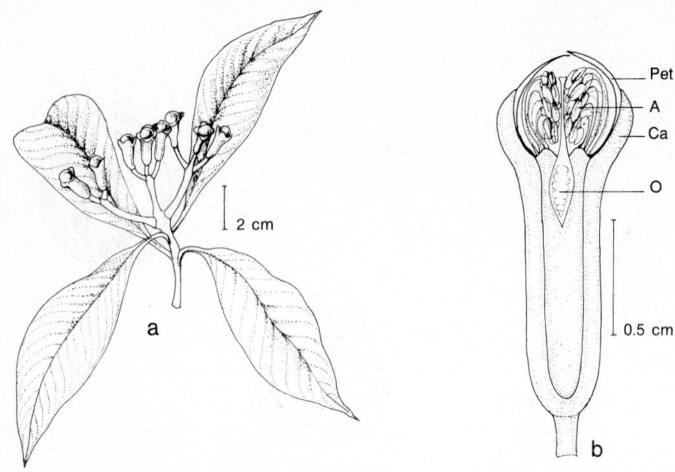


Fig. 85. *Syzygium aromaticum*. (a) shoot tip with flower buds, (b) flower bud in longitudinal section. A = anthers, Pet = petals, Ca = calyx, O = ovary

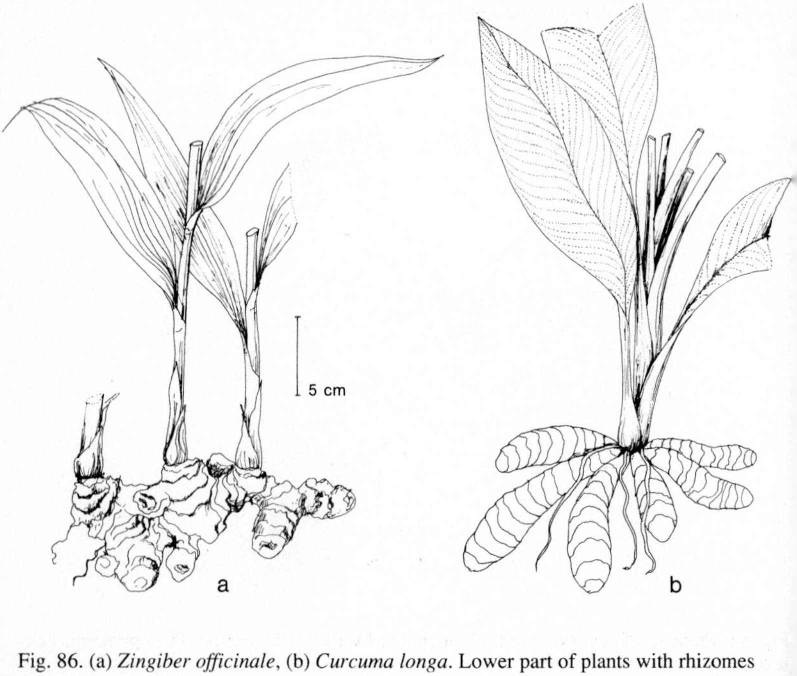

Fig. 86. (a) *Zingiber officinale*, (b) *Curcuma longa*. Lower part of plants with rhizomes

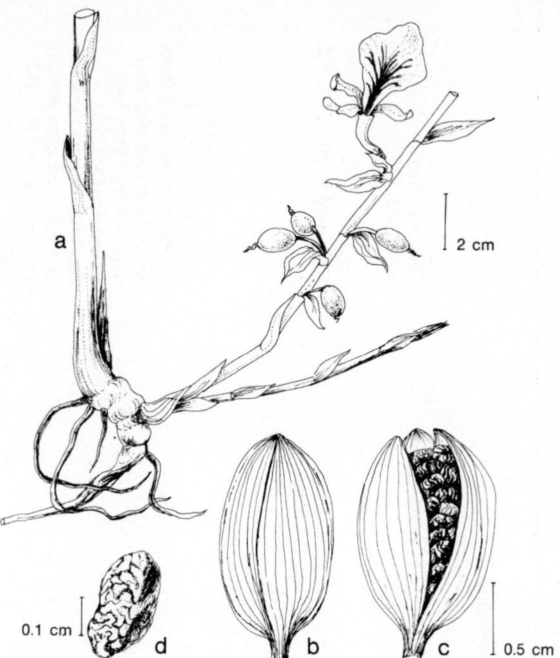


Fig. 87. *Elettaria cardamomum*. (a) lower part of plant with inflorescence, (b and c) closed and opened capsule, (d) seed

Table 39. Further spices (explanation of symbols, see footnotes page 55)

| Botanical name | Vernacular names | Used part | Climatic region | Distribution | Econ. value | Remarks and literature references |
|---|---|---|---|---|---|---|
| | | | **Dicotyledonae** | | | |
| **Annonaceae** | | | | | | |
| *Xylopia aethiopica* (Dun.) A. Rich. | Guinea pepper, African pepper, malagueta da Guiné | fruit | Tr | l: W Africa | + | Only used locally (262, 893, 934, 1442, 1477) |
| *Monodora myristica* (Gaertn.) Dunal | Jamaica nutmeg, calabash nutmeg, muscade de calabash | seed | Tr | l: W Africa, W Indies | + | Only used locally (200, 262, 1477) |
| **Anacardiaceae** | | | | | | |
| *Schinus molle* L. | Peruvian mastic, California pepper, molle seed, pimentero falso, aroeira | fruit | S, Tr | w | + | Used in Peru for the preparation of a mildly alcoholic drink. Dried fruits used like pepper, regularly exported from Peru and Ecuador to the USA. Essential oil from the leaves and fruit used as an aromatic (81, 661, 981, 1378, 1382, 1444) |

| | | | | | | |
|---|---|---|---|---|---|---|
| **Boraginaceae** | | | | | | |
| *Borago officinalis* L. | borage, bourrache, borraja, borragem | leaf | Te, S | w | + | Culinary herb used as seasoning for gherkins and salads. Bee food (see p. 402) (200, 376) |
| **Cannabidaceae** | | | | | | |
| *Humulus lupulus* L. | hop, houblon, lúpulo | female inflores-cence | Te, S | w | +++ | Mostly used as a beer flavouring. The resin contains bitter compounds of the lupulon and humulon types, which are also utilized medicinally because of their soothing effect ("lupulin"). The essential oil (hops oil) is extracted by distillation and used as an aromatic. World production (dried inflorescen-ces) 110,000 tons, of which 70,000 t grown in Europe (255, 376, 1383) |
| **Capparidaceae** | | | | | | |
| *Capparis spinosa* L. | capers, câprier, alcaparra | flower-buds | S | w | ++ | Cultivated especially in S France and Spain, but also in other countries in the Mediter-ranean region, the USA, and elsewhere. Roots utilized medicinally (1382, 1442) |

Table 39. Further spices, cont'd (explanation of symbols, see footnotes page 55)

| Botanical name | Vernacular names | Used part | Climatic region | Distribution | Econ. value | Remarks and literature references |
|---|---|---|---|---|---|---|
| **Compositae** | | | | | | |
| *Artemisia absinthium* L. | absinthe, wormwood, ajenjo | leaf | Te, S | w | + | Herb or its aqueous extract used in the production of vermouth. Essential oil utilized in liqueurs and medicinally. Main cultivation regions Mediterranean region, USA, S Siberia, Kashmir (406, 661) |
| *A. dracunculus* L. f. *dracunculus* | German tarragon, sweet tarragon, estragon | leaf | Te, S | w | + | Propagated only vegetatively. Vinegar and salad seasoning (642, 643, 1358, 1442, 1538) |
| f. *redowskii* hort. | Russian tarragon, false tarragon | leaf | Te, S, | w | + | Propagated by seed. Bitter, less aromatic than sweet tarragon (1358, 1538) |
| **Cruciferae** | | | | | | |
| *Armoracia rusticana* Ph. Gaertn., B Mey. et Scherb. | horseradish, cran, cranson, mérédic, rábano picante | root | Te, S, TrH | w | + | Propagated using pieces of root. Very sensitive to *Meloidogyne*. Traded fresh or prepared with vinegar (381, 1442, 1538) |

| | | | | | | |
|---|---|---|---|---|---|---|
| **Illiciaceae** | | | | | | |
| *Illicium verum* Hook.f. | Chinese anise, star anise, anis étoilé, badiane | fruit | Tr | r: SW China | ++ | (Fig. 81) For spicing of baked goods and plum jam, for production of the Egyptian drink "er'e'souz". To a large extent used for the distillation of the essential oil, which is used for liqueurs (anisette), in perfumery, and medicinally (661, 796, 1358, 1383, 1442) |
| **Labiatae** | | | | | | |
| *Hyssopus officinalis* L. | hyssop, hyssope | leaf | Te, S | w | + | Fresh or dried as a sauce seasoning, likewise the distilled oil (1442) |
| *Melissa officinalis* L. | lemon balm, melisse | leaf | Te, S | w | + | To a small extent used as a salad and soup seasoning. Aromatic in liqueurs. Medicinal uses (661, 1442, 1842) |
| *Ocimum basilicum* L. | sweet basil, basilic commun, albahaca | leaf | Te, S, Tr | w | ++ | Includes a variety of forms (some authors have divided it into seperate botanical species). Is used fresh for salad seasoning, or dried for sausages and meat dishes. The distilled oil is used in perfumery (403, 642, 643, 1442, 1538) |
| *Origanum majorana* L. (*Majorana hortensis* Moench) | sweet marjoram, marjolaine | leaf | Te, S | w | ++ | Cultivated to a large extent in the Mediterranean region (642, 643, 1442, 1538) |

Table 39. Further spices, cont'd (explanation of symbols, see footnotes page 55)

| Botanical name | Vernacular names | Used part | Climatic region | Distribution | Econ. value | Remarks and literature references |
|---|---|---|---|---|---|---|
| *O. onites* L. (*Majorana onites* (L.) Benth.) | pot marjoram, Cretian dost | flowering shoots | S | r: Mediterranean countries | + | Weaker aroma than *O. majorana* (642, 643, 1538, 1875) |
| *O. syriacum* L. (*O. maru* L.) | Egyptian marjoram, biblical hyssop | leaf | S | r: E Mediterranean countries | + | Used similarly to sweet marjoram (642, 643) |
| *O. vulgare* L. | wild marjoram, European oregano, marjolaine sauvage, orégano | leaf | Te, S | r: Mediterranean countries | + | Mostly gathered from wild plants in the Mediterranean region. Often traded mixed with other labiates (e.g. *Thymus capitatus*) as "pizza spice" (642, 643, 1127, 1442, 1538) |
| *Plectranthus amboinicus* (Lour.) Spreng. (*Coleus amboinicus* Lour.) | Indian borage, oreille | leaf | Tr | r: S and SE Asia, West Indian Islands | + | Fresh leaves as seasoning for fish and meat, also as a leaf vegetable (259, 347, 376, 731, 1296, 1383) |

| | | | | | | |
|---|---|---|---|---|---|---|
| *Rosmarinus officinalis* L. | rosemary, romero, alecrim | leaf | Te, S | w | + | Originates in the Mediterranean region. Used in Italy, France and England to a great extent for meat dishes. The essential oil is more important, used not only as a seasoning, but as an aroma for soaps, and for technical products (642, 643, 661, 1442, 1538) |
| *Salvia officinalis* L. | sage, sauge, salvia | leaf | Te, S | w | ++ | Widespread in the Mediterranean region, used there to a great extent for oil distillation. Fresh or dried leaves used as a meat seasoning (642, 643, 661, 1442, 1538, 1837, 1842) |
| *Satureja hortensis* L. | summer savory, sedrée, ajedra | leaf | Te, S | w | ++ | Annual. Much used as a seasoning for beans, but also for sauces and meat dishes (642, 643, 1442, 1538) |
| *S. montana* L. | winter savory, sarriette vivace | leaf | Te, S | r: Europe, Mediterranean countries | + | Perennial, woody bush. Used as for *S. hortensis* (642, 643, 1442) |
| *Thymus capitatus* Hoffmgg. et Link (*Coridothymus capitatus* (L.) Rchb.f.) | conehead thyme, Spanish origanum | leaf | S | r: Mediterranean countries | + | Component of European oregano, with other labiates (see *Origanum vulgare*). One of the sources of origanum oil (661, 1127) |

Table 39. Further spices, cont'd (explanation of symbols, see footnotes page 55)

| Botanical name | Vernacular names | Used part | Climatic region | Distribution | Econ. value | Remarks and literature references |
|---|---|---|---|---|---|---|
| *Thymus vulgaris* L. | garden thyme, frigoule, tomillo | leaf | Te, S | w | ++ | Cultivation of the traded herb is primarily in Spain, France, Italy and Bulgaria. Traditional herb for meat, component of "mixed herbs". Also medicinal, due to antiseptic properties (thymol) (642, 643, 661, 1127, 1383) |
| **Lauraceae** | | | | | | |
| *Cinnamomum aromaticum* Nees (*C. cassia* Bl.) | cassia, Chinese cassia, camelier casse, koriza | bark, unripe fruits | Tr | r: S China, SE Asia | ++ | From this and the following two species, the bark is won from 6-10 year old trees or branches. The outside of the bark is mostly not or only partially scraped off, the bark pieces do not roll up tightly like Ceylon cinnamon. The aroma is stronger, the essential oil content higher than with Ceylon cinnamon. It provides the majority of cinnamon which is traded. The names "cassia buds" and "flores cassiae" refer to the dried, unripe fruits, which are likewise used as a spice |

| | | | | | | |
|---|---|---|---|---|---|---|
| | | | | | | (1107, 1123, 1358, 1383, 1442, 1446, 1513, 1538, 1688) |
| *C. burmanii* (Nees) Bl. | Padang cassia, Batavia cinnamon | bark | Tr | r: Indonesia, Philippines | + | see *C. aromaticum* |
| *C. loureirii* Nees | Saigon cassia | bark | Tr | r: Indochina | + | see *C. aromaticum* |
| *C. verum* J.S.Presl. (*C. zeylanicum* Bl.) | cinnamon, canelle de Ceylan, canela | bark | Tr | w | ++ | (Fig. 82) Bark from 1-2 year old suckers, the outer side is scraped of, with drying forms tight rolls. Fine aroma, more valued than cassia in Europe. Main producers Sri Lanka, Madagascar, the Seychelles, and Brazil (406, 466, 1107, 1123, 1358, 1383, 1442, 1444, 1446, 1688) |
| *Laurus nobilis* L. | laurel, sweet bay, laurier | leaf | S, TrH | w | ++ | Homeland is in the Mediterranean region, the largest production is also there. Has spread to all subtropical countries as an ornamental tree (642, 643, 1358, 1442, 1538) |

Table 39. Further spices, cont'd (explanation of symbols, see footnotes page 55)

| Botanical name | Vernacular names | Used part | Climatic region | Distribution | Econ. value | Remarks and literature references |
|---|---|---|---|---|---|---|
| **Leguminosae** | | | | | | |
| PAPILIONOIDEAE | | | | | | |
| *Dipteryx odorota* (Aubl.) Willd. | tonka bean, cumaru, fava-tonca | seed | Tr | w | + | The seeds contain up to 10% coumarin. Mostly used for the aromatizing of tobacco and for perfumery. Cultivated in Venezuela, W Indian Islands, and Nigeria (444, 1444) |
| **Myristicaceae** | | | | | | |
| *Myristica fragrans* Houtt. | nutmeg, mace, noix muscade, nuez moscada | seed | Tr | w | ++ | (Fig. 83) Dioecious tree, especially on W Indian Islands and in SE Asia. Exports mostly from Indonesia and Grenada, some from Sri Lanka (total 8500 t). Mace is the dried arillus, nutmeg is the endosperm freed from the seed coat. The fruit flesh is used locally for preserves and pickles (797, 907, 1107, 1358, 1442, 1446, 1538, 1688, 1825) |

| | | | | | | |
|---|---|---|---|---|---|---|
| **Myrtaceae** | | | | | | |
| *Pimenta dioica* (L.) Merr. | allspice, Jamaican pepper, piment, pimenta de Jamaica | fruit | Tr | r: Central America, West Indian Islands | ++ | (Fig. 84) Known under many names - ger. Nelkenpfeffer; fr. quatre-épices, tout-épices, piment giroflé, poivre giroflé (the french piment is the name for *Capsicum annuum*). Exported mostly from Jamaica, Honduras, Guatemala and Mexico (2760 t). Predominantly used in sausages and fish conserves (1358, 1442, 1446, 1538, 1825) |
| *Syzygium aromaticum* (L.) Merr. et L.M.Perry (*Eugenia aromatica* (L.) Baill. non Berg) | clove, girofle, árbol del clavo | flower-buds | Tr | w | +++ | (Fig. 85) World production about 80,000 t, exports about 40,000 t, mostly from Tanzania, Madagascar and Brazil, some also from Sri Lanka. The largest producer is Indonesia, where annually over 40,000 t is used for kretek cigarettes. Clove oil is primarily distilled from leftovers, stems, young shoots, and leaves (12, 797, 1107, 1147, 1358, 1383, 1442, 1446, 1538, 1688) |

Table 39. Further spices, cont'd (explanation of symbols, see footnotes page 55)

| Botanical name | Vernacular names | Used part | Climatic region | Distribution | Econ. value | Remarks and literature references |
|---|---|---|---|---|---|---|
| **Ranunculaceae** | | | | | | |
| *Nigella sativa* L. | black cumin, black caraway, cumin noir | seed | S | r: SW Asia, S Europe | + | Used in the Middle East and Balkans as seasoning for baking and cheese (376, 1442, 1564) |
| **Rutaceae** | | | | | | |
| *Murraya koenigii* (L.) Spreng. | curry leaf, ghandela | leaf | S, Tr | r: S Asia | + | Cultivated all over India. Leaves used for curry mixtures. Ornamental tree (376, 1383, 1392, 1442, 1538) |
| **Umbelliferae** | | | | | | |
| *Anethum graveolens* L. | dill | fruit, leaf | Te, S, Tr | w | ++ | Annual. Seeds have a multitude of uses, leaves fresh for salads, dried for sauces (642, 643, 798, 1358, 1442, 1538) |
| *Ammodaucus leucotrichus* Coss. et Dur. | cafoun | fruit | S, Tr | r: Sahara | + | Small, annual plants, cultivated as a seasoning and healing plant in the oases of the Sahara (200, 1846) |

| | | | | | | |
|---|---|---|---|---|---|---|
| *Anthriscus cerefolium* (L.) Hoffm. | chervil, cerfeuil, cerafolio | leaf | Te, S | w | + | Annual. Cultivated especially in Europe, the Mediterranean region and the USA. Mostly fresh in soups and salads, dried also with fish and sauces (642, 643, 1442) |
| *Bunium bulbocastanum* L. | black zira, earth nut, noix de terre | fruit, root | S, TrH | l: Kashmir | + | Cultivated in the Himalayas as a spice (fruits). Easy to propagate using tubers, which are eaten in times of shortage (200, 376, 798, 864, 1665) |
| *Carum carvi* L. | caraway, carvi, alcaravia | fruit | Te, S | w | +++ | Biennial. Apart from in Europe, cultivated especially in N Africa, USSR, and N America. Utilized in bread, with potatoes, cabbage, in cheese. Essential oil for spirits and liqueurs (376, 798, 1442) |

Table 39. Further spices, cont'd (explanation of symbols, see footnotes page 55)

| Botanical name | Vernacular names | Used part | Climatic region | Distribution | Econ. value | Remarks and literature references |
|---|---|---|---|---|---|---|
| *Coriandrum sativum* L. | coriander, cilantro, coriandre | fruit, (leaf) | Te, S, Tr | w | +++ | Annual. Fruits harvested when not quite ripe, and dried in the sun. Important spice in most tropical and subtropical countries. In S and SE Asia, the leaves also used as a soup seasoning. World production about 200,000 t, exports about 35,000 t. USSR and Morocco are the major exporters, India the major producer (376, 710, 798, 1236, 1442, 1446, 1635, 1687) |
| *Cuminum cyminum* L. | cumin, comino, cominho, jira, zeera | fruit | S, TrH | r: W Asia to India | ++ | Annual. Important spice in E Asia, India, and considerable exports from there and also Iran, especially to SE Asian countries. The USA annually imports more than 3000 tons (376, 437, 798, 1442, 1538, 1687) |

| | | | | | | |
|---|---|---|---|---|---|---|
| *Foeniculum vulgare* Mill. | fennel, fenouil, hinojo | fruit, (leaf) | Te, S, TrH | w | ++ | Many forms. Fruits for baking and liqueurs, leaves sometimes for salads. Fruit also used medicinally and for distillation of the oil. Forms with enlarged leaf bases cultivated as a vegetable (316, 376, 798, 1442, 1812) |
| *Petroselinum crispum* (Mill.) Nym. ex A. W. Hill | parsley, persil, perejil | leaf, roots | Te, S, Tr | w | +++ | Biennial. Cultivated every-where in many varieties. The fresh leaves are mostly used, but also dried, rubbed leaves, roots, and the seeds (for oil) (642, 643, 661, 1442, 1538) |
| *Pimpinella anisum* L. | anise, anis | fruit | Te, S, TrH | w | +++ | Annual. Exports mainly from Spain, Turkey and Mexico. Fruits especially used in baking. The main product is the essential oil, used for liqueurs (Pernod, ouzo, raki) (798, 1442) |
| *Trachyspermum ammi* (L.) Sprague (*Carum copticum* (L.) Benth. et Hook.f.) | ajowan, bishop's weed | fruit | S, Tr | r: E Mediter-ranean countries to India | ++ | Drought tolerant. Cultivated especially in India (exported). Also used medicinally (376, 798, 1383, 1442) |

Table 39. Further spices, cont'd (explanation of symbols, see footnotes page 55)

| Botanical name | Vernacular names | Used part | Climatic region | Distribution | Econ. value | Remarks and literature references |
|---|---|---|---|---|---|---|
| *T. roxburghianum* (DC.) Craib. (*Carum roxburghianum* Benth.) | ajmud | fruit, leaf | S, Tr | r: SE Asia | + | Much cultivated in India. Leaves used like parsley, fruits used as spice, and medicinally (259, 376, 1383) |
| **Verbenaceae** | | | | | | |
| *Lippia graveolens* H.B.K. | Mexican sage, orégano | leaf | S, Tr | r: Mexico, Central America | ++ | Apart from this species, also *L. micromera* Schau. and others. Important spice in its homeland, and in the USA (annual imports 1800 t). Not to be confused with *Origanum vulgare* from the Mediterranean (642, 643, 1183, 1538, 1825) |

## Monocotyledonae

| | | | | | | |
|---|---|---|---|---|---|---|
| **Alliaceae** | | | | | | |
| *Allium sativum* L. | garlic, ail blanc, ajo, alho, | bulb | Te, S, TrH | w | +++ | Drought and heat tolerant. World production 1.3 million t (India 250,000 t, Egypt 145,000 t, Spain 135,000 t). Utilized as fresh cloves and dehydrated, also the essential oil (659, 903, 1358, 1383, 1442) |
| **Iridaceae** | | | | | | |
| *Crocus sativus* L. | saffron, safran, azafrán | stigmas, style-tops | S | r: Mediter-ranean countries to India | + | World trade is small due to the high price (saffron costs 20 times as much as vanilla), the production is much reduced. Cultivated in Spain, S France, Italy, Middle East and India, where there is still a regular demand for it. The annual production in Kashmir amounts to 9000 kg (150, 376, 406, 816, 1096, 1383, 1442, 1538, 1671) |
| **Zingiberaceae** | | | | | | |
| *Aframomum mele-gueta* (Rosc.) K. Schum. | grains of paradise, melegueta pepper | seed | Tr | r: W Africa, Surinam | + | In W Africa generally cultivated in house gardens, and traded there. In world trade still only of slight importance (709, 1070, 1442, 1477) |

Table 39. Further spices, cont'd (explanation of symbols, see footnotes page 55)

| Botanical name | Vernacular names | Used part | Climatic region | Distribution | Econ. value | Remarks and literature references |
|---|---|---|---|---|---|---|
| *Alpinia galanga* (L.) Willd. (*Languas galanga* (L.) Stuntz) | greater galangal, galanga de l'Inde | rhizome | Tr | r: SE Asia | ++ | One of the most important spices in Malaysia and Java. Little used outside this region (259, 731, 1296, 1383, 1442) |
| *A. officinarum* Hance | lesser galangal, galanga de la Chine | rhizome | Tr | l: S China | + | Sharp taste, used less as a spice, more as a medicinal plant (259, 1383) |
| *Amomum aromaticum* Roxb. | Bengal cardamom | seed | Tr | l: India | + | Cultivated in N Bengal and the Khasi hills. Outside India, occasionally traded as "false cardamom" (376, 1383, 1442) |
| *A. compactum* Soland. ex Maton (*A. kepulaga* Sprague et Burk.) | round cardamom, amome à grappe, kapol, kapul | fruit | Tr | l: Java, Sumatra | + | Locally of considerable importance, also exported to other SE Asian countries, but seldom outside this region (259, 1296, 1442) |
| *A. krervanh* Pierre | Cambodia cardamom, krervanh | seed | Tr | l: Cambodia, Thailand | + | Of local importance in Indochina (200, 259, 1383, 1442) |

| | | | | | | |
|---|---|---|---|---|---|---|
| *A. subulatum* Roxb. | Nepal cardamom, greater cardamom | seed | Tr | l: India | + | Cultivated in swampy places in Bengal, Sikkim, Assam, and Madras (376, 797, 1442, 1666, 1671) |
| *A. xanthioides* Wall. | bastard cardamom, tavoy cardamom | seed | Tr | l: Thailand | + | Traded in SE Asia, seldom outside (259, 1383) |
| *Boesenbergia rotundata* (L.) Mansf. (*B. pandurata* (Roxb.) Schlecht.) | Chinese keys, temu kontji | rhizome | Tr | r: SE Asia | + | Much cultivated in the whole region as a spice for rice and pickles (259, 731, 1296, 1383, 1411) |
| *Curcuma amada* Roxb. | mango ginger, amada | rhizome | Tr | l: India | + | Cultivated in Bengal, Andhra Pradesh, and Madras, especially used in pickles. Smells like mango (259, 376, 1446) |
| *C. longa* L. (*C. domestica* Val.) | turmeric, safran des Indes, curcuma | rhizome | Tr | w | +++ | (Fig. 86) Cultivated in S, SE, and E Asia, and on the W Indian Islands. Used in all curry recipes. India alone produces 120,000 t annually, of which about 10,000 t are exported. Also used as a dyestuff (see p. 386) (376, 797, 912, 987, 1296, 1340, 1393, 1442, 1446, 1481, 1687) |

Table 39. Further spices, cont'd (explanation of symbols, see footnotes page 55)

| Botanical name | Vernacular names | Used part | Climatic region | Distribution | Econ. value | Remarks and literature references |
|---|---|---|---|---|---|---|
| *C. mangga* Val. et van Zijp | temu pauh, temu mangga | rhizome | Tr | r: Malaysia, Java | + | Cultivated as a spice in Malaysia and Java. Smells like mango (259, 1296, 1383, 1446) |
| *Elettaria cardamomum* (L.) Maton | cardamom cardamomier, cardamomo | fruit | Tr | w | ++ | (Fig. 87) Cultivated in a range of varieties. Most important producer is India. Total exports about 3000 t, of which 2000 t from India, 650 t from Guatemala, the remainder from Thailand and Sri Lanka. The greatest users are the Arabian countries, where cardamom is used as a coffee spice. The most expensive spice after saffron and vanilla (376, 1107, 1346, 1442, 1446, 1515, 1538, 1567, 1582, 1801) |
| *Kaempferia galanga* L. | galanga, bataki, kentjur | rhizome | Tr | r: S and SE Asia | + | Used from India to Java as a rice spice. To a small extent exported. The related *K. rotundata* L. is only used medicinally and as an ornamental plant (259, 376, 1296, 1383) |

| | | | | | | |
|---|---|---|---|---|---|---|
| *Zingiber mioga* (Thunb.) Rosc. | Japanese ginger, mioga ginger, myoga | rhizome | S | l: Japan | ++ | Used as with *Z. officinale*, mostly peeled and covered with lime powder when traded, of lesser quality (less sharp). In Japan, the young shoots and inflorescences also eaten (200, 1383) |
| *Z. officinale* Rosc. | ginger, gingembre, gengibre | rhizome | S, Tr | w | +++ | (Fig. 86) Cultivated in many countries. Main producers India and China. Exports about 16,000 t annually, especially from India, Malaysia, and Nigeria. Traded in many forms: fresh, as white and black ginger (dried after different preparation processes), candied (in pieces), powdered. Further important products are the extract (resinoid) and the essential oil (53, 259, 376, 797, 912, 1107, 1296, 1383, 1411, 1442, 1446, 1538, 1825) |

# Medicinal Plants

A very large number of medicinal plants are found in the tropics and subtropics; in India alone, the figure of 2500 has been named (954). By far the largest number of medicinal plants are gathered from the wild, and only a few are cultivated to a noteworthy extent (Table 40). Again, of these, only a few are of great interest for commerce and agricultural development. The list of plants which are also used in European pharmacy has to be increased by the plants whose major usage is in other fields (starch plants, vegetables, fruit, stimulants, spices, essential oils, fatty oils, resins, tanning materials, gums) (1703). On the other hand, there are some species which are traditionally considered medicinal plants, the production of which is now mostly used for other purposes, such as quinine as the bitter principle in tonic water and vermouth, or liquorice as flavouring for tobacco. Comprehensive listings are found in (6, 327, 445, 1216, 1382, 1383, 1842, 1962). For the agriculturalist, the most important medicinal plants are those containing active constituents which cannot be synthesized by the pharmaco-chemical industries (or only at very high cost), and which cannot be substituted by other compounds. The compounds which so far cannot be replaced are particularly alkaloids (from *Cinchona*, *Datura*, *Hyoscyamus*, *Papaver*, *Rauvolfia*, and others), *Digitalis* glycosides, and also flavonoids and mucilages.

The cultivation of medicinal plants has been strongly influenced by the chemical industry in recent years. Several medicinal plants have been practically driven off the market, latterly the *Hydnocarpus* spp. (Flacourtiaceae), which provided chaulmoogra oil, once the most important means of controlling leprosy until the development of the sulphonamides and thalidomide. On the other hand, industry has also promoted the cultivation of plants which it needs (in part organized by the chemical industry itself), because it could not work with the supplies of plants gathered from the wild, which vary too greatly in quality and quantity. This cultivation has actually led to the breeders selecting cultivars which are especially rich in the compounds which are required. The wild plants can then no longer compete with these selected lines. The classic example of this is the selection of *Cinchona* species in the second half of the 19th century by the Dutch in Indonesia (684), on whose results the present day *Cinchona* cultivars are based. Moreover, the chemists have succeeded in transforming crude plant material with little pharmacological effect into important new drugs, which led to the introduction and cultivation of new "medicinal plants", or which unlocked new markets for the old plants. The transformation of steroid sapogenins into cortisone and sex hormones is the outstanding example of this, but there are also some other compound groups to which this applies (codeine from thebaine from *Papaver bracteatum*, Table 40).

In general the economic importance of medicinal plants is slight for the cultivator. In only a few exceptional cases is the market for drugs so open that the cultivation of medicinal plants can be carried out independently by a farmer.

As a rule, cultivation takes place on the basis of contracts with the industry or with a wholesaler, who handles the supplies to the export or internal markets. As with spices, the cultivation of drugs declined in the industrial countries, due to the high labour costs, and, wherever it is climatically possible, it has shifted to the subtropical developing countries. Many medicinal plants originate from the tropics, and will only thrive there (e.g. *Rauvolfia*, *Cephaelis*, Zingiberaceae, and many of the spices and stimulants which are also used medicinally).

For most of the medicinal plants the data provided in Table 40 have to suffice. Only three groups need additional explanation.

## Quinine

Until the 1930s, quinine was the only agent against malaria. While this is no longer the case, it is still an important drug against fevers and cardiac arrhythmias (quinidine against heart flutter). The biggest demand comes from the drinks industry. The botanical classification is difficult for the cultivated species of *Cinchona*, Rubiaceae, particularly as nowadays mostly vegetatively propagated hybrids are cultivated (684, 1657). The most important are *C. calisaya* Wedd., with up to 3% quinidine and 10% total alkaloids in the bark, *C. ledgeriana* Moens ex Trim., with up to 16% total alkaloids but with low quinidine content, and *C. pubescens* Vahl (*C. succirubra* Pav. ex Klotsch), which is the original source of cinchona bark, and is often used as a rootstock for the vegetative propagation of selected types. All forms need humid tropical highland climates, or corresponding subtropical climates. The major producers are Zaire, Bolivia, Ecuador, and Peru, and minor producers are India, Indonesia, and the USSR (6, 106, 376, 684, 1216, 1590, 1960).

## Rauvolfia Alkaloids

The old Indian medicinal plant, *Rauvolfia serpentina* (L.) Benth., Apocynaceae, has been introduced into modern medicine since 1947, and nowadays is indispensable for reducing blood pressure and for the treatment of mental illnesses. The part of the plant which contains most alkaloids is the root, in which more than 30 different alkaloids have been identified. Of these, reserpine is the most active. *Rauvolfia* alkaloids are also found in similar concentrations in *R. vomitoria* Afzel. (Africa, exploited in Zaire) and *R. tetraphylla* L. (*R. canescens* L.) (originating on West Indian islands, also introduced to India). For cultivation, *R. serpentina* comes first into the picture, as this is easy to propagate vegetatively (the seeds often do not germinate well). It has been cultivated in India, Thailand and the Philippines for many years. India has now forbidden the export of the root to encourage the expansion of its own pharmaceutical industry (6, 106, 376, 455, 850, 1104, 1105, 1216, 1383, 1563, 1989).

## Steroid Hormones

The most valuable of the chemical compounds of plant origin which are available from the pharmaceutical industry are the steroid hormones, with an annual production of much more than 1000 t of pure substance. This development was initiated in the 1940s, after the discovery that cortisone could be obtained from diosgenin from *Dioscorea* species, Dioscoreaceae, through the action of fungi and by chemical transformations. Then, chemists and microbiologists discovered methods, whereby sex hormones could be created from the same raw materials which have formed the basis of hormonal contraception (307, 794). A small part of the steroid hormones are synthesized from cholic acid or cholesterol, or by total chemical synthesis, but over 90% originate from plant steroid saponins and steroid alkaloids. At present, the most important sources are diosgenin from *Dioscorea* species, mainly from Mexico and Central America ("barbasco", *D. macrostachya* Benth., *D. floribunda* Martens et Gal., *D. spiculiflora* Hemsl.), and Asia (*D. deltoidea* Wall., *D. prazeri* Prain et Burk.), stigmasterol from soya (USA), hecogenin from agaves (Tanzania), smilagenin from *Smilax*, *Yucca*, and *Agave* (Mexico), and solasodine from *Solanum laciniatum* Ait. and *S. viarum* Dun. (101, 106, 362, 378, 695, 1120, 1186, 1216, 1590). *Dioscorea* spp. are normally gathered from the wild, but are also cultivated for the purpose of diosgenin extraction, though the yields per ha and per year are not satisfactory. The most suitable are *D. macrostachya* and *D. floribunda*, or hybrids. At present, the interest of the agriculturalists is concentrated on annual plants which can be cultivated as field crops, such as *Solanum laciniatum*, *S. viarum* (both with about 5% solasodine), and *Trigonella foenum-graecum* (diosgenin, Table 25) (6, 196, 409, 523, 928). The *Solanum* species and *Trigonella foenum-graecum* are already utilized commercially for solasodine and diosgenin production respectively.

Table 40. Further medicinal plants (explanation of abbreviations, see footnotes page 55)

| Botanical name | Vernacular or product name | Used part | Climatic regions | Distribution | Econ. value | Remarks and literature references |
|---|---|---|---|---|---|---|
| | | | **Ascomycetes** | | | |
| **Clavicepitaceae** | | | | | | |
| *Claviceps purpurea* (Fr.) Tul. | ergot | sclerotium of the fungus | Te, S, TrH | w | + | Active substances: ergot alkaloids. For migraine and hypertonia, aid in giving birth. Strains with high alkaloid content bred, cultivated by artificial inoculation of rye fields. Also industrially in saprophytic culture of the fungus, similarly *C. paspali* Stev. et Hall (106, 863, 1216, 1884) |
| | | | **Dicotyledonae** | | | |
| **Apocynaceae** | | | | | | |
| *Catharanthus roseus* (L.) G. Don (*Vinca rosea* L.) | rose periwinkle, pervenche de Madagascar | leaf | S, Tr | w | ++ | Active substances: vinca-blastine, and many others. Anti-leukemia, anti-neo-plastic (oncolytic). Exporters: India, S Africa and other countries (6, 106, 376, 1216, 1350, 1382, 1590, 1784) |

Table 40. Further medicinal plants, cont'd (explanation of abbreviations, footnotes page 55)

| Botanical name | Vernacular or product name | Used part | Climatic regions | Distribution | Econ. value | Remarks and literature references |
|---|---|---|---|---|---|---|
| *Vinca minor* L. | common periwinkle, running myrtle | leaf | Te, S | r: SW Asia, SE Europe | + | Active substances: vincamine, vincine, and others. Against eczema and catarrh (6, 376, 1350, 1382) |
| **Araliaceae** | | | | | | |
| *Panax ginseng* C.A. Meyer | ginseng, Asiatic ginseng, jen shen | root | Te, S | r: Korea, China, Japan | ++ | Active substances: triterpene-saponins, ginsenosides. Overall tonic. In the USA, *P. quinquefolius* L. is cultivated (141, 763, 1383, 1441, 1479, 1884) |
| **Chenopodiaceae** | | | | | | |
| *Chenopodium ambrosioides* L. var. *anthelminticum* (L.) A. Gray. | wormseed, Mexican tea, ambroisine, paico | branches after flowering | S, TrH | r: S, Central and N America | + | Active substances: ascaridole, and others. Oil used against hookworm, and in veterinary medicine. Rarely used now (1383) |
| **Compositae** | | | | | | |
| *Anthemis nobilis* L. (*Chamaemelum nobile* (L.) All.) | Roman chamomile, camomille Romaine, manzanilla | flower | Te, S | r: S Europe, France, USA, Argentina | + | Active substances: chamazulene, esters of angelic acid. For rinsing mouths and wounds. Essential oil to a small extent used for liqueurs and perfumes (6, 376, 1382) |

| | | | | | | |
|---|---|---|---|---|---|---|
| *Matricaria recutita* L. (*Chamomilla recutita* (L.) Rauschert) | German chamomile, camomille vraie, camomila | flower | Te, S | w | ++ | Apigenin in the extract, chamazulene in the essential oil. Utilized as for *Anthemis nobilis* (6, 106, 376, 1382) |
| *Saussurea costus* (Falc.) Lipschitz (*S. lappa* C.B.Clarke) | costus, kuth, kut | root | TrH | r: India, Pakistan | + | Active substance: saussurine alkaloid. Resin and essential oil for perfumes and against skin diseases (376, 406, 661, 954, 1383, 1630) |
| *Silybum marianum* (L.) Gaertn. | holy thistle, milk thistle, chardon Marie, cardo de Maria | fruit, leaf, root | S | w | + | Active substances: flavonoids ("silymarin"). For illnesses of the spleen, liver and gall, and against seasickness (6, 376, 1382, 1842) |
| **Convolvulaceae** | | | | | | |
| *Ipomoea purga* (Wender.) Hayne (*Exogonium purga* (Wender.) Benth.) | jalap | root | Tr | r: Mexico Jamaica India | + | Active substances: convolvulin, jalapin. Purgative, especially in veterinary medicine. Other species used as well as *I. purga* (6, 376, 1382, 1444, 1957, 1960) |
| **Cucurbitaceae** | | | | | | |
| *Citrullus colocynthis* (L.) Schrad. | colocynth, bitter apple, altandal, tumba | fruit | S, Tr | r: SW and Central Asia, N Africa | + | Active substances: elaterinide (cucurbitacin E-glucoside). Mostly gathered from wild plants. Purgative (106, 1494, 1590) |

Table 40. Further medicinal plants, cont'd (explanation of abbreviations, footnotes page 55)

| Botanical name | Vernacular or product name | Used part | Climatic regions | Distribution | Econ. value | Remarks and literature references |
|---|---|---|---|---|---|---|
| **Labiatae** | | | | | | |
| *Orthosiphon aristatus* (Bl.) Miq. (*O. stamineus* Benth.) | Java tea, kidney tea | leaf | Tr | r: Indonesia, Philippines, Malaysia | + | Essential oil, bitter principle and tanning material. Diuretic for gout and kidney problems (259, 684, 1383) |
| **Leguminosae** | | | | | | |
| CAESALPINIOIDEAE | | | | | | |
| *Cassia angustifolia* Vahl | Tinnevelly senna, Arabian senna, Indian senna | leaf, fruit | S, Tr | r: India, Arabia, Pakistan | + | Active substances: sennoside A+B, sennidin A+B, kaempferol and kaempferin. Purgative. Cultivated in India (6, 376, 444, 1216, 1351, 1352, 1370, 1382, 1590) |
| *C. senna* L. (*C. acutifolia* Del.) | Alexandrian senna | leaf, fruit | S, Tr | l: Sudan | + | as for *C. angustifolia* |
| PAPILIONOIDEAE | | | | | | |
| *Glycyrrhiza glabra* L. | common licorice, liquorice réglisse | root | S | w | ++ | Active substance: glycyrrhizin. Remedy for coughs and relief of cramps. Mainly used for aromatizing tobacco, also for sweets, and in England as an addition to beer (6, 106, 376, 444, 1216, 1382, 1383, 1443, 1590) |

| | | | | | | |
|---|---|---|---|---|---|---|
| **Papaveraceae** | | | | | | |
| *Papaver bracteatum* Lindl. | Oriental poppy | whole plant | Te, S | l: Hungary | + | Active substance: thebaine (for codeine synthesis). Cannot be used for opium or the production of heroin (see p. 246) (6, 106, 443, 499, 1048, 1216, 1295, 1619, 1796) |
| *P. somniferum* L. | opium poppy, pavot somnifère, adormidera | leaf fruit seed | Te, S, Tr | w | ++ | Active substances: morphine, codeine, etc. Apart from the dried latex, leaves, ripe capsules, seeds, and the fatty oil won from the seeds (Table 18) are medicinal. For soothing and pain-killing preparations. Opium is also extracted from the straw of harvested plants. Main producers: Turkey, India, Afghanistan, SE and E Asia (6, 106, 376, 443, 986, 1216, 1442, 1796) |
| **Plantaginaceae** | | | | | | |
| *Plantago arenaria* Waldst. et Kit. (*P. psyllium* L.) | black psyllium, kala isabgol (India) | seed | S | l: India | + | Seeds traded as black psyllium. Low mucilage content. Also other *Plantago* species are used, but seldom cultivated now (376, 954, 1194, 1216, 1382) |

Table 40. Further medicinal plants, cont'd (explanation of abbreviations, footnotes page 55)

| Botanical name | Vernacular or product name | Used part | Climatic regions | Distribution | Econ. value | Remarks and literature references |
|---|---|---|---|---|---|---|
| *P. ovata* Forssk. | blonde psyllium, isabgol (India), ispaghula | seed, testa | S | l: India, China | ++ | Most important source of the commercial drugs (blonde psyllium, psyllium husks). Main production in India (about 4500 t/year), important export drug. Mucilage containing (see p. 373). Mild laxative, against dysentry and inflamations (376, 954, 995, 1170, 1194, 1216, 1382, 1470) |
| **Rhamnaceae** | | | | | | |
| *Rhamnus purshianus* DC. | cascara buckthorn, Western buckthorn, cáscara | bark | Te, S, TrH | l: N America, Kenya, India | + | Active substances: anthrachinones. Purgative. The bark as a drug is called "cáscara sagrada" (6, 376, 1216, 1382) |
| **Rubiaceae** | | | | | | |
| *Cephaëlis ipecacuanha* (Brot.) A. Rich. | ipecac, ipecacuanha | root | Tr | l: Brazil, India, Indonesia, Bangladesh | ++ | Active substances: emetine, cephaeline, psychotrine. Against whooping cough, bronchial asthma, amoebic dysentry, also tuberculostatic (6, 106, 376, 542, 1216, 1383, 1590, 1960) |

| | | | | | | |
|---|---|---|---|---|---|---|
| **Rutaceae** | | | | | | |
| *Agathosma betulina* (Bergius) Pillans (*Barosma betulina* Bartl. et Wendl.) | buchu | leaf | S | l: S Africa | + | Active substance: diosphenol. Diuretic, inflammation restraint. *A. crenulata* (L.) Pillans also utilized. Only *A. betulina* is cultivated. Essential oil (buchu oil) from the leaves used in liqueurs, perfumes, aromatics (190, 191, 590, 1905) |
| *Pilocarpus microphyllus* Stapf | jaborandi | leaf | Tr | r: S America | ++ | Active substances: pilocarpine and related alkaloids. Stimulant of heart and parasympathicus, and against glaucoma. Essential oil from leaves used in perfumery. Propagated by cuttings. Main exporter Brazil. Other *P.* species used for the same purposes (6, 376, 551, 758, 1216, 1382, 1518, 1625) |
| *Ruta graveolens* L. | common rue, rue puante, arruda | leaf | Te, S, Tr | w | + | Active substances: aromatic oil, methyl-n-nonylketone, rutin. Sedative material, mostly in folk medicine. Essential oil produced in Spain and Algeria. In small quantities used as an aromatic and in perfumes (259, 327, 376, 661, 1382) |

Table 40. Further medicinal plants, cont'd (explanation of abbreviations, footnotes page 55)

| Botanical name | Vernacular or product name | Used part | Climatic regions | Distribution | Econ. value | Remarks and literature references |
|---|---|---|---|---|---|---|
| **Scrophulariaceae** | | | | | | |
| *Digitalis lanata* Ehrh. | Grecian foxglove, digitale laineuse | leaf | Te, S | l: S Europe | +++ | Active substances: digitoxin, digoxin. Better suited for cultivation in warm countries than *D. purpurea* L. Cultivars bred with high content of glycosides (6, 106, 376, 1216, 1382, 1590) |
| **Solanaceae** | | | | | | |
| *Atropa acuminata* Royle ex Lindl. | Indian belladonna, mait-brand | leaf, root | TrH | l: India | + | Used like *A. bella-donna* (106, 376, 1216, 1590) |
| *A. bella-donna* L. | deadly nightshade, banewort, morel | leaf, root | Te, S | r: Europe, Asia, N America, N Africa | | Active substances: l-hyoscy-amine, atropine. Utilized as narcotic and nerve drug (6, 376, 1216, 1382) |
| *Datura metel* L. | Hindu datura, thornapple, dhatura (India) | leaf, seed | S, Tr | w | + | Active substances: predomin-antly hyoscine. Sedative (106, 327, 376, 1383) |

| | | | | | | |
|---|---|---|---|---|---|---|
| *D. stramonium* L. | Jimson weed, common thorn apple, stramoine | leaf, seed | Te, S, Tr | w | ++ | Active substances: l-hyoscyamine, hyoscine, atropine. Affects the central nervous system, for asthma cigarettes and fumigating materials (6, 106, 327, 376, 1216, 1382) |
| *Duboisia leichhardtii* F. v. Muell. | duboisia, corkwood | leaf | S | l: Australia | + | Active substances: hyoscyamine, hyoscine. Cultivated in Australia for the commercial extraction of the alkaloids. Also *D. myoporoides* R. Br. is utilized (6, 143, 275, 795, 1079, 1216) |
| *Hyoscyamus muticus* L. | Egyptian henbane | leaf | S, Tr | w | + | Most important species for the extraction of hyoscyamine. Primarily cultivated in Egypt (327, 376, 475, 1216) |
| *H. niger* L. | black henbane, jusquiame, beleño | leaf, root, seed | Te, S, Tr | w | ++ | Active substances: l-hyoscyamine, hyoscine. Painkilling, cramp relief, used against eye problems (6, 106, 327, 376, 1216, 1382) |
| **Umbelliferae** | | | | | | |
| *Ammi visnaga* (L.) Lam. | toothpick ammi, bishop's weed, khella (Egypt) | seed | S | w | + | Active substances: khellin, visnagin. Khellin used against angina pectoris and bronchial asthma. Mostly gathered from the wild. Small scale cultivation in Egypt (6, 106, 376, 1216, 1382, 1453) |

Table 40. Further medicinal plants, cont'd (explanation of abbreviations, footnotes page 55)

| Botanical name | Vernacular or product name | Used part | Climatic regions | Distribution | Econ. value | Remarks and literature references |
|---|---|---|---|---|---|---|
| **Monocotyledonae** | | | | | | |
| **Liliaceae** | | | | | | |
| *Aloë vera* (L.) N.L.Burm. (*A. barbadensis* Mill.) | Barbados aloe, true aloe, sábila | leaf | Tr | w | ++ | Active substances: aloin (barbaloin), aloe-emodin. Purgative and for injuries of the skin. Main exporters are the West Indian Islands. South African aloe (*A. ferox* Mill.) is gathered from wild plants (6, 656, 744, 1216, 1382, 1383, 1975) |
| **Zingiberaceae** | | | | | | |
| *Curcuma xanthorrhiza* Roxb. | temu lawak | rhizome | Tr | r: SE Asia | + | Active substances: curcumin, essential oil, p-toluylmethyl-carbinol. Gall-stimulating. Also used as a spice in SE Asia (6, 259, 684, 731, 1296, 1382, 1383) |
| *C. zedoaria* (Bergius) Rosc. | zedoary, zédoaire, temu puteh | rhizome | Tr | r: India, Sri Lanka | + | Active substances: gingerol, essential oil. Stomach treatments. Used like ginger (6, 259, 327, 376, 406, 946, 1296, 1383, 1446) |

| | | | | | | |
|---|---|---|---|---|---|---|
| *Zingiber cassumunar* Roxb. | cassumunar ginger, bangle | rhizome | Tr | r: SE Asia | + | Active substance: gingerol. Stomach treatments. Also as a spice (200, 259, 327, 1383, 1446) |
| *Z. zerumbet* (L.) Rosc. ex Sm. | zerumbet ginger, lempoyang | rhizome | Tr | r: India, Indochina | + | Active substances: gingerol, zingiberene. Stomach treatments (200, 259, 327, 376, 1296, 1383, 1446) |

# Essential Oils

Essential oils are volatile substances. They can be distinguished from the fatty oils (see p. 76) in that a droplet placed on filter paper does not leave a grease spot. Their volatility explains their distinctive odours and their use as thinners for paints (turpentine oil, camphor oil). Chemically, they are totally different from the fatty oils. Their characteristic compounds are monoterpenes, which are present as hydrocarbons, alcohols, aldehydes, ketones and esters with short-chained fatty acids. In some, there occur also indole and S-containing compounds (allyl isothiocyanate). All essential oils are mixtures of such compounds. Of the approximately 2500 types that are known, about 100 are used, the majority only occasionally and in small amounts in perfumery, about 40 regularly, and sometimes in considerable quantities. They are utilized:

- especially in the perfume industry and for cosmetic articles (soap, ointments, powder);
- in the food industry (lemonades, liqueurs, confectionery);
- for technical purposes (solvents, flotation agents);
- for masking smells in plastics, artificial leather, rubber, floor wax, household sprays;
- and in pharmaceutical preparations because of their specific effects (anise oil, fennel oil, camomile oil), their antiseptic properties, and to improve taste (toothpaste).

Because of their manifold uses, the literature on essential oils is scattered through pharmaceutical, cosmetic, chemical, and technical journals. There is basic information for the agronomist in the encyclopaedia of GUENTHER (661). A compendium of all aromatic substances (not only essential oils) is given in (758).

Essential oils are formed in the cytoplasm, and mostly stored in special organs (epidermal hairs, secretion glands, resin ducts). Their concentration in the plant organs is seldom so high that they can be extracted by mechanical means (citrus peel oils). As a rule they are separated by steam distillation, but some valuable perfume oils, especially from flowers, are won by solvent extraction (hexane or benzol). After evaporation of the solvent, the "concrete" remains, in which the concentration of the essential oil is usually only a few percent; its main constituents are then fats and waxes. The alcohol extraction from the concrete leaves the "absolute", which consists of 20-25% essential oil.

In terms of quantity and value, the citrus oils lead: peel oils from lemon, lime, mandarin, grapefruit, bergamot; leaf oils (petitgrain oils), and flower oils (orange flower oil, neroli oil from bitter orange). They are followed by the group of grass oils (citronella oil, lemongrass oil, palmarosa oil, vetiver oil). Among the individual oils which are required in relatively greater quantities, the following should be mentioned: clove oil, eucalyptus oil (1371), oils from the various *Mentha* species, patchouli oil, rosewood oil, lavender and lavandin oils, nutmeg oil, geranium oil, rosemary oil, sandalwood oil, and cassia oil. In terms of price, the aromatic oils fall into the very expensive category (neroli, orris root, rose, and some seldom used flower oils), or the middle price range (geranium, sandalwood, bergamot, fine peppermint oil, ylang-ylang, vetiver). There are also cheap oils, such as lavandin, patchouli, cassia, rosewood, and very cheap ones (citronella, bitter almond, eucalyptus, camphor).

There are few groups of useful plants which have been planted in the developing countries as often as the essential oil plants for "crop diversification" and as a potential source of foreign exchange, but frequently with disappointing results. The reasons for these failures are (1521, 1822):

1. expensive oils have a limited market, and mostly require considerable technical experience to compete with the established producers;
2. cheap bulk oils provide very little profit for the producers, and competition with the chemical industry is especially intense, so that only very efficient production methods can lead to economic success;
3. the purchasers are a relatively small number of large firms;
4. climate, soil and cultivar have a great influence on the quality, and oil that is below the standard quality for trading is unsaleable, or saleable only at a large discount;
5. the demand is always subject to large fluctuations (e.g. the citral from lemongrass oil was very much sought after in the 1950s for the production of vitamin A; the price of lemongrass oil has fallen dramatically since citral and vitamin A can be derived synthetically from pinene) (1822).

For developing countries, the oils which are of special interest are those in the middle price range with a proportionately large volume of trading (see above), and apart from these, the oils whose procurement requires a lot of manual work (most flower oils). Other positive factors can be: plants which are cultivated for protection against soil erosion, and those where the residues can be used for animal fodder (*Cymbopogon* spp.). The position is different where essential oils are a by-product (citrus, spices, medicinal plants); with these oils, which sometimes fetch a good price, it is possible for some developing countries to enter the world markets (778, 1526, 1766, 1822).

In Table 41, only those plants are included which are cultivated solely or mainly for the production of essential oils. It has already been indicated in the relevant sections where spices, medicinal plants and fruit species are additional important sources of essential oils. To complete the picture, there is a relatively large number of essential oils which are exclusively extracted from wild plants or forest trees, and which can sometimes be an important source of income for the producer. Among these are West Indian sandalwood oil from *Amyris balsamifera* L. (Haiti), cabreuva oil from *Myrocarpus frondosus* Allem. and *M. fastigiatus* Allem. (Brazil), cajuput oil from *Melaleuca leucadendron* (L.) L. (Southeast Asia), guaiak wood oil from *Bulnesia sarmienti* Lorentz (Argentina, Paraguay), may chang oil from *Litsea cubeba* Pers. (China), muhuhu oil from *Brachylaena hutchinsii* Hutch. (East Africa), myrtle oil from *Myrtus communis* L. (Mediterranean countries), niaouli oil from *Melaleuca viridiflora* Gaertn. (New Caledonia), rosewood oil from *Aniba duckei* Kosterm. (Brazil), sassafras oil from *Ocotea pretiosa* Mez (Brazil), spike oil from *Lavandula latifolia* (L.f.) Medik. (Spain), and the large group of the various conifer oils (turpentine, pine needle, cypress, thuja oils, etc.).

Table 41. Essential oils. Further species in (661, 758) (explanation of symbols, see footnotes page 55)

| Botanical name | Vernacular or product name | Used part | Climatic regions | Distribution | Econ. value | Remarks and literature references |
|---|---|---|---|---|---|---|
| **Dicotyledonae** | | | | | | |
| **Annonaceae** | | | | | | |
| *Canangium odoratum* (Lam.) Baill. | ylang-ylang, cananga, cadmia | flower | Tr | r: SE Asia, Madagascar, Réunion, Comores | ++ | Ylang-ylang is a high-value oil, and is the first produced in the distillation. Cananga oil is won in the further course of distillation; this name is also used in trade for an oil extracted in Java from a different form of *C. odoratum*. Ylang-ylang is twice as expensive as cananga. Both oils are used in perfumery (376, 661) |
| **Burseraceae** | | | | | | |
| *Bursera penicillata* (Sessé et Moç. ex DC.) Engl. | elemi gum, linaloe oil | fruit, wood | Tr | r: Mexico, India | + | In India (Mysore), cultivated for the extraction of the essential oil from the berry peels. In Mexico, the wood of wild-growing trees is also distilled. Oil for cosmetics and soap (376, 410) |

| | | | | | | | |
|---|---|---|---|---|---|---|---|
| **Geraniaceae** | | | | | | | |
| *Pelargonium* hybrids | geranium, rose geranium, géranium rosat | leaf, green shoots | S, TrH | w | | +++ | All of the *Pelargonium* strains which are cutivated for oil production are amphidiploid hybrids of South African species, probably *P. capitatum* (L.) L'Hérit. and *P. radens* H.E.Moore (419). None of the true species originating from Cape Province yield a usable oil. The botanical names which are found in the literature for oil geraniums should be deleted. The most important producers are Réunion, Algeria, Morocco, Zaire. The oil has a range of uses in perfumery (419, 661, 1184, 1525) |
| **Labiatae** | | | | | | | |
| *Lavandula angustifolia* Mill. (*L. officinalis* Chaix, *L. vera* DC.) | lavender, lavande, lavander, | flowering shoots | Te, S | w | | ++ | Produced mostly in S France and other Mediterranean countries, in small amounts also in Australia, Argentina, S Africa and other countries. Important perfume oil (661, 662) |

Table 41. Essential oils, cont'd . Further species in (661, 758) (explanation of symbols, see footnotes page 55)

| Botanical name | Vernacular or product name | Used part | Climatic regions | Distribution | Econ. value | Remarks and literature references |
|---|---|---|---|---|---|---|
| *L.* x *intermedia* Emeric ex Loisel. (*L. angustifolia* Mill. x *L. latifolia* (L.f.) Medik.) | lavandin, lavande bâtarde | flowering shoots | Te, S | w | +++ | Yield per surface area substantially higher than for lavender. The oil has a definite smell of camphor and cineole, and thus is less valuable. Used to a large extent for soap and cosmetic preparations. From wild stands of spike lavender (*L. latifolia* (L.f.) Medik.), the cheap spike oil is extracted on the Iberian peninsula (661, 662) |
| *Mentha arvensis* L. | cornmint, Japanese mint, baume des champs | leaf | S, TrH | w | +++ | Was previously very sought after because of its high menthol content (82-86%). After separation of the majority of the menthol (nowadays also synthesized from pinene), the oil is used like peppermint oil, but is of lower value. Main production in E Asia and Brazil (637, 661, 1442) |

| | | | | | | |
|---|---|---|---|---|---|---|
| *M.* x *piperita* L. | peppermint, menthe poivrée | leaf | Te, S, TrH | w | +++ | The main production is in the temperate zone. The yield is lower in the subtropics, and the oil is of lower quality due to the higher menthol and menthofuran contents. Utilized in the food industry, cosmetics and pharmacy (641, 661, 1822) |
| *M. pulegium* L. | pennyroyal, pouliot | leaf | S | r: W Mediter-ranean countries | + | Production slight. Main constituent of the oil is pulegone. Mostly for soaps and for masking the odour of plastics and detergents (661) |
| *M. spicata* L. | spearmint, menthe verte | leaf | Te, S | w | ++ | Cultivated predominantly in the temperate zone, especially in the USA. Chemically totally different from peppermint oil. Especially used as an aromatizer in the food industry, for chewing gum and toothpaste. The herb is often used as a seasoning (661) |
| *Pogostemon cablin* (Blanco)Benth. | patchouli | leaf | Tr | r: SE Asia | ++ | 90% of the oil traded comes from Indonesia (Sumatra). Small quantities also from the Seychelles and China. Used in perfumery (661, 1383, 1520) |

Table 41. Essential oils, cont'd . Further species in (661, 758) (explanation of symbols, see footnotes page 55)

| Botanical name | Vernacular or product name | Used part | Climatic regions | Distribution | Econ. value | Remarks and literature references |
|---|---|---|---|---|---|---|
| *Salvia sclarea* L. | clary sage, sclarée | inflores-cence | Te, S | w | + | Main producers: Mediter-ranean countries, USSR. Utilized in perfumery and as an aroma in drinks and sweets (661, 1442) |
| **Lauraceae** | | | | | | |
| *Cinnamomum camphora* (L.) J.S.Presl | camphor tree, camphrier, alcanfor, chang shu | wood, leaf | S, Tr | r: E Asia | ++ | Oil from the wood of the camphor tree and several related species. The camphor separates from the oil during distillation. The oil is cheap, and is mostly used for technical purposes (lacquers, insecticides, ore flotation), and also pharmaceutically. Since the synthetic production of camphor, the oil is the most important product. The leaf oil of *C. camphora* is used in Taiwan (Ho leaf oil) for the production of linalool (661, 1383, 1446, 1590) |

| | | | | | | |
|---|---|---|---|---|---|---|
| **Leguminosae**<br>MIMOSOIDEAE<br>*Acacia farnesiana* (L.) Willd. | sweet acacia, cassie ancienne, aromo | flower | S | r: Mediterranean countries | + | For perfume, extracted as the concrete and absolute (333, 376, 406, 444, 661, 1383) |
| **Magnoliaceae**<br>*Michelia champaca* L. | champaca, champa | flower | Tr | r: SE Asia, Réunion, Madagascar | + | Mostly extracted as both concrete and absolute. Perfumes, and perfuming of tea (406, 661, 1383) |
| **Malvaceae**<br>*Abelmoschus moschatus* Medik. (*Hibiscus abelmoschus* L.) | ambrette, musk mallow, abu-el-misk, habb el mosk | seed | Tr | w | + | The aromatic oil is won from the seeds by steam distillation or by extraction. Used in the drinks industry and perfumery. The seeds are themselves traded (musk grains, musk pods). Production in Java, Martinique, W Indian islands. A fibre which is used locally is won from the stems (376, 406, 661) |
| **Myrtaceae**<br>*Eucalyptus citriodora* Hook. | lemon-scented gum | leaf | S, Tr | w | + | Valued for perfume purposes. For the production of citronellal (contains 60-80%). Mainly cultivated in: USSR, India, Zaire, Brazil, China (512, 1234, 1235, 1371, 1522) |

Table 41. Essential oils, cont'd . Further species in (661, 758) (explanation of symbols, see footnotes page 55)

| Botanical name | Vernacular or product name | Used part | Climatic regions | Distribution | Econ. value | Remarks and literature references |
|---|---|---|---|---|---|---|
| *E. dives* Schau. | broadleaf peppermint | leaf | S, Tr | w | + | For technical purposes (flotation processes), in perfumery, and as an antiseptic, contains 40% piperitone, which can be converted to thymol. Mainly cultivated in Zaire and Australia (512, 1371, 1522, 1822) |
| *E. globulus* Labill. | Tasmanian blue gum, eucalipto azul | leaf | S, Tr | w | +++ | The oil contains over 70% cineole, which has a strongly antiseptic effect. Utilized in medicine and perfumery. Mainly cultivated in Brazil, USSR, Spain, Ecuador, Portugal and India. The tree is often used in shelterbelts (376, 512, 1371, 1522, 1822) |
| ssp. *maidenii* (F.v.Muell.) Kirkp. (*E. maidenii* F.v.Muell.) | Maiden's gum | leaf | S, Tr | w | + | Used like *E. globulus* (higher oil yield). Mainly cultivated in Brazil, Zaire, and USSR (512, 1371, 1522) |

| | | | | | | |
|---|---|---|---|---|---|---|
| *E. macarthurii* Dean et Maiden | camden wollybut | leaf | S, Tr | w | + | For perfume, contains 60-70% geranyl acetate. Mainly cultivated in Zaire and USSR (512, 1371, 1522) |
| *E. radiata* Sieb ex DC. | narrowleaf peppermint | leaf | S, Tr | w | + | Depending on the chemical composition of the oil, 4 types are distinguished, which to some extent are valued perfume oils. Production in Australia (512, 1371) |
| *E. smithii* R.T.Bak. | gully gum | leaf | S, Tr | w | + | Used like *E. globulus* (higher oil yield). Mainly cultivated in Guatemala, Zaire, USSR, and Brazil (512, 1371, 1522) |
| *E. staigeriana* F.v.Muell. | lemon-scented iron bark | leaf | S, Tr | w | + | For perfume and technical purposes, contains 60% l-limonene and 30% citral. Mainly cultivated in Brazil, Guatemala and Zaire (512, 1371) |
| *Leptospermum flavescens* Sm. (*L. citratum* Challinor, Cheel et Penfold) | yellow tea tree, tea bush, tea tree oil | leaf | S, TrH | r: Australia, Zaire, Kenya, S Africa, Guatemala | + | The oil smells refreshingly like lemons. Was in demand in the 50s due to its high citral content. Nowadays used to a small extent in perfumes (3) |

Table 41. Essential oils, cont'd . Further species in (661, 758) (explanation of symbols, see footnotes page 55)

| Botanical name | Vernacular or product name | Used part | Climatic regions | Distribution | Econ. value | Remarks and literature references |
|---|---|---|---|---|---|---|
| *Pimenta racemosa* (Mill.) J.W.Moore (*P. acris* (Sw.) Kostel) | bay rum tree, bay oil, bois bay-rum | leaf | Tr | l: Dominica, Puerto Rico | + | In perfumery, and for the production of bay rum, smells of cloves. Also used as a spice (50, 68, 661) |
| **Oleaceae** | | | | | | |
| *Jasminum officinale* L. (*J. grandiflorum* L.) | royal jasmine, jasmin commun | flower | S, Tr | r: Mediter-ranean countries | + | Flower oil won by extraction. Also other species, especially *J. sambac* (L.) Ait., cultivated for production of the flower oil. Utilized in perfumery (376, 661, 1822) |
| **Rosaceae** | | | | | | |
| *Rosa* x *centifolia* L. | cabbage rose, rose de mai, rose oil | flower | Te, S | w | ++ | Oil mostly won by extraction. Used in perfumery and cosmetics. Main producers France and Morocco (376, 661, 1822) |

| | | | | | | |
|---|---|---|---|---|---|---|
| *R.* x *damascena* Mill. f. *trigintipetala* (Dieck) Keller | Damask rose, otto of rose, attar of rose | flower | Te, S | w | +++ | Oil predominantly extracted by steam distillation. Apart from the oil, the rosewater which is formed in the distillation is also traded. Perfumery and cosmetics, confectionery (marzipan). Main producers Bulgaria, Turkey, USSR and India (376, 661, 1673, 1941) |
| **Santalaceae** | | | | | | |
| *Santalum album* L. | white sandalwood, chandana | wood | S, Tr | r: India, SE Asia | ++ | The plant is semi-parasitic, and is reared in the nursery on suitable host plants. The world market is dominated by Indian oil. Attempts to cultivate it in other regions have been successful, particularly in Indonesia. Utilized in perfumery and pharmacy (273, 376, 406, 684, 1887) |

Table 41. Essential oils, cont'd . Further species in (661, 758) (explanation of symbols, see footnotes page 55)

| Botanical name | Vernacular or product name | Used part | Climatic regions | Distribution | Econ. value | Remarks and literature references |
|---|---|---|---|---|---|---|
| | | | **Monocotyledonae** | | | |
| **Agavaceae** | | | | | | |
| *Polianthes tuberosa* L. | tuberose, tubéreuse, jacinthe des Indes, omixochitl | flower | Te, S, Tr | w | + | Found worldwide as a garden flower. The strongly scented flower oil is produced as the concrete by extraction in France, Morocco, Réunion, India. One of the most expensive perfume oils, only used in small amounts due to the strong scent (376, 1383, 1822, 1831) |
| **Araceae** | | | | | | |
| *Acorus calamus* L. | sweetflag, acore odorant | rhizome | Te, S, Tr | w | + | Oil won from the root by steam distillation. Most important producer is India. Used for drinks, in perfumery and pharmacy (376, 406, 661, 1383) |

| | | | | | | |
|---|---|---|---|---|---|---|
| **Gramineae** | | | | | | |
| *Cymbopogon citratus* (DC.) Stapf | West Indian lemongrass, verveine des Indes, capim cidreira | leaf | Tr | w | ++ | Produced in Guatemala, Haiti, Brazil, China. Contains 70-85% citral, from which ionone (violet scent) can be derived. For perfumes, cosmetics, an aromatic in drinks. Leaves used as a herb in soups in SE Asia (376, 859, 1522) |
| *C. flexuosus* (Nees ex Steud.) W.Wats | East Indian lemongrass, herbe de Malabar | leaf | Tr | l: India | ++ | Used like the oil from *C. citratus* (376, 661, 859, 1522) |
| *C. martinii* (Roxb.) W.Wats. | gingergrass, rosha grass, palmarosa | leaf | Tr | r: S and SE Asia | ++ | Palmarosa oil is won from the form "motia" particularly in India from wild-growing and cultivated plants. Because of its high geraniol content, it is also called "East Indian geranium oil" when traded. Used in perfumery and the food industry. Gingergrass oil is a cheap oil, which is distilled from the wild-growing form "sofia" in India. Used for cheap perfumes (376, 661, 859, 1349) |

Table 41. Essential oils, cont'd . Further species in (661, 758) (explanation of symbols, see footnotes page 55)

| Botanical name | Vernacular or product name | Used part | Climatic regions | Distribution | Econ. value | Remarks and literature references |
|---|---|---|---|---|---|---|
| *C. nardus* (L.) Rendle | Ceylon citronella, lenabatu | leaf | Tr | r: Sri Lanka, India, Taiwan | ++ | Climatically demanding. Used in perfumery, and for the production of geraniol (376, 661, 859, 1522) |
| *C. winterianus* Jowitt | Java citronella, mahapangiri | leaf | Tr | w | +++ | Morphologically similar to *C. nardus* (also combined with this in the species *C. confertiflorus* (Steudt.) Stapf; the separation is preferred because of the different nature of the oils). Most important producers: Indonesia, Guatemala, Honduras, Malaysia, Taiwan, Brazil. Utilized in perfumery, and for extraction of citronellal. Also planted for the control of soil erosion (661, 859, 1429, 1522) |

| | | | | | | |
|---|---|---|---|---|---|---|
| *Vetiveria zizanioides* (L.) Nash | vetiver, khas-khas | root | Tr | w | +++ | Main production in India, Indonesia, Réunion, Zaire, Haiti, Brazil. The oil is extracted by steam distillation of roots which have been dried in the sun. Much used for perfumes, cosmetics and soaps. Also planted for erosion control (see p. 439) (406, 661, 1520, 1822) |
| **Iridaceae** | | | | | | |
| *Iris pallida* Lam. | oris root | rhizome | S | l: Italy | + | Won by extraction. Next to rose and neroli oils, one of the most expensive perfume oils (376, 661) |

# Fibre Plants

Vegetable fibres have served mankind since our beginnings. Early in the prehistoric epoch man learned to process them into textiles by weaving. Some of the fibre plants were among the first cultivated plants (flax, sunn hemp, cotton, kenaf). The most important fibres are those that can be processed to make the finest fabrics for clothing and bed-linen; of these, cotton has been by far the most important since the 19th century (Fig. 88). Other fibres are of great economic importance too, such as those that are used for the production of bags and other packaging materials, for curtain materials and floor coverings (jute, kenaf, agave fibre), and those fibres which are indispensable for tear-proof paper (halfa grass, and many others). Finally, some plant fibres are irreplaceable for doormats (coir), brooms (piassave, sorghum and others), and for basket-work and binding materials (raffia and other palm fibres, agave fibres). The tropics and subtropics have almost a monopoly in the production of plant fibres. In terms of value they provide 96% of the world's production.

In spite of competition from the development of the synthetic fibre industry, plant fibres have retained their dominant role. Cotton alone supplies 46% of all fibres, the total for all plant fibres is 65%, animal fibres (wool and silk) supply 5%, and artificial fibres (cellulose and synthetic fibres) provide 30% of the world's production. The quantitatively dominant role of plant fibres is based on their low price compared to animal and some synthetic fibres, as well as on their technical characteristics (e.g high capacity for the absorption of moisture, which is indispensable for underwear and bedding, or for packaging materials). Other fibres are superior for other purposes (ease of dyeing and warming effect of wool and silk; tensile strength, shrink resistance and non-creasing qualities of many synthetic fabrics). From the viewpoint of environmental effects, the biodegradability of natural fibres is to their advantage, especially for packing materials. The major share of the market will always remain with plant fibres. In the fields of utilization, where they are in competition with the synthetic fibres, their market position is also relatively good, as long as efficient production methods are used, and as long as they are available at a maintained quality and in sufficient quantity. Since 1973, the oil crisis and the rise in the price of wood have had a favourable effect on the agriculturally produced fibres.

The only fibre for which the production has risen further in the last decade is cotton (Fig. 88), which provides 78% of the plant fibres. Jute, kenaf and similar fibres provide 14%, sisal 2%. Then follow the two fibres which are predominantly produced in the temperate zone: flax, 3%, and hemp, 1%. The remainder is extracted from a large number of other fibre plants (Table 44), most of which are of only local importance. There is an export market only for coir and abaca (Manila hemp), which are of economic interest, though only on a small scale. Only the fibres of worldwide economic importance can be fully described here; for the others, Table 44 and the literature must be referred to (176, 420, 972, 1444, 1445).

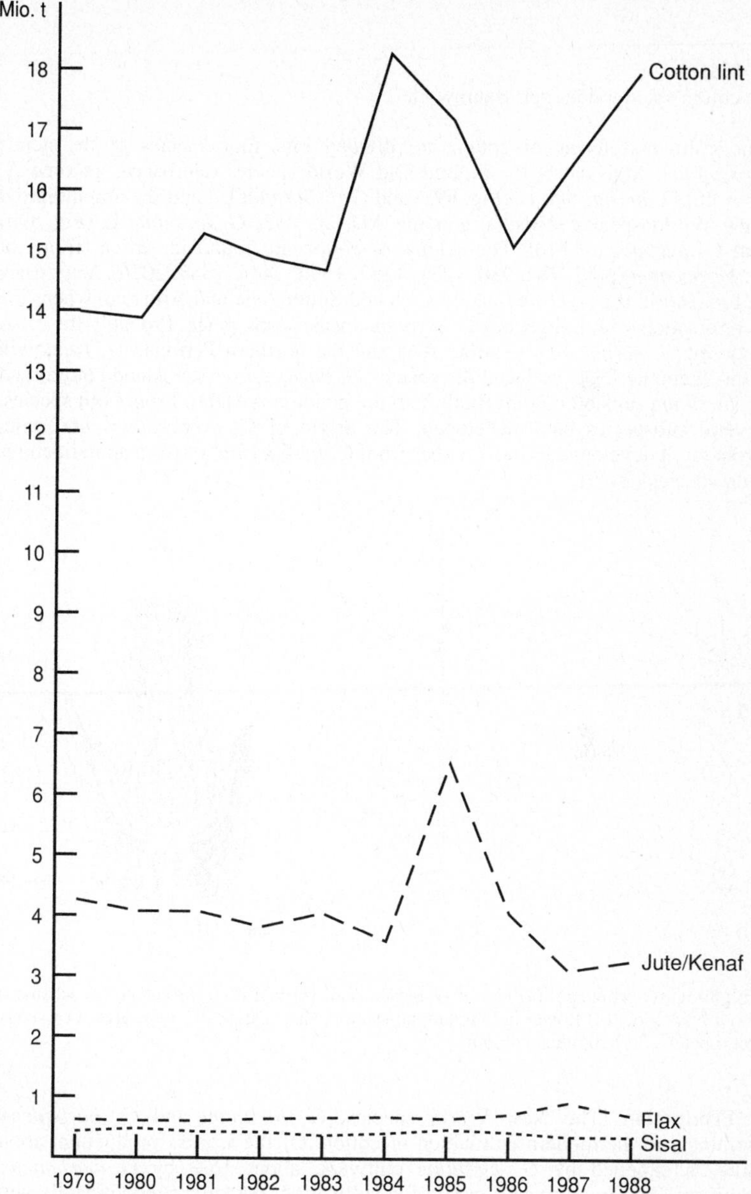


Fig. 88. World production of vegetable fibres, 1979-1988

## Cotton

fr. coton; sp. algodón; ger. Baumwolle

The cultivated forms of cotton are divided into four species of the genus *Gossypium*, Malvaceae, the diploid Old World species (Herbacea, genome A, 2n = 26) *G. herbaceum* L. (Fig. 89a) and *G. arboreum* L., and the amphidiploid New World species (Hirsuta, genome AD, 2n = 52) *G. hirsutum* L. (Fig. 89b) and *G. barbadense* Mill. The origins of cotton are found in Africa (forms of *G. herbaceum*) (782, 784, 980, 1231, 1357, 1396, 1444, 1657, 2026). Very early in prehistoric times, *G. herbaceum* reached South Asia and America, where the hybridisation with indigenous *Gossypium* species took place. Probably the cross between *G. herbaceum* (genome AA) and the northern Peruvian *G. raimondii* Ulbr. (genome DD) produced the species *G. barbadense* (sea island cotton) and *G. hirsutum* (upland cotton) (both with the genome AADD). From both species, several subspecies have developed. The origin of *G. arboreum* is uncertain. Probably it developed in India, either from *G. herbaceum*, or from an indigenous wild species (2026).

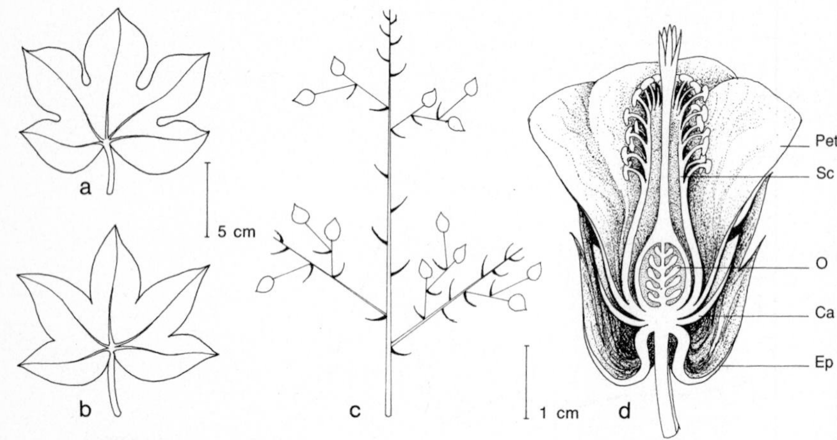


Fig. 89. *Gossypium* spp. (a) leaf of *G. herbaceum*, (b) leaf of *G. hirsutum*, (c) scheme of shoot branching, (d) flower in longitudinal section. Ca = calyx, Ep = epicalyx, O = ovary, Pet = petals, Sc = staminal column

**Production**. The New World species, *G. hirsutum* and *G. barbadense* dominate in the modern cultivation of cotton. Of the world's production, about 80% is supplied by *G. hirsutum* cultivars, about 15% by *G. barbadense* cultivars, the former mostly in rain-fed cultivation, the latter predominantly with irrigation. The Old World species are occasionally still cultivated in Africa, and to a considerable extent in South and East Asia. In India, *G. arboreum* takes 28% of the surface area cultivated with cotton, and *G. herbaceum* 19%. Two technical advances were responsible for developing cotton into by far the most important fibre plant: the invention of the ginning machine by Whitney (USA) in 1794, and the use of modern insecticides, without which the cultivation of cotton would no longer be thinkable in most countries.

The most conspicuous event in the last years has been the increase in cotton production in China, based on *G. hirsutum* (1979: 2.2, 1988: 4.2 million t lint) (1064, 2003). India still has the largest surface area under cotton (7.5 million ha), China now takes second place (5.5 million ha), the USA (4.8 million ha), the USSR (3.4 million ha), Brazil and Pakistan (each 2.6 million ha) follow. In production of lint, the USA, for decades the largest cotton producer, is now second (3.4 million t), the USSR third (2.7 million t), followed by India and Pakistan (1.5 million t each), while Brazil, with its low yield per ha, produces only 0.7 million tons of lint. The USA is still the largest cotton lint exporter (1.2 million t); the USSR exported 0.8 million t in 1988, Pakistan 0.5, China 0.5, the Sudan 0.2, Australia 0.2, Paraguay 0.2, Argentina and Turkey 0.1 million tons each. However, the figures for lint exports give an incomplete picture of the importance of cotton for the economies of a considerable number of developing countries. Cotton has been the basis of a local textile industry in these countries, which not only provides for the local needs, but has become the supplier for an expanding export business in yarn, cloth and finished products.

**Morphology and Anatomy**. All cotton species are potentially perennial, even though they are normally grown for only one year in modern agriculture They form a strong tap-root, which develops even at the seedling stage, and which can reach a depth of 3 metres. The shoot system is dimorphic: main axis and lower branches are monopodial and vegetative (that is, no flowers form in the leaf axils); the fruiting branches are sympodial (each flower stands at the end of the shoot, further growth occurs from the axil of the subtending leaf, so that the fruit appears to be inserted opposite the leaf) (Fig. 89c) (341). Leaves and stems are mostly hairy. All parts of the plant usually bear glands which are visible as dark spots, and gossypol is formed in these. Each flower is surrounded by three deeply divided bracts (epicalyx, Fig. 89d). The fibres themselves are single-celled hairs, which develop from the outer epidermis cells of the integument (Fig. 90a). Some of the hairs remain short, and form the fuzz covering the ripe seeds (linters). The most useful ones are the long hairs (lint), which are more than 20 mm long in modern cultivars, in primitive types more than 9.5 mm (660). The fruits (bolls) grow very quickly after pollination. After about 20 days, they reach their final size, and they are ripe after a further 25-45 days. With most species, the dry boll walls open in the middle of each carpel, and the fibre-mass emerges. However, the seeds remain clinging to the placenta, and are first separated from the fruit by picking (in unfavourable weather, this can also occur with strong winds). When the seeds ripen, the hairs die, and their wall collapses, so that only a narrow cavity remains inside, which still contains the remainder of the protoplasm. The wall of the hairs is composed of many layers of cellulose fibres, which run spirally. The direction of the spiral bands reverses at certain points and changes from layer to layer. This explains the twisting which is characteristic for dry cotton fibres (Fig. 90b). As parts of the epidermis, the hairs are covered with a cuticle. Because of this layer of wax, unprocessed cotton fibres feel fatty to the touch and repel water.

Species and cultivar differences are found in the type of branching, in the leaf shape (Fig. 89a and b), in the length and shape of the bracts, in the length and shape of the bolls, and in the amount and quality of the fibre. Modern cultivars are often difficult to identify on the basis of vegetative characteristics. Reliable seed propagation methods are therefore especially important for the cultivation of cotton. For descriptions of the cultivars, the local literature must be referred to (240, 821, 1002, 1432, 1629).

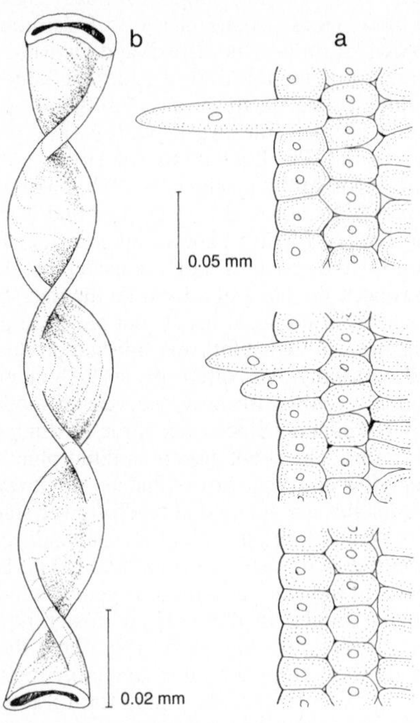


Fig. 90. *Gossypium* spp. (a) development of hairs from epidermal cells of the outer integument, (b) part of a ripe, dry hair with reversal point

**Breeding.** The breeding of cotton was limited until the 1930s to crossing within the diploid and tetraploid groups. Polyploidisation of the diploids has greatly widened the potentialities for the breeders. Crosses between the Old World and New World cultivars have become important, particularly for resistance breeding, and for the breeding of better performance cultivars for the Asiatic regions where *G. arboreum* and *G. herbaceum* grow well, but *G. hirsutum* to a lesser degree. In China, mutations, which have been obtained by irradiation, have been successfully incorporated into new cultivars (2027). Hybrid seed has played a role in India for several years. However, since male sterile lines are not yet available, hybrid seed has to be produced by hand pollination, a costly procedure even under Indian conditions (407). The general goals of breeding are primarily fibre yield and quality. Both have a variety of component factors, and require a high degree of technical equipment in the laboratory of the breeders. Because cotton suffers more than almost any other cultivated plant from diseases and pests, breeding for resistance is second in importance. It has had little success as yet with insect pests. Progress regarding resistance against fusarium and verticillium wilts, bacterial blight and nematodes is quite different; there is now a large number of cultivars which are resistant to one or more of these diseases or pests, or at least tolerant of them. After this, breeding focusses locally on tolerance of cold, drought and high soil salinity. The introduction of mechanical harvesting in the USA and USSR has lead to the breeding of cultivars which allow a high degree of efficiency in these harvesting procedures (low growth, little branching, short flowering time, seeds which are not too firmly attached, leaves which are not or only slightly hairy). Cultivars without gossypol in the cotyledons have been bred (glandless) to improve the suitability of the oilcake as a source of protein for humans and animals. Several commercial cultivars which are gossypol-free have been available since 1970 in a variety of countries (216, 1002, 1131, 1364, 1634).

**Ecophysiology**. Cotton is a decidedly warmth-loving plant. The seed should not be sown at soil temperatures under 18°C, and 35°C is optimal. For further development, 27°C is the optimum. At temperatures over 40°C, and with strong insolation, the bolls will be damaged and fall off. Cotton is extremely sensitive to frost. Its cultivation is only possible where 200 frost-free days can be relied on. *G. hirsutum* is mostly day-neutral, the flowering time being primarily governed by the temperature. But short days accelerate the development if the temperature lies substantially below the optimum. A lot of sunshine promotes flowering and fruit-setting. Therefore, the highest yields are achieved in dry areas under irrigation (southern Russia, Egypt). Cotton originated in the semi-arid summer rainfall region. Cultivation is possible with rainfall of between 600 and 1500 mm. The ripening time should occur in a rainless period, because rainfall after the opening of the bolls damages the quality of the fibres, and can lead to considerable losses. The plants are drought tolerant due to their deep-reaching root systems. However, prolonged dryness during flowering and boll development leads to noticeable decreases in the yield. Strong winds can damage the seedlings, and can blow away the fibres after the opening of the bolls.

Cotton needs deep soil with sufficient drainage. Otherwise, its demands are slight. The pH should lie between 6-8. It is relatively salt tolerant, and a salt content of 0.5-0.6% generally causes no damage, although there are considerable differences among cultivars with regard to sensitivity to salt. The nutrient uptake ability is strong, and the nutrient requirements are moderate. Too much N fertilization encourages vegetative growth, and extends the vegetation period. Sufficient supplies of K are important for attaining good fibre quality and for disease resistance. The requirement for Ca is decidedly high. Deficiencies of B have been reported from various countries, and can be eliminated by spraying at the correct time. On the other hand, cotton withstands relatively high B concentrations in the soil (176, 459, 474, 495, 595, 682, 1793).

**Cultivation**. Good preparation of the land is especially important before sowing cotton, because the seedlings, which germinate epigeally, can only penetrate hard or crusted soils with difficulty, and, until they are three weeks old, they have little ability to compete with weeds. Sowing is carried out by hand in many countries. For mechanical sowing, the fuzz must be removed from the seeds either mechanically or chemically, because the seeds otherwise cling together. The density of sowing varies within wide limits, depending on the cultivar, soil fertility, cultivation and harvest methods. The row-spacing lies between 50 and 120 cm, the spacing within the rows between 20 and 60 cm (55). For mechanical harvesting, types which are weakly branching are densely sown, in order to achieve an early uniform ripening in the crop. Here, the spacings are reduced to 15-20 cm between the rows, and 8-10 cm within the rows. Such "short-season" cultivars are available only of *G. hirsutum* (391). The seed should be sown no deeper than 5 cm.

Cotton can be sown on level soil, in furrows (protection against driftsand), or on ridges. Sowing on ridges is necessary with poorly drained soil. It makes irrigation easier, and assists the entry of water into the soil. However, in regions with irregular rainfall, it makes the control of weeds and mechanical harvesting more difficult.

In the USA, the most economical method of weed control has proved to be the application of a strip of soil herbicide over the row of seeds, and later, the flaming of the weeds between the rows. At later stages, weeds are not a serious problem where the seed is sown thickly. With cultivars which have a strong tendency to shed their young bolls (up to 10 days after flowering), a spraying with NAA (napthylacetic acid) is recommended (97). Considerable increases in yield have been reported from various countries where CCC (cycocel) and other growth inhibitors (e.g. mepiquat) have been utilized (487, 1284, 1913).

To decrease the infestation by pests, and to control the various soil-borne diseases, cotton is seldom cultivated in monoculture. In many countries, not only is crop rotation prescribed (the cotton is usually grown in the same field only every third year), but also the destruction of the harvested plants. For this, the plants should be pulled up with their roots, to eliminate the disease carriers as much as possible (*Fusarium*, *Verticillium*, *Xanthomonas*) (1334, 1357).

**Diseases and Pests**. Protection of the plants is of such tremendous importance for a successful harvest of cotton, that all the handbooks cover it in detail (821, 980, 1002, 1231, 1357, 1423, 1629, 1793). Diseases and pests are dealt with in specialist texts (417, 982, 1605, 1901), and pests alone in (1157, 1368 and 1512).

Of the diseases, bacterial blight, anthracnose, and the wilt diseases are widespread. Bacterial blight, also called "angular leaf spot" and "black arm", is caused by *Xanthomonas malvacearum* (E.F.Sm.) Dowson. For its control, seed treatment with Hg compounds and destruction of the remains of the harvest are the most important; there are *G. hirsutum* cultivars with good resistance. Anthracnose, caused by *Glomerella gossypii* Edg., can be controlled by the same measures; resistant cultivars have not yet been found. The fusarium wilt (caused by *Fusarium oxysporum* Schlecht. f. sp. *vasinfectum* (Atk.) Snyder et Hansen) occurs especially with rain-fed cultivation. The verticillium wilt (caused by *Verticillium albo-atrum* Reinke et Berth. and *V. dahliae* Kleb.) occurs mostly in cultivation with irrigation. For a long time there have been cultivars of *G. hirsutum* which are resistant against fusarium wilt, and good resistance against verticillium wilt is found in several Egyptian cultivars of *G. barbadense*. Agricultural precautions against the wilt diseases are crop rotation, sufficient fertilization with K, and the control of nematodes.

The number of insects which appear as pests on cotton is unusually large. Their control always requires the use of insecticides, in addition to the sanitation measures (destruction of harvest remains, weed control, crop rotation). The greatest damage is caused by the bollworms (moths of the genera *Pectinophora*, *Earias*, *Heliothis*, etc.), and their control is usually achieved by repeated spraying with contact insecticides. In Egyptian cotton cultivation, the cotton leafworm, *Spodoptera littoralis* Boisd., is much feared; in America, the boll weevil (*Anthonomus grandis* Boh.) is the most dangerous pest, and in some regions of tropical Africa, the plant bugs (*Dysdercus* spp., cotton stainers, *Lygus* spp.) are the greatest threat. Apart from these, aphids, thrips, red spiders, whiteflies, leafhoppers, and cicadas can cause major losses of yield; their control has been definitely made easier by the use of modern systemic insecticides.

**Harvest and Processing**. The majority of the world's harvest is picked by hand. It is a labour-intensive operation (a picker gathers 20-80 kg of seed cotton a day), but it produces the cleanest cotton and the highest yields per surface area (repeated picking 3, 4 or even more times). There is still no other procedure nowadays for fine, long-fibred cotton. In the USA, USSR and Australia, cotton is almost exclusively harvested by machines, and in other countries partially. Of the two types of machine, spindle pickers and strippers, the spindle picker works more slowly, but it delivers a more uniform and less impure product than the stripper (1691). For mechanical harvesting, only low growing, weakly branching cultivars can be considered. The main sources of impurities are pieces of leaf. By spraying with defoliants, the leaves are therefore made to fall off before harvesting. In spite of this, a special cleaning of the fibres is necessary with mechanical harvesting, and this is carried out in the gin, mostly after the seed removal process.

The yield of cotton (seed cotton) can reach 4 t/ha under optimal conditions, but in practice it is seldom over 2.5 t, and the global average is only about 1.6 t, because the yield in many countries is still very low (India, 0.6 t/ha). With primitive cultivars, the yield of fibres (ginning out-turn) is 20-25%, good upland cultivars nowadays yield at least 35%, and the best more than 40%. The ginning is carried out mechanically in all countries. In the grading of the fibres, apart from length and fineness, cleanness (colour and freedom from foreign matter) plays a decisive role. The cultivator can considerably increase the value of his product by careful picking at the correct time, and by sorting before sale.

The seeds which are left over after ginning are also a valuable product. They provide linters (5%), cottonseed oil (24%), oilcake (33%), and hulls (34%). The most valuable product is the oil (world production 3.3 million tons, Fig. 19, Table 15), which after refining (which also destroys the gossypol) is usable as a salad oil and for other purposes. The cake and meal contain over 40% crude protein; they are a valuable animal fodder for ruminants, but, because of their high gossypol content, they are not without dangers for monogastrics. There are chemical and mechanical methods to remove the gossypol, or to render it harmless (1379). This problem does not occur with the gossypol-free cultivars (see p. 341). The linters are mechanically separated, and are used particularly for cellulose fibres and other cellulose products, but also for coarse threads, as a material for cushions, and in paper-making. The seed hulls are used as roughage for livestock, as bedding, fertilizer, or fuel.

# Jute

fr. jute; sp. yute; ger. Jute

The true jute which is traded comes from two *Corchorus* species, *C. capsularis* L. and *C. olitorius* L., Tiliaceae (420, 598). The most important differences between the species are summarized in Table 42, and the characteristic differences in the fruit shape are shown in Fig. 91. *C. capsularis* originated as a fibre plant in the region of East India - Burma; wild ancesters are not known. It provides 75% of the jute in India and Bangladesh. *C. olitorius* is widespread as a low-growing weed in the tropics of the Old World, and it is an ancient and popular vegetable in Africa, the Middle East and India (Arabic meloukhia or molokhia) (Table 23) (480, 659, 997, 1657); the tall-growing fibre form was developed in East Asia.

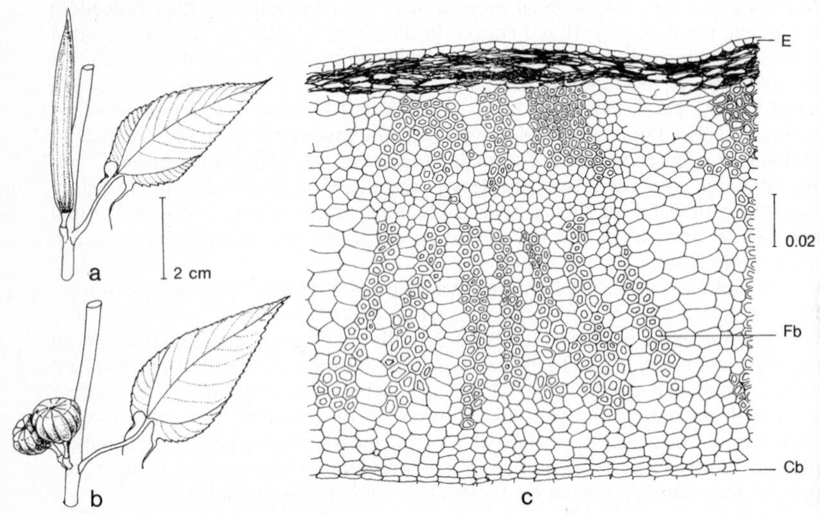


Fig. 91. *Corchorus* spp. Leaf and fruit of (a) *C. olitorius* and (b) *C. capsularis*. (c) cross section through bark. E = epidermis, Fb = fibre bundle, Cb = cambium

**Production**. In the statistics, the production figures for jute are often given together with those for kenaf and its related fibres, which are used for the same purposes. Of the total quantity of 3.3 million tons (Fig. 88), 2.6 million tons are from jute. The major jute producers are India (1.2 million t) and Bangladesh (0.9 million t); the production does not reach 100,000 tons in any other country (Burma, Nepal, etc.). Of the total world production, 98% comes from Asia; of the countries outside Asia, only Brazil (about 30,000 tons) produces a significant amount.

**Morphology and Anatomy**. Jute is densely sown in order to suppress branching. On good soil, the plants of both species grow to 4 m tall, and sometimes even higher with *C. olitorius*. The flowers stand in cymes with few flowers (mostly 2-3 flowers, occasionally up to 5), opposite the alternate leaves. The fibres consist of the sclerenchyma bundles of the bark (Fig. 91). They are separated from the wood and the bark parenchyma by retting in water.

**Breeding**. In India and Bangladesh, about 100 cultivars are distinguished. These differ in their ecological adaptation, vegetation time, fibre quality, and morphological characteristics. The two species cannot be crossed with each other, thus breeding is limited to crossing and selection within each species (149, 598, 1055).

Table 42. Differences between the two jute species. After (420, 997)

| Characteristic | White jute<br>*C. capsularis* | Tossa jute<br>*C. olitorius* |
|---|---|---|
| Leaves | Up to 12 cm long<br>mostly bitter | Up to 20 cm long<br>not bitter, edible |
| Fruit shape | Round,<br>1-1.5 cm diameter | 5-10 cm long<br>0.3-0.8 cm diameter |
| Seeds | Brown<br>3.3 g /1000 seeds | Grey to black, or greenish<br>2 g /1000 seeds |
| Fibres | Whitish, coarser and slightly weaker than *C. olitorius* | Fine, white, shining yellowish or reddish coloured, good resistance to tearing |
| Habitat | Withstands occasional flooding when the plants are over 1 m high | Does not endure water-logging |
| Resistance against *Macrophomina phaseolina* | Little resistance | Cultivars with good resistance |

**Ecophysiology**. Jute has a narrow adaptation to the Asiatic monsoon climate. Uniformly high temperatures (27-32°C), a rainfall of about 1500 mm and high air humidity are the requirements for quick growth and high fibre quality. Both species are short-day plants. Flowering is not wanted for fibre production. Jute therefore is sown when the day-length exceeds 12.5 hours (1587), but there are also breeding lines which are not sensitive to photoperiod (908). The typical soils for jute are heavy alluvial soils (loams to sandy loams), with a pH of 6-7. In regions of cultivation in India and Bangladesh, mineral fertilizers are very seldom used, but organic remnants and animal manure are applied. Because the amount of nutrients removed by a good stand of jute is high (65 kg N, 13 kg P, 130 kg K, 60 kg Ca for 2 tons of fibres), a profitable increase in yield is often achieved with mineral fertilizers (595, 921).

**Cultivation**. The small-seeded jute requires a thorough soil preparation. This is also desirable for making weed control easier, because this is very important in the first weeks after sowing. In India and Bangladesh, jute was mostly sown broadcast. Today drilling in rows predominates, and in Taiwan it is also transplanted from seedbeds (420). With broadcasting, the seed quantities are 6-9 kg/ha for *C. olitorius*, and 11-13 kg/ha for *C. capsularis*. After germination, the stand is thinned out to 10 x 10 cm. With drilling, the row-spacing is 20-30 cm, and the spacing within the rows is 6-8 cm; the seed requirement is thus reduced to about 5 to 7 kg/ha, respectively.

Because the vegetation period for jute only lasts 100-150 days, a second and even a third crop is generally planted after the jute. For *C. capsularis*, rice usually follows, and after this a pulse, rape or wheat (1473); for *C. olitorius* (dry sites) potatoes or a pulse can follow.

**Diseases and Pests**. Jute suffers relatively little from diseases and pests. The seedling blight or stem rot (*Macrophomina phaseolina* (Tassi) Goid.) can cause considerable losses in the South Asian regions of cultivation. With unfavourable weather and soil conditions, several fungi can appear (*Corticium rolfsii* Sacc., *Diplodia corchori* Syd., and others), and lead to the death of the seedlings (damping-off). Disinfection of the seeds and Cu spraying can be used against these.

Of the animal pests, the jute semilooper (*Anomis sabulifera* Gn.) is the most frequent, followed by the caterpillars of a number of moths (*Dicrasia obliqua* Walk., *Prodenia litura* F.). Considerable damage can also be caused by the larvae of the jute apion (*Apion corchori* Mshl., Curculionidae) and the ring pest (*Nupserha bicolor postbrunnea* Breun, Lamiidae). For controlling these, contact insecticides are used (417, 420, 2001).

**Harvest and Processing**. The best fibre quality is achieved when jute is harvested early: with *C. olitorius* at 90 days after sowing, *C. capsularis* at 100 days (1108). Higher yields are achieved by harvesting after 100 or 110 days, but the fibres are then of lesser quality. There are considerable differences among cultivars with regard to vegetation time. The plants are harvested with a sickle, or mechanically. For the extraction of the fibres, the whole plant is retted in water for 8-10 days (water temperature at least 25°C; with lower temperatures, the retting lasts longer). After this, the fibres are separated by hand, washed, and hung in the sun to dry (140, 420, 996, 997, 1546).

The fibre content varies, depending on the cultivar and time of harvesting, between 4.5 and 7.5% (average = 5.5%) of the fresh weight of stems. Maximum yields are 3.5-4 t fibres/ha, but as a good average, 1.5 t/ha are achieved.

The fibres will not permit fine spinning, but they are outstandingly well-suited for usage as packing materials because of their low elasticity and their high capacity to absorb water (24% of the dry weight). 75% of the production is used to make sacks, the remainder for coarse yarns, which are used as the backing material for carpets and linoleum. Because of its excellent dying properties, jute is a favourite material for decorative purposes (curtains).

The stems, which are the residue after retting (about 3.5 million tons in India), are used for fencing and as firewood, but they have also been found suitable for paper making and building panels (138).

# Kenaf and Roselle

fr. kénaf; sp. kenaf; ger. Kenaf
fr. roselle; sp. rosella; ger. Roselle

Several Malvaceae provide bark fibres which are similar to jute (Table 44) (212, 972). Among these kenaf, *Hibiscus cannabinus* L., and roselle, *H. sabdariffa* L. var. *altissima* Wester, are by far the most important (176, 420). In the recent literature, the two common names which are given above are predominantly used, because their cultivation has spread to all parts of the world. The numerous local names (gambo hemp, dekkan hemp, etc.) no longer have any international meaning. Because the species are difficult to differentiate from each other, and the fibres are practically identical, kenaf is used collectively for both (376, 1648). Some distinctive characteristics are compiled in Table 43. It must, however, be noted that both species have many cultivars, particularly *H. cannabinus*, which differ greatly from each other, e.g. in leaf shape or flower colour. In the usual botanical descriptions of roselle, the characteristics of the var. *sabdariffa* (Table 30) with its fleshy, bright red-coloured calyces are frequently included. Both species were first brought under cultivation in Africa; roselle was developed as a fibre plant in Southeast Asia (213).

**Production**. While the production of jute has altered little since the Second World War, the production of kenaf and roselle has increased about sixfold. They are the fibre plants with the highest growth in production (cotton production since 1945 has about tripled), and for the last 20 years they have been in third place on the list of fibre plants. In the last decade, kenaf production has shown no further increase, while cotton production has risen further (Fig. 88). Of the total figure of about 1 million tons of "jute-like fibres" given in (514), about 90% comes from kenaf and roselle, the remainder being supplied by *Urena lobata*, *Abutilon theophrasti*, other Malvaceae, and *Crotalaria juncea* (Table 44). The rapid spread in the cultivation of both plants is based on their wide ecological adaptability, the possibility of mechanizing the fibre production, and on the attempt of many developing countries to be independent of jute imports from Asia. Kenaf is appreciably more important than roselle in Africa and America; but roselle takes a strong share of the market, for example, in the considerable production of these fibres in Thailand (1648). The main producers of roselle, each with about 300,000 tons, are India, China and Thailand (503). The information given in (514) is very incomplete, because many countries, which cultivate kenaf only for the provision of their domestic needs are not included (129).

**Breeding**. Breeding work has been carried out especially in India, Cuba and Florida. In the foreground are ecological adaptation, fibre content and quality, suitability for mechanical fibre extraction, resistance against *Colletotrichum hibisci* and nematodes (*Meloidogyne* spp.). Nematode resistance is found only in roselle cultivars. Cross-breeding *H. cannabinus* and *H. sabdariffa*, is an attempt to introduce nematode resistance into kenaf cultivars, although the $F_1$ hybrids are triploid and sterile. This difficulty can be overcome by polyploidization (171, 420, 548, 1963).

**Ecophysiology**. Both species are very adaptable, and thrive on the most widely varying soil types, and in a wide climatic range (Table 43). This is especially true for kenaf, which is cultivated from the warm temperate zone to the equator, while roselle is limited to the tropics. Both species are decidedly short-day plants. Because flowering and fruit formation bring growth in height to a standstill, the day-length during the main growing season should be over 12.5 hours.

Table 43. Differences between kenaf and roselle

| Characteristic | Kenaf | Roselle |
|---|---|---|
| Chromosome number | 36 | 72 |
| Stem | Rough-haired | Soft-haired |
| Leaves at medium height | Heart-shaped to deeply lobed | Finger-shaped lobes with narrow, sharp-pointed tips |
| Calyx | Woolly haired; long, awl-shaped tips; dry when seeds ripen | Glabrous; short, broad tip; still green when seeds ripen but not fleshy as in var. *sabdariffa* |
| Epicalyx | Almost free, about half as long as the calyx, tip is serrated | Attached to the calyx, about 1/4 as long as the calyx, tip is smooth |
| Vegetation period | 70-140 days | 130-180 days |
| Rainfall | 500-700 mm | 1000-2000 mm |
| Most favourable temperature range | 15-27°C | 20-30°C |
| Limit of cultivation | 45°N, 30°S | About 20°N and S |
| Mechanized fibre extraction | Easy | Difficult |
| Nematode resistance | Poor | Good |
| Main regions of cultivation | Africa, Central and S America, Europe, Asia. | SE Asia |

The optimal pH range is 6-7. Water-logging is tolerated by neither kenaf nor roselle (although this is more adapted to a high rainfall climate). Kenaf grows very well on sandy soils, however *Meloidogyne* is more dangerous there.

In spite of the low demands made by both species, adequate fertilization is necessary for high yields (up to 6 t/ha of fibre is possible), especially with N; 60-100 kg N/ha is recommended (176, 420, 595).

**Cultivation**. The cultivation of both species offers no particular problems. They are sown broadcast, and in mechanized agriculture they are drilled with a row-spacing of 20-30 cm, and a spacing within the row of 5-10 cm (1566). The stand develops so quickly that weed control is no problem. One hoeing after germination may be necessary.

**Diseases and Pests.** By observing a regular crop rotation, by the choice of suitable growing sites, and by the correct choice of cultivars, the losses through diseases and pests are usually slight. Seed disinfection is recommended against anthracnose (*Colletotrichum hibisci* Poll.) and other fungus diseases (*Rhizoctonia solani* Kühn, *Macrophomina phaseolina* (Tassi) Goid., etc.) (417, 420). The damage by leaf-eating insects normally stays within acceptable limits, and only exceptionally is the application of contact insecticides necessary. The danger of nematodes for kenaf has already been indicated.

**Harvest and Processing.** Both species are best harvested at the beginning of flowering, either by hand or mechanically (1984). The fibre is then still easy to separate from the wood; also, the fibre content does not increase significantly after this point. With roselle, the fibre is obtained by retting the whole stem after stripping off the leaves (420, 684, 1648). With kenaf, the fibre can be mechanically separated from the wood. There are machines which extract the finished fibre from the fresh stems. A substantially better fibre is obtained if a retting process is carried out on the fibres, for both mechanically or manually extracted fibres. For management efficiency, it is very advantageous that the separated fibres can be first dried (best in the shade), and then later retted at any convenient time. This is a particular advantage in the drier regions of cultivation, where there is insufficient water to ret large quantities all at once. The harvest can be carried out at the optimum time, and the transport of just the fibres to a central retting station is no great problem (34, 35, 420).

The fibre content of the fresh stems is 5-6%, and 18-22% of the dry weight. In practical cultivation, the yield is mostly 1-2 tons fibre/ha, and 3-3.5 t/ha under favourable conditions.

The fibre is somewhat coarser than jute. Well-retted fibre can, however, be used for the same applications as jute, or for weaving combined with jute. The main utilization of kenaf is found in the production of sacks (129). The stems or broken stems which remain after the removal of the fibres can be used as firewood, as well as in the paper industry. Also, the whole plant can be used for paper making; the yield is 20 tons pulp/ha, which is much greater than for tree plantations (65, 1527). The young leaves are used as a vegetable in many countries (1317).

# Sisal

fr. sisal; sp. sisal; ger. Sisal

The leaves of several *Agave* species, Agavaceae, are used for the extraction of the sclerenchyma and vascular bundle fibres. The fibres are hard, and only permit spinning into coarse threads, which are used for twine and ropes, and also for coarse sacks. Some species are suitable for the production of brushes and paint brushes.

Only sisal, *A. sisalana* Perrine, has worldwide importance. In Mexico (Yucatan), henequen, *A. fourcroydes* Lem., is an important cultivated plant (fibre production 80,000 tons). It is also cultivated to a limited extent in Guatemala, Honduras and Nicaragua. The cultivation of *A. letonae* F.W.Taylor is limited to El Salvador, and its fibres are finer and softer than those of henequen ("Salvador henequen", "letona fibre"). The production amounts to 3,500 tons. *A. cantala* Roxb. is a cultivated form developed in the Philippines from a wild Mexican species. It tolerates more rainfall than other fibre agaves. Its production has decreased considerably (2,400 t). Fibres which are suitable for brushes and paint brushes are provided by *A. lecheguilla* Torr. (ixtle fibre) and *A. xylonacantha* Salm-Dyck (jamauve fibre) (684, 972, 1039, 1069, 1657, 2026). The following description applies only to sisal.

**Production**. Sisal occupies only sixth place among the fibre plants (Fig. 88). It is produced in many countries of the tropics and subtropics, in limited amounts to supply internal needs. There are significant exports only from Brazil (65,000 t), Kenya (31,000 t), Tanzania (18,000 t), and Madagascar (9,000 t). These are also the countries with the highest productions (Brazil 199,000 t, Kenya 40,000 t, Tanzania 28,000 t, and Madagascar 20,000 t).

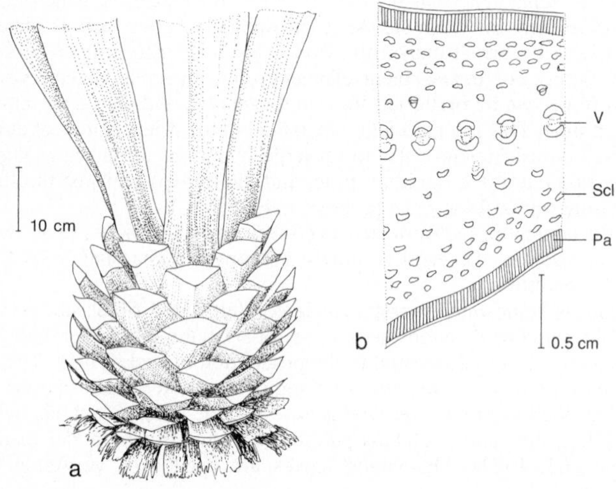


Fig. 92. *Agave sisalana*. (a) bole, lower leaves harvested, (b) leaf in cross section. V = vascular bundle, Pa = palisade tissue, Scl = sclerenchyma fibre bundle

**Morphology**. Sisal is a perennial plant, which with good growing conditions, forms an inflorescence after 6-9 years, after having produced 200-250 leaves, and then dies. The leaves, which average 120 cm in length, are arranged in a spiral around the thick stem (Fig. 92a). Two types of fibre bundles are found in the leaf; about 75% are purely sclerenchyma bundles, situated mainly near the upper and lower surfaces, the remainder consists of vascular tissue bundles, which prevail in the centre of the leaf (Fig. 92b). The root system is shallow (up to 35 cm deep), and extends up to 3.5 m from the stem. Stolons are formed from the axils of basal leaves, which are mostly removed, but can also be used for propagation. The usual planting material is the bulbils which are formed in the axils of the flower bracts, 1000-4000 for each inflorescence.

**Breeding**. Breeding of agaves is a tedious process due to the long time between generations. Because of this, yield, disease resistance, etc., can only be evaluated after a long time. *A. sisalana* types are aneuploid pentaploids (x = 30, 2n = 138-149). Planned cross-breeding has provided no new cultivars as yet. The cultivars are selections from existing populations (208). So far, the best results have been achieved by the crossing of the diploid species *A. amaniensis* Trel. et Nowell (blue sisal) x *A. angustifolia* Haw. (dwarf sisal), and the back-crossing of the $F_1$ hybrids with *A. amaniensis*. Hybrid No. 11648 which forms more than 500 leaves before it flowers and dies has become the most important. In the lowlands of East Africa, yields have been achieved with this hybrid which are more than double those with *A. sisalana* (535, 1069, 1325, 1445, 1657).

**Ecophysiology**. Sisal is a succulent (Crassulacean acid metabolism type of photosynthesis) which can survive long dry periods. For good yields, 1000-1250 mm rain are necessary. However, it is also cultivated in regions with substantially lower rainfall (500 mm or less). Under such conditions the plants form fewer leaves each year, and so have a longer lifetime (up to 20 years). Sisal reacts in the same way to temperatures which are below the optimum (under 25°C; subtropics and tropical highlands up to 2000 m). A lot of sunshine benefits production. The demands on the soil are slight as long as the soil is well-drained. The pH should lie between 5.5 and 7.5, and the soil must contain enough Ca. The nutrient removal per ton of fibre is about 30 kg N, 5 kg P, 80 kg K, 65 kg Ca and 40 kg Mg. Because the fibres themselves contain few minerals, the majority of the nutrients can be returned to the land with the pulp. In conditions with good rainfall, the N can be supplied by the use of leguminous cover crops. K fertilization has proved to be profitable in many trials (176, 595, 1069).

**Cultivation**. The bulbils are planted in nursery fields with a spacing of 50 x 25 cm, usually shortly before the beginning of the rainy season. After 12-18 months, the plants are large enough (25-40 cm) to be transplanted into the production fields. All of the roots are cut back before planting. The normal planting method is in double rows with a spacing of 4 x 1 x 0.8 m. If the rainfall is sufficient, ground cover plants are sown between the double rows, such as *Pueraria phaseoloides*, *Centrosema pubescens*, etc. (Table 53). Apart from hoeing the weeds at the end of the rainy season, and the removal of stolons, the plants do not usually need any other maintenance work, which would also not be justified from an economic viewpoint.

In the first (and second) year after planting out, maize, cotton or beans can be sown between the double rows. If an appropriate amount of fertilizer is supplied, the later yields of sisal will not be affected (1445).

**Diseases and Pests**. Fungus pests only appear in humid places. *Aspergillus niger* v. Tiegh. can enter through cuts at the leaf base, and causes bole rot; *Phytophthora* spp. are parasitic on the leaves (zebra disease), as is *Colletotrichum agaves* Cav. Spraying with Cu compounds helps with these, and is economically justified in high-yielding plantations.

The only serious animal pest is the sisal weevil (*Scyphophorus interstitialis* Gyll.), which originates from Mexico, but is now present in all areas of cultivation. Treatment of the planting holes with contact insecticides has been recommended as a counter-measure (1069, 1445).

**Harvest and Processing**. The harvest usually begins when the plants have about 100 leaves, normally in the third year (in very favourable situations, it can even begin in the second year, but it is later with unfavourable conditions - dryness, low temperatures). The leaves are cut off with a knife about 2.5 cm from the bole. About 20 of the upright leaves are left at the top of the bole, so that the plant still has enough surface area for photosynthesis (Fig. 92a). The sharp spine tips of each leaf are cut off immediately to make subsequent handling easier. About 40 leaves are taken in the first cut, and in each following year about 25. The fibre content increases in the course of the years from 2.2 to 4.5-5%, and so the fibre yield increases from 0.8 t/ha in the first year to 2.5 t/ha in the fifth. The cut-off leaves are bundled and taken to the factory either by light railway or by truck.

The decortication now takes place mechanically everywhere. All types of fibre-removal machines follow the pattern of the raspador (developed in Mexico in 1839 by CERON), in which the leaves are crushed and beaten by a rotating wheel set with blunt knives, so that only the fibres remain. All other parts of the leaf are so chopped up that they can be washed away by water. The fibres are dried in the sun or by hot air. Sun-dried fibre is yellowish, oven-dried is white. Apart from the long fibres (grade A, over 90 cm long), which are used for ropes and binder twine, the decortication produces a quantity of shorter fibres (flume tow), as does the brushing of the dried fibre (tow fibre) (about 25% of the total fibre). These are used especially for padding, for mats and stair carpets, and also for paper and building panels (972, 1069).

The utilization of the great quantity of material that remains after fibre extraction (95-96% of the leaves' weight) has been given a great deal of consideration (1069). The best application is usage as fertilizer, however this requires a large expenditure on transport. The dried pulp can serve as a fuel. Equipment for the production of methane gas is still in the experimental stage. The production of hecogenin (see p. 308) (362), of pectin and wax is impossible or uneconomic in normal processing due to their great dilution in the wash water. Small quantities of pulp can be used as animal fodder, also after silaging (708, 733). The dry poles can be used as structural timber.

## Further Fibre Plants

Table 44 includes several fibres which are exclusively or predominantly extracted from wild plants, as they are of substantial economic importance, and are regularly exported.

The numerous fibres from wild plants which are of only local importance (376, 972) have been omitted, even if there have been occasional attempts to cultivate them (e.g. *Sansevieria trifasciata* Prain in Mexico), or when their cultivation is slight and local. All these plants have little prospect of becoming economically important crops; the competition of the highly developed fibre plants and the synthetic fibres has become too intense for this.

It must be pointed out that there is a variety of cultivated plants which are dealt with in other sections whose leaf or stem fibres are used as by-products. Of these the only ones which have worldwide economic importance are the fibres from sugar cane (see p. 68), whereby the bagasse is used for the manufacture of paper and building panels, and coconut fibre (coir, Fig. 22, see p. 94) which is extracted from the carpels by a dry method, or after retting, and is used for doormats, carpets, ropes, and as padding and stuffing material (324, 376, 430, 462, 873, 874, 1824). It is an important export product from India and Sri Lanka.

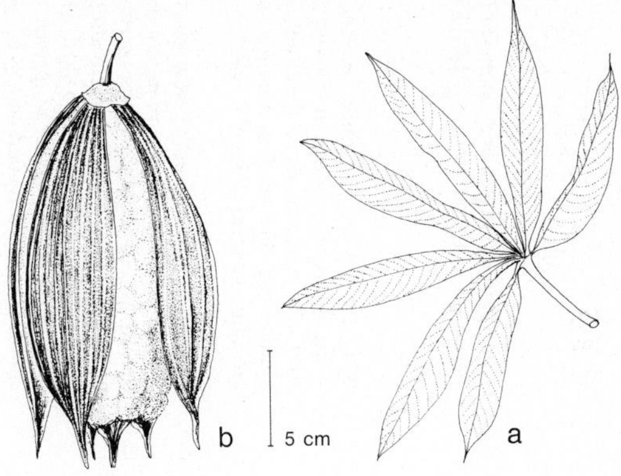


Fig. 93. *Ceiba pentandra.* (a) leaf, (b) ripe, bursting fruit

Fig. 94. *Crotalaria juncea.* Shoot tip with flowers and young fruits

Table 44. Further fibre plants (explanation of symbols, see footnotes page 55)

| Botanical name | Vernacular names of plants or products | Used part | Climatic region | Remarks and literature references |
|---|---|---|---|---|
| | | **Dicotyledonae** | | |
| **Araliaceae** | | | | |
| *Tetrapanax papyrifer* (Hook.) K. Koch (*Fatsia papyrifera* Benth. et Hook.f.) | rice-paper tree, tung tsao chin, tsuso, fatsi | pith | S, Tr | Cultivated in Taiwan and China for Chinese rice paper and artificial flowers (259, 1375, 1383) |
| **Bombacaceae** | | | | |
| *Bombax ceiba* L. (*B. malabaricum* DC., *Salamalia malabarica* (DC.) Schott et Endl.) | red silk-cotton tree, Indian kapok, purani, simbal | fruithair | Tr | Main regions of cultivation are India and SE Asia, the world production is about the same as for *Ceiba pentandra*, and the fibres are used in a similar way. The seed oil is edible, the oil cakes are animal fodder (259, 376, 972) |
| *Ceiba pentandra* (L.) Gaertn. (*Eriodendron anfractuosum* DC.) | kapok, silk-cotton tree, arbre kapok, ceiba | fruithair | Tr | (Fig. 93) Cultivated in the New World, Africa, and SE Asia. The fibres are predominantly used locally as a filling material for mattresses and cushions, and as a water-repellent insulation material. The seed oil is also used. The fruit hairs of other Bombacaceae species are used similarly to those named here (176, 259, 406, 684, 972, 1290, 1297, 1383, 1774) |

| | | | | |
|---|---|---|---|---|
| **Cannabidaceae**<br>*Cannabis sativa* L. | hemp,<br>chanvre,<br>cáñamo | bast | Te, S | World production of fibres is 250,000 tons. Main production in the USSR, China and India, also cultivated in the Balkan countries. Seeds provide an edible oil (176, 406, 420, 1052, 1115, 1383, 1490) |
| **Cucurbitaceae**<br>*Luffa cylindrica* (L.) Roem.<br>(*L. aegyptiaca* Mill.) | luffa,<br>loofah,<br>sponge gourd,<br>courge torchon | fruit | S, Tr | Cultivated worldwide, an important crop in Japan, India and Egypt. The network of vascular bundles in the ripe fruit is used for bath sponges, shoe padding, and wallpaper. The young fruits of non-bitter forms are eaten as a vegetable (Table 25). The seeds are used for oil production and as cattle fodder (259, 376, 731, 1136, 1895) |
| **Leguminosae**<br>PAPILIONOIDEAE<br>*Crotalaria juncea* L. | sun hemp,<br>san hemp,<br>sunnhemp,<br>crotalaire | bast | S, Tr | (Fig. 94) Cultivated as a fibre plant only in India, Pakistan and Brazil. Production about 80,000 t. Used for fish nets, twine, thick cloth, and cigarette papers. Cultivated worldwide as a green manure plant (see p. 429) (376, 420, 444, 972, 1261, 1565, 1933) |
| **Linaceae**<br>*Linum usitatissimum* L. | flax,<br>lin,<br>lino | bast | Te, S | As a rule special fibre cultivars grown but also forms with double usage (fibre and oil (Table 18)). World production 700,000 tons. Main producer USSR, 300,000 t, otherwise noteworthy production only in China (200,000 t). Recent recovery of production in Europe (France) 1988: 73,000 t (176, 376, 420, 663, 972) |

Table 44. Further fibre plants, cont'd (explanation of symbols, see footnotes page 55)

| Botanical name | Vernacular names of plants or products | Used part | Climatic region | Remarks and literature references |
|---|---|---|---|---|
| **Malvaceae** | | | | |
| *Abutilon indicum* (Torner) Sweet | country mallow, jhampi | bast | Tr | Cultivated in S and SE Asia, especially India (Bombay), for ropes. Found worldwide as a weed (259, 376, 972) |
| *A. theophrasti* Medik. (*A. avicennae* Gaertn.) | China jute, Indian mallow, velvetleaf, chingma | bast | S | In India, China and the USSR for ropes and paper. Other *Abutilon* species are cultivated as fibre plants, especially in India and Pakistan (376, 406, 420, 972, 1714) |
| *Malachra capitata* L. | bhanbhendi | bast | Tr | Cultivated in India, Central and S America. Quick growing, withstands flooding. Fibre similar to jute. Found worldwide as a weed, locally also a medicinal plant (376, 972) |
| *Urena lobata* L. | Congo jute, cousin, aramina, paka, malva | bast | Tr | Yields up to 3 t fibre/ha. Fibre is water and termite proof, utilized like kenaf. Largest producer Brazil (75,000 t) (404, 405, 406, 420, 477, 972, 1383) |
| **Moraceae** | | | | |
| *Artocarpus elasticus* Reinw. ex Bl. | wild breadfruit benda terap | bast | Tr | For fish nets and paper. Latex for bird-lime. The fibres from other *Artocarpus* species are used especially for rope making (259, 972, 1383) |

| | | | | |
|---|---|---|---|---|
| *Broussonetia papyrifera* (L.) Vent. | paper mulberry, mûrier à papier, kodzu | bast | S, Tr | This species and *B. kazinoki* Sieb. are cultivated in E and SE Asia for the finest papers (kodzo paper) (259, 376, 972, 1383) |
| **Sterculiaceae** | | | | |
| *Abroma augustum* (L.) L.f. | perennial Indian hemp, ramia sengat | bast | Tr | Widespread in SE Asia, used for fish nets, washing lines, and false hair. Low demanding, recommended as an ancillary plant for soil reclamation (200, 376, 406, 972) |
| **Urticaceae** | | | | |
| *Boehmeria nivea* (L.) Gaud. | ramie, China grass, ramio | bast | Tr | Production in E Asia about 100,000 t, of great local importance, largely replaced on the world market by synthetic fibres. Is of interest for other regions due to the possibility of mechanical processing. Yields up to 2 t/ha crude fibre. High demands on the soil and fertilizer. The fibre is strong, fine and spinnable, for tablecloths, serviettes, curtains and clothing. Fodder plant (34, 35, 176, 406, 420, 640, 872, 972, 998, 1093, 1258) |
| **Monocotyledonae** | | | | |
| **Agavaceae** | | | | |
| *Furcraea cabuya* Trel. | cabuya castilla | leaf | Tr | Cultivated and wild from Costa Rica to Ecuador. Fibres for ropes and sacks (287, 515, 1377, 1960) |
| *F. foetida* (L.) Haw. (*F. gigantea* Vent.) | Mauritius hemp, piteira, pita | leaf | Tr | Of less value than sisal, for sacks and ropes. Production 2000 t, also hedge plant (972, 1039, 1297, 1445) |

Table 44. Further fibre plants, cont'd (explanation of symbols, see footnotes page 55)

| Botanical name | Vernacular names of plants or products | Used part | Climatic region | Remarks and literature references |
|---|---|---|---|---|
| *F. macrophylla* (Hook.) Bak. | fique<br>wild sisal | leaf | Tr | Wild and cultivated in Colombia. Annual production 39,000 t, particularly for coffee and sugar bags. Other *Furcraea* species used locally for fibre production (515, 972, 1039, 1377) |
| *Phormium tenax* J.R. et G.Forst. | New Zealand flax<br>formio | leaf | S | Annual production 9000 t. Cultivated in a few countries to a limited extent. Fibres utilized like jute (515, 972) |
| *Samuela carnerosana* Trel. | palma ixtle,<br>palma istle,<br>palma barreta | leaf | Tr | Production from wild plants in Mexico 4-8000 tons. Fibres utilized for brushes and rope products (972) |
| **Bromeliaceae** | | | | |
| *Neoglaziovia variegata* (Arr. Cam.) Mez | caroá<br>makimbeira | leaf | Tr | Large wild stands in the dry regions of NE Brazil. Production about 4000 t. For coarse cloths and mats. Other Bromeliacean species, also the pineapple (see p. 190) are used to obtain fibres (972, 1960) |
| **Cyclanthaceae** | | | | |
| *Carludovica palmata* Ruiz et Pav. | Panama hat palm,<br>toquilla,<br>palma de sombrero | leaf | Tr | Cultivated from Central America to Ecuador. This, together with *C. incisa* Wendl. and *C. jamaicensis* Lodd. are used for Panama hats and as roofing material (972, 1445, 1960) |

| | | | | |
|---|---|---|---|---|
| **Gramineae** | | | | |
| *Bambusa vulgaris* Schrad. ex J.C.Wendl. | golden bamboo<br>yellow bamboo<br>feather bamboo<br>bambú común | shoot | Tr | Purely a cultivated plant. Many other species of the Bambusoideae are utilized for paper, basket and building materials. Young shoots as a vegetable (Table 25) (259, 376, 510, 1296, 1376, 1383, 1445) |
| *Lygeum spartum* Loefl. ex L. | esparto<br>albardín | leaf | S | Utilized like *Stipa tenacissima* in N Africa and Spain (972) |
| *Stipa tenacissima* L. | halfa<br>alfa grass<br>esparto | leaf | S | Covers large areas in Spain and NW Africa from Tunisia to Morocco. This species and *Lygeum spartum* are not always clearly distinguished. The total production amounts to about 100,000 t. Utilized for paper making and for basketwork. Slight cultivation in Tunisia (74, 250, 972) |
| **Juncaceae** | | | | |
| *Juncus effusus* L. var. *decipiens* Buchenau (*J. decipiens* Nakai) | mat rush,<br>soft rush,<br>igusa, naga-i | stem | S | Cultivated in Japan on 11,000 ha, production 120,000 t for making mats (692, 1383) |
| **Musaceae** | | | | |
| *Musa textilis* Née | Manila hemp,<br>abacá,<br>cáñamo de Manila | leaf sheath | Tr | 60% of the production in the Philippines. To a small extent also in other countries (Ecuador). Annual production about 90,000 t. There are about 100 cultivars. Yields up to 4 t fibre per ha. Producing the fibre requires a lot of manual labour. Mechanical processes have been developed for large plantations. Fibre for coarse yarn and paper (176, 259, 684, 972, 1297, 1445, 1828) |

Table 44. Further fibre plants, cont'd (explanation of symbols, see footnotes page 55)

| Botanical name | Vernacular names of plants or products | Used part | Climatic region | Remarks and literature references |
|---|---|---|---|---|
| **Palmae** | | | | |
| *Attalea funifera* Mart. ex Spreng. | Bahia piassava<br>coquilla nut | leaf sheath | Tr | Piassava fibres are the brushlike vascular bundles of the leaf sheath, which are used for hard brooms. Apart from *Attalea* species, *Leopoldinia piassaba* Wallace is also used. Brazilian annual production 52,500 t. Similar fibres are obtained in W Africa from palms of the genus *Raphia* (972, 1518) |
| *Calamus rotang* L. | rattan cane<br>bet<br>bent | stem | Tr | Wild-growing climbing plant of SE Asia. To a small extent also cultivated. The stems are mechanically stripped, and provide the rattan cane from the inner part, and the cane for chairs from the outer part. By shredding the stem, a fibre for ropes and mats is obtained. Several other *Calamus* species are used in a similar manner, and sometimes also cultivated (259, 406, 827, 1445, 1452) |
| *Chamaerops humilis* L. | dwarf fan palm<br>European fan palm<br>palmier nain | leaf | S | The fibres are mechanically obtained from wild-growing palms, especially in Morocco and Algeria. They are traded under the name "crin vegetal", and are used primarily for upholstery, and locally for ropes. The world production is about 100,000 t (972) |

| | | | | |
|---|---|---|---|---|
| *Raphia hookeri* Mann et Wendl. | raphia<br>giant raffia<br>wine palm<br>Ivory Coast raffia<br>mimbo wine | leaf | Tr | Raffia bast is the easily removable upper surface of young leaves. It is used for tying in gardening and basketmaking. Apart from the W African *R. hookeri*, bast is obtained from other *Raphia* species. The largest exporter is Madagascar. The material comes from *R. farinifera* (Gaertn.) Hyl. (*R. ruffia* (Jacq.) Mart.). The raffia palms otherwise provide palm wine, wax, and African piassava fibre. The leaf stalks are used as building material and for furniture (200, 972, 1445, 1559) |
| *Trachycarpus fortunei* (Hook.) H.Wendl. | windmill palm<br>chusan palm | leaf sheath, leaf | S, Tr | Cultivated in E Asia for the production of fibres. For raincoats, hats, brooms and string. In other parts of the world as ornamental palm (259, 376, 492, 1383) |
| **Pandanaceae** | | | | |
| *Pandanus utilis* Bory | common screw pine | leaf | Tr | Originates from Madagascar, and is planted there and in other countries to obtain basket making material. Many other *Pandanus* species are used for basketry and as fruit (Table 30) (1445) |

# Elastomers

Natural and synthetic polymers with rubber-elastic properties are classified as elastomers. The natural elastomers are polyisoprenes. They occur in two isomeric forms: cis-1,4 polyisoprene is caoutchouc with highly elastic properties, trans-1,4 polyisoprene is gutta-percha, with slight elasticity, but with strongly thermoplastic properties (softening at high temperatures, re-hardening at room temperature).

Polyisoprene occurs in small quantities in many plants. It is technically usable only in the species where it is concentrated in the vacuoles of cells, but especially in the latex tubes. Such latex tubes are characteristic of several families (Apocynaceae, Asclepiadaceae, Compositae, Moraceae, Sapotaceae). The most commercially important of all is *Hevea brasiliensis*, as its latex contains only caoutchouc. There are also several species in which gutta-percha occurs predominantly in the latex, and this provides the basic ingredient of chewing gum (866). Pure gutta-percha (from *Palaquium* and *Payena* species) has lost much importance, because it has been replaced by the numerous thermoplastic synthetics. Locally, the latex from many plants is utilized in primitive ways for tools (knife handles), as a caulking material for boats and containers, and for making toys.

## Hevea

fr. hévéa; sp. hevea, jebe; ger. Hevea

All other caoutchouc plants have been driven from the world market by the new high-yielding cultivars and the improvements in cultivation methods of hevea, but even more by the competition with synthetic rubbers. More than 99% of the world's production of natural rubber comes from cultivated *Hevea brasiliensis*, the remainder from wild trees of this species, and from a few other *Hevea* species in tropical South America (especially *H. benthamiana* Muell. Arg.). *Hevea brasiliensis* (H.B.K.) Muell. Arg., Euphorbiaceae, originates from the Amazon region. Several botanical varieties of the species have been described; the cultivated forms belong to the var. *brasiliensis* (142, 267, 535, 1405, 1610, 1657, 1910).

**Production**. The Second World War brought production and trade of natural rubber almost to a standstill, and this was the cause of the rapid development of the synthetic rubber industry in the USA. However, the pre-war level of production was reached again a few years after the end of the war. Since then, the production has risen steadily, and continues to rise, by 44% from 1968 to 1978, by 24% from 1978 to 1988. Of the world's production of 4.7 million tons, 4.4 million t (93%) comes from Southeast Asia, of which 1.61 million is from Malaysia, 1.09 million from Indonesia, 0.86 million from Thailand, and 0.13 million from Sri Lanka. Since 1945, the production in West Africa (Liberia,

Côte d'Ivoire, Nigeria, Cameroon, Zaire) has increased more than fivefold. In 1988, it reached 264,000 t, with the highest output in Liberia (85,000 t) and Nigeria (73,000 t). The production in South America remains almost unaltered, because the production there is made very difficult by the leaf blight (*Microcyclus ulei*).

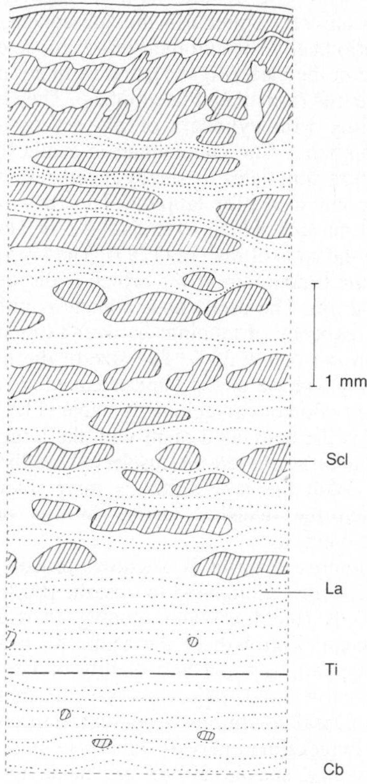


Fig. 95. *Hevea brasiliensis.* Cross section through bark. Cb = cambium, La = latex vessels, Scl = sclerenchyma fibres, Ti = maximum depth of tapping incision (564)

No other natural raw material has had such a vacillating history as rubber - from fabulous beginnings, to economically destructive catastrophes. Due to the increases in the price of crude oil, the raw material for synthetic rubbers, and owing to the decisive improvements in the production techniques for natural rubber, favourable conditions have been created for the further growth of the commodity markets for natural rubber. On a global average, natural rubber takes a 32% share of the total rubber production at the moment. This share will probably increase, because natural rubber is preferred for vehicle tyres, especially for trucks and aeroplanes, and its use will probably increase as the number of motor vehicles increases - in any event in the developing countries (589, 610, 1124).

**Morphology and Anatomy**. Hevea is a tree which grows up to 30 m tall, with a tap root 4.5 m long. The side roots are 7-10 m long. The trifoliate leaves are formed in flushes, particularly on young trees. In each flush, the first four to six leaves are large, with long petioles, and are separated by long internodes, while the internodes are very stunted in the subsequent smaller leaves (1297). Fully grown trees periodically shed all their leaves (once or twice a year).

The latex tubes lie in the bark (Fig. 95), and form an interconnecting network throughout the whole tree (articulated latex tubes). They differentiate in the bark parenchyma alternating with cylinders of phloem tissue, both of which are derived from the cambium. The caoutchouc is synthesized in the latex tubes themselves, and tapping stimulates its steady regeneration; this is most intensive in the region of the cut made for tapping. The flow of latex which follows tapping comes from an area which extends 1-1.5 m along the trunk, and 10-15 cm laterally from the ends of the cut (1297). The latex contains an average of about 30% caoutchouc (maximum 40%), which is dispersed as globules with a diameter of about 1-2 µm. The caoutchouc content is influenced by the weather conditions and the frequency of tapping; by over-extracting (ethrel stimulation, daily tapping), it can fall below 25%. The size of the globules is then smaller than usual. The aqueous phase of the latex is called serum. It has a pH of 7.1-7.2, at which the suspension of caoutchouc is stable (112, 267, 1087, 1910).

**Breeding**. Hevea is the cultivated plant with the largest increase in yields in the last 50 years. Plantations based on unselected seedlings yielded 300-500 kg caoutchouc/ha/year. With the new cultivars more than 3000 kg is achieved. Breeding has created the decisive factors in this progress, supported by improved cultivation methods.

The selection of improved forms is a tedious process, which takes 10 years under favourable conditions. Compared to this, the propagation of new cultivars by vegetative methods (budding) and cloning is relatively quick. General selection criteria are quick growth of the trunk, thick and smooth bark, good bark renewal after tapping, amount and quality of latex (caoutchouc content, purity), and resistance to wind damage (low growth, small crown). Additionally, there is adaptation to local conditions, reaction to latex stimulation, high yields also when tapped at longer time intervals (labour-saving).

In resistance breeding, special efforts are made to find forms which are resistant to *Microcyclus ulei*. The resistance comes from *H. benthamiana*, *H. pauciflora* (Spruce) Muell. Arg., and others, but as yet there are no hybrids which combine the high yields of the Malaysian *H. brasiliensis* cultivars with *Microcyclus* resistance. Also, new races of *Microcyclus* are always appearing, against which the hybrids are not resistant. Powdery mildew (*Oidium heveae*) is a serious disease, especially in Sri Lanka. High-yielding cultivars with tolerance to this disease are available today. Obviously, the resistance of new clones is also tested against other diseases (*Glomerella cingulata*, *Phytophthora* spp., *Corticium salmonicolor* B. et Br.). The selection of suitable rootstocks also receives attention (1, 530, 535, 743, 1548, 1767, 1900, 1910).

**Ecophysiology**. Hevea is a tree of the tropical rainforest. Its cultivation is limited to the equatorial region between 15° North and South. An average temperature of 28°C, and 2000-4000 mm rain are optimal. During the dry season, hevea sheds all of its leaves, and it tolerates longer dry periods in the leafless state; however, the yield decreases markedly under these conditions.

Because of the extended root system, the soil should be deep. The optimal pH is 5-6, however even a pH of 4 or 8 will be tolerated. Although hevea endures periodic flooding in the Amazon region, it regularly produces high yields only on well-drained soil.

The nutrient removal amounts to about 25 kg N, 5 kg P and 25 kg K per ton of caoutchouc (921). Mineral fertilizers, particularly P, are used especially to aid the rapid development of the legumes which are sown as ground cover plants, and are also generally given to encourage a quick growth of the trees in the first 2-3 years after planting. Obviously, the new high-yielding clones make more demands on soil fertility and fertilizer usage than the old seedling plantations, and they require regular fertilizing with N, P, K, and if necessary other elements (44, 163, 165, 595, 621, 1380).

**Cultivation**. For propagation, fresh seeds are usually used (the ability to germinate decreases rapidly after 1 month). They are germinated on the surface of seed-beds, and then planted in the tree nursery or in plastic bags. After 3-6 months ("green budding") or 1-2 years ("brown budding"), they are budded. With green budding, the trees are planted out 4-5 months after budding, with brown budding planting out is done immediately after budding, and before the sprouting of the scion. Other procedures for propagation are also used (cloning, cuttings with mist propagation) (1042, 1297, 1369, 1380, 1548, 1553). New techniques of propagation predominantly aim at shortening the time interval between planting and the beginning of tapping (1042, 1553). In South America, trials have been carried out to bud the budded trees a second time in the region of the crown (crown budding), when the trees have reached a height of about 2 m, this time with a *Microcyclus*-resistant species (*H. guianensis* Aubl., *H. spruceana* (Benth.) Muell. Arg., *H. benthamiana* Muell. Arg.) in order to create a resistant crown. The procedure is intricate, creates a time delay of 2 years, the yield is often somewhat lower than for trees with their own crown, and the danger of wind breakage to the crown is substantially increased. In Malaysia too, trials are carried out to test and improve crown budding (2020).

Planting distances in the stand depend, among other factors, on economic considerations of the farm management. Dense planting (300-400 trees/ha) gives higher yields per hectare, but less yield per tree, and thus a higher labour cost than planting wider apart (150-200 trees/ha). Planting in wide rows (with 20 m separation) with narrower spacing within the rows (1-2 m) is also labour saving. With this, there is a shorter distance to walk when tapping, and other crops can be cultivated between the rows (coffee, cocoa, food crops of all types), as is normal with small-holdings. Such wide planting distances have also been reported to be favourable where diseases endanger the cultivation (*Microcyclus* in Brazil); with the clone plantations, this aspect will become important in many regions (1333, 1550).

For many years, thorough soil preparation (in Southeast Asia: eradication of alang-alang grass) and immediate sowing of legumes as ground cover (see p. 434) have been obvious actions to take before establishing new hevea plantations (236, 302, 1068). With the choice of the right combination of ground cover plants (including species which stand a lot of shade, such as *Centrosema pubescens*), the harmful weeds (weedy grasses) are kept down, and later only the paths and areas around the trees need to be kept free from growth. Under favourable conditions, the tapping can begin 4 years after planting out, but often only after 6-9 years. The economic lifetime of a well-tended plantation is 30-35 years. In recent years, new planting has deliberately been done earlier (after 20-25 years) in order to utilize the benefits of the new high-yielding cultivars.

The latex production can be controlled by stimulation with a variety of growth regulators; nowadays mainly ethrel (ethephon) is utilized for this purpose. The hormone solution is applied to a 4 cm wide band beneath the tapping cut. The treatment is carried out only at time intervals of several months, so that the trees do not become too weakened. The individual clones react differently to the treatment, on average, an increase in yield of 30% is achieved. Latex stimulation makes high demands on the production capacity of the trees, and can lead to a weakening of the trunk (danger of wind breakage) and to reduced disease resistance. Therefore, it should be used cautiously on young trees, and fully only on trees which are older than 15 years; as a rule it necessitates additional supplies of fertilizer, especially K (4, 112, 156, 1549, 2009). Because of this, latex stimulation is by no means generally used; however, it presents an outstanding method of making the caoutchouc production more flexible, and of increasing production in times of high demand (and higher prices).

**Diseases and Pests**. The most serious disease of hevea, the South American leaf blight, caused by *Microcyclus ulei* (P.Henn.) Arx, is limited to Central and South America, and there it makes cultivation on plantation-scale almost impossible. It can only be controlled by resistance breeding, crown budding, and by very wide planting distances (42, 318, 715, 748, 1900).

The other leaf diseases, colletotrichum leaf disease (*Glomerella cingulata* (Ston.) Spaut. et Schr.), mildew (*Oidium heveae* Steinm.), and bird's eye spot (*Drechslera heveae* (Petch) M.B.Ellis, *Helminthosporium heveae* Petch), appear more or less regularly in the Southeast Asian regions of cultivation, and must be controlled in the tree nurseries and young plantations with suitable fungicides, wherever tolerant clones are not available (1552, 1724, 1910). On unfavourable growing sites, and with insufficient care, the usual root and trunk diseases of tropical tree cultivation can cause considerable damage (*Rosellinia* spp., *Ganoderma* spp., *Phytophthora* spp.), and they can make control measures necessary (uprooting, soil disinfection).

Damage by pests mostly plays a limited role, and does not require regular prevention measures. An exception is *Erinnyis ello* L., whose caterpillars can cause considerable damage in the Amazon region (42). A peculiarity which must be mentioned is the damage caused to young and old trees in Indonesia, where the bark is eaten by the giant snail *Achatina fulica* Fer. (1723).

**Harvest and Processing**. When the stem has reached a circumference of 45-50 cm at a height of 1 m above the ground, the tapping can begin. The tapping cut is made at an angle of 30° from the horizontal, from left downwards to the right. Special knives are used for this, so that the bark is cut as deeply as possible, but without damaging the cambium. Using the standard method, the cut has a length of 1/2 the tree circumference, and is repeated every 2 days (this is denoted by S/2, d/2. S/4, d/3 would be a less intensive tapping, cutting only 1/4 of the circumference, and repeated only every third day). The latex flows immediately after cutting. It runs into a vertical groove to a spout, and is collected in a vessel which is hung on the tree (a plastic vessel usually nowadays). The tapping is carried out in the early morning (high turgor pressure in the latex tubes). Several hours later, the latex (which has not yet coagulated) is collected in buckets, and brought immediately to the factory for processing. If the quantity of latex decreases noticeably, or if the tree shows signs of weakness (reduced increase in girth of the trunk, poor development of the renewed bark), a break of several months is made, and also during dry periods. With each subsequent tapping cut, 1.5 mm of bark are removed. This involves an annual bark consumption of about 25 cm. When one side of the tree is consumed, the tapping changes to the other side; when this is also used up, tapping begins on the renewed bark of the first tapping panel. In the intervening time period of about 8 years, the newly formed bark has regrown so much that it provides as much latex as it did originally. A new, labour-saving method of tapping is puncture tapping, whereby a needle is stuck into the bark which has been previously stimulated with ethephon. This can be combined with the conventional tapping method. Good results have been achieved with this technique in Malaysia and Africa (716, 1551). Towards the end of their productive lives, the trees are very intensively tapped (tapping at two heights, upward tapping), because it is no longer necessary to maintain their vigour (79, 303, 1851).

The tapping system allows so many variations, that the production level can be adapted within quite wide limits to the market conditions and labour costs. In several production regions, the labour costs have limited the profitability in times of low rubber prices (1272). One method, which drastically reduces the costs of collection, is the polybag technique, where polythene bags are used as the initial collection vessels, in which the caoutchouc coagulates naturally as it is collected only once a month. There are problems with dilution and contamination with dirt by rainwater which runs down the tree trunk, and this must be kept out using a rain protector made of plastic sheet. Caoutchouc technology has developed procedures which can process this raw material which has coagulated in an uncontrolled way into a standardizable product of high quality (crumb rubber, see p. 110).

The classical processing procedure consists of diluting the latex with water, and lowering the pH so much by the addition of acids (usually formic acid) that the caoutchouc globules coagulate (1369). Other coagulation procedures have been developed in recent years (67, 360, 890). The caoutchouc has a specific gravity of 0.91, and that of the serum is 1.02, so that the coagulate gathers on the surface. It is lifted as a thick curd, which is then pressed into sheets by passing it a number of times between pairs of rollers. During the rolling process, the coagulate is thoroughly washed, the finished sheets are dried and smoked (ribbed smoked sheets, RSS). Crepe rubber is produced in machines which tear and macerate the sheets.

The marketing of natural rubber has gained much from the introduction of a standard in Malaysia. The grade of Standard Malaysian Rubber (SMR) is primarily decided by the degree of soiling. The type of processing (sheet or crumb) is without relevance. Indonesia has followed this example (Standard Indonesian Rubber, SIR), with very favourable results, especially for the rubber produced by smallholders, which is processed using the crumb rubber procedure, and comes on the market as block rubber.

In order to meet the requirements of the processing industries, numerous techniques have been developed which through additions or treatments yield a supply of natural rubber which is easy to process, and which possesses special physical or chemical properties, such as CV rubber with constant viscosity, rubber with a defined plasticity retention index (PRI), rubber "thinned" with mineral oil (oil extended natural rubber, OENR), rubber which vulcanizes rapidly, or is even pre-vulcanized (special or superior processing rubber, SPR), and de-proteinized rubber (enzyme deproteinized natural rubber, DPNR). About 15% of the natural rubber is now shipped as latex. For this, the natural latex is thickened by centrifuging (caoutchouc content 60%), and stabilized by the addition of ammonia and other chemicals. With the increasing demand for rubber goods used as protection against the AIDS virus, the latex industry in Malaysia and other Southeast Asian countries has grown greatly in recent years. In 1989, Malaysia alone exported protective rubber-goods worth US $ 250 million. The technology of natural rubber has been developed into a special science under the pressure of competition from synthetic rubbers (32, 142, 1222, 1380, 1910, 2010).

A by-product is the profusion of seeds which are formed (about 0.5 t/ha), which contain 40-50% of an oil which dries well and which is suitable for use as food and for technical purposes. Its utilization is made difficult by the high labour costs of gathering the seeds, which are quick to decay. In Sri Lanka, about 2500 tons of the oil are produced each year (a substitute for linseed oil). The seed meal contains about 28% protein, and is a suitable addition to animal fodder (1294, 1464, 1744). The wood of exhausted trees can be used for paper pulp and timber (611, 884, 1777).

## Guayule

The name guayule is used in all languages for the shrub *Parthenium argentatum* A. Gray (Compositae), which originates from the arid regions of northern Mexico, and neighbouring parts of Texas. Guayule contains 5-10% caoutchouc in its tissue. During the Second World War, 12,000 ha of it were cultivated in the USA. The availability of synthetic and hevea rubbers after the war brought its cultivation to an end. In the 70s, cultivation attempts were resumed in several countries, but they have not yet led to renewed commercial production of any size. Types with higher caoutchouc content (plants with 25% caoutchouc in the dry mass have been reported), cultivars which are adapted to local climates (in their home region, caoutchouc is not synthesized during the hot summer months, but in late autumn and winter), and improved extraction and purification procedures could make the cultivation of guayule profitable even under present-day economic conditions, because the production of guayule rubber can be fully mechanized, whereas the production of hevea rubber will always remain labour-intensive. This can be especially valuable for marginal dry regions where agricultural usage is limited. If *Microcyclus ulei* (see p. 366) spreads to Southeast Asia, and endangers hevea cultivation there, then guayule could become an important source of natural rubber (192, 438, 493, 577, 1189, 1250, 1258, 1259, 1513, 1804).

## Further Elastomers

Table 45 includes only species whose latex is still nowadays extracted and exported in noteworthy quantities. They are collectively called "balata" in the American literature, in order to distinguish them as a group from caoutchouc

(866). Their latex is characterized by the presence of gutta-percha, either alone or predominating, a high proportion of resins (sometimes over 20%), and other compounds, e.g. saturated and unsaturated long-chain alcohols derived from isoprene (waxes, see p. 394). Because of these constituents, they sometimes have a distinctly individual taste, which does not interfere with their main use as a base for chewing gums. Chewing gum contains 20-25% elastomer base, as well as carbohydrates (sorbitol, glucose, etc.), softeners, emulsifiers (gum arabic or others), and flavourings. The world's production of chewing gum amounts to many thousand tons. The demand for natural gutta-percha elastomers will therefore remain strong, even if fully synthetic chewing gum bases have already been produced, or synthetic compounds are added to the natural base (butyl rubber, polyisobutylene, polyvinyl esters, etc.).

Table 45. Further elastomers

| Botanical name | Vernacular names of plants or products | Origin | Remarks and literature references |
|---|---|---|---|
| **Apocynaceae** | | | |
| *Couma macrocarpa* Barb.-Rodr. | sorva gum, leche caspi, palo de vaca, sorva grande | Amazonas Territory | Annual production 2800 t. Wild-growing trees, tapped every 1-2 years. Thickened by boiling for chewing gum. Latex tastes good, used like milk for coffee, manioc dishes, and other meals. Fruits edible. Used similarly to the latex from *C. rigida* Muell. Arg. (mucugê) and *C. utilis* Muell. Arg. (sorva) (42, 423, 550, 866, 1518, 1952, 1960) |
| *Dyera costulata* (Miq.) Hook.f. | jelutong, pontianak | W Malaysia, Indonesia | One of the most important sources for chewing gum. Is tapped at short intervals, and coagulated by the addition of acid, and brief boiling (259, 376, 866, 1095, 1953) |
| **Moraceae** | | | |
| *Brosimum utile* (H.B.K.) Pittier (*B. galactodendron* D.Don) | cow tree, árbol de la leche | Central and northern S America | Latex used like milk, and for chewing gum and wax production, fruit flesh and seeds eaten (550, 1444, 1960) |
| **Sapotaceae** | | | |
| *Ecclinusa balata* Ducke | coquirana ucuquirana | Brazil | Annual production 1600 t. Tapped from wild trees, which are found in large numbers in the Amazon region. Tapped at long time intervals. Coagulates spontaneously, used for chewing gum, seldom now for driving belts and other technical purposes. Further *Ecclinusa* species are used in the same way as this one (423, 550) |

| | | | |
|---|---|---|---|
| *Manilkara bidentata* (A. DC.) A.Chev. (*Mimusops balata* Crueg. ex Griseb.) | balata, bullet tree | Northern S America | Annual production in Brazil 400 tons. Obtained by tapping and boiling. Addition to chewing gum. Previously used for insulation and for driving belts, fruit flesh edible. Has greatly decreased in importance (423, 550, 1518) |
| *M. elata* (Allem.) Monac. | massaranduba, maçaranduba | Brazil | Annual production 800 t. Obtained by tapping wild trees in the Amazon region. After boiling down, used for chewing gum (423, 1039, 1425, 1518) |
| *M. zapota* (L.) van Royen | chiclé tree, sapodilla | Central America, Mexico | Cultivated as a fruit tree (Table 30) and for latex production; the majority of the chicle is collected from wild trees. The most important base in the chewing gum industry. The trees are tapped at long time intervals, and the juice thickened by boiling (550, 866, 1039, 1297, 1444, 1518, 1960) |
| *Palaquium gutta* (Hook.)Baill. | gutta-percha | SE Asia | Previously an important source of gutta-percha, which was used for insulation, for golfballs, and in medicine. Obtained after felling the trunk, or by boiling the leaves and twigs. Nowadays of little importance (259, 684, 866, 1444, 1954) |
| *Payena leerii* (Teijsm. et Binn.) Kurz | gutta sundek | Java | Used like gutta-percha (259) |

# Gums and Mucilages

fr. gommes, mucilages; sp. gomas, mucinas; ger. Gummen, Schleime

Gums and mucilages are polymeric carbohydrates, whose main chain consists of sugars (mannose, galactose), linked with side chains of other sugar molecules or sugar derivatives (galacturonic acid, glucuronic acid). They are chemically related to hemicelluloses (xylans) and pectins (main chain of galacturonic acid). Physically they are characterized by the fact that they swell or dissolve in warm water, while they are insoluble in alcohol, benzene, etc. The gums which are utilized as raw materials are constituent parts of the cell wall (seed gums), or are formed from cell-wall material (bark gums).

Depending on the site of their formation in the plant, the following groups can be differentiated: bark gums, endosperm gums, and seed coat gums or mucilages.

Bark gums are mostly obtained from wild-growing trees and shrubs. Only gum arabic is produced in the Sudan in a type of semi-cultivation of *Acacia senegal* (L.) Willd. (*A. verek* Guill. et Perrott.), Leguminosae-Mimosoideae (in the "gum-gardens" of Kordofan). The propagation is natural, by seeds and suckers, the maintenance work consists in the removal of unwanted trees. The growing trees are exploited for 15-25 years, then they are cut down, and the land is used for several years for food crops. The proportion of agricultural land to land where yielding trees are grown is about 1:2. To stimulate the exudation of gum, the bark is injured by tearing off a strip 3-5 cm wide and 30-100 cm long at the end of the rainy season (tapping). The gum solidifies in the air, and is collected several times during the dry season (444, 511, 587, 760, 1125, 1261, 1928). Gum arabic is also gathered in other regions, especially in western Sudan from *A. senegal*. In the Sahel zone, its systematic cultivation as a windbreak for preventing erosion is recommended (273, 1160). Apart from *A. senegal*, *A. nilotica* (Table 55) and *A. seyal* Del. produce a product which is worth exporting (638). Locally, the gums from numerous other *A.* species are used. Gum arabic appears on the market under a variety of trade names, some of which indicate the geographical, and some the botanical origin (babul gum, zedou gum from *A. nilotica*; hashab gum, vereck gum from *A. senegal*; talha gum, Suakim gum from *A. seyal* etc.). World production amounts to about 60,000 tons, and a stable market can be found for good qualities (11). 55-60% of the gum arabic is utilized in the food industry (highest quality requirements), as a stabilizer, emulsifier, and for increasing viscosity. Following this, it is used technically for inks and paints, as a binder in pharmaceutical preparations, and in the textile industry (1928).

Other regular exports are tragacanth, obtained from wild shrubs of *Astragalus gummifer* Labill. and other *A.* species, Leguminosae-Papilionoideae, in the Middle East, and karaya gum (Indian tragacanth), obtained from *Sterculia urens* Roxb., Sterculiaceae, in India. Tragacanth is the most expensive plant gum (6-10 times more costly than gum arabic), and is used almost exclusively for medicinal purposes (444, 760, 1261, 1928). Karaya is nearly twice as expensive as gum arabic, and is used for pharmaceutical preparations and in the food industry. The production of tragacanth has sharply receded (400 t); 1000 tons/year of karaya are exported. Ghatti gum is obtained in India from *Anogeissus latifolia* Wall., Combretaceae; it is exported on a small scale (a few 100 t/year). The gum from the Indian-Southeast Asian plant *Lannea coromandelica* (Houtt.) Merill (L. grandis (Dennst.) Engl.), Anacardiaceae, "jhingan gum", is only locally used in the textile industry, and for the clarification of sugar cane juice (376) (see also cashew, p. 238).

Among the endosperm gums, guar is by far the most important; its production exceeds even that of gum arabic, and the USA alone requires about 50,000 tons annually. The plant, *Cyamopsis tetragonoloba* (L.) Taub., Leguminosae-Papilionoideae, is an old Indian vegetable and fodder plant, which was cultivated previously in other countries, mostly as a green manure and fodder plant. Since the start of the 1950s, guar has been used primarily as a source of gum. The plant is quite drought tolerant, and is a good $N_2$ fixer; its cultivation can be fully mechanized. The separation of the endosperm from the seed coat and embryo can be carried out using wet or dry milling processes. The yield of seeds is about 800 kg/ha, and the endosperm is 35-42% of the seed weight. The product which is traded is the finely milled endosperm. The main exporter is India, but guar is also produced in many other countries. Guar gum is somewhat cheaper than gum arabic, and can usually be used for the same purposes as this. However the bulk of it does not go to the food industry, but is used for technical purposes (376, 444, 608, 885, 947, 1258, 1928, 1929, 1930). The two other sources of endosperm gum of commercial importance are carob and tamarind. Carob gum is traded internationally also as locust bean gum. The seed contains 35-45% endosperm, the gum is utilized for all purposes, particularly in the food industry, and in pharmaceutical and cosmetic preparations (488). The production of tamarind seed gum (tamarind seed powder) amounts to about 20,000 tons in India. Most of this is used locally in the textile industry, and only a small part is exported. The seeds of *Trigonella foenum-graecum* (Table 25), *Crotalaria* spp., and *Sesbania* spp. provide usable gums, which however play very little role in trading (376, 444, 608, 935, 1928). A new candidate on the market is tara gum from *Caesalpinia spinosa* (Table 47) (916).

Mucilages are obtained particularly from *Plantago ovata* (Table 40) and from quince seeds (see p. 205). *P. ovata* seeds are predominantly used medicinally (psyllium seed gum). They have almost replaced linseed gum. Quince seeds are harvested 75% from wild plants in Iran, and 25% are a byproduct from canning factories. Quince seed gum is almost as expensive as gum tragacanth, and is used medicinally and in cosmetics (wave setting lotions). The production amounts to about 500 tons/year (1928).

The gums extracted from the higher plants face a lot of competition in many areas of usage, especially in the food industries; their competitors are algal mucilages (algins from brown algae, agar and carrageenan from red algae), starch derivates, cellulose ethers, and bacterial mucilages (dextran, xanthan) (608, 1125, 1928). With the steadily increasing demand (drinks industry, ice cream, yoghourt, frozen foods, etc.), the plant gums have kept their place in the market well, although in the future, firm price levels may need to be set if they are to remain competitive (1928).

# Resins and Gums

fr. résines; sp. resinas; ger. Harze

Resins are complicated mixtures of diterpenes (type: abietic acid) with volatile terpenes (e.g. pinene), coniferyl esters, gums and aromatic compounds. They are not in all cases clearly separable from gums, but generally they are more or less soluble in alcohol and other organic solvents, and scarcely soluble in water; they mostly burn well. In the plants, they are exuded in a liquid state into secretory ducts in the bark and the wood. Many resins solidify in the air due to the evaporation of their volatile components, and also due to oxidation and polymerisation. Some resins are called soft resins and balsams, as they form no solid, glass-like products.

Most resins are obtained from trees or shrubs growing wild (Table 46). The following are cultivated: *Pistacia terebinthus*, *Toxicodendron verniciflua*, and the forest trees *Pinus pinaster*, *Shorea robusta*, *Styrax benzoin*, and others, with which the extraction of resin is often a by-product (644). In some plants, the formation of resin is so strong that its collection is worthwhile without any special measures (copal, also fossil). More frequently however, the resin flow must be stimulated by wounding the bark or the wood. The volatile aromatic components of many resins are separated by distillation, and are used as solvents (turpentine oil) or in perfumery. The remains from the distillation (rosins) are especially hard, and are used technically (colophony, etc.). The multitude of applications for resins can be seen in Table 46. The plant resins are irreplaceable in pharmacy and in perfumery as aromatics. In their technical applications as lacquers (colourless and coloured) and putty, strong competition has developed from the resin from tall oil (see p. 78) and the synthetic resins, but for some particular purposes, the natural resins are still preferred. The same is true for shellac, the resin from the lac insects (see p. 402), which occupies a leading place with a production of 60,000 t (376, 760).

Table 46. Resins (explanation of symbols, see footnotes page 55)

| Botanical name | Vernacular names of plants or products | Origin | Climatic region | Remarks and literature references |
|---|---|---|---|---|
| | | **Gymnospermae** | | |
| **Araucariaceae** | | | | |
| *Agathis dammara* (Lamb.) L.C. Rich. (*A. alba* (Bl.) Foxw.) | Manila copal, batjan gum, batu gum | SE Asia | Tr | Cultivated in forests in Java, the resin is used for lacquers and linoleum (259, 355, 760, 1937) |
| **Cupressaceae** | | | | |
| *Tetraclinis articulata* (Vahl) Mast. | sandarac gum, arar tree, cyprès de l'Atlas | Algeria, Morocco | S | Cultivated in forests in Morocco, utilized for lacquer, cement, fumigation products, in perfumery, and for liqueurs (758, 760) |
| **Pinaceae** | | | | |
| *Pinus pinaster* Ait. (*P. maritima* Lam. non Mill.) | turpentine, maritime pine, pin maritime | S Europe, Mediterranean countries | S | Cultivated in forests for resin production. For oil of turpentine and colophonium. Resins also obtained from other *Pinus* species (644) |

Table 46. Resins, cont'd (explanation of symbols, see footnotes page 55)

| Botanical name | Vernacular names of plants or products | Origin | Climatic region | Remarks and literature references |
|---|---|---|---|---|
| | | **Dicotyledonae** | | |
| **Anacardiaceae** | | | | |
| *Pistacia lentiscus* L. var. *chia* Desf. | Chios mastic, mastic tree, lentisque | Mediterranean countries | S | Cultivated on Chios. Resin for picture and photograph lacquers, aromatics used especially for chewing gum and in medicine. Leaves of this and other *Pistacia* species are used for the production of tanning materials and colourants (758, 760) |
| *Toxicodendron verniciflua* (Stokes) Barkl. (*Rhus verniciflua* Stokes) | Japanese lacquer, Chinese lacquer, urushi | Central and W China | S | Cultivated in China and Japan, resin obtained by tapping the bark, utilized for lacquer (760, 1904) |
| **Burseraceae** | | | | |
| *Boswellia sacra* Flückiger (*B. carteri* Birdw.) | olibanum, frankincense | Somalia, Iran, Iraq | S, Tr | Utilized for fumigation products, perfume, cosmetics, and also for lemonade and baking (10, 758, 760, 1836) |
| *Canarium luzonicum* Miq. | elemi, Manila elemi | Philippines | Tr | Applications in oil lacquers, skin creams, fumigation products, perfume, dressing of felt cloths, and as a flavouring in lemonade, bakery and soups. Other Burseraceae species are used for the production of resin, and are utilized similarly (259, 758, 760) |

| | | | | |
|---|---|---|---|---|
| *Commiphora abyssinica* Engl. | Abyssinian myrrh, hotai | NE Africa, S Arabia, Somalia | S, Tr | Bitter compound in pharmacy (against inflammation of oral cavities), perfumery, used for sweets, lemonade and chewing gum (10, 758, 1836) |
| *C. erythraea* Engl. var. *glabrescens* Engl. | opopanax gum, sweet myrrh | Somalia | Tr | Oriental spice for liqueurs, in perfumery "perfumed bdellium" (10, 758, 1836) |
| *C. opobalsamum* (Le Moine) Engl. | Mecca myrrh, harobol myrrh, Gilead balsam | Arabia, Somalia | S, Tr | Previously cultivated in Egypt and Palestine, utilized like the other Burseraceae resins (758, 760, 1836) |
| **Cistaceae** | | | | |
| *Cistus ladanifer* L. | ladanum, gum cistus | Mediterranean countries | S | Obtained by boiling the leaves and twigs, utilized in perfumery and as an aromatic for sweets, baking and chewing gum. The resin from *C. creticus* L. is used similarly (10, 760, 1382) |
| **Dipterocarpaceae** | | | | |
| *Dipterocarpus costatus* Gaertn.f. (*D. alatus* Roxb.) | gurjun | Burma, Thailand, Indochina | Tr | Cultivated in forests in tropical Asia. For lacquer and in perfumery; utilized for adulterating patchouli oil. In India, *D. indicus* Bedd. and *D. turbinatus* Gaertn.f. are also called gurjun (259, 376, 758, 1383) |
| *Shorea robusta* Gaertn.f. | sal tree, sal dammar, Indian sal, dammar de l'Inde | India, SE Asia | Tr | Important forest tree in SE Asia, supplies resin for cheap paints and lacquers, utilized for "attar", incense and salves. Bark provides tanning material and black dye, seeds for vegetable oil and press cakes (234, 376, 406, 760, 807, 1471) |

Table 46. Resins, cont'd (explanation of symbols, see footnotes page 55)

| Botanical name | Vernacular names of plants or products | Origin | Climatic region | Remarks and literature references |
|---|---|---|---|---|
| *S. wiesneri* Schiffn. | Batavian dammer | Sumatra, Kalimantan, Philippines | Tr | Utilized for high value lacquers, and especially for embedding microscopic samples. A multitude of other *Shorea* species are used similarly (259, 760) |
| *Vateria indica* L. | piney varnish, white dammar, Indian copal | India | Tr | Resin for fumigation goods, dyes and lacquer. Seed fat (piney tallow, dhupa fat) used for lighting, for soaps, and also as a foodstuff. The fruit shells provide a tanning material (234, 376, 406, 760, 1471) |
| **Guttiferae** | | | | |
| *Garcinia morella* Desr. | gamboge, Indian gamboge tree | India, Sri Lanka, Thailand, Indochina | Tr | Used as an additive to painting colours and lacquers, strong purgative in veterinary medicine. *G. hanburyi* Hook.f. is used similarly in Indochina (259, 376, 406, 760) |
| **Hamamelidaceae** | | | | |
| *Liquidambar orientalis* Mill. | Oriental storax, Oriental sweet gum, Levant storax | SW Asia | S | In pharmacy and perfumery, also as aromatic for sweets and baking, lemonade and chewing gum. Turkey almost has an export monopoly of this resin. The resin from *L. styraciflua* L., American storax or sweet gum, is used similarly in N and Central America (10, 661, 758, 760) |

| | | | | |
|---|---|---|---|---|
| **Leguminosae** | | | | |
| CAESALPINIOIDEAE | | | | |
| *Copaifera officinalis* (Jacq.) L. | copaiba balsam, baume de copahu, Maracaibo balsam | Central and S America, Africa | Tr | Obtained by tapping the trees, used for soap, lacquer, varnish, tracing paper, perfume. Other *Copaifera* species are used similarly (10, 760, 1518) |
| *Daniellia oliveri* Hutch. et Dalz. | African copaiba, Niger copal, gum copal | W Africa | Tr | Obtained by tapping, utilized for lacquer. Other *Daniellia* species also provide copal (760) |
| *Guibourtia demeusii* (Harms) Léonard | copal noir, copalier | W Africa | Tr | Fossil copal for lacquer, for linoleum production, for caulking boats, and in medicine. The fossil resins of other *Guibourtia* species are used in a similar way (1477) |
| *Hymenaea courbaril* L. | West Indian locust, jutaicica, guapinol, jutaí | Central and S America | Tr | Obtained by tapping, and semi-fossilized from the base of the tree, used for lacquering wood and metal, and for fumigating materials. Other *Hymenaea* species are exploited similarly (760, 1425, 1518, 1960) |
| *Trachylobium verrucosum* (Gaertn.) Oliv. | Zanzibar copal East African copal | E Africa | Tr | Predominantly a fossil resin, utilized for lacquers and varnishes (376, 760) |
| PAPILIONOIDEAE | | | | |
| *Myroxylon balsamum* (L.) Harms | | | | |
| var. *balsamum* | tolu balsam, bálsamo de Tolú | Colombia, Venezuela, Peru | Tr | Utilized in medicine (pectoral), perfumery, for lemonade, baking, and chewing gum (10, 444, 758, 760, 1378, 1518) |
| var. *pereirae* (Royle) Harms | balsam of Peru, El Salvador balsam, bálsamo blanco | Central America | Tr | Obtained by tapping, utilized in perfumery and medicine (skin diseases), also as an aromatic for chewing gum (10, 444, 758, 760, 1518, 1960) |

Table 46. Resins, cont'd (explanation of symbols, see footnotes page 55)

| Botanical name | Vernacular names of plants or products | Origin | Climatic region | Remarks and literature references |
|---|---|---|---|---|
| **Styracaceae** | | | | |
| *Styrax benzoin* Dryand. | styrax, Sumatra benzoin, benjamin gum | Sumatra, Java, Indochina, Thailand | Tr | Easily grown forest tree, cultivated in Sumatra for re-afforestation and for resin production, utilized in perfumery, medicine and as an aromatic for lemonade, baking, and chewing gum, an expectorant and component of "friar's balsam", textile printing and spirit lacquers. Considerable exports from SE Asia. Other *Styrax* species are similarly exploited for resin production (10, 376, 684, 758, 760) |
| **Umbelliferae** | | | | |
| *Ferula assa-foetida* L. | asafetida gum, asant | Iran<br>Afghanistan<br>Indus territory | S, Tr | The gum resin from the roots is utilized, collected from wild plants. In perfumery and medicine (sedative, antispasmodic, laxative), spice for meat (Worcester sauce), and pickles. The resins from *F. foetida* (Bunge) Regel and *F. narthex* Boiss. are used similarly (376, 1383, 1442, 1846) |
| *F. gummosa* Boiss. (*F. galbaniflua* Boiss. et Buhse) | galbanum, boridschah, kasnih | Iran<br>Afghanistan | S | Utilized in perfumery and medicine, and as an aromatic for lemonade, sweets and baking (10, 758, 1846) |

| | | | | |
|---|---|---|---|---|
| **Zygophyllaceae**<br>*Guajacum officinale* L. | guaiac resin,<br>lignum vitae,<br>lignum sanctum,<br>bois de vie,<br>guayacán | West Indies,<br>Northern<br>S America | Tr | Resin content of wood is 18-25%, utilized in medicine (arthritis, rheumatism, kidney and skin diseases), as a preservative and an aromatic for sweets, baking and chewing gum. *G. sanctum* L. is used in the same way (760, 1960) |
| | | **Monocotyledonae** | | |
| **Palmae**<br>*Daemonorops draco* Bl. | dragon's blood palm,<br>rattan palm,<br>sanguis draconis | Sumatra,<br>Kalimantan,<br>Malaysia | Tr | The resin is an exudation from the fruit, and is separated mechanically. It is utilized for coloured lacquers and in medicine (astringent for dysentery), for toothpaste and lemonade. Numerous other *Daemonorops* species are used similarly (259, 376, 760, 1383) |
| **Xanthorrhoeaceae**<br>*Xanthorrhoea hastilis* R. Br. | yacca gum,<br>grass-tree gum,<br>Botany Bay gum,<br>blackboy | Australia | S | The resin is exuded from the base of the leaves, and is used for lacquers, varnishes and paper. Other *Xanthorrhoea* species are used in a similar way (760, 1106) |

# Tanning Materials

fr. substances tannantes; sp. materias curtientes; ger. Gerbstoffe

Tanning materials are complex phenolic compounds with a large number of oxygen functions. The most important basic molecules are gallic acid and catechins, both together in mixed tannins. One can differentiate between hydrolysable tannins, predominantly from gallic acid, which is esterified with sugar (e.g. oak gall, chestnut), and condensed tannins, which are composed of the basic substances flavandiols, flavones, and catechins, and their oxidation and condensation products (e.g quebracho, mimosa bark). Tanning agents which are technically usable are soluble in water. Their tanning effect (leather formation) consists in the precipitation of protein, which afterwards no longer swells in water, and resists rotting. Their medicinal effect is also attributed in part to precipitation of proteins, especially firming the protein in the body tissues (478, 1071).

Tannins are found in the parenchymatic tissues of many plants, in usable concentrations especially in the bark, in the wood of the trunk, in the roots, and in fruits and galls. The number of plants which have been used by man since ancient times in all parts of the earth amounts to many hundreds (478, 761). Those which are regularly used commercially are presented in Table 47. At present, the most important are mimosa bark, quebracho, chestnut wood and mangrove.

The vegetable tanning materials now face strong competition from chemical tanning compounds (chrome tanning, synthetic tanning materials). In spite of this, the demand for plant tannins is still considerable, because they give weight to the leather and characteristic colouring. An important part of the tannins are used for other technical purposes, e.g. for reducing the viscosity of drilling slurry from deep-drilling projects; in the USA alone, 40,000 t of quebracho extract are used for this each year (478). The amount of tannins which find uses in pharmaceutical preparations is small compared to this. The majority of tanning products have for many years come on the market in the form of extracts (moist or dried), and only material which is very high in tannin is traded in dried or powdered form (e.g. tara powder).

Table 47. Tanning materials (explanation of symbols, see footnotes page 55)

| Botanical name | Vernacular names of plants or products | Origin | Climatic region | Remarks and literature references |
|---|---|---|---|---|
| **Anacardiaceae** | | | | |
| *Astronium balansae* Engl. | urunday | N Argentina, Paraguay | S, Tr | Obtained from wild-growing trees containing about 12% tannin in the wood. Production much decreasing, utilized like quebracho (478, 1382) |
| *Rhus coriaria* L. | Sicilian sumach, sumac, zumaque | SW Asia and Mediterranean countries | S | Cultivated in S India, leaves contain 25-30% tannin (478) |
| *Schinopsis balansae* Engl. | willowleaf red quebracho, quebracho colorado chaqueño | N Argentina, S Brazil, Paraguay | S, Tr | Wood contains 20-25% tannin, from wild stands in the western Gran Chaco, annual tannin production from both *Schinopsis* species about 130,000 tons (478) |
| *S. quebracho-colorado* (Schlechtend.) Barkl. et T. Mey | red quebracho, quebracho colorado, quebracho vermelho | N Argentina, S Brazil, Paraguay | S, Tr | From wild stands of the eastern Gran Chaco, only 16-17% tannin in the wood, used like *S. balansae* (478) |
| **Combretaceae** | | | | |
| *Terminalia chebula* Retz. | myrobalans, myrabolans | India | Tr | Fruits of mostly wild-growing trees have a tannin content of 25-35%, annual production of about 100,000 t fruit, of which 10,000 t exported; India (a monopoly) recently also exports the myrobalan extract. The fruits of other *Terminalia* species are used similarly (259, 376, 406, 478, 1383) |

Table 47. Tanning materials, cont'd (explanation of symbols, see footnotes page 55)

| Botanical name | Vernacular names of plants or products | Origin | Climatic region | Remarks and literature references |
|---|---|---|---|---|
| **Fagaceae** | | | | |
| *Castanea sativa* Mill. (*C. vesca* Gaertn.) | European chestnut, châtaigne, castaño | Italy, S France, Spain | S | Annual production about 50,000 t tannin, the wood contains about 8.5% tannin. Seeds eaten as nuts (Table 33) (478) |
| *Quercus infectoria* Olivier | Aleppo oak, gall oak, Mecca galls | SW Asia and E Mediter-ranean countries | S | The leaf galls contain 36-78% tannin, their usage is much decreased. Galls from other *Quercus* species are used like the species named, originating mostly from the Middle East (478) |
| **Leguminosae** | | | | |
| MIMOSOIDEAE | | | | |
| *Acacia mearnsii* De Wild. (*A. decurrens* (Wendl.) Willd. var. *mollis* Lindl.) | black wattle, acacia noir, acacia negra | w | S, TrH | Annual world production 130,000 t, main producers S Africa and Brazil, 35-39% tannin in the dried bark. Other acacias are to a small extent used locally for tannin (44, 174, 376, 478, 684, 1642) |
| CAESALPINIOIDEAE | | | | |
| *Caesalpinia brevifolia* (Clos.) Baill. | algarrobilla, algarroba | Chile | S | The fruits contain 40-50% tannin (478) |
| *C. coriaria* (Jacq.) Willd. | divi-divi, cascalote, nacascol | Central and S America | Tr | Main producers Venezuela, Colombia, and the W Indian Islands (about 10,000 t), pods contain 40-50% tannin (406, 444, 478, 1960) |

| | | | | |
|---|---|---|---|---|
| *C. spinosa* (Mol.) O.Kuntze | tara, guarango, huarango | Brazil Peru Chile | S, Tr | The main producer Peru exports up to 20,000 t, the product traded is the milled pods (tara ) powder, which contain 35-55% tannin and black colourant. Pods of other *C.* species are used similarly only locally. Tara gum (p. 373) is obtained from the seeds (81, 444, 478) |
| **Myrtaceae** | | | | |
| *Eucalyptus astringens* Maiden | brown mallet | W Australia | S | About 1000 t bark with 40-55% tannin content are obtained in Australia, also the bark of other *Eucalyptus* species is used for tannin production to a small extent (478, 512, 1371) |
| **Rhizophoraceae** | | | | |
| *Rhizophora mucronata* Lam. | mangrove cutch, red mangrove | tropical coasts | Tr | The bark contains 25-40% tannin, the wood 9-14%. In the tropics, the bark of many other mangrove species is used for tanning, e.g. *Bruguiera eriopetala* Wight et Arn., *Ceriops candolleanum* Arn., *Sonneratia* spp., and other *Rhizophora* spp. (376, 406, 478, 1383) |

# Dyes, Colourings

fr. colorants; sp. tintes; ger. Farbstoffe

From earliest times to the highest civilizations, man has used pigments of plant origin for beautifying the body, for clothing, for decorating the home, and in foods and drinks. Colourant additions also play a major role in the modern food and cosmetic industries. Since the second half of the 19th century, the products of the chemical industry have almost driven the natural organic pigments from the world market. However, since the discovery of the carcinogenic effects of many synthetic colourings, vegetable pigments again play an increasing role in all products which are eaten or put on the skin because legislation in all countries does not require proof of the harmlessness of plant products, while it is required for synthetic colourings. The procedure for proving their safety is very tedious and costly (451, 469, 623, 728).

The plant pigments which are used in the food industry and in cosmetics belong to a variety of chemical groups. Reds come from anthocyanins, betacyans (betanin), some carotenoids (bixin from annatto, capsanthin from paprika, lycopene from tomatoes, and carthamin from safflower), yellows come from the carotenoids xanthophyll, zeaxanthin and azafrin (*Escobedia* spp.), and curcurmin from *Curcuma longa* (Table 39). As a green colouring, chlorophyll is obtained from the leaves of a variety of plants. The Cu-containing complexes of chlorophyll and chlorophyllin are more stable than chlorophyll, and are allowed by the food regulations. A blue colour is obtained from the flowers of *Clitoria ternatea* (Table 52) in Asia, and is used for the colouring of rice dishes (1082), and many anthocyanins are also blue at alkaline reaction (red cabbage) (988).

Pigments are found in the plant cells either in the plastids or in vacuoles. The plastid pigments (chlorophyll, carotenoids) are soluble in fats, and are therefore used in fatty foods and cosmetic preparations, yet they can also be used in fat-free foodstuffs with the help of dispersing agents and emulsifiers (gums, starch derivatives). The vacuole pigments (anthocyanins, betacyanins) are water-soluble, and are therefore of primary importance in the drinks and confectionery industries.

In Table 48, apart from the pigments which are used in the food industries, some are included which are locally cultivated for the dyeing of cloth (e.g. indigo, which has recently regained some commercial importance with the fashion for blue jeans). Colourings which are exclusively gathered from wild plants, and which have no commercial value are not included in the Table (complete listing in (988)). Some plants, which are classed as suppliers of pigments, have already been dealt with in other sections (cereals: sorghum, see p. 29; vegetables: beetroot, Table 25, and tomato, p. 133; fruit: skins of blue grapes (see p. 175); beverages: gambir, Table 38; spices: saffron, sweet paprika, turmeric, Table 39; oil plants: safflower, p. 105; fodder plants: *Clitoria ternatea*,

Table 52). It should also be remembered, that the juice of many fruits is used as a colouring additive to drinks and other foodstuffs (cherry, mulberry, etc.).

Of the plants in Table 48, *Bixa orellana* deserves special mention (Fig. 96). It is spread throughout all tropical regions as an ornamental and hedge bush. Its pigments have become an important export product for a number of developing countries (India, East Africa, Peru, Brazil, Panama, Ecuador, Jamaica). The world's production amounts to some 1000 tons (451). The seeds are covered with papillae, in which the pigments bixin and norbixin are stored (pigment content of the whole seed is 3.4-5.3%). Bixin is soluble in fat, the alkaline salt of norbixin is water-soluble. As a result, it has unusually versatile application possibilities in the food and cosmetic industries (54, 376, 550, 684, 758, 819, 1301, 1960).

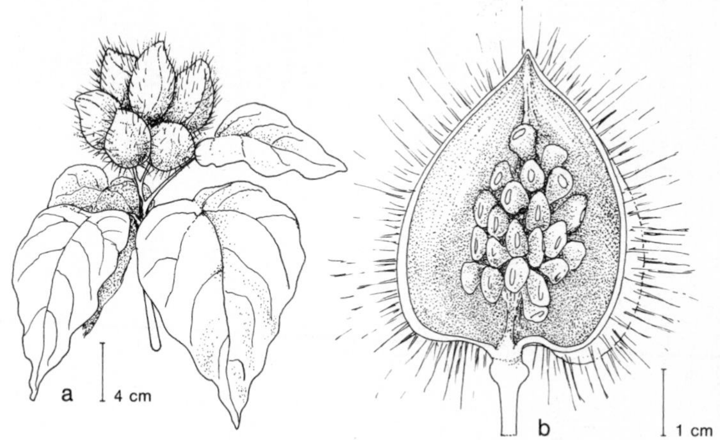


Fig. 96. *Bixa orellana*. (a) fruiting branch, (b) opened fruit with parietal placentation

Table 48. Dyes and colourings (explanation of symbols, see footnotes page 55)

| Botanical name | Vernacular names of plants or products | Origin | Climatic region | Remarks and literature references |
|---|---|---|---|---|
| **Bixaceae** | | | | |
| *Bixa orellana* L. | annatto, urucúm, achiote | Central and S America E Africa, India | S, Tr | (see p. 387, Fig. 96) |
| **Boraginaceae** | | | | |
| *Alkanna tuberculata* (Forssk.) Meikle (*A. tinctoria* Tausch) | alkanna root, alkanet | S and E Europe, Turkey | S | Still cultivated in the Middle East, for dyeing foodstuffs, salves and cloth (758) |
| **Leguminosae** | | | | |
| MIMOSOIDEAE | | | | |
| *Acacia catechu* (L.f.) Willd. | catechu, black cutch | India | Tr | Dark-coloured, tannin-containing gum, which is utilized in India for colouring and preserving, and chewed with betel. Exported as an aromatic for sweets and baking, lemonade and chewing gum (54, 376, 406, 1125, 1383, 1444) |
| CAESALPINIOIDEAE | | | | |
| *Caesalpinia sappan* L. | sappan, Indian redwood | India | Tr | The heartwood, "sappan wood", supplies a red pigment, bark and roots are likewise used for colour; in India, the pods are used for producing a black dye and ink. Other *Caesalpinia* species, especially *C. echinata* Lam. (in Brazil) are used for the same purposes. Also planted as a hedge shrub (54, 376, 406, 444, 1383, 1518) |

| | | | | |
|---|---|---|---|---|
| *Haematoxylum campechianum* L. | logwood, bois de campeche, palo campeche | Central America, S Asia | Tr | The heartwood is used for the production of haematoxylin, which is used in microscopy and locally for dyeing (54, 758, 1518, 1857, 1960) |
| PAPILIONOIDEAE | | | | |
| *Indigofera arrecta* Hochst. ex A. Rich. | Java indigo, Bengal indigo, Natal indigo | India, W Africa | S, Tr | Apart from *I. arrecta*, *I. tinctoria* L., and *I. sumatrana* Gaertn. are cultivated locally in India, and *I. articulata* Gouan in W Africa. Some species are used as ground covers (Table 53) and fodder plants (259, 376, 444, 1444) |
| **Lythraceae** | | | | |
| *Lawsonia inermis* L. (*L. alba* Lam.) | henna, henné, alheña | India, Egypt | S, Tr | Cultivated in India, the Middle East, and N Africa. The colouring is obtained from the leaves, for dyeing leather, hair, fingernails, etc. For leather dye and modern hair shampoos, considerable quantities are exported from Egypt, the Middle East, and especially India. Hedge plant (54, 259, 376, 406, 444, 1382, 1383) |
| **Phytolaccaceae** | | | | |
| *Phytolacca americana* L. | poke, pokewood, raisin d'Amérique | N America | Te, S | Stable red pigment betanin, primarily obtained from plants which have run wild. The berries of *P. acinosa* Roxb. and *P. chilensis* Miers are also used (200, 376, 451) |
| **Rubiaceae** | | | | |
| *Morinda citrifolia* L. | Indian mulberry, nuna, al | India | Tr | Several species cultivated in India for colourant production from their roots. Has lost importance (259, 376, 1383) |

Table 48. Dyes and colourings, cont'd (explanation of symbols, see footnotes page 55)

| Botanical name | Vernacular names of plants or products | Origin | Climatic region | Remarks and literature references |
|---|---|---|---|---|
| *Rubia cordifolia* L. | Indian madder<br>manjistha<br>manjit | India | Tr | Red pigment from stem and roots, used in India for dyeing cotton cloths. Other species, especially *R. tinctorum* L., are used similarly in India (376, 406, 1383) |
| **Scrophulariaceae** | | | | |
| *Escobedia scabrifolia* Ruiz et Pav. | Color azafrán,<br>raíz de color,<br>açafrão do campo | S America | TrH | The yellow pigment azafrin is obtained from the roots of this and other *E.* species, and is used for colouring rice, manioc, and other foodstuffs (1378, 1518, 1958) |

# Pesticides

fr. pesticides; sp. pesticidas; ger. Pestizide

Many plants contain natural protective compounds against insect pests. They were already used by primitive man for the control of ectoparasites (lice, bugs). The production of highly active extracts (pyrethrum in household sprays) and their application in crop protection is relatively new. Especially for crop protection, an intensive research effort was begun in the 1930s, when, on the one hand, the market for insecticides had expanded very rapidly, while on the other hand, the dangers became recognized of the great persistence or high acute toxicity of the heavy metal insecticides (As, Pb, Hg, etc.), and of some of the synthetic insecticides (537, 631, 854, 1618, 1625, 1827).

The insecticidal plants that are cultivated to a noteworthy extent are set out in Table 49; there is a great number of other species which are being examined in a variety of places (neem (see p. 117); (1606)). Of the species named, pyrethrum is by far the most important which is cultivated nowadays in many countries (1524). This is followed by derris and cube which contain rotenone, and third place is taken by nicotine and the chemically related anabasine. Vegetable insecticides are traded as a highly purified substance (nicotine), concentrated extract (pyrethrum), and dried powder (cube). The proportion of plant products in the total insecticide market is slight. In spite of this, some have become important export products for the producer countries, and the market for them shows a tendency to increase. The discovery of plant hormones which disturb the development of insects indicates a new possibility for the use of plant products in the battle against pests (222). Attention has also been attracted to plants with molluscicidal or nematidicidal properties (755, 976, 1134, 1208, 1649).

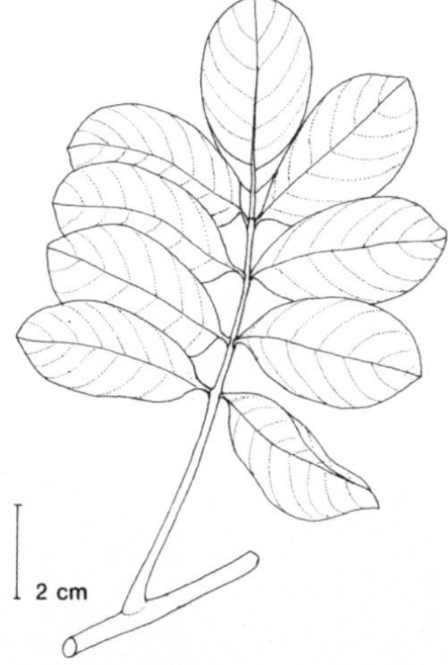


Fig. 97. *Derris elliptica*. Leaf

Table 49. Pesticides (explanation of symbols, see footnotes page 55)

| Botanical name | Vernacular names of plants or products | Origin | Climatic region | Remarks and literature references |
|---|---|---|---|---|
| **Celastraceae** | | | | |
| *Tripterygium wilfordii* Hook.f. | Thundergod vine, lei kung tung | China, Taiwan | S | Active constituents: alkaloids (wilfordine, etc.). The root bark is used. Cultivated in China (537) |
| **Chenopodiaceae** | | | | |
| *Anabasis aphylla* L. | leafless anabasis | S USSR, Iran | S | Active constituent: anabasine (537, 1846) |
| **Compositae** | | | | |
| *Chrysanthemum cinerariifolium* (Trev.) Vis. | pyrethrum, piretro | Kenya, S America, New Guinea, S Africa, India | S, TrH | Active constituents: pyrethrins. Annual world production 22,000 t dried flowers, of which 14,000 t from Kenya. The pyrethrin content is considerably improved by breeding (to 2%). Now marketed as an extract with 25-30% pyrethrin (279, 537, 1444, 1511, 1524) |
| **Flacourtiaceae** | | | | |
| *Ryania pyrifera* (L. C. Rich.) Uittien et Sleumer (*R. speciosa* Vahl) | ryania | S America, Trinidad | Tr | The finely ground stem wood and the roots provide the alkaloid ryanodine. The product traded is ryania powder, which is used against fruit maggots, and in the USA against maize stem-borers (537, 1716) |

| | | | | |
|---|---|---|---|---|
| **Leguminosae** | | | | |
| PAPILIONOIDEAE | | | | |
| *Derris elliptica* (Sweet) Benth. | derris, tuba root, akar tuba | S and SE Asia, E Africa | Tr | (Fig. 97) Active constituent: rotenone. The dried and ground roots are utilized. Other *Derris* species likewise provide rotenone (259, 376, 406, 444, 684, 1465) |
| *Lonchocarpus nicou* (Aubl.) DC. | cubé, timbó | Guayana | Tr | see *L. utilis* |
| *L. urucu* Kill. et A. C. Sm. | timbó vermelho | Brazil | Tr | see *L. utilis* |
| *L. utilis* A. C. Sm. (*Derris utilis* Ducke) | cubé, barbasco, timbó | Peru | Tr | Ancient cultivated plant of the Indios. The product traded, the dried pulverized root, contains 5-12% rotenone. Also traded as an extract with 25-40% rotenone. Peru is the largest exporter (444, 1039, 1518) |

# Waxes

fr. cires; sp. ceras; ger. Wachse

Waxes are fatty substances, mostly with a melting point of 50-90°C. The characteristic chemical constituents are esters of long chain fatty acids ($C_{18}$-$C_{32}$) with long chain primary alcohols ($C_{22}$-$C_{34}$). In addition, there are bivalent alcohols, hydroxy fatty acids, and paraffins ($C_{27}$-$C_{31}$), and as unwanted impurities, resins. Liquid waxes are found in the seeds of *Simmondsia chinensis* (see below). Under the term waxes are also included the long chain alcohols in the latex of Moraceae (*Ficus* spp., *Brosimum utile*, Table 45 and p. 368), as well as some vegetable fats which consist of triacylglyceroles of long chain fatty acids with a correspondingly high melting point (Japan tallow, bayberry wax, Table 50).

Most of the waxes which are of economic importance are exudations from the epidermis of leaves, stems and fruits. Japan wax comes from the fruit pulp, jojoba wax from the seeds. The most important sources of wax are given in Table 50. In terms of production quantities, carnauba leads, followed by candelilla, jojoba, ouricury, and Japan wax. Wax is also extracted as a by-product from the processing of sugar cane, rice and sorghum (stems of sugar sorghum, processing of grain sorghum), from the production of raffia fibres (Table 44), and recently also from the bark of Douglas fir (*Pseudotsuga menziesii* (Mirb.) Franco). The wax from other plants is used locally (*Scheelea martiana* (Table 17), *Bulnesia retama* (Gill.) Griseb., Zygophyllaceae). It only occasionally appears on the world market. Hydrated castor oil ("opal wax", "castor wax") is an artificial plant wax, which is used to adulterate carnauba wax.

Plant waxes are used particularly for the wax coating of fruits (citrus, apples), in cosmetics (especially lipsticks), carbon paper, chewing gum, and pharmaceutical preparations. Lesser qualities are used as polishing materials (floors, furniture, cars, shoes), and are utilized in the textile, leather and paper industries. For technical applications, the products of the synthetics industry (polystyrene, styrene-acrylcopolymers) and the mineral oil industry (paraffin wax, microwax) have become extremely competitive. There is a fossil plant wax, montan wax (extracted from brown coals using organic solvents), which is produced especially in the German lignite industry. This currently costs about half as much as plant waxes, and can be used to replace them for many purposes (carbon paper, shoe creams, floor polish).

# Jojoba

Jojoba (*Simmondsia chinensis*) is a dioecious shrub from the southwestern USA and the neighbouring regions of Mexico. As a rule, the fruits enclose one seed (a maximum of 3), containing 47-62% wax, which is liquid at 7°C. One can expect the first yields in the 4th or 5th year, and full yields are reached after about 10 years.

The product which is traded is called jojoba oil. The present production amounts to about 1,000 t oil per year. The USA is the largest producer, followed by Mexico, and on a smaller scale by Costa Rica, Australia, Brazil, Paraguay, and a few others. Due to the good yielding potential of the plant (up to 4.5 t seed per ha has been reported) and the high price of the product (jojoba oil is 2-4 times more expensive than the other vegetable waxes), cultivation has been tried in many localities, frequently with disappointing results, due to mainly unsuitable climatic conditions. Depending on the future price, world demand is estimated at 20,000 to 200,000 t. Production is still far from even the lower figure, mainly because there are two conditions which often have not been met in the past: though jojoba is very tolerant of drought, 750 mm of rain or irrigation are necessary for good yield, and the plant needs a cool period of about 2 months (mean temperature around 15°C) for breaking the dormancy of the flower buds (at constantly high temperatures it never flowers) (534).

Breeding is in progress in several places for high-yielding, monoecious, and early bearing cultivars, and for types which are adapted to local climate. However, the breeding of advanced and uniform cultivars will take decades. In the meantime, the vegetative propagation of highly productive genotypes is the most promising approach (1027). In countries where manual labour is expensive, the high cost of collecting the small fruits is a limiting factor. The solution is either to breed cultivars which are suited to mechanical picking, or to let the ripe fruits or seeds fall to the ground and to collect them afterwards using suction or other means.

Jojoba oil is a high quality substitute for sperm oil, which is hardly available any more due to the decrease in whaling. It is especially needed in the cosmetic industry, and as a high pressure lubricant. If the price were lower, numerous other applications would be available (173, 534, 577, 1027, 1269, 1513, 1519, 1807, 1894, 1915, 1976, 2018, 2019).

Table 50. Waxes (explanation of symbols, see footnotes page 55)

| Botanical Name | Vernacular names of plants or products | Origin | Climatic region | Remarks and literature references |
|---|---|---|---|---|
| | | **Dicotyledonae** | | |
| **Anacardiaceae**<br>*Toxicodendron succedaneum* (L.) O.Kuntze (*Rhus succedanea* L.) | Japan tallow, wax tree, sumac cirier | Japan, China, India | S | One of the most expensive waxes, double the price of beeswax, the fruit mesocarp contains 65% wax, the kernel a fatty oil. Utilized for floor polish, shoe polish, furniture polish, candles, and matches, and also in the cosmetic and leather industries (376, 406, 1856) |
| **Euphorbiaceae**<br>*Euphorbia antisyphilitica* Zucc. | candelilla | Mexico | Tr | Gathered from wild plants, the wax is formed as a covering on the shoots, and is obtained by boiling. Annual production about 3000 t, utilized for polishes and painting materials, for waxing of fruits, and for chewing gum (619, 746, 1095, 1258) |
| *Pedilanthus pavonis* Boiss. | candelilla | Mexico | Tr | Obtained and utilized like the candelilla wax from *Euphorbia antisyphilitica*, other Euphorbiaceae species are likewise used for wax production (746, 1095, 1444) |

| | | | | |
|---|---|---|---|---|
| **Myricaceae** | | | | |
| *Myrica cerifera* L. | wax myrtle, bayberry, arrayán, cera vegetal | Central America, Northern S America | S, TrH | Obtained from wild plants, regularly exported from Colombia to the USA, several other *Myrica* species likewise provide bayberry wax (1378, 1856, 1956, 1960) |
| **Simmondsiaceae** | | | | |
| *Simmondsia chinensis* (Link) Schneid. | jojoba, goat nut | w | S | (see p. 395) |
| | | **Monocotyledonae** | | |
| **Palmae** | | | | |
| *Ceroxylon alpinum* Bonpl. ex DC. (*C. andicola* Humb. et Bonpl.) | South American wax palm, palma cera | S America | TrH | The stem is covered by a thick, whitish layer of wax. Used locally for the production of candles and matches (1378, 1445) |
| *Copernicia alba* Morong (*C. australis* Becc.) | caranday, carandá | Brazil | Tr | Large stands in the Mato Grosso, as yet little used (1128, 1518) |
| *C. prunifera* (Mill.) H.E.Moore (*C. cerifera* (Arr. da Cam. ex Koster) Mart.) | carnauba wax palm | Brazil | Tr | In large stands in NE Brazil, annual production, 12,000 tons. The wax covers the leaves, has a very high melting point (83-86°C), and is especially suitable for the production of carbon paper (891, 1095, 1445, 1518, 1783) |
| *Syagrus coronata* (Mart.) Becc. | ouricury palm, ouricuru, uricuri, licuri, nicuri | Brazil | Tr | Mainly from the state of Bahia, on dry sites. Wax from the leaves, obtained like carnauba wax, annual production about 3000 t. The palms also supply oil (Table 17) (132, 1518) |

# Forage and Pasture Plants

A large part of the production from plants is used for the nourishment of the various domestic animals, which contribute greatly to human nutrition as providers of meat, milk, and eggs, serve us for draught and transport, and also provide wool, silk and leather. Animal husbandry is the only practical form of agriculture possible where crop cultivation is impossible due to low rainfall, steepness of the land, and shallowness, wetness or salinity of the soil. However it also plays a large role in the tropical and subtropical summer rainfall regions which have sufficient rainfall for arable cultivation. Even in the humid tropics, it gains more and more in importance, because it offers one possibility for permanent land usage, particularly where the provision of sufficient protein for the population is a problem (see p. 120). However, a serious warning must be given against the clearance of trees from large areas to create grazing land with low productivity, as has become the fashion in Latin America, because in the long term this bears great ecological and socio-economic dangers (1617).

The presentation which follows must be limited in substance to the nutrition of ruminants, because these are economically the most important domestic animals. Some indications for the food supply of domesticated insects are given at the end of this section.

The farmer has three possibilities to improve and ensure the basic supplies of fodder:

- sowing higher yielding and more nutritious species in natural pastures,
- establishing seeded pastures,
- and cultivating forage crops on arable land.

Sowing in existing pastures serves the purpose of closing any gaps in the vegetation cover caused by overgrazing, or improving the botanical composition of the pasture by the introduction of more productive grasses or legumes. For this, special measures are mostly necessary to encourage the germination and first growth of the sown plants (partially killing the old plants using herbicides; breaking up the soil using ploughs or crust breakers; fertilizers). The success is often uncertain, because it depends to a high degree on factors over which the farmer has no control. In spite of this, more attention should be paid to this technique, especially with severely degraded pastures, where it can take decades for the natural growth of the indigenous flora to regenerate the previous level of productivity (92, 266, 1626, 1718, 1776, 1865).

Where the rainfall is sufficient for arable farming, the cultivation of grasses and legumes for grazing or for fodder production can be carried out, as well as the cultivation of food crops. In regions with only a few rainy months, mostly only annual species can be considered, such as tef or other millets (Table 6), sudangrass, oats, barley, *Trifolium subterraneum*, *Medicago truncatula*, and other legumes (Table 52). The choice of good pasture plants is much wider if the rainfall permits the growth of perennial plants (Tables 51 and 52). For permanent pastures, there are nowadays suitable grass and legume species available even for relatively dry and for very humid regions, for the establishment of either pure or mixed pastures (195, 211, 390, 775, 785, 810, 1040, 1206, 1347, 1579, 1638, 1684, 1684a, 1935). In summer rainfall regions, permanent pastures, which are used for 3-6 years in rotation with arable crops (ley farming), have a special importance for maintaining the productivity (build-up of humus under grass), and the reduction of harmful organisms (nematodes, noxious soil fungi, etc.) (1560).

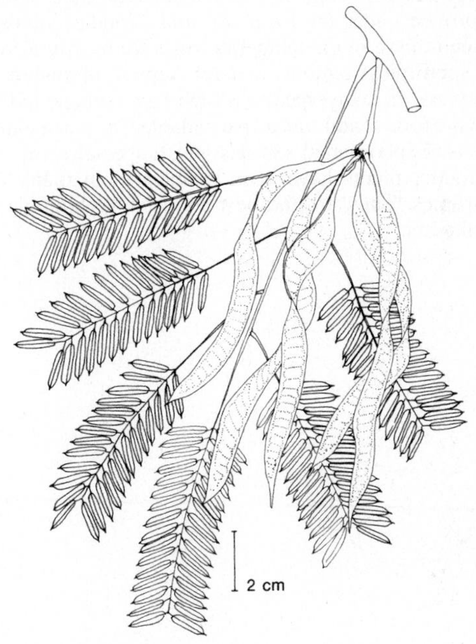


Fig. 98. *Leucaena leucocephala*. Leaf and fruits

In arable fodder cropping, the highest yields are obtained in the humid tropics with sufficient N fertilization with grasses such as Napier fodder and Guinea grass. Yields of 30-50 t dry matter/ha/year are normal, and under favourable conditions, up to 80 t can be achieved (1872). In the subtropics, lucerne, berseem and some other species (Table 52) can give excellent yields of high protein fodder under irrigation. There is a range of tropical forage legumes (*Pueraria phaseoloides*, *Neonotonia wightii*, etc.) which are very high-yielding. Arable cultivation of fodder has special importance in regions with seasonal rainfall, because pastures there possess very little nutritional value during the dry period. Supplements of hay or silage are then necessary to prevent growth cessation and even weight loss in the animals. Apart from grasses and legumes, attention should be paid to those plants whose fruits or roots make succulent fodder available also in the dry season (fodder melons, *Citrullus lanatus*; fodder gourds, *Cucurbita pepo* and other species; fodder beets, *Beta vulgaris*; Japanese fodder radish, *Raphanus sativus*; spineless forms of *Opuntia ficus-indica* (Tables 25 and 30)) (389).

The number of species which are cultivated as fodder plants is so great, that only the most important species can be named in Tables 51 and 52. More complete lists can be found in (92, 195, 718, 773, 785, 917, 1040, 1188, 1206, 1347, 1537, 1626, 1684, 1684a, 1871, 1939).

In the literature referred to above, further details of the ecological adaptation of the different species are given. In individual cases, their suitability has to be experimentally proven under the local soil and climatic conditions (786). The fodder plants which thrive in the temperate zone are mostly also suitable for the cool seasons in subtropical regions, and for tropical highlands. Plant breeding has selected cultivars of many species which have substantially higher fodder value than the wild forms, and are in part adapted to particular ecological and farming conditions. Experimental stations which specialize in the improvement of pasture and fodder plants have been established in many countries of the tropics and subtropics. They are the best source for obtaining seed of locally recommended cultivars.

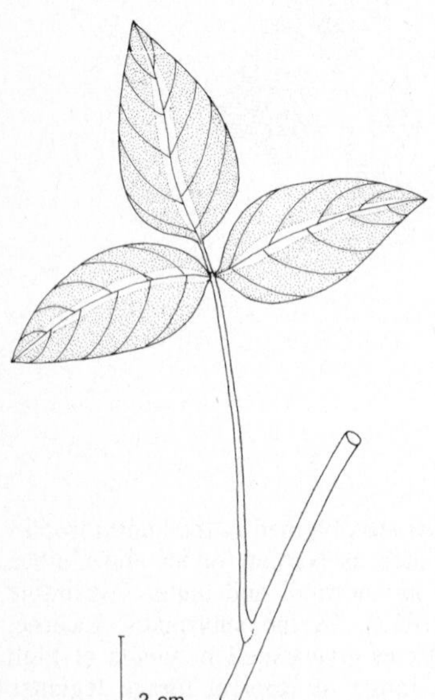


Fig. 99. *Desmodium uncinatum*. Leaf

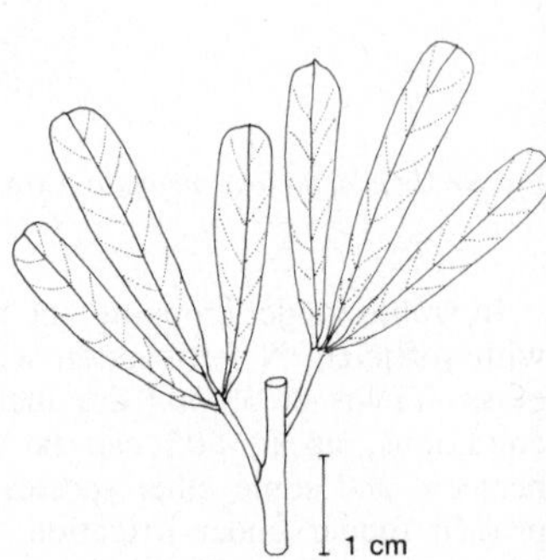


Fig. 100. *Lespedeza cuneata*. Piece of shoot with leaves

Some of the best modern fodder plants belong to species which are important as cereals or vegetables. The following should be especially mentioned: the fodder types of *Sorghum bicolor* (Table 5) (434, 1890) and *Pennisetum americanum* (see p. 29 and Table 6), the hybrids of *Pennisetum americanum* x *P. purpureum*, forms of triticale (see p. 7) (665, 1233), forms of *Eragrostis tef*, the fodder beets (*Beta vulgaris* L. var. *alba* DC.), and the various fodder forms of *Brassica oleracea* L. and *B. rapa* L. (Table 25) (1252).

Apart from grasses and legumes, only a few families provide plants which are regularly cultivated for fodder production. To mention some: of the Dicotyledons Boraginaceae (*Symphytum asperum* Lepech., comfrey), Caryophyllaceae (*Spergula arvensis* L. corn spurrey), Chenopodiaceae (*Atriplex* spp., saltbush, in dry regions, and primarily important on saline soils (1258); *Beta vulgaris* L., fodder beet), Compositae (*Helianthus tuberosus* L., topinambur), Cruciferae (*Brassica* spp., fodder kale, *Raphanus sativus* L., fodder radish), and of the Monocotyledons, the Araceae (*Alocasia* spp., *Xanthosoma* spp., Table 12) (301, 718, 1300).

Hay, straw and residues are very important for animal feeding, and these are obtained from almost all of the plants which are cultivated for other purposes. They often form the only additions to the natural pasture foods. This usage as a by-product is mentioned in the chapters dealing with the various plant groups. The following should be remembered as examples: sunflower, soya, peanuts, sweet potato, sugar cane, carob bean, ramie, and citrus (1826).

The listing of fodder plants would be incomplete without mentioning that many plants are used as much for human nutrition as for animal feeding. Often the same cultivars are used for both human and animal nutrition (e.g. maize, sorghum, cassava, cowpea, hyacinth bean, breadfruit, dates), but with other species, special cultivars have been bred for fodder purposes (fodder melons, fodder gourds). The export of calorie- and protein-rich fodder products is an important source of foreign exchange for many countries; for this, maize, sorghum, cassava, and soya can primarily be mentioned.

The usage of trees and shrubs for animal feeding in the dry regions is controversial. Tree growing in these places is primarily for erosion control, and for supplying the population with firewood. Great damage is often caused to the plants by animals eating young trees, or by the pruning of older trees to obtain fodder. On the other hand, fallen leaves and fruit can definitely improve the feed of animals in dry times, when there are always insufficient supplies of protein. These tree species originate mostly in the indigenous flora, and so cannot be named individually (for exceptions, see Table 52, e.g. *Prosopis* spp.) (612, 1035, 1247, 1262, 1277, 1818).

The nutritional value of fodder plants (264, 612, 948, 1167) varies to such a great extent, depending on the growing conditions and the stage of development, that examples for the individual species which are listed here would have little value. It is generally true that legumes exceed grasses, not only in protein content, but also in mineral and vitamin contents. The digestibility of grasses decreases quickly with the development of stems and flowers. To obtain high fodder value, the grass must be cut at short time intervals (2-3 weeks) during the time of active growth. The digestibility of grass which has grown at high temperatures is less than when the development takes place at lower temperatures (415). Therefore, tropical grasses can have an energy content which is too low for the optimal growth of the animals (431).

It is very common that the mineral content of the fodder is insufficient, so that deficiency diseases develop in the animals. This is especially true for P, and also for Ca, Na, Fe, Co, Cu, and other elements. With extensive pasture systems, it is generally not profitable to improve the mineral deficiency through the usage of fertilizers. It is simplest to increase the mineral supply to the animals using licks. Exceptions are the surprisingly good successes achieved in some regions with trace element fertilization, especially with Mo and Cu (1206). With intensive pastures, it often pays to supply large amounts of N fertilizer (274, 1206). The establishment of mixed pastures of grasses and legumes can be economically more profitable than N fertilization of pure grass pasture, even if the yields are lower.

Three groups of domesticated or semi-domesticated insects are of considerable economic importance in some regions: silkworms, bees, and lac insects.

Of the silkworms (70, 209, 376, 645, 1878), the mulberry silkworm (*Bombyx mori* L.) is by far the most important, and provides 97% of the world's production. It has been inbred so much that the adult moths have lost their ability to fly. The many breeds have been assigned to individual species by some taxonomists. The main food plant is *Morus alba* L., Moraceae, of which there are numerous cultivars in East Asia. It is grown in a variety of ways (tree, bush, espalier), always with the aim of obtaining the maximum number of leaves which can be easily harvested. A yield of 25 tons of fresh leaves per ha per year is considered good. Several other *Morus* species are likewise suitable as food plants (*M. laevigata* Wall. ex Brandis, *M. nigra* L., Table 30, *M. serrata* Roxb.). The other silkworms provide coarser silk. The eri silkworm is fully domesticated in India (*Philosamia ricini* Boisd.). Its main fodder plant is *Ricinus communis* (see p. 108), but it can also be fed with the leaves of many other plants (other Euphorbiaceae; *Heteropanax fragrans* Seem., *Carica papaya*) (1803). The muga silkworm is also domesticated to some extent (*Antheraea assamensis* Westw.) and is raised on *Machilus bombycina* King ex Hook.f., which is cultivated for this purpose. The tasar or tussah silk is obtained in China from *Antheraea proylei* Guér. et Méne., fed on *Quercus* spp.; the tasar silkworm utilized in India, *Antheraea mylitta* Drury, is found wild; its fodder plants are wild-growing trees (*Shorea* spp., *Terminalia* spp., *Ziziphus jujuba*, Table 30) which are regularly cut back to achieve as many young leaves as possible (899). In Nigeria, silk is obtained on a small scale from *Anaphe* spp. (1878).

Bees are valued domesticated animals, not only for the production of honey (see p. 68) and wax, but also for ensuring the pollination of many cultivated plants (fruit trees, sunflowers and other Compositae, Cucurbitaceae, lucerne and other legumes for seed production, rape and other Cruciferae, coconut palms and other palms) (559, 1472). "Bee pastures" are specially planted only in exceptional cases, but this by-product can be considered when choosing ground cover plants (*Fagopyrum esculentum*, Table 7, *Phacelia tanacetifolia* Benth., *Cleome monophylla* L.). The following are exceptionally good plants for feeding bees: *Eucalyptus*, *Acacia*, and *Prosopis* species, and Proteaceae such as *Grevillea robusta* (Table 55). Beekeeping plays an important role in many countries of the tropics and subtropics (25, 376, 383, 384, 385, 386, 514, 559, 792, 1879). It can be much improved by the introduction of European honey bees, and crossing these with indigenous species.

Shellac (376, 803) is produced in large quantities only in India and Thailand, and is an important export product from these countries. It is the resinous exudation from a scale insect, *Laccifer lacca* Kerr. (there are other generic names used for the species by other authors, *Lakshadia*, *Kerria*, and *Tachardia*;

a number of types can be identified, which have also been classified as species). A range of species are cultivated as their food plants, and wild plants are also used in part (1627). The best lac comes from *Schleichera oleosa* (Lour.) Oken, Sapindaceae. However, *Butea monosperma* (Lam.) O.Kuntze, and *Flemingia macrophylla* (Willd.) O.Kuntze, Leguminosae, are said to provide equally good lac, and are easier to cultivate. *Cajanus cajan* (Table 25), *Ficus* spp., *Croton* spp., *Acacia* spp., *Albizia* spp., and other species are also used as food plants, but they provide less valuable lac. The trees must be pruned and allowed rest periods to ensure regular and high yields.

Carmine is the colouring matter from the dried bodies of the cochineal insect (*Dactylopius coccus* Costa) and is produced to a considerable extent in Peru and on the Canary Islands. The colourant has a market in cosmetics, for foodstuffs (Campari, salame), and in microscopy (54). The food plants of the insects are *Nopalea cochenillifera* (L.) Salm-Dyck and *Opuntia* species, Cactaceae (54, 451, 1007).

The Chinese wax scale insect, *Ericerus pe-la* Char., supplies "white wax" which is used for candles and for polishing. For its food, *Fraxinus chinensis* Roxb. and *Ligustrum lucidum* Ait. (both Oleaceae) are cultivated (1050).

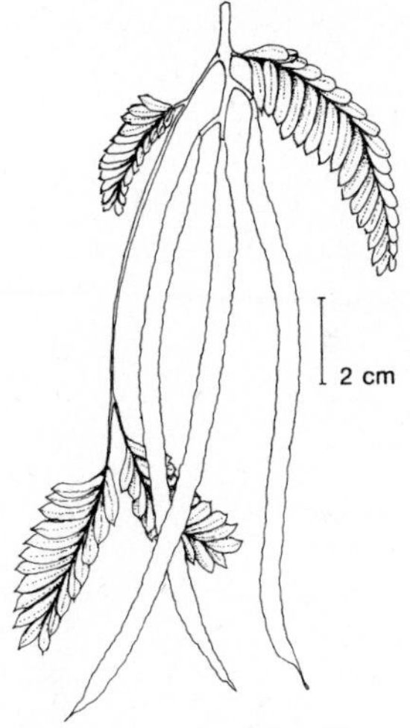


Fig. 101. *Sesbania sesban*. Tip of branch with fruits

Table 51. Fodder grasses (explanation of symbols, see footnotes and footnotes page 55)

| Botanical name | Vernacular names | Climatic region | Vegetation period[1] | Propagation[2] | Tolerances[3] of | | | Remarks and literature references |
|---|---|---|---|---|---|---|---|---|
| | | | | | drought | wet soil | salt | |
| *Acroceras macrum* Stapf | Nile grass | S, Tr | per | se | - | + | - | High value for pastures, green forage, hay and silage (195, 1180, 1626, 1684a) |
| *Agropyron desertorum* (Fisch.) Schult. | standard crested wheatgrass | S | per | se | + | - | + | Valuable and high-yielding pasture and fodder grass, also soil stabilizer. Other *Agropyron* species are important pasture plants in N America (426, 619, 626, 741, 773, 1871) |
| *Andropogon gayanus* Kunth | Gamba grass, sadabahar | Tr | per | se | + | - | - | Grows well on various soil types, even on acid soils. For hay, pastures and soil stabilizing (195, 718, 902, 1174, 1684a) |
| *Arrhenatherum elatius* (L.) P.Beauv. ex J.S.et K.B.Presl | tall oatgrass, fromental | Te, S | per | se | - | - | - | Good fodder grass for pastures and hay production. For the cooler parts of the subtropics (1626, 1871) |

| | | | | | | | | |
|---|---|---|---|---|---|---|---|---|
| *Axonopus compressus* (Swartz) Beauv. | savannah grass, tropical carpet grass | S, Tr | per | se | + | + | - | Tolerates shade, forms a cover which is resistant to trampling. Several other *Axonopus* species are used likewise as pasture and lawn grasses (195, 626, 917, 1149, 1626, 1684a, 1694, 1871) |
| *Bothriochloa ischaemum* (L.) Keng (*Andropogon ischaemum* L.) | Turkestan bluestem, yellow bluestem | S, Tr | per | se | + | - | - | Useful for re-colonizing eroded pastures. Other *Bothriochloa* species are used similarly (626, 1626, 1684a, 1871) |
| *Bouteloua eriopoda* Torr. | black grama | S | per | se | ++ | - | ++ | Important pasture grass in semi-arid regions. Numerous other *Bouteloua* species are used in America for pasture improvement (626, 718) |
| *Brachiaria brizantha* (A. Rich.) Stapf | palisade grass, signalgrass | S, Tr | per | v | - | - | - | High forage value also during the dry season. Especially suitable for rotation pastures in the tropics (195, 1505, 1626, 1684a) |

[1] Vegetation period: an = annual, bi = biennial, per = perennial
[2] Propagation: se = by seed, v = vegetative
[3] Tolerances: - = not tolerant, + = tolerant, ++ = very tolerant

Table 51. Fodder grasses, cont'd (explanation of symbols, see footnotes page 55 and 405)

| Botanical name | Vernacular names | Climatic region | Vegetation period | Propa-gation | Tolerances of | | | Remarks and literature references |
|---|---|---|---|---|---|---|---|---|
| | | | | | drought | wet soil | salt | |
| *B. decumbens* Stapf | Suriname grass, signalgrass | Tr | per | v | - | - | - | Resistant to trampling, gives very high yields with high N supplies. Voluntarily eaten at all stages. Several eco-types are known (785, 917, 1067, 1505, 1684a, 1694) |
| *B. humidicola* (Rendle) Schweickerdt | creeping signalgrass, quicuyu da Amazônia | Tr | per | v | - | + | - | Important pasture grass in the Amazon region because it also thrives on poor ultisols (195, 1180, 1505, 1684a, 1694) |
| *B. mutica* (Forsk.) Stapf | Pará grass, hierba de Pará | Tr | per | v | - | ++ | + | Grows in water. Of good fodder value, suitable also for hay and silage (195, 718, 1206, 1505, 1684a, 1871) |
| *B. ruziziensis* Germain et Evrard | ruzi grass, Congo grass | TrH | per | v | - | + | - | Good, moisture-tolerant pasture grass for tropical highlands of 1000-2000 m (195, 414, 718, 1505, 1684a) |

| | | | | | | | | |
|---|---|---|---|---|---|---|---|---|
| *Bromus unioloides* H.B.K. (*B. catharticus* Vahl) | rescue grass, cebadilla criolla | Te, S, TrH | an bi | se | - | - | - | Good pasture grass in the subtropical winter rainfall region during the cool season, and in tropical highlands. Other annual *Bromus* species are used similarly (195, 254, 741, 1871) |
| *Cenchrus ciliaris* L. | African foxtail, buffel grass, dhaman grass, anjan grass | S, Tr | per | se | ++ | - | - | Very resistant to trampling, important fodder grass for dry regions, good suitability for mixtures with legumes, soil stabilizer (195, 619, 718, 785, 917, 1684a, 1871) |
| *Chloris gayana* Kunth | Rhodes grass, grama Rhodes, zacate gordura | S, Tr | per | se | + | - | + | Various eco-types, some are salt-tolerant, very good pasture grass, and for hay production, high yields with good fertilization (195, 626, 718, 785, 1684a, 1871) |

Table 51. Fodder grasses, cont'd (explanation of symbols, see footnotes page 55 and 405)

| Botanical name | Vernacular names | Climatic region | Vegetation period | Propa-gation | Tolerances of | | | Remarks and literature references |
|---|---|---|---|---|---|---|---|---|
| | | | | | drought | wet soil | salt | |
| *Cynodon dactylon* (L.) Pers. | Bermuda grass, Bahama grass, couch grass, dhubgrass | Te, S, Tr | per | v | ++ | - | + | Very resistant to trampling, many different cultivars, "coastal Bermuda" especially suitable as pasture grass. Lawn grass and soil stabilizer, frequently a weed (92, 195, 619, 626, 718, 785, 917, 1684a, 1871) |
| *C. plectostachyus* (K.Schum.) Pilg. | stargrass, Naivasha stargrass estrella | S, Tr | per | v | + | - | + | Suitable for warmer sites and grows more strongly than *C. dactylon* (195, 917, 1180, 1684a) |
| *Dactylis glomerata* L. | cocksfood, orchard grass, pasto ovillo | Te, S, TrH | per | se | - | - | - | Many cultivars, for winter pastures in the subtropics, and for tropical highlands (92, 195, 626, 848, 1871) |
| *Digitaria decumbens* Stent | Pangola grass, pasto pangola | S, Tr | per | v | - | - | - | Resistant to trampling, one of the best fodder grasses. Makes good use of high N supplies. Many other *Digitaria* species |

| | | | | | | | | |
|---|---|---|---|---|---|---|---|---|
| | | | | | | | | have proved to be good pasture grasses (195, 785, 917, 1180, 1684a, 1871) |
| *Echinochloa pyramidalis* (Lam.) Hitchc. et Chase | antelope grass, pasto limpapo | S, Tr | per | v | - | ++ | - | Grows in water. Good fodder in the younger stages, especially suitable for water buffalo in flood plains. Seeds eaten in times of need in W Africa (195, 718, 1180, 1684a) |
| *E. stagnina* (Retz.) Beauv. | bourgou, koempai | Tr | per | se, v | - | ++ | - | Grows in open water and in swamps, is voluntarily eaten by cattle, along the Niger it is cultivated for sugar production (see p. 74). Other species are often weeds, especially in rice (718, 1180, 1537) |
| *Eragrostis curvula* (Schrad.) Nees | weeping lovegrass, pasto llorón | S | per | se | + | - | - | Remains green for a long time in the dry season, and maintains its high fodder value. Important grass for ley farming in summer rainfall regions (Ermelo strain). Many other *Eragrostis* species are valuable fodder grasses (195, 1180, 1684a, 1753) |

Table 51. Fodder grasses, cont'd (explanation of symbols, see footnotes page 55 and 405)

| Botanical name | Vernacular names | Climatic region | Vegetation period | Propa-gation | Tolerances of | | | Remarks and literature references |
|---|---|---|---|---|---|---|---|---|
| | | | | | drought | wet soil | salt | |
| *Festuca arundinacea* Schreb. | tall fescue, fétuque élevée, festuca alta | Te, S, TrH | per | se | + | - | - | Resistant to trampling, valuable pasture grass, grows even on poor soils. Sometimes poisonous due to phytotoxins (92, 195, 619, 848, 1180, 1871) |
| *Hyparrhenia rufa* (Nees) Stapf | jaragua grass, yaragua grass | Tr | per | se, v | - | + | - | Very hardy grass for hay and pastures, good fodder only in the young stages. When fully mature, especially suitable for thatching roofs (195, 718, 917, 1684a) |
| *Lolium multiflorum* Lam. | Italian ryegrass, annual ryegrass | Te, S | an bi | se | - | - | - | Very valuable fodder plant, makes high demands on the soil, belongs to cooler climates than *L. perenne* (626, 1871) |

| | | | | | | | | |
|---|---|---|---|---|---|---|---|---|
| *L. perenne* L. | English ryegrass, perennial ryegrass | Te, S, TrH | per | se | - | - | - | Good pasture grass for the cooler seasons in the subtropics, and for tropical highlands over 2000 m (195, 626, 848, 1871) |
| *L. rigidum* Gaud. | Wimmera ryegrass | S | an | se | + | - | + | Important pasture grass in the Mediterranean region and in Australia (619, 1206, 1626, 1871, 1939) |
| *Melinis minutiflora* Beauv. | molasses grass, gordura, capim melado | Tr | per | se | - | + | - | Especially suitable for mixtures with legumes. Cattle must first become used to its taste (195, 626, 718, 785, 917, 1537, 1684a) |
| *Panicum antidotale* Retz. | blue panic, giant panic, pánico azul | S, Tr | per | se | - | - | - | Especially suitable for highly fertilized and irrigated pastures, high protein content (195, 773, 1626, 1684a, 1871) |
| *P. coloratum* L. | coloured Guinea grass, small buffalo grass, kleingrass | S, Tr | per | se | + | - | - | For regions with summer rainfall, withstands long dry seasons, many varieties, 'Makarikari' is salt tolerant (195, 1066, 1180, 1626, 1684a) |

Table 51. Fodder grasses, cont'd (explanation of symbols, see footnotes page 55 and 405)

| Botanical name | Vernacular names | Climatic region | Vegetation period | Propa-gation | Tolerances of | | | Remarks and literature references |
|---|---|---|---|---|---|---|---|---|
| | | | | | drought | wet soil | salt | |
| *P. maximum* Jacq. | Guinea grass, herbe de Guinée | S, Tr | per | se | - | + | - | One of the most important fodder grasses for pastures and hay in the tropics, newly bred types and selections with high proportion of leaves and outstanding fodder value (195, 785, 917, 1164, 1180, 1505, 1537, 1684a, 1871) |
| *Paspalum dilatatum* Poir. | Dallis grass, watergrass | S, Tr | per | se | - | + | - | Resistant to trampling, high-value pasture grass (195, 718, 917, 1180, 1684a, 1871) |
| *P. notatum* Flügge | Bahia grass, grama dulce | Tr | per | se | - | - | - | Pasture grass, less demanding than *P. dilatatum*. Also for protection of water ways against erosion (195, 785, 1180, 1684a, 1871) |

| | | | | | | | | |
|---|---|---|---|---|---|---|---|---|
| *P. vaginatum* Swartz | biscuit grass, silt grass, salt water couch | S, Tr | per | se | - | + | ++ | For sand stabilization on beaches and for salt marshes (195, 718, 1180, 1258) |
| *Pennisetum clandestinum* Hochst. ex Chiov. | Kikuyu grass | S, TrH | per | v | - | - | - | Used worldwide for pastures and lawns, produces well only with high rainfall. Well suited for soil stabilization (195, 626, 718, 1180, 1684a, 1871) |
| *P. purpureum* Schum. | elephant grass, Napier fodder grass, pasto gigante | S, Tr | per | v | - | + | - | Good fodder grass (for green forage, hay, and silage) only in the young stages, under good conditions gives the highest yields. Likewise utilized for mulching, soil stabilization, and shelterbelts (195, 532, 718, 785, 917, 1110, 1180, 1684a, 1706, 1985) |
| *Phalaris aquatica* L. (*P. tuberosa* L.) | bulbous canary grass, Harding grass | S, TrH | per | se, v | + | - | - | Especially for winter pastures in the subtropics and for tropical highlands over 2000 m. Roots survive long dry periods, requires good soils (195, 718, 848, 1871) |

Table 51. Fodder grasses, cont'd (explanation of symbols, see footnotes page 55 and 405)

| Botanical name | Vernacular names | Climatic region | Vegetation period | Propagation | Tolerances of | | | Remarks and literature references |
|---|---|---|---|---|---|---|---|---|
| | | | | | drought | wet soil | salt | |
| *P. arundinacea* L. | reed canarygrass, phalaris roseau, alpiste, caniço malhado | Te, S | per | se, v | - | ++ | - | Grows in water. Good fodder grass for hay and silage production in flood plain meadows. Cut before flowering (195, 626, 718, 1871) |
| *Poa pratensis* L. | Kentucky bluegrass, pâturin des prés | Te, S, TrH | per | se | - | - | - | Lawn grass, and good fodder grass in the cool seasons (626, 741, 1871) |
| *Puccinellia airoides* (Nutt.) Wats. et Coult. | Nuttal alkali grass | Te, S | per | se | - | - | ++ | Several *Puccinellia* species are of importance because of their salt and alkali tolerance (741) |
| *Setaria sphacelata* (Schumach.) Stapf et C.E.Hubbard | golden timothy, napierzinho | S, Tr | per | se | - | - | - | Nutrient-rich grass for pastures, hay and silage, well suited for ley farming with sufficient rainfall. Many forms, most important cultivars 'Nandi' and 'Kazungula'. Several other *Setaria* species (some annuals) are cultivated as fodder |

| | | | | | | | | |
|---|---|---|---|---|---|---|---|---|
| | | | | | | | | plants (195, 674, 718, 785, 1180, 1684a) |
| *Sorghum* x *almum* Parodi | Columbus grass, black sorgo, pasto negro | S, Tr | per | se | - | - | - | One of the highest-yielding fodder grasses, especially for hay and silage. Suitable for a wide climatic range (195, 434, 1684a, 1871) |
| *S. halepense* (L.) Pers. | Johnson grass, sorgo de Alepo | S, Tr | per | se | + | - | - | Very hardy, for pastures and cutting. Because of its tough rhizomes, it is a troublesome weed in many countries (195, 434, 718, 785, 1684a, 1871) |
| *Stenotaphrum secundatum* (Walter) O.Kuntze | St. Augustine grass, buffalo couch, chiendent de boef | S, Tr | per | v | - | + | ++ | Hardy pasture grass, also for sand stabilization on beaches; tough but coarse lawn grass (195, 261, 626, 917, 1505, 1626, 1684a, 1694, 1871) |
| *Tripsacum fascicu-latum* Trin. ex Aschers. (*T. laxum* Nash) | Guatemala grass, pasto Guatemala | Tr | per | v | - | + | - | Robust grass for hay and silage, for the production of mulch material, as soil stabilizer, and as catch crop for new plantations of tea in Sri Lanka (nematode control) (195, 718, 1684a, 1871, 1922) |

Table 51. Fodder grasses, cont'd (explanation of symbols, see footnotes page 55 and 405)

| Botanical name | Vernacular names | Climatic region | Vegetation period | Propa-gation | Tolerances of | | | Remarks and literature references |
|---|---|---|---|---|---|---|---|---|
| | | | | | drought | wet soil | salt | |
| *Urochloa mosambicensis* (Hack.) Dandy | sabi grass | S, Tr | per | se | + | + | - | Low demanding, nutrient-rich pasture grass. Remains green a long time in dry season. Especially suitable for sowing in degraded pastures and for soil stabilization in waterways. Also a weed in arable crops (195, 670, 1180, 1206, 1684a) |
| *Zea mays* L. ssp. *mexicana* (Schrad.) Iltis (*Euchlaena mexicana* Schrad.) | teosinte | S, Tr | an | se | - | - | - | Climatic and soil needs similar to maize, but less susceptible to diseases and pests. Grains are usable as a food and feed (195, 376, 1684a, 1871, 2026) |

Table 52. Fodder legumes (for explanation of symbols, see footnotes page 55 and 405)

| Botanical name | Vernacular names | Climatic region | Vegetation period | Tolerances of | | | Remarks and literature references |
|---|---|---|---|---|---|---|---|
| | | | | drought | wet soil | salt | |
| MIMOSOIDEAE | | | | | | | |
| *Calliandra calothyrsus* Meissn. | calliandra | Tr | per | - | - | - | Origin: Central America. Spread first as ornamental, then for fuel wood, recently as forage crop (Indonesia). Quick growing, cut stumps coppice readily. Versatile tree in agroforestry: mixed cropping, living fences, nurse crop, soil improver (124, 473, 600, 1247, 1261, 1262, 1267) |
| *Leucaena leucocephala* (Lam.) de Wit (*L. glauca* (Moench) Benth.) | leucaena, horse tamarind, ipil-ipil, koa haole, | S, Tr | per | + | + | + | (Fig. 98) Low demanding, capacity to suppress weeds, different growth types. When too much is given in fodder, there is a danger of mimosine poisoning. Shade tree and soil stabilizer. Young shoots and leaves are eaten as vegetables in SE Asia (195, 229, 444, 473, 785, 904, 1046, 1247, 1261, 1262, 1265, 1431, 1513, 1684) |

Table 52. Fodder legumes, cont'd (for explanation of symbols, see footnotes page 55 and 405)

| Botanical name | Vernacular names | Climatic region | Vegetation period | Tolerances of | | | Remarks and literature references |
|---|---|---|---|---|---|---|---|
| | | | | drought | wet soil | salt | |
| *Prosopis juliflora* (Sw.) DC. | mesquite algarrobo | S, Tr | per | ++ | - | - | Cattle eat leaves, pods and seeds. Very hardy, can easily lead to formation of uncontrollable thicket. Several other *Prosopis* species are cultivated in dry regions as fodder plants and for soil stabilization. A low-value gum can be obtained from the bark (376, 427, 1247, 1262, 1513, 1658, 1684) |
| PAPILIONOIDEAE | | | | | | | |
| *Alysicarpus vaginalis* DC. | alyce clover, one-leaf clover | Tr | per | - | - | - | Also cultivated as a ground cover plant and for hay production (195, 259, 444, 718, 1684) |
| *Arachis glabrata* Benth. | perennial peanut, rhizoma peanut, arb | S, Tr | per | + | - | - | Rich in protein, often cultivated mixed with grasses and for hay production. Other perennial *Arachis* species are also used as fodder plants (166, 195, 602, 1437, 1684) |

| | | | | | | | |
|---|---|---|---|---|---|---|---|
| *Centrosema pubescens* Benth. | centro, fleur languette, campanilla, jetirana | Tr | per | + | - | - | Important fodder and pasture plant. Remains green a long time in dry periods, well suited for haymaking. One of the most used ground cover plants, often in combination with *Calopogonium*, *Pueraria*, and others (see p. 434, Tab. 53) (195, 655, 787, 1297, 1684, 1762) |
| *Chamaecytisus proliferus* (L.f.) Link (*Cytisus proliferus* L.f.) | tagasaste, tree lucerne | S | per | ++ | - | + | Origin: Canary Islands. As browse and forage shrub in Mediterranean countries and Australia. Foliage relished by all grazing animals. Also planted as ornamental and hedge shrub (169, 712, 1611, 1699, 2026) |
| *Clitoria ternatea* L. | butterfly pea, Kordofan pea, campanilla | Tr | per | - | - | - | Good fodder and ground cover plant. Flowers and young pods are eaten as vegetables in the Philippines. The flowers provide a blue food colouring for meals and drinks (see p. 386) (195, 444, 686, 1082, 1684) |
| *Desmodium intortum* (Mill.) Urb. | greenleaf desmodium, beggarlice, desmodio verde | S, Tr, TrH | an per | - | - | - | Wide climatic adaptability (from the coasts to the highland), relatively high need for water, good in grass mixtures (195, 476, 785, 1206, 1626, 1684) |

Table 52. Fodder legumes, cont'd (for explanation of symbols, see footnotes page 55 and 405)

| Botanical name | Vernacular names | Climatic region | Vegetation period | Tolerances of | | | Remarks and literature references |
|---|---|---|---|---|---|---|---|
| | | | | drought | wet soil | salt | |
| *D. tortuosum* (Sw.) DC. | Florida clover, giant beggarweed | S, Tr | an per | - | - | - | Tolerates shade, good in grass mixtures and for hay production, also for green manuring and for ground cover (195, 376, 1684) |
| *D. uncinatum* (Jacq.) DC. | silverleaf desmodium, Spanish tickclover | S, Tr | per | - | + | - | (Fig. 99) Good pasture and fodder plant for poor sites, suitable for grass mixtures. Numerous other *Desmodium* species can be considered for fodder, pasture and ground cover (195, 785, 1206, 1684) |
| *Hedysarum coronarium* L. | French honeysuckle, sulla, zulla | S | bi | + | - | - | Forage and pasture plant of the Mediterranean region, very rich in protein, and high-yielding. For calcareous, deep soils, especially with irrigation, many local varieties. Also for hay and silage (444, 619, 1871) |

| | | | | | | | |
|---|---|---|---|---|---|---|---|
| *Lathyrus ochrus* (L.) DC. | ochrus vetch, gesse ocre, alverjana, ervilha dos campos | S | an | + | - | - | Cultivated especially in the eastern Mediterranean region for green forage and hay, the seeds are also eaten or used as fodder. Green manure and ground cover plant. Many other *Lathyrus* species are similarly utilized. With some, the seeds are poisonous (2, 1871) |
| *Lespedeza cuneata* (Dum. Cours.) G. Don | perennial lespedeza, sericea lespedeza | Te, S | per | ++ | - | - | (Fig. 100) Low demanding, likes acid soils and withstands frosts to -17°C, cultivated for pastures, hay, and soil stabilization (444, 1146) |
| *L. stipulacea* Maxim. | Korean lespedeza | Te, S | an | + | - | - | Cultivated on light soils, for pastures, hay, and for soil stabilization, self-propagating from its own seed (444, 1146) |
| *L. striata* (Thunb. ex Murr.) Hook. et Arn. | Japanese lespedeza, common lespedeza | S | an | + | - | - | For light soils, for pastures and for hay production, sensitive to cold, self-propagating from its own seed. Also as ground cover and stabilizer (1146) |
| *Lotononis bainesii* Baker | Miles lotononis, creeping indigo | S | per | + | - | - | Inoculation important, due to species-specific *Bradyrhizobium* (Table 22). Requires heavy grazing in grass mixtures. Good for ground cover (195, 266, 785, 1206, 1684) |

Table 52. Fodder legumes, cont'd (for explanation of symbols, see footnotes page 55 and 405)

| Botanical name | Vernacular names | Climatic region | Vegetation period | Tolerances of | | | Remarks and literature references |
|---|---|---|---|---|---|---|---|
| | | | | drought | wet soil | salt | |
| *Lotus corniculatus* L. | bird's foot trefoil, trèfle cornu | Te, S | per | - | + | - | Cultivated in winter rainfall regions during the cool season, good for mixed pastures (376, 444, 1871) |
| *Lupinus angustifolius* L. | blue lupin, lupin bleu, altramuz silvestre, tremoço azul | Te, S | an | - | - | - | For light soils. Only the forms free of bitter compounds for fodder. Bad suitability for hay (juicy stems), can be silaged in grass and maize mixtures, green manuring. Seed for feed exported from Australia (165, 376, 444, 828, 1762, 1871) |
| *L. luteus* L. | yellow lupin, lupin jaune, tremoço amarelo | Te, S, TrH | an | - | - | - | This and other species are used similarly to *L. angustifolius* (see also Table 25) (376, 444, 828, 1762, 1871) |
| *Macroptilium atropurpureum* (DC.) Urb. (*Phaseolus atropurpureus* DC.) | siratro, purple bean, conchito | S, Tr | per | + | - | - | Important pasture and fodder plant for sandy soils, especially in Australia, also for mixed pastures (195, 444, 785, 788, 1206, 1626, 1639, 1684) |

| | | | | | | | |
|---|---|---|---|---|---|---|---|
| *M. lathyroides* (L.) Urb. (*Phaseolus lathyroides* L.) | phasemy bean, phasey bean, frijol de los arrozales | S, Tr | an bi | + | - | - | Quick growing, utilizes large supplies of fertilizer, good in grass mixtures and for hay, also good pig food when green. Other bean species (Table 25) likewise used as fodder plants (195, 211, 376, 1206, 1684) |
| *Medicago lupulina* L. | hop clover, black medic, minette | Te, S, Tr | an bi | - | - | - | For light soils, especially for the Mediterranean region in winter. Good in mixtures with *Festuca arundinacea* and *Lolium* species (376, 444, 1045, 1871) |
| *M. sativa* L. | alfalfa, lucerne, mielga | Te, S, Tr | per | + | - | - | Most important fodder plant for good soils with irrigation in arid regions, especially for hay production, but also for green fodder and grazing. Crosses with ssp. *falcata* (L.) Arcang. are less demanding and withstand more cold. Bee food, especially if grown for seed production (195, 329, 376, 444, 693, 718, 1045) |
| *M. truncatula* Gaertn. (*M. tribuloides* Desv.) | barrel medic, barrel clover | S | an | ++ | - | + | Pasture plant, in Australia cultivated as a fodder plant wherever the rainfall is not sufficient for *Trifolium subterraneum* (92, 1206, 1762, 1871) |

Table 52. Fodder legumes, cont'd (for explanation of symbols, see footnotes page 55 and 405)

| Botanical name | Vernacular names | Climatic region | Vegetation period | Tolerances of | | | Remarks and literature references |
|---|---|---|---|---|---|---|---|
| | | | | drought | wet soil | salt | |
| *Melilotus alba* Medik. | Bokhara clover, white sweet clover, mélilot blanc, meliloto branco | Te, S | an bi | + | - | - | Very low demanding, coumarin-free breeds and the new US cultivars have much increased the fodder value. For hay, silage, green manure, and bee food (376, 444, 1871) |
| *M. indica* All. (*M. parviflora* Desf.) | Indian sweet clover, trêvo de cheiro, senji | Te, S, Tr | an | + | - | + | For fodder and pastures, especially in the cool season. Soil improver. Other *Melilotus* species are used similarly (195, 376, 444, 1871) |
| *Mucuna pruriens* (Stickm.) DC. | velvet bean, Bengal bean | S, Tr | an per | + | - | - | Strongly growing pasture and fodder plant, ground cover and green manuring plant. Seeds with *utilis* (Wall. ex Wight) Baker ex Burck., also leaves and young pods are used as foodstuff. Previously, several species were separated, e.g. *M. aterrima* (Piper et Tracey) Holland, Mauritius bean, and *M. deeringiana* (Bort) Merr., Florida velvet bean (195, 259, 376, 444, 718, 947, 1296, 1611, 1684) |

| | | | | | | | |
|---|---|---|---|---|---|---|---|
| *Neonotonia wightii* (Arn.) Lackey (*Glycine wightii* (Arn.) Verdc.) | glycine, perennial soybean, soja perene, fundo-fundo | S, Tr | per | + | - | - | One of the most important fodder and ground cover plants of the tropics, well suited for mixed pastures with grasses, prefers heavy soils (195, 444, 785, 791, 1206, 1625, 1626, 1684, 1871) |
| *Onobrychis viciifolia* Scop. | sainfoin, cock's head, esparceta | Te, S | per | + | - | - | For calcareous soils, for hay production and pastures, several varieties (444, 1033, 1632, 1871) |
| *Ornithopus sativus* Brot. | serradella | Te, S | an | + | - | - | Especially suitable for acid, poor, sandy soils in winter rainfall regions, a pasture and fodder plant there (202, 444, 606, 1871) |
| *Pueraria lobata* (Willd.) Ohwi (*P. thunbergiana* (Sieb. et Zucc.) Benth.) | kudzu | S, Tr | per | + | - | - | Pasture and fodder plant, one of the most important soil stabilizers, ground cover and green manuring plants. Other *Pueraria* species are likewise cultivated, especially as ground cover (195, 376, 444, 946, 1684, 1780, 1871) |
| *P. phaseoloides* (Roxb.) Benth. (*P. javanica* Benth.) | tropical kudzu, puero | Tr | per | - | - | - | Better pasture plant for the tropics than *P. lobata*, can be silaged. Also ground cover and green manure plant (see p. 429) (195, 259, 376, 444, 718, 1261, 1684) |

Table 52. Fodder legumes, cont'd (for explanation of symbols, see footnotes page 55 and 405)

| Botanical name | Vernacular names | Climatic region | Vegetation period | Tolerances of | | | Remarks and literature references |
|---|---|---|---|---|---|---|---|
| | | | | drought | wet soil | salt | |
| *Sesbania cannabina* (Retz.) Pers. (*S. aculeata* Pers.) | prickly sesban, dhunchi, dhaincha, danchi | Tr | an bi | - | ++ | + | Grazing and fodder plant for wet sites, particularly in India, ground improver and green manuring. Bark fibres are used for nets in India. Weed, especially in rice. Other *Sesbania* species (Table 55) planted as soil improvers, windbreaks, and shade plants (376, 444, 494, 596, 718, 1247, 1261, 1262, 1611, 1664) |
| *Stylosanthes guianensis* Sw. (*S. gracilis* H.B.K.) | Brazilian stylo, Brazilian lucerne, alfalfa do nordeste | S, Tr | per | - | + | - | Adapted to various also acid soils. One of the most important tropical pasture plants, sensitive to over-grazing and fire. Used also as a soil stabilizer (195, 278, 348, 444, 785, 1206, 1626, 1684, 1727, 1762) |
| *S. hamata* (L.) Taub. | Caribbean stylo | Tr | an per | + | - | - | Thrives well on sandy soils. Important pasture plant in N Australia (195, 266, 278, 348, 1684, 1726, 1727, 1762) |

| | | | | | | | |
|---|---|---|---|---|---|---|---|
| *S. humilis* H.B.K. | Townsville stylo, Townsville lucerne | S, Tr | an | + | - | - | Modest demands on nutrients (poor soils), sensitive to shade conditions, must be sufficiently grazed in mixture with grasses, good seed formation (self-propagating) (195, 278, 348, 444, 785, 1206, 1626, 1684, 1727, 1762) |
| *Trifolium alexandrinum* L. | berseem, Egyptian clover, trèfle d'Alexandrie, bersim | S, TrH | an | - | - | + | Most important fodder plant in Egypt and other N African countries, cultivars for different requirements, only for fertile soils, frequently in irrigated cultivation, for grazing, green forage and hay, also for green manuring (195, 444, 633, 1871) |
| *T. fragiferum* L. | strawberry clover, trébol fresero | Te, S | per | - | ++ | + | Good pasture plant for swampy and salty sites (444, 619, 718, 1871) |
| *T. incarnatum* L. | crimson clover, trèfle incarnat | Te, S | an | - | - | - | Quick growing fodder plant for light soils, many cultivars, suitable for grazing and hay (444, 1871) |
| *T. pratense* L. | red clover, trèfle commun, trébol violeta | Te, S, TrH | per | - | - | - | For good soils in cool sites as a pasture and fodder plant (195, 444, 1206, 1383, 1626, 1871) |
| *T. repens* L. | white clover, ladino clover, trèfle rampant, trébol blanco | Te, S | per | - | - | - | Good pasture plant for cool sites, less demanding than red clover (130, 195, 444, 849, 1206, 1626, 1871) |

Table 52. Fodder legumes, cont'd (for explanation of symbols, see footnotes page 55 and 405)

| Botanical name | Vernacular names | Climatic region | Vegetation period | Tolerances of | | | Remarks and literature references |
|---|---|---|---|---|---|---|---|
| | | | | drought | wet soil | salt | |
| *T. resupinatum* L. | Persian clover, reversed clover, trébol persa | Te, S | an | + | - | - | In the cool season in winter rainfall regions, withstands more cold than berseem (444, 718, 992, 1871) |
| *T. subterraneum* L. | subclover, subterranean clover, trèfle souterrain, trébol subterráneo | S | an | ++ | - | - | Self-propagating due to large seed production, protects the soil from wind erosion. Sheep pasture. Many other *Trifolium* species are cultivated, and used similarly to those named (195, 444, 870, 871, 1541, 1652, 1762) |
| *Vicia sativa* L. | common vetch, vesce commune, veza común | Te, S | an | - | - | - | Pasture and fodder plant, several subspecies and many cultivars. In mixtures with cereals, especially oats, in the cooler season (2, 444, 1871) |
| *V. villosa* Roth | Russian vetch, hairy vetch, vesce velue, veza vellosa | Te, S | an | - | - | - | For poorer (also more acid) soils than *Vicia sativa*, also in the cooler season (cold tolerant), green forage, hay, soil improver. Several other *Vicia* species are cultivated for fodder and green manuring, especially in the Mediterranean region (444, 1871) |

# Ancillary Plants

The name ancillary plants is used here for those plants which are cultivated in agriculture and forestry not to reap a crop from them, but because they encourage the growth and yields of other plants. The most important goals in the cultivation of ancillary plants are soil improvement, ground covering, prevention of erosion, wind protection and shade provision. To this can be added plants grown as supports for climbing plants (black pepper, betel pepper, vanilla), and also plants which form "living fences" (hedges). Obviously, these goals cannot be sharply divided; ground cover, erosion prevention and soil improvement overlap frequently. Support plants not only give support to climbers, but also shade and wind protection. The grouping of plants in Tables 53, 54, and 55 should only be taken as an indication of the pre-eminence of the individual plants for particular purposes. Some of the species named in other chapters can also be used as ancillary plants. However, there is a large number of cultivated plants which are primarily used as ancillary plants. This does not exclude their use for other purposes, especially for yielding firewood and fodder.

## Green Manure Plants

The value or otherwise of green manuring in agriculture in the tropics and subtropics has been much discussed (1911). It can be taken as certain that green manuring in the warm zones is not capable of increasing the humus content of the soil to a substantial extent, and also that the improvement in soil structure from the plant remains disappears after a few weeks, due to their quick decomposition (591). Old, lignified material decays more slowly, but its fertilizing effect is slight, or even negative due to the binding of N. Green manuring can be used sensibly only in places where the rainy season is long enough to ensure that there is no harvest loss. In practice, there are four particular situations where the cultivation of green manure plants really brings advantages: fruit farms in the winter rainfall regions, alley cropping in the tropics, fertilization of wet rice, and irrigated cultivation.

A basic difficulty exists everywhere if great quantities of green material must be worked into the soil using simple implements. There is a considerable investment of labour, and in spite of that, the results are often unsatisfactory. Also, even a tractor-drawn plough is not always sufficient, so that the use of slashing or cutting machines must precede the ploughing. Powerful rotavators have been strongly recommended, but they require a high power input, and wear out rapidly.

It is usual in all winter rainfall regions where the rainfall is sufficient, that, during the cool humid winter months, the weeds are allowed to grow in fruit orchards and vineyards, and these are worked into the soil in spring, before the start of the dry season. In advanced agriculture, suitable green manure plants are

Table 53. Green manure and ground cover plants (explanation of symbols, see footnotes page 55)

| Botanical name | Vernacular names | Climatic region | Remarks and literature references |
|---|---|---|---|
| | | | **Pteridophyta** |
| **Azollaceae** | | | |
| *Azolla pinnata* R. Browm | azolla, water fern, water velvet | S, Tr | Traditional green manure plant for rice in Indochina. Used as source of nitrogen and organic matter for rice, arrowhead and cocoyam. Feed for carps and pigs. Several other species besides *A. pinnata* (76, 867, 1000, 1089, 1394, 1498, 1664) |
| | | | **Angiospermae** |
| **Compositae** | | | |
| *Ageratum houstonianum* Mill. (*A. mexicanum* Sims) | ageratum, goatweed | Tr | Useful as ground cover in Indonesia, together with *Centrosema* (684, 1297) |
| *Clibadium surinamense* L. var. *asperum* Baker | jakass breadnut, cunambi | Tr | Together with *Centrosema*, quickly forms a dense ground cover (1297, 1518) |
| *Eupatorium inulifolium* H.B.K. (*E. pallescens* DC.) | kirinjoeh | Tr | Upright bush, ground cover in combination with creeping legumes. Tolerates acid soil. Also a weed (684, 1297) |
| *Mikania micrantha* H.B.K. | guaco | Tr | Quick growing annual climbing plant, as ground cover to suppress unwanted vegetation (smother crop). Often a difficult weed to control (259, 376, 1095, 1355) |

| | | | |
|---|---|---|---|
| *Tithonia diversifolia* (Hemsl.) A.Gray | Mexican sunflower, Mexican tango flower | S, Tr | As green manure for rice fields, also weed and ornamental plant (376, 1095) |
| **Hydrophyllaceae** | | | |
| *Phacelia tanacetifolia* Benth. | tansy phacelia, valley vervenia, fiddleneck | Te, S | Fodder plant, ground cover, and green manure plant, good bee food (see p. 402) (1611) |
| **Leguminosae** | | | |
| MIMOSOIDEAE | | | |
| *Mimosa invisa* Mart. var. *inermis* Adelb. | sensitiva trepadora, malícia de mulher, raspacanilla | Tr | The spineless form is a great improvement on the wild form. Very low demanding. As ground cover especially with rubber, and for the control of alang-alang (see p. 434) (259, 376, 1297) |
| CAESALPINIOIDEAE | | | |
| *Cassia didymobotrya* Fresen. | candelabra tree | S, Tr | This and other low growing *Cassia* species used for green manuring in Asia (259, 1343, 1444) |
| PAPILIONOIDEAE | | | |
| *Aeschynomene americana* L. | joint vetch, pega pega, pulmón | Tr | Ground cover and green manure plant, also fodder (195, 259, 266, 1192, 1343, 1684) |
| *Calopogonium mucunoides* Desv. | calopo, frisolilla | Tr | One of the most important ground cover plants. Also used as a fodder plant in the Philippines and in India (259, 444, 718, 1297, 2017) |
| *Centrosema plumieri* (Pers.) Benth. | butterfly pea | Tr | Traditional ground cover plant, nowadays mostly replaced by *C. pubescens* (Table 52). Also fodder plant (259, 1343, 1444, 1684) |
| *Crotalaria anagyroides* H.B.K.(*C. micans* Link) | rabbit bells, maromera, maruga | S, Tr | Perennial, much used in SE Asia for green manuring. Serves also for animal fodder. In 3 months becomes 2.5 m tall (684, 1297) |
| *C. pallida* Ait. (*C. mucronata* Desv., *C. striata* DC.) | smooth crotalaria, stripped crotalaria | S, Tr | Annual to perennial, one of the oldest green manuring plants in Java. In spite of a certain alkaloid content also used as fodder. Seeds eaten in Java (444, 684, 1296) |

Table 53. Green manure and ground cover plants, cont'd (explanation of symbols, see footnotes page 55)

| Botanical name | Vernacular names | Climatic region | Remarks and literature references |
|---|---|---|---|
| *C. zanzibarica* Benth. (*C. usaramoensis* Bak.f.) | curara pea | Tr | Annual to biennial, grows exuberantly even on very poor soils. One of the most important green manuring plants in SE Asia. Reputed not to be poisonous, and yields protein-rich fodder. Apart from the species named, *C. juncea* (Table 44) and other *C.* species are used as green manure and ground cover plants (684) |
| *Indigofera hirsuta* L. | hairy indigo, tomtoman | S, Tr | Drought tolerant, annual, for light soils (259, 376, 444, 923, 1561) |
| *I. spicata* Forsk. (*I. endecaphylla* Jacq.) | creeping indigo, trailing indigo | S, Tr | Non-climbing, perennial ground cover plant especially for tea plantations, also fodder (259, 444, 1444) |
| *Sesbania rostrata* Brem. et Oberm. | rostrate sesbania | Tr | Green manure mainly for rice. Grows in water or wet soil, forming rhizobial nodules on roots and stem with active $N_2$ fixation (*Azorhizobium caulinodans*, Table 22). Sown as an intercrop between two rice crops with a vegetation period of less than two months, but also in rows mixed with the cash crop (210, 596, 922, 1478) |
| *Tephrosia candida* (Roxb.) A. DC. | white tephrosia, boga medeloa | Tr | Perennial, good production of green material also on poor soils with sufficient water supply. Ground cover, up to heights of 1200 m (259, 444, 1308) |
| *T. purpurea* (L.) Pers. | purple tephrosia, wild indigo | Tr | Perennial, mainly for ground cover, in SE Asia used as green manure plant, before flowering in limited quantities also as fodder, but danger of poisoning (259, 619, 1343, 1383) |

| | | | |
|---|---|---|---|
| *T. vogelii* Hook.f. | fish poison bean, igongo | Tr | Perennial, good green manure plant for heavy soils with sufficient rainfall; high rotenoid content, potential insecticide plant (see p. 391), in E Africa used as a fish poison (259, 376, 444, 537, 1297, 1444) |
| *Vigna hosei* (Craib) Backer (*Dolichos hosei* Craib) | Sarawak bean | Tr | Especially used as a ground cover plant in tea plantations, also fodder (259, 376, 718, 1444, 1684) |

sown in autumn (605). Excellent results are obtained with cereal-vetch mixtures, peas, various vetches, yellow lupins, and others. Where the soil is fertile, and enough rain falls, a dual use as fodder and green manure is possible if suitable plants are chosen.

Under humid tropical conditions, a new system of simultaneous green manuring has been developed: alley cropping. It consists in sowing annual crops such as maize between rows of perennial legumes (e.g. *Leucaena leucocephala* (Table 52) or *Gliricidia sepium* (Table 55)). The young branches of the hedges are pruned at more or less regular intervals and left in the field as green manure and soil cover (555).

In India, it has been normal since ancient times to use green plant matter for fertilizing wetland rice fields. Research has proved the value of this procedure (1343). The green manure is produced on the rice fields as a catch crop, or is obtained from the shrubs and trees which grow around the rice field. Annual plants suitable for this cultivation include: *Crotalaria juncea* (sun hemp, Table 44), *Tephrosia candida*, *Vigna trilobata* (pillipesara) and other *Vigna* species (Table 25), *Sesbania cannabina* (Table 52), *Sesbania rostrata* (Table 53), *Indigofera* spp. and other low-demanding legumes. Among the woody plants, which grow outside the rice field, are *Gliricidia sepium*, *Sesbania grandiflora*, *Butea monosperma*, and many others. In recent years, the utilization of the water fern, *Azolla pinnata*, has received special attention as a green manure plant for wet rice cultivation (Table 53), as this fixes considerable amounts of $N_2$ through its symbiosis with *Anabaena* (see p. 14).

On irrigated land with badly drained soil, the regular supply of organic material is necessary to make the soil permeable. Thus, here, green manuring does not directly nourish the plants, much more, it encourages the development of bacteria whose mucilage binds the clay particles in crumbs, so that cavities are formed, through which drainage and the washing out of harmful salts can take place. Instead of green manuring, animal dung, or any sort of organic remains can be used (1772, 1909). Quick-growing green manure plants are often the simplest and cheapest source of organic materials for this purpose.

## Ground Cover Plants

The introduction of suitable herbaceous and shrubby cover plants in the low-yielding hevea plantations of Indonesia in the first decades of this century was a definite turning point in the cultivation techniques of tropical tree culture in the humid regions (684, 1902, 1903). The ground cover plants carry out their most important function immediately after forest clearance. They protect the soil from the impact of raindrops, and from the scorching effects of direct sunlight, and also prevent the development of an unwanted weed flora (e.g. alang-alang, *Imperata cylindrica* (L.) Raeusch.). Legumes are most suitable, as they simultaneously contribute on N supply to the main crop due to their $N_2$ fixing. Ground cover plants should develop quickly, and should be efficient in the acquisition of mineral nutrients, so that they themselves demand little or no fertilizers, while at the same time mobilizing the mineral resources of the soil for the main crop. They should be deep-rooted, so as not to compete with the main crop for water and nutrients, they should also remain green through several months of low rainfall in order to reduce the risk of fires, and they should tolerate a certain amount of shade (218, 1903, 1911, 2000). Not all ground cover plants fulfill all of these postulates. Because of this, mixtures of different species are often used, e.g. *Mimosa invisa*, which is extremely undemanding, but dies back in dry periods, with *Pueraria phaseoloides* and *Centrosoma pubescens*, of which the latter tolerates more shade than the former. A mixture of upright growing shrubs and herbaceous perennials (e.g. *Eupatorium inulifolium*, *Tephrosia candida*, *Leucaena leucocephala*, Table 52) with creeping plants can be useful, since it achieves a high production of biomass and a long lifetime of the ground cover. The choice of species depends not only on the local soil and climatic conditions, but also on the growth habit of the main crop. For tea, strongly growing creepers should be avoided, such as *Mucuna* spp., *Pueraria* spp., or *Centrosema pubescens*. These would quickly overrun the tea plants. However less aggressive species can be used such as *Indigofera spicata* or *Vigna hosei*.

Ground cover plants can have a decisive role as a means of controlling weeds. Where the land is left fallow, even if only for a few weeks, in humid regions or in irrigated cultivation, the weeds can propagate without hindrance. This situation is especially favourable for the vegetatively propagating and light-loving grasses and sedges (e.g. *Sorghum halepense*, *Cynodon dactylon* (Table 51), *Cyperus esculentus* (Table 52), *C. rotundus* L.). If a quick growing ground cover plant is sown immediately after harvesting, the possibility of weed propagation is reduced. By far the best plant for this purpose is thickly sown buckwheat (Table 7). The soil is shaded within a week of sowing, and after several weeks, enough mass has been formed that it is worthwhile to work it into the soil. Quick growing legumes such as *Lablab purpureus*, *Vigna* and *Phaseolus* species (Table 25) are also suitable for keeping down weeds in the short term. However, vigourously growing ground cover plants are not only used to hinder the appearance of unwanted weeds, but also to suppress existing populations of weeds. Thus for the control of alang alang, *Leucaena leucocephala*, *Mimosa invisa*, *Centrosema pubescens*, *Pueraria phaseoloides*, *Mucuna pruriens*, and others are employed (1360, 1760).

For afforestation in the tropics, a great range of ancillary ground cover plants are available, but these cannot be elaborated on here.

## Soil Stabilizing Plants

Severe erosion damage occurs where man has damaged or destroyed the vegetation. The most endangered are arid regions (wind erosion) and mountainous terrain (water erosion). Because the exposure of the soil surface is the primary cause of erosion, the prevention and repair of erosion damage is based on the preservation or regeneration of vegetation. Plants protect the soil from the direct effects of rain and wind, improve the ability of the soil to absorb water, reduce the speed of flowing water, and hold the soil particles in place with their roots. Depending on the soil type, amount of rainfall and gradient of slope, one of these protection effects will be in the foreground, and on this will depend the choice of the most suitable protection plants.

In arid regions, the most difficult problem is the stabilization of loose sand. Because the movement of grains of sand is the greatest obstacle to new growth, the movement of the sand grains must first be prevented by mechanical means.

Table 54. Soil stabilizers (explanation of symbols, see footnotes page 55)

| Botanical name | Vernacular names | Climatic region | Remarks and literature references |
|---|---|---|---|
| | | | **Dicotyledonae** |
| **Aizoaceae** | | | |
| *Carpobrotus edulis* (L.) L. Bolus | Hottentot fig, sour fig | S,Tr | Drought and salt tolerant, for stabilization of open sand surfaces and slopes. Fruits eaten (1905) |
| **Amaranthaceae** | | | |
| *Alternanthera bettzickiana* (Rgl.) Nichols. (*A. amoena* (Lem.) Voss) | kanchari, nadarang, busbusi | Tr | For stabilization of slopes, on terraces and water ditches. Leaves used as a vegetable (684, 1296, 1297) |
| **Casuarinaceae** | | | |
| *Casuarina equisetifolia* J.R. et G.Forst. | horsetail tree, Polynesian ironwood, Australian beefwood, Australian pine | S, Tr | Salt tolerant, for stabilizing sandy coastal regions and shifting dunes, likewise used as a windbreak in dry regions, other *Casuarina* species are used similarly (259, 406, 1095, 1185, 1262, 1266, 1383, 1591) |
| **Chenopodiaceae** | | | |
| *Haloxylon ammodendron* (C.A.Mey.) Bunge | saxoul, saksaul | S | Salt and drought tolerant, planted in Iran for afforestation and stabilization of sandy soils and dunes (1611) |
| *H. aphyllum* (Minkw.) Iljin | black saxoul | S | This and several other *H.* species are used like *H. ammodendron* (1262) |
| **Compositae** | | | |
| *Eupatorium riparium* Regel | mist flower | Tr | Low growing, especially for stabilizing slopes on ditches and roads (684, 1297) |

| | | | |
|---|---|---|---|
| *E. triplinerve* Vahl | ayapana | Tr | On poor soils, for stabilizing terrace benches, tolerates shade. Ground cover, medicinal and tea plant, leaves used as a vegetable (200, 376, 684, 1297, 1383) |
| **Convolvulaceae** | | | |
| *Ipomoea pes-caprae* (L.) R.Br. | beach morning-glory, seaside morning-glory | S, Tr | Salt tolerant, planted for stabilizing sand beaches, dunes and slopes, leaves as fodder, especially for pigs (259, 376, 1444, 1383) |
| **Leguminosae** | | | |
| MIMOSOIDEAE | | | |
| *Acacia saligna* (Labill.) Wendl. | golden wreath wattle, blue-leaved wattle, Port-Jackson wattle | S, Tr | This and other evergreen *Acacia* species are shrubs or small trees which are planted in many countries for stabilizing dunes, especially near the sea (1261, 1262, 1335) |
| **Passifloraceae** | | | |
| *Passiflora foetida* L. | running pop, tagua passionflower, granadilla de culebra | Tr | Planted for ground cover and to prevent erosion because of its quick growth, fruits edible (259, 550, 1444) |
| **Polygonaceae** | | | |
| *Calligonum polygonoides* L. | phog, balonja, tirni | S, Tr | For stabilizing sand and dunes, in Iran on soils after stabilization by petroleum, several other *Calligonum* species cultivated in the Middle East for stabilizing dunes. Emergency food in India (200, 1095) |
| **Salicaceae** | | | |
| *Salix acutifolia* Willd. | Caspic willow, osier violet | Te, S | Protection of river banks, also cultivated for dune stabilization and windbreaks, twigs for basket making, many other *Salix* species are used similarly (2026) |
| **Tamaricaceae** | | | |
| *Tamarix aphylla* (L.) Karst. | tamarisk salt tree, athel tamarisk, salt cedar | Te, S, Tr | For stabilization of dunes and eroded surfaces, also in shelterbelts; all *Tamarix* species withstand high salt contents in the soil, several other species are used in arid regions in a similar way to *T. aphylla* (153, 376, 406, 580, 1262) |

Table 54. Soil stabilizers, cont'd (explanation of symbols, see footnotes page 55)

| Botanical name | Vernacular names | Climatic region | Remarks and literature references |
|---|---|---|---|
| **Umbelliferae** | | | |
| *Centella asiatica* (L.) Urb. (*Hydrocotyle asiatica* L.) | Indian pennywort, Asiatic pennywort | S, Tr | From sea level up to 2500 m high; good for soil stabilization in humid regions due to strong formation of runners, tolerates dense shade. Locally used as a vegetable and medicinal plant (259, 376, 684, 1296, 1297, 1383) |
| | | **Monocotyledonae** | |
| **Gramineae** | | | |
| *Ammophila arenaria* (L.) Link (*A. arundinacea* Host.) | European beach grass, marram grass, roseau des sables | Te, S | Used for the stabilization of dunes and sandy surfaces in many countries. Stalks utilized for basket making (1846) |
| *Arundo donax* L. | Spanish reed, giant reed, canne de Provence, caña de Castilla | Te, S | For stabilizing river banks. The reeds have numerous uses, for basket making, as supports, for paper. Leaves also used as fodder (259, 376, 406, 964, 1374) |
| *Calamovilfa gigantea* (Nutt.) Scribn. et Merr. | sandreed | S | Planted for dune stabilzation in dry regions of the USA (741) |
| *Phragmites australis* (Cav.) Trin. ex Steud. (*P. communis* Trin.) | common reed grass, roseau commun, carrizo común | Te, S, Tr | Salt tolerant, planted for protection of river and lake shores and for water purification. Good basket material, young shoots a vegetable. Fodder, weed (259, 376, 1980) |

For this, hedges of straw or dry shrubbery are used, and the soil stabilizing plants are sown or planted in the protection of this barrier. Otherwise sprays of bitumen solution, rubber latex, or plastic preparations can be used (33). Deep-rooted shrubs are primarily suited for permanent growth, as these can use the moisture which may be available deep in the sand. The most important species are *Acacia salicina* Lindl., *A. saligna* and other acacias, *Prosopis* spp., *Tamarix* spp., *Casuarina equisetifolia* and *Haloxylon* spp. The choice of suitable species depends on the rainfall and the salt content of the soil.

In more humid regions, the greatest concern is to diminish the effect of flowing water. For the protection of terraced slopes, grasses which form a thick root mass are preferred (*Pennisetum clandestinum* (Table 51), *Vetiveria zizanioides* (1991) (Table 41), *Cymbopogon* spp.). The banks of ditches can be secured with creeping plants (*Eupatorium triplinerve*, *E. riparium*, *Alternanthera bettzickiana*, *Centella asiatica*, *Passiflora foetida*, and others (Table 54)). For vegetation cover of waterways, low-growing grasses which form runners and rhizomes are preferred, such as *Urochloa* spp., *Paspalum* spp., *Cynodon dactylon* (Table 51).

To protect surfaces which are used for arable purposes from wind and water erosion during the times with little or no rainfall, cultivation without ploughing (no-tillage farming) is increasing in importance (1397). In this, the stubble is allowed to remain to protect the soil. It can also be useful, in order to achieve an effective stubble, to sow a quick growing grass such as teff at the end of the vegetation time, which then serves no purpose other than to provide sufficient ground cover during the dry period (1914).

## Windbreak Plants

Some crop plants, such as banana and cassava, are rather sensitive to strong winds. On sites which are exposed to the wind, they can only be cultivated when the strength of the wind is broken by trees. In arid and semi-arid regions, windbreaks can be necessary in order to reduce the danger of soil erosion. In arid regions, the windbreak can also be favourable for the development of the crops, primarily because the reduction of the wind speed reduces the transpiration rate. However, it should not be overlooked that windbreaks can not only increase the temperature above the optimum, they can also increase the danger of night frosts by hindering the airflow and preventing the outflow of cold air. Whether planting a windbreak is necessary or advantageous, and which type of plant is the most suitable is determined by many factors, and can only be decided with a thorough knowledge of the local conditions (226, 367, 468, 1756). Therefore, it is impossible to deal with details here. Also, Table 55 can only include the windbreak plants which are most used, particularly since numerous indigenous plants have been proved to be especially suitable, e.g. in dry regions of South Africa, the combination of *Acacia dealbata* with *Rhus lancea* and *Euclea undulata*.

In general, it can only be stated that a good windbreak lies across the prevailing wind direction, and should not be totally impermeable to the wind to prevent the formation of turbulent eddies behind the trees. The effective reduction of the wind speed extends up to 5 times the tree height on the windward side, and up to 20 times the tree height on the lee side (15, 502, 1163, 1741, 1911). The species

Table 55. Windbreak, shade, support, and hedge plants (explanation of symbols see footnotes page 55)

| Botanical name | Vernacular names | Climatic region | Remarks and literature references |
|---|---|---|---|
| | | **Gymnospermae** | |
| **Cupressaceae** | | | |
| *Cupressus arizonica* Greene | Arizona cypress | S, TrH | Used in many countries for shelterbelts and for afforestation. Other *Cupressus* species are also used like *C. arizonica* (1846) |
| *Juniperus virginiana* L. | eastern red cedar | Te, S, Tr | Cultivated in forests and in shelterbelts, especially in the central and southern states of the USA. Other *Juniperus* species are likewise cultivated for these purposes (1846, 2026) |
| **Pinaceae** | | | |
| *Pinus halepensis* Mill. | Aleppo pine, pin blanc, pino carrasco | Te, S | Drought tolerant, used especially in the Mediterranean region and Argentina for afforestation and shelterbelts. Some other *Pinus* species are likewise suitable for subtropical regions (644, 1262) |
| | | **Dicotyledonae** | |
| **Anacardiaceae** | | | |
| *Rhus lancea* L.f. | willow rhus, karreeboom | S | Shrub or small tree, very low demanding, as lowest storey in shelterbelts. Other evergreen *Rhus* species are likewise used (1335, 1336) |
| **Bignoniaceae** | | | |
| *Parmentiera aculeata* (H.B.K.) L. Wms. (*P. edulis* DC.) | food candletree, cuachilote | Tr | Shade tree, also cultivated for fruit and medicinal purposes (550, 1444) |

| | | | |
|---|---|---|---|
| *Spathodea campanulata* Beauv. | African tuliptree, spathodea | S, Tr | Cultivated as shade, road, and ornamental tree (259, 1095, 1383) |
| **Ebenaceae** | | | |
| *Euclea undulata* Thunb. | gwarri | S, Tr | As lowest storey in shelterbelts. Also other *Euclea* species are used for this (1336) |
| **Euphorbiaceae** | | | |
| *Croton tiglium* L. | purging croton | Tr | Hedge plant and shade tree. Purgative seeds and seed oil are used medicinally (259, 1095, 1383) |
| *Euphorbia tirucalli* L. | milkbush, milk-hedge | S, Tr | As a hedge plant in Africa and SE Asia. Other, especially spiny, *Euphorbia* species are often planted as hedges. Cultivation attempted as a source of hydrocarbons (259, 1095, 1383) |
| *Jatropha curcas* L. | physic nut, purging nut, frailejón, pinhão do Paraguai | Tr | Hedge and fencing plant in many tropical countries. Seed oil used for technical purposes and medicinally. Dark blue colouring and tanning compounds from the bark. *J. gossypiifolia* L. used similarly (259, 376, 406, 1383, 1384, 1620) |
| **Guttiferae** | | | |
| *Calophyllum brasiliense* Cambess. | maria, aceite de María, jacareúba | Tr | In the W Indies, shade tree and windbreak for cocoa. Also provides an aromatic balsam (1373, 1444, 1518) |
| **Leguminosae** | | | |
| MIMOSOIDEAE | | | |
| *Acacia dealbata* Link. | silver wattle | S | Windbreak, tannin, and acacia gum of lower quality (259, 580, 760, 1095) |
| *A. decurrens* (Wendl.) Willd. | green wattle | S, TrH | In windbreak hedges and for tannin production, also on poor soils (259, 1095, 1262, 1444) |
| *A. nilotica* (L.) Del. (*A. arabica* (Lam.) Willd.) | Egyptian acacia, babul | S, Tr | Windbreak and soil stabilizer in arid regions, salt tolerant. Provides low quality gum arabic and tannin from its bark (see p. 372) (406, 444, 511, 638, 760, 1262, 1684) |

Table 55. Windbreak, shade, support, and hedge plants, cont'd (explanation of symbols see footnotes page 55)

| Botanical name | Vernacular names | Climatic region | Remarks and literature references |
|---|---|---|---|
| *A. tortilis* (Forssk.) Hayne | Israeli babul, umbrella thorn, sayal, talha | S, Tr | Very drought tolerant, planted for soil protection and windbreak (273, 411, 444, 585, 639, 1247, 1261, 1684) |
| *Albizia chinensis* (Osb.)Merr. | siran | Tr | As windbreak for coffee and tea in S and SE Asia. Leaves used as animal fodder (259, 273, 1261) |
| *A. falcataria* (L.) Fosb. | Molucca albizia, banda | Tr | Quick growing shade tree in SE Asia (259, 1247, 1261, 1262, 1297) |
| *A. lebbeck* (L.) Benth. | lebbek, shak-shak, woman's tongue | S, Tr | Widespread as a shade tree. Several other *Albizia* species are planted locally as shade trees (259, 273, 406, 580, 1081, 1247, 1261, 1262, 1438, 1684) |
| *Andira inermis* (W. Wright) H.B.K. ex DC. | bastard mahogany, chigo, angelim | Tr | Windbreak and shade tree on the W Indian Islands. Bark used medicinally (259, 1095, 1247, 1960) |
| *Enterolobium cyclocarpum* (Jacq.) Griseb. | earpod tree, parota, corotu, guanacaste | Tr | Quick growing shade tree. Young pods and seeds as cattle fodder and vegetable (1247, 1261, 1373, 1960) |
| *Inga edulis* Mart. | food inga, ice cream bean, inga espada, inga sipó | Tr | Support plant for vanilla in Brazil, shade tree for coffee in Colombia. Young pods eaten as vegetable (444, 1247, 1262, 1518, 1960) |
| *I. laurina* (Sw.) Willd. | sackysac, pois doux, guamo | Tr | Windbreak and shade tree on the W Indian Islands. Other *Inga* species are also planted in Central and S America as shade trees (1444, 1960) |

| | | | |
|---|---|---|---|
| *Pithecellobium dulce* (Roxb.) Benth. | Manila tamarind, guamuchil, dinde | Tr | Planted as a hedge in many countries, also generally used as fruit (259, 376, 1262, 1960) |
| *P. saman* (Jacq.) Benth. (*Enterolobium saman* (Jacq.) Prain) | rain tree, cow tamarind, cenícero, saman | Tr | Shade tree for crops in the humid tropics (cocoa, pepper, vanilla). Pods are high value animal fodder (259, 376, 528) |
| CAESALPINIOIDEAE | | | |
| *Parkinsonia aculeata* L. | Jerusalem thorn, palo de rayo, espinheiro de Jerusalem | S, Tr | Low-demanding, drought tolerant hedge shrub, also planted to protect against erosion. Young twigs as fodder for goats and sheep (376, 580, 1262, 1684) |
| PAPILIONOIDEAE | | | |
| *Erythrina fusca* Lour. (*E. glauca* Willd.) | coral bean, bucare, bucago | Tr | Shade tree for cocoa in Central America (990, 1444, 1591) |
| *E. senegalensis* DC. | coral flower, arbre corail | Tr | Much planted for fences and hedges (1381, 2026) |
| *E. variegata* L. var. *orientalis* (L.) Merr. (*E. indica* Lam.) | Indian coral tree, mandara | Tr | Cultivated as hedges and support plant for pepper. Locally, many other *Erythrina* species are used similarly. Leaves serve as fodder and for green manuring (376, 406, 1383) |
| *Gliricidia sepium* (Jacq.) Steud. | quick stock, madre de cacao, mata ratón, cacaute | Tr | Hedge shrub and shade tree, leaves much used in India for green manuring and mulching (376, 473, 1247, 1262, 1591, 1684, 1695, 1960) |
| *Sesbania grandiflora* (L.) Pers | agathi, corkwood tree, fagotier, turi | Tr | For light shade and as support for pepper and betel pepper. Flowers, leaves and young pods are eaten as vegetables. Fodder crop (259, 273, 376, 473, 494, 596, 731, 1095, 1247, 1261, 1383, 1684) |
| *S. sesban* (L.) Merr. (*S. aegyptiaca* Pers.) | common sesban, Egyptian sesban, djanti | Tr | (Fig. 101) Shade, windbreak and hedge plant. Twigs are utilized for green manuring. Tolerates flooding (376, 494, 596, 1095, 1262, 1684) |

Table 55. Windbreak, shade, support, and hedge plants, cont'd (explanation of symbols see footnotes page 55)

| Botanical name | Vernacular names | Climatic region | Remarks and literature references |
|---|---|---|---|
| *S. speciosa* Taub. | seemai agathi | Tr | Shade plant in nurseries. Windbreak and green manure plant for rice and other crops in India. Other *Sesbania* species are used like those named, and for obtaining fibres (376, 494, 596, 1262) |
| **Malvaceae** | | | |
| *Thespesia populnea* (L.) Sol. ex Corr. | Portia tree, umbrella tree, false rosewood | Tr | Shade tree and windbreak, suitable for sites near the sea. Valuable wood. Seed oil and bast fibres used locally (259, 273, 376, 406, 1383) |
| **Myrtaceae** | | | |
| *Eucalyptus camaldulensis* Dehnhardt | Australian kino, river red gum | S, Tr | After *E. globulus* (Table 41) the most cultivated species. Outstanding for shade and windbreaks. Withstands long dry periods (511, 512, 1262, 1371) |
| *E. cladocalyx* F.v.Muell. | sugar gum | S, Tr | Especially well suited for shelterbelts in dry regions (512, 1371) |
| *E. cornuta* Labill. | yate | S, Tr | Small compact tree, especially suitable for dry regions as a windbreak (512, 1371) |
| *E. gomphocephala* A. DC. | white gum, tuart | S, Tr | Tall growing, good in shelterbelts, also growing near the coast (512, 1262, 1371) |
| *E. sargentii* Maiden | salt river gum | S, Tr | Medium sized, as a windbreak on sandy and salty soils. Several other *Eucalyptus* species are planted as windbreak, shade and street trees (512, 1371) |
| **Proteaceae** | | | |
| *Grevillea robusta* A. Cunn. ex R.Br. | silver oak, silky oak | S, Tr | Planted as a shade tree and windbreak for coffee and tea, withstands mild frosts, and can be cultivated to over 2000 m (259, 376, 714, 1262, 1591) |

| | | | |
|---|---|---|---|
| **Salicaceae** | | | |
| *Populus euphratica* Oliv. | Indian poplar, bahan, hodung | Te, S | Used in India and Iran as windbreak and for soil stabilization. Other *Populus* species are used in many countries for shelterbelts (376, 1262) |
| **Salvadoraceae** | | | |
| *Salvadora indica* Royle (*S. oleoides* Decne.) | jal, diar, mithi, khakan | S | Drought and salt tolerant, cultivated for shelterbelts in India and Pakistan. Fruits edible, shoots serve as camel fodder. Seed oil usable for technical purposes (234, 376, 580, 619) |
| | | **Monocotyledonae** | |
| **Agavaceae** | | | |
| *Agave americana* L. | century plant, agave d'Amérique | S, Tr | Much used as a fencing plant in dry regions (200, 406, 964, 1383, 1445) |
| **Gramineae** | | | |
| *Dendrocalamus strictus* (Roxb.) Nees | Calcutta bamboo, male bamboo, bans | S, Tr | This and other bamboo species (Tables 25, 44) serve also as windbreaks for other crops. *D. strictus* is the most important raw material for paper production in India (376, 580, 406) |
| *Saccharum spontaneum* L. | thatch grass, kans, bagberi | Tr | Polymorphic species. Used in the breeding of sugar cane (see p. 69). Forms dense windbreak hedges, and serves for sand stabilization. Supplies a fibre for basketwork, ropes and paper, thatching material and fodder. Young shoots as a vegetable (376, 406, 1644) |

chosen should not make high demands on the soil, and in dry regions should compete as little as possible for water with the pasture and arable plants. The only plants which can be considered here are succulents (e.g. *Agave americana*) and deep-rooted species with little horizontal growth of their roots.

Good shelterbelts can have a positive side-effect by forming an effective barrier against grass fires. They can serve as nesting sites for birds, but this can also be a disadvantage (e.g. red-billed quelea in Africa). An important by-product is the provision of firewood (964, 1262). The usage of leaves from windbreak trees for the fertilization of rice fields in India has already been mentioned (see p. 434). Plants which are themselves useful can also be planted primarily as windbreaks, such as *Anacardium occidentale* (Table 33) (511). Shade for animals in the fields in hot regions is also an important side effect provided by windbreaks.

## Shade Trees

A certain amount of shade is necessary for all tropical tree crops in the first months or years after planting out, primarily to reduce the transpiration of the young saplings until their root system is well-developed. For this purpose, individual trees are allowed to remain from the previous forest clearing, or planting is done in glades, quick and tall-growing legumes are sown (*Crotalaria* spp., *Tephrosia* spp.), or tall-growing food plants (banana, cassava) are planted between the rows.

The extent to which shade is an advantage for adult trees and bushes depends on the local climatic conditions and the availability of mineral fertilizers. Shade trees reduce the yields and make working the soil more difficult. They are especially useful in places where temperature differences between day and night are large (highlands), where protection against strong winds is necessary, and where mineral fertilizers are not available or are too expensive (7, 1911, 1946).

Good shade trees should have a high crown which is not too dense. The reduction in sunlight should be 10-15%, and not more than 25%, except when the shade creators themselves produce economically important products (e.g. bananas shading coffee). Leguminosae are preferred due to their $N_2$ fixing. Apart from them, species such as *Grevillea robusta* are used, which have deep roots, a good fertilizing effect due to their abundant leaf-fall, and provide a usable wood.

## Fencing Plants

A chapter about ancillary plants would be incomplete without a mention of the widespread use of plants for the boundaries of fields, especially with the purpose of keeping both domestic and wild animals away from cultivated fields (161, 1591). For this, it is best to use undemanding plants which can be vegetatively propagated and form hedges so thick and ideally thorny that they can not easily be crossed by animals. The most important species belong to the genera *Agave*, *Aloe*, *Euphorbia*, *Opuntia*, and *Yucca*, which also grow in very dry regions. With more rainfall, shafts of *Ficus*, *Erythrina*, and *Dracaena* species and others are set as hedges, as they quickly develop roots and form a long-lasting living fence. It is common in South America to use trees, such as *Gliricidia sepium*, as living fence posts, which serve as supports for the usual barbed wire (1591, 1960). *Pithecellobium dulce*, which is propagated by seeds, is a frequently used hedge plant in India (376). Plants with bad tasting or smelling leaves, such as *Croton tiglium*, can also provide useful protective hedges.

# Literature


1 Abbot, A.J., and Atkin, R.K. (eds.), 1987. Improving Vegetatively Propagated Crops. Academic Press, London.

2 Abd el Moneim, A.M., Khair, M.A., and Cocks, P.S., 1990. Growth analysis, herbage and seed yield of certain forage legume species under rainfed conditions. J. Agron. Crop Sci. 164, 34-41.

3 Abeele, M. van den, et Vandenput, R., 1956. Les Principales Cultures du Congo Belge. 3rd ed. Direction de l'Agriculture des Forêts et de l'Elevage, Bruxelles.

4 Abraham, P.D., Anthony, J.L., and Arshad, N.L., 1981. Stimulation practices for smallholdings. Planters' Bull. No. 167, 51-66.

5 Abu-Irmaileh, B.E., 1982. Weeds of Jordan. University of Jordan, Amman.

6 Achtnich, W., 1989. Arzneipflanzen. In: Rehm S., (Hrsg.) 1989. Spezieller Pflanzenbau in den Tropen und Subtropen. Handbuch der Landwirtschaft und Ernährung in den Entwicklungsländern, 2. Aufl. Bd. 4, 521-540. Ulmer, Stuttgart.

7 Acland, J.D., 1971. East African Crops. Longman, London.

8 Adam, A.V., and Rodriguez, A., 1970. "Clean seed" and "certified seed" programmes for bananas in Mexico. FAO Plant Prot. Bull. 18, 57-63.

9 Adams, C.D., Kasasian, L., and Seeyave, J., 1968. Common Weeds of the West Indies. Univ. West Indies, Trinidad.

10 Adamson, A.D., 1971. Oleoresins. Production and Markets with Particular Reference to the United Kingdom. G 56, Trop. Prod. Inst., London.

11 Adamson, A.D., and Bell, J.-M.K., 1974. The Market for Gum Arabic. G 87, Trop. Prod. Inst., London.

12 Adamson, A.D., and Robbins, S.R.J., 1975. The Market for Cloves and Clove Products in the United Kingdom. G 93, Trop. Prod. Inst., London.

13 Adrian, J., et Jacquot, R., 1968. Valeur Alimentaire de l'Arachide et de ses Dérivés. Maisonneuve et Larose, Paris.

14 Aggarwal, P.K., Liboon, S.P., and Morris, R.A., 1987. A Review of Wheat Research at the International Rice Research Institute. IRRI Res. Paper Ser. 124, IRRI, Los Baños, Philippines.

15 Ahmed, I., 1972. The role of forestry in correct land use with emphasis on shelterbelts in Kano State. Samaru Agric. Newsletter 14 (1), 5-7.

16 Ahmed, S., Bamofleh, S., and Munshi, M., 1989. Cultivation of neem (*Azadirachta indica*, Meliaceae) in Saudi Arabia. Econ. Bot. 43, 35-38.

17 Ahmed, S., and Khalid, A., 1971. Why did Mexican dwarf wheat decline in Pakistan?. World Crops 73, 211-216.

18 Ahn, P.M., 1970. West African Soils. Oxford Univ. Press, London.

19 Aiyer, A.K.Y.N., 1966. Field Crops of India. 6th ed. Bangalore Printing and Publishing, Bangalore.

20 Akehurst, B.C., 1981. Tobacco. 2nd ed. Longman, London.

21 Akinola, J.O., Whiteman, P.C., and Wallis, E.S., 1975. The Agronomy of Pigeon Pea (*Cajanus cajan*). Commonwealth Agric. Bureaux, Farnham Royal.

22 Akobundu, I.O., 1978. Weeds and Their Control in the Humid and Subhumid Tropics. IITA, Ibadan, Nigeria.

23 Akobundu, I.O., 1987. Weed Science in the Tropics. Principles and Practices. Wiley Interscience, New York.

24 ALAMU, S., and MCDAVID, C.R., 1985. Genetic variability in tannia, *Xanthosoma sagittifolium* (L.) Schott. Trop. Agric. (Trinidad) 62, 30-32.

25 ALAMUDDIN, S.N., 1977. Apiculture in tropical climates. World Crops 29, 12-15, 66-71.

26 ALAZARD, D., NDOYE, I., and DREYFUS, B., 1988. *Sesbania rostrata* and other stem-nodulated legumes. In: BOTHE, H., DE BRUIJN, F. J., and NEWTON, W. E. (eds.): Nitrogen Fixation: Hundred Years after, 765-850. Fischer, Stuttgart.

27 ALBUQUERQUE, M. DE, e PINHEIRO, E., 1970. Tuberosas Feculentas. IPEAN, Série Fitotécnica, 1 (3), 1-115.

28 ALEXANDER, A.G., 1973. Sugarcane Physiology. A Comprehensive Study of the *Saccharum* Source-to-Sink System. Elsevier, Amsterdam.

29 ALLEN, D.J., 1984. The Pathology of Tropical Food Legumes. Disease Resistance in Crop Improvement. Wiley, Chichester.

30 ALLEN, J.L., 1970. Lime Juice and Lime Oil Production and Markets. G 45, Trop. Prod. Inst., London.

31 ALLEN, O.N., and ALLEN, E.K., 1981. The Leguminosae - A Source Book of Characteristics, Uses, and Nodulation. Macmillan, New York.

32 ALLEN, P.W., 1972. Natural Rubber and the Synthetics. Crosby Lockwood, London.

33 ALLES, W.S., 1971. Chemical sprays for controlling soil erosion. Trop. Agriculturist (Ceylon) 127, 179-185.

34 ALLISON, R.V., and BOOTS, V.A., 1970. A three-way look at the mechanical harvesting of two long-fiber crops, ramie and kenaf. Soil Crop Sci. Soc. Flor. Proc. 30, 100-112.

35 ALLISON, R.V., TRADEWELL, G.E., and BOOTS, V.A., 1973. A further look at the use of mechanical harvesting for the long-fiber crop, ramie, and the in-line reception and processing of the fiber and other products. Soil Crop Sci. Soc. Flor. Proc. 32, 66-73.

36 ALMEYDA, N., and MARTIN, F.W., 1978. Tropical fruits: The mangosteen. World Farm. 20 (8), 10, 12, 20-23.

37 ALMEYDA, N., and MARTIN, F.W., 1980. Cultivation of neglected tropical fruits with promise. Part 8. The pejibaye. USDA, New Orleans.

38 ALSINA, E., VALLE-LAMBOY, S., and MENDEZ-CRUZ, A.V., 1975. Preliminary evaluation of ten sweet sorghum varieties for sugar production in Puerto Rico. J. Agric. Univ. Puerto Rico 59, 5-14.

39 ALVARENGA, A.A., and VALIO, I.F.M., 1989. Influence of temperature and photoperiod on flowering and tuberous root formation of *Pachyrrhizus tuberosus*. Ann. Bot. 64, 411-414.

40 ALVIM, P. DE T., 1966. Factors affecting flowering of the cocoa tree. Cocoa Growers Bull.No. 7, 15-19.

41 ALVIM, P. DE T., 1977. Factors affecting flowering of coffee. J. Coffee Res. 7, 15-25.

42 ALVIM, P. DE T., 1981. A perspective appraisal of perennial crops in the Amazon basin. Interciencia 6, 139-145.

43 ALVIM, P. DE T., et ALVIM, R., 1979. Sources d'énergie d'origine végétale: hydrates de carbon, huiles et hydrocarbures. Oléagineux 34, 465-472.

44 ALVIM, P. DE T., and KOZLOWSKI, T.T. (eds.), 1977. Ecophysiology of Tropical Crops. Academic Press, New York.

45 ALVIM, R., ALVIM, P. DE T., and LEITE, R.M. DE O., 1978. Mechanical injury of wind to recently transplanted cacao seedlings as related to the shade problem. Rev. Theobroma 8, 117-124.

46 AL-ZAND, O.A., 1974. The economics of olive oil and oilseeds in the Mediterranean region. World Crops 26, 24-27.

47 AMEFIA, Y.K., CILAS, C., DJIEKPOR, E.K., et PARTIOT, M., 1985. La multiplication végétative du cacaoyer. Note sur une méthode d'obtention de boutures à enracinement orthotrope. Café Cacao Thé 29, 83-88.

48 AMERICAN PHYTOPATHOLOGICAL SOCIETY, 1976. A Compendium of Corn Diseases. Amer. Phytopath. Soc., St. Paul, Minnesota.

49 AMERICAN SOCIETY OF AGRICULTURAL ENGINEERS, 1982. Vegetable Oil Fuels. Amer. Soc. Agric. Eng. (ASEA).
50 AMES, G.R., BARROW, M., BORTON, C., CASEY, T.E., MATTHEWS, W.S., and NABNEY, J, 1971. Bay oil distillation in Dominica. Trop. Sci. 13, 13-15.
51 AMUTI, K., 1980. Geocarpa groundnut (*Kerstingiella geocarpa*) in Ghana. Econ. Bot. 34, 358-361.
52 ANAND KUMAR, K., 1989. Pearl millet: current status and future potential. Outlook on Agriculture 18 (2), 46-53.
53 ANAND, N., 1982. Selected Markets for Ginger and Its Derivatives with Special Reference to Dried Ginger. G 161, Trop. Prod. Inst., London.
54 ANAND, N., 1983. The market for annatto and other natural colouring material, with special reference to the United Kingdom. G 174, TDRI, London.
55 ANASTASSIOU-LEFKOPOULOU, S., and SORTIRIADIS, S.E., 1984. Effect of plant population and spacing on cotton. I. Plant characters/density relationship and production stability. Cot. Fib. Trop. 39, 15-21.
56 ANDERSON, A.B., and BALICK, M.J., 1988. Taxonomy of the babassu complex (*Orbignya* spp.: Palmae). Syst. Bot. 13, 32-50.
57 ANDERSSON, G., HALL, O., and LÖÖF, B., 1969. Oil crops as a source of protein. Sver. Utsädesfören. Tidskr. 3-4, 249-255.
58 ANDREW, C.S., and KAMPRATH, E.J. (eds.), 1978. Mineral Nutrition of Legumes in Tropical and Subtropical Soils. CSIRO, Melbourne, Australia.
59 ANDREWS, J., 1984. Peppers, the Domesticated Capsicums. University of Texas Press, Austin, USA.
60 ANELLI, G., FIORENTINI, R., and LEPIDI, A.A., 1982. Food and feed from banana by-products. Riv. Agric. Subtrop. Trop. 76, 67-75.
61 ANGLADETTE, A., 1966. Le Riz. Maisonneuve et Larose, Paris.
62 ANGLADETTE, A., et DESCHAMPS, L., 1974. Problèmes et Perspectives de l'Agriculture dans les Pays Tropicaux. Maisonneuve et Larose, Paris.
63 ANGUS, J.F., NIX, H.A., RUSSELL, J.S., and KRUIZINGA, J.E., 1980. Water use, growth and yield of wheat in a subtropical environment. Austr. J. Agric. Res. 31, 873-886.
64 ANNECKE, D.P., and MORAN, V.C., 1982. Insects and Mites of Cultivated Plants in South Africa. Butterworths (South Africa), Durban.
65 ANONYM, 1970. A new career for kenaf?. World Farming 12 (10), 12-13.
66 ANONYM, 1970. Weed control in rice: herbicides herald a new area. World Farming 12 (4), 12-13, 16, 29.
67 ANONYM, 1971. Assisted biological coagulation. Planters' Bull. R.R.I.M. 112, 52-53.
68 ANONYM, 1973. Dominican bay oil industry. Trop. Sci. 15, 100-101.
69 ANONYM, 1973. Peanuts - culture and uses. A Symposium. Amer. Peanut Res. and Education Assoc., Raleigh, N. C.
70 ANONYM, 1973. Sericulture: a note on the present situation and prospects in developing countries. Trop. Sci. 15, 157-165.
71 ANONYM, 1973. The protein problem - a UN statement. World Agric. 22 (3), 18-21.
72 ANONYM, 1973. Trends in the world mushroom market. Monthly Bull. Agric. Econ. Statistics 22 (12), 12-17.
73 ANONYM, 1974. Latest IRRI rice shows resistance to pests and diseases. World Crops 26, 23.
74 ANONYM, 1978. Tunisia to improve esparto grass. World crops 30 (1), 17.
75 ANONYM, 1984. Keen rivalry faces newest entry to sweetener market. Ceres 17 (4), 10.
76 ANONYM, 1985. Utilization of Azolla in China. Intern. Rice Comm. Newsletter 34, 265-273.
77 ANONYM, 1986. African plant looms as major contender in sweetener market. Ceres No. 113 (Vol. 19, No. 5), 6-7.

78 ANOSIKE, E.O., and EGWUATU, C.K., 1981. Biochemical changes during the fermentation of castor oil (*Ricinus communis*) seeds for use as a seasoning agent. Qual. Plant. 30, 181-185.

79 ANTHONY, J.L., and ABRAHAM, P.D., 1981. Approaches to minimise constraints with upward tapping on smallholdings. Planters' Bull. No. 167, 67-75.

80 ANTONIO, A.A., JULIANO, B.O., and DEL MUNDO, A.M., 1975. Physicochemical properties of glutinous rice in relation to "suman" quality. Phil. Agric. 58, 351-355.

81 ANTUNEZ DE MAYOLO, K.K., 1989. Peruvian natural dye plants. Econ. Bot. 43, 181-191.

82 APPELQVIST, L.-A., and OHLSON, R. (eds.), 1972. Rapeseed. Elsevier, Amsterdam.

83 APPERT, J., et DEUSE, J. (eds.), 1982. Les Ravageurs des Cultures Vivrières et Maraîchères sons les Tropiques. Maisonneuve et Larose, Paris.

84 ARAMBOURG, Y., 1975. Les Insectes Nuisibles à l'Olivier. Sém. Oléic. Intern. 6-17 Oct., Cordue, Spain.

85 ARAULLO, E.V., DEPADUA, D.B., and GRAHAM, M., (eds.) 1976. Rice: Post-Harvest Technology. Intern. Dev. Res. Centre, Ottawa.

86 ARAULLO, E.V., NESTEL, B., and CAMPBELL, M., (eds.) 1974. Cassava Processing and Storage. Intern. Dev. Res. Centre, Ottawa.

87 ARE, L.A., and GWYNNE-JONES, D.R.G., 1974. Cacao in West Africa. Oxford Univ. Press, Ibadan.

88 ARKCOLL, D., 1988. Lauric oil resources. Econ. Bot. 42, 195-205.

89 ARNAUD, F., 1979. La pollinisation assistée dans les plantations de palmiers à huile. Oléagineux 34, 117-122.

90 ARNOLD, A.J., and HARRIES, H.C., 1979. Hybrid coconut seed production. World Crops 31, 12-16.

91 ARNOLDO, M., 1971. Gekweekte en Nuttige Planten van de Nederlandse Antillen. Natuurwetensk. Werkgroep Nederl. Antillen, Curaçao.

92 ARNON, I., 1972. Crop Production in Dry Regions. 2 Vols. Leonard Hill, London.

93 ARNON, I., 1975. Mineral Nutrition of Maize. Intern. Potash Institute, Bern.

94 ARORA, R.K., 1977. Job's-tears (*Coix lacryma-jobi*) - a minor food and fodder crop of Northeastern India. Econ. Bot. 31, 358-366.

95 ARORA, R.K., CHANDEL, K.P.S., JOSBI, B.S., and PANT, K.C., 1980. Rice bean: tribal pulse of eastern India. Econ. Bot. 34, 260-263.

96 ARYA, P.S., and SAINI, S.S., 1979. Tips for chicory growing. Indian Farming 28 (10), 17-18, 21.

97 ASANA, R.D., 1975. Physiological research in cotton - past and present. Indian Farming 24 (10), 17-19.

98 ASHBY, D.G., and PFEIFER, R.K., 1956. Weeds, a limiting factor in tropical agriculture. World Crops 8, 227-229.

99 ASHLEY, J., 1980. The culture of vanilla in Uganda. World Crops 32, 124-129.

100 ASHRI, A., 1971. Evaluation of the world collection of safflower, *Carthamum tinctorius* L. I. Reaction to several diseases and associations with morphological characters in Israel. Crop Sci. 11, 253-257.

101 ASOLKAR, L.V., and CHADHA, Y.R., 1979. Diosgenin and Other Steroid Drug Precursors. C. S. I. R., New Delhi.

102 ASSAF, R., et RIVALS, P., 1977. Néoculture du néflier du Japon (*Eriobotrya japonica* Lindl.). Fruits 32, 237-251.

103 ASSOCIATION OF JAPANESE AGRICULTURAL SCIENTIFIC SOCIETIES, 1975. Rice in Asia. Univ. Tokyo Press, Tokyo.

104 ASSOCIATION SCIENTIFIQUE INTERNATIONAL DE CAFE, 1979. Huitième Colloque Scientifique International sur le Café. ASIC, Paris.

105 ASTRIDGE, S.J., 1975. Cultivars of Chinese Gooseberry (*Actinidia chinensis*) in New Zealand. Econ. Bot. 29, 357-360.

106 ATAL, C.K., and KAPUR, B.M., 1977. Cultivation and Utilization of Medicinal and Aromatic Plants. Regional Res. Lab., Jammu-Tawi, India.

107 Atherton, J.G., and Rudich, J., 1986. The Tomato Crop. The Scientific Basis for Improvement. Chapman and Hall, London.

108 Atwal, A.S., 1976. Agricultural Pests of India and South-East Asia. Kalyani, New Delhi.

109 Aubert, B., 1972. Viticulture en région tropicale pour la production de raisin de table. Aspects et possibilités. Fruits 27, 513-537.

110 Auld, B.A., and Medd, R.W., 1987. Weeds, An Illustrated Guide to the Weeds of Australia. Inkata Press, Australia.

111 Austin, D.F., 1978. The *Ipomoea batatas* complex - I. Taxonomy. Bull. Torrey Bot. Club 105, 114-129.

112 Auzac, J.d', 1989. Physiology of Rubber Tree Latex. CRC Press, Boca Raton, Florida.

113 Awada, M., I-pai Wu, Suehisa, R.H., and Padgett, M.M., 1979. Effects of Drip Irrigation and Nitrogen Fertilization on Vegetative Growth, Fruit Yield, and Mineral Composition of the Petioles and Fruits of Papaya. Techn. Bull. 103, Hawaii Agric. Exp. Sta., Univ. of Hawaii, Honolulu.

114 Awasthi, Y.C., Bhatnagar, S.C., and Mitra, C.R., 1975. Chemurgy of Sapotaceous Plants: *Madhuca* species of India. Econ. Bot. 29, 380-398.

115 Aya, F.O., 1978. A preliminary assessment of the influence of age and the time of transplanting on the performance of polybag seedlings in the field. J. Nigerian Inst. Oil Palm Res. 5 (20), 7-14.

116 Ayanaba, A., and Dart, P.J. (eds.), 1977. Biological Nitrogen Fixation in Farming Systems of the Tropics. Wiley, New York.

117 Ayensu, E.S., 1972. Morphology and anatomy of *Synsepalum dulcificum* (Sapotaceae). Bot. J. Linn. Soc. 65, 179-187.

118 Ayensu, E.S., Doggett, H., Keynes, R.D., Marton-lefevre, J., Musselman, L.J., Parker, C., and Pickering, A. (eds.), 1984. *Striga*: Biology and Control. Intern. Council of Scientific Unions, Paris.

119 Azam, B., Lafitte, F., Obry, F., et Paulet, J.L., 1981. Le feijoa en Nouvelle-Zélande. Fruits 36, 361-384.

120 Azariah, M.D., and Rai, R.P., 1960. Breaking the dormancy of seed potatoes in the Nilgiris. Ind. Pot. J. 2, 100-101.

121 Azoulay, E., 1972. Le marché des fruits tropicaux en France et en Europe connaît une expansion considérable. Marchés Trop. Médit. No. 1402.

122 Baagøe, J., 1974. The genus *Guizotia* (Compositae). A Taxonomic Revision. Bot. Tidsskrift (Kopenhagen) 69, 1-39.

123 Badillo, V.M., 1971. Monografia de la familia Caricaceae. Asoc. de Profesores, Maracay, Venezuela.

124 Baggio, A., and Heuveldop, J., 1984. Initial performance of *Calliandra calothyrsus* Meissn. in live fences for the production of biomass. Agrofor. Systems 2, 19-29.

125 Bahre, C.J., and Bradbury, D.E., 1980. Manufacture of mescal in Sonora, Mexico. Econ. Bot. 34, 391-400.

126 Bailey, L.H., 1951. Manual of Cultivated Plants. Macmillan, New York.

127 Bajwa, M.S., Singh, P., and Singh, R., 1972. Ber cultivation in Punjab. Indian Hortic. 16 (4), 7-12.

128 Baker Castor Oil Company, 1968. Castor Oils and Chemical Derivatives. Techn. Bull. 123, Baker Castor Oil Co., Bayonne, N. J.

129 Baker, E.F., 1970. Kenaf and roselle in Western Nigeria. World Crops 22, 380-386.

130 Bakker, M.J., and Williams, W.M. (eds.), 1987. White Clover. C. A. B. International, Wallingford, Oxon, England.

131 Baldev, B., Ramanujam, S., and Jain, H.K. (eds.), 1988. Pulse Crops (Grain Legumes). Oxford and IBH Publ., New Delhi.

132 Balick, M.J., 1979. Amazonian oil palms of promise: a survey. Econ. Bot. 33, 11-28.

133 BALICK, M.J., 1979. Economic botany of the Guahibo. I. Palmae. Econ. Bot. 33, 361-376.

134 BALICK, M.J., 1981. Une huile comestible de haute qualité en provenance des éspèces *Jessenia* et *Oenocarpus*: un complexe de palmiers natifs de la Vallée de l'Amazone. Oléagineux 36, 319-326.

135 BALICK, M.J., 1986. Systematics and Economic Botany of the *Oenocarpus-Jessenia* (Palmae) Complex. Sci. Publ. Dep., New York Botanical Garden, Bronx, NY.

136 BALICK, M.J., and GERSHOFF, S.N., 1981. Nutritional evaluation of the *Jessenia bataua* palm: source of quality protein and oil from tropical America. Econ. Bot. 35, 261-271.

137 BALLESTREM, C.GRAF, and HOLLER, H.-J., 1977. Potato Production in Kenya. Experiences and Recommendations. Schriftenreihe der GTZ No. 50. TZ Verlagsges., Roßdorf.

138 BANDYOPADHYAY, S.B., and SANYAL, A.K., 1974. The wasted jute-stick has many industrial uses. Indian Farming 23 (12), 31-35.

139 BANERJEE, A., BAGCHI, D.K., and SI, L.K., 1984. Studies on the potential of winged bean as a multipurpose legume cover crop in tropical regions. Exp. Agric. 20, 297-301.

140 BANERJEE, B., 1955. Jute - especially as produced in West Bengal. Econ. Bot. 9, 151-174.

141 BARANOV, A., 1966. Recent advances in our knowledge of the morphology, cultivation and uses of ginseng (*Panax ginseng* C.A. Meyer). Econ. Bot. 20, 403-406.

142 BARLOW C., 1978. The Natural Rubber Industry. Its Development, Technology and Economy in Malaysia. Oxford Univ. Press, Kuala Lumpur.

143 BARNARD, C., 1952. The Duboisias of Australia. Econ. Bot. 6, 3-17.

144 BARNES, D.E., and CHANDAPILLAI, M.M., 1972. Common Malaysian Weeds and Their Control. Ansul (M) Sdn. Berhad, Kuala Lumpur.

145 BARRAU, J., 1962. Les Plantes Alimentaires de l'Océanie. Origins, Distribution et Usages. Ann. Musée Colon. Marseille.

146 BARRIOS, B.I. DE, 1977. Mil delicias de la quinoa. Editoria Quelco, Oruro, Bolivia.

147 BARRITT, B.H., BRINGHURST, R.S., and VOTH, V., 1982. Inheritance of early flowering in relation to breeding day-neutral strawberries. J. Amer. Soc. Hort. Sci. 107, 733-736.

148 BARUA, D.N., 1969. Seasonal dormancy in tea (*Camellia sinensis* L.). Nature 224, 514.

149 BASAK, S.L., JANA, M.K., and PARIA, K., 1974. Approaches to genetic improvement of jute. Indian J. Genet. 34A, 891-900.

150 BASKER, D., and NEGBI, N., 1983. Uses of saffron. Econ. Bot. 37, 228-236.

151 BATTAGLINI, M., 1969. Les méthodes traditionelles de propagation de l'olivier. Inf. Oléic. Intern. 48, 51-69, Madrid.

152 BAUDET, J.C., 1981. Les Céréales Mineures. Bibliographie Analytique. Agence de Coopération Culturelle et Technique, Paris.

153 BAUM, B.R., 1978. The Genus *Tamarix*. Israel Academy of Sciences and Humanities, Jerusalem.

154 BAVAPPA, K.V.A., 1980. Breeding and genetics of arecanut, *Areca catechu* L. - a review. J. Plantation Crops 8, 13-23.

155 BAVAPPA, K.V.A., NAIR, M.K., and KUMAR, T.P. (eds.), 1982. The Arecanut Palm (*Areca catechu* L.). Central Plantation Crops Res. Inst. Kasaragod, India.

156 BEAGLE, E.C., 1970. Rice hulls, the largest agricultural opportunity in the world. Amer. Ass. Cereal Chemists, 55th Ann. Meeting, 1-10.

157 BECK, N.G., and LORD, E.M., 1988. Breeding system in *Ficus carica*, the common fig. I. Floral diversity. Amer. J. Bot. 75, 1904-1912.

158 BECK, N.G., and LORD, E.M., 1988. Breeding system in *Ficus carica*, the common fig. II. Pollination events. Amer. J. Bot. 75, 1913-1922.

159 BEEN, B.O., 1981. Observations on field resistance to lethal yellowing in coconut varieties and hybrids in Jamaica. Oléagineux 36, 9-11.

160 BEER, J., 1987. Advantages, disadvantages and desirable characteristics of shade trees for coffee, cacao and tea. Agrofor. Systems 5, 3-13.

161 BEER, J.W., CLARKIN, K.L., SALAS, G. DE LAS, and GLOVER, N.L., 1979. A case study of traditional agro-forestry practices in a wet tropical zone: the "La Suiza" project. CATIE, Turrialba.

162 BEGLEY, B.W., 1981. Taro - the flood-irrigated root crop of the Pacific. World Crops 33 (2), 28-30.

163 BELLIS, E., 1971. The evolution of current manuring practices in *Hevea brasiliensis* plantations. Fertilité 38, 29-43.

164 BELOTTI, A., and SCHOONHOVEN, A. VAN, 1978. Cassava Pests and Their Control. CIAT, Ser. 09EC-2, Cali, Colombia.

165 BELTEKY, B., and KOVACS, I., 1984. Lupin: The New Break. Heffers, Cambridge, England.

166 BELTRANENA, R., BREMAN, J., and PRINE, G.M., 1981. Yield and quality of Florigraze rhizoma peanut (*Arachis glabrata* Benth.) as affected by cutting height and frequency. Soil and Crop Sci. Soc. Florida Proc. 40, 153-156.

167 BENERO, J.R., 1972. A mechanical method for extracting tamarind pulp. J. Agric. Univ. Puerto Rico 56, 185-186.

168 BENERO, J.R., COLLAZO DE RIVERA, A.L., and DE GEORGE, L.M.I., 1974. Studies on the preparation and shelf-life of soursop, tamarind, and blended soursop-tamarind soft drinks. J. Agric. Univ. Puerto Rico 58, 99-104.

169 BENGE, M.D., 1987. Multipurpose uses of contour hedgerows in highland regions. World Animal Review 64, 31-39.

170 BENJAMIN, C., 1977. A survey of food gardens in the Hoskins oil palm scheme. Papua New Guinea Agric. J. 28, 57-71.

171 BENJASIL, V., 1973. THS 30, a new variety of kenaf. Thai J. Agric. Sci. 6, 45-55.

172 BENTES, M.H., SERRUYA, H., ROCHA FILHO, G.N., GODOY, R.L.O., CABRAL, J.A.S., e MAIA, J.G.S., 1986. Estudo químico das sementes de bacuri. Acta Amazonica 16/17, 363-368.

173 BENZIONI, A., and DUNSTONE, R.L., 1986. Jojoba: adaptation to environmental stress and the implications for domestication. Quart. Rev. Biol. 61, 177-199.

174 BERENSCHOT, L.M., FILIUS, B.M., and HARDJOSOEDIRO, S., 1988. Factors determining the occurrence of the agroforestry systems with *Acacia mearnsii* in Central Java. Agrofor. Systems 6, 119-135.

175 BERGER, J., 1962. Maize Production and the Manuring of Maize. Centre d'Etude de l'Azote, Genf.

176 BERGER, J., 1969. The World's Major Fibre Crops. Their Cultivation and Manuring. Centre d'Etude de l'Azote, Zürich.

177 BERTHAUD, J., 1978. L'hybridation interspécifique entre *Coffea arabica* L. et *Coffea canephora* Pierre. Obtention et comparaison des hybrides triploides, Arabusta et hexaploides. Café Cacao Thé 22, 3-12, 87-112.

178 BERTIN, Y., 1976. La culture de la grenadille au Sri-Lanka (Ceylan). Fruits 31, 171-176.

179 BEZUNEH, T., and FELEKE, A., 1966. The Production and Utilization of the Genus *Ensete* in Ethiopia. Econ. Bot. 20, 65-70.

180 BHANTHUMNAVIN, K., and MCGARRY, M.G., 1971. *Wolffia arrhiza* as a possible source of inexpensive protein. Nature 232, 495.

181 BHAT, K.S., 1978. Agronomic research in arecanut - a review. J. Plantation Crops 6, 67-80.

182 BHATNAGAR, P.S., 1985. Soybean in India: problems and prospects. Indian J. Agric. Sci. 55, 709-722.

183 BIRD, J., and MARAMOROSCH, K. (eds.), 1975. Tropical Diseases of Legumes. Academic Press, New York.

184 BISCHOF, F., 1978. Common Weeds from Iran, Turkey, the Near East and North Africa. GTZ, Eschborn.
185 BLAAK, G., 1976. Pejibaye. Abstr. Trop. Agric. 2 (9), 9-17.
186 BLAAK, G., 1980. Vegetative propagation of pejibaye (*Bactris gasipaes* H.B.K.). Turrialba 30, 258-161.
187 BLACKBURN, F.H., 1984. Sugar-Cane. Longman, London.
188 BLANCHET, P., 1989. Description et comportement d'espèces d'*Actinidia* à fruits glabres dans le Sud-Ouest de la France. Fruits 44, 543-552.
189 BLENCOWE, J.W., 1971. Recent advances in the stimulation of rubber. World Crops 23, 126-132.
190 BLOMMAERT, K.L.J., 1972. Buchu - difficult to cultivate, but very valuable. Farming S. Afr. 48 (12), 10-11, 16.
191 BLOMMAERT, K.L.J., and BARTEL, E., 1976. Chemotaxonomic aspects of the buchu species *Agathosma betulina* (Pillans) and *Agathosma crenulata* (Pillans) from local plantings. J. South Afr. Bot. 42, 121-126.
192 BLOSS, H.E., and PFEIFFER, C.M., 1984. Latex content and biomass increase in mycorrhizal guayaule (*Parthenium argentatum*) under field conditions. An. appl. Biol 104, 175-183.
193 BOCK, K.R., 1982. Kenya: avoiding cassava mosaic disease. Span 25 (1), 11-13.
194 BOERMA, A.H., 1973. Forward planning: meeting the requirements. The world food and agricultural situation. Phil. Trans. R. Soc. Lond. B. 267, 5-12.
195 BOGDAN, A.V., 1977. Tropical Pasture and Fodder Plants (Grasses and Legumes). Longman, London.
196 BOHANNON, M.B., HAGEMANN, J.W., EARLE, F.R., and BARCLEY, A.S., 1974. Screening seed of *Trigonella* and three related genera for diosgenin. Phytochem. 13, 1513-1514.
197 BOHLEN, E., 1978. Crop Pests in Tanzania and Their Control. 2nd ed. Parey, Berlin.
198 BOHM, B.A., GANDERS, F.R., and PLOWMAN, T., 1982. Biosystematics and evolution of cultivated coca (Erythroxylaceae). Systematic Bot. 7, 121-133.
199 BOHS, L., 1989. Ethnobotany of the genus *Cyphomandra* (Solanaceae). Econ. Bot. 43, 143-163.
200 BOIS, D., 1927-37. Les Plantes Alimentaires. 4 Vols. Lechevalier, Paris.
201 BOKUCHAVA, M.A., and SKOBELEVA, N.I., 1969. The chemistry and biochemistry of tea and tea manufacture. Adv. Food Technol. 17, 215-292.
202 BOLLAND, M.D.A., and GLADSTONES, J.S., 1987. Serradella (*Ornithopus* spp.) as a pasture legume in Australia. J. Austr. Inst. Agric. Sci. 53, 5-10.
203 BOMPARD, J.M., 1986. Arboriculture fruitière en Indonesie occidentale: Traditions et perspectives. Fruits 41, 531-551.
204 BONAVIA, E., 1976. Cultivated Oranges and Lemons of India and Ceylon. 2 Vols. Today and Tomorrow's Book Agency, New Delhi.
205 BOND, D.A. (ed.), 1980. *Vicia faba*. Feeding Value, Processing and Viruses. Martinus Nijhoff, The Hague.
206 BOND, D.A., 1987. Recent developments in breeding field beans (*Vicia faba* L.). Plant Breeding 99, 1-26.
207 BORNEMISZA, E., and ALVAREDO, A. (eds.), 1975. Soil Management in Tropical America. Soil Sci. Dep., North Carolina State Univ., Raleigh, N. C.
208 BOS, J.J., and LENSING, F.H.G., 1973. A new cultivar in sisal from East Africa: *Agave sisalana* Perr. ex Engelm. cv. Hildana. East Afr. Agric. For. J. 39, 17-25.
209 BOSE, P.C., 1989. Sericulture in India, its progress and prospects. Indian Farming 38 (12), 28-31.
210 BOTHE, H., BRUIJN, F.J. DE, and NEWTON, W.E. (eds.), 1988. Nitrogen Fixation Hundred Years after. G. Fischer, Stuttgart.
211 BOUDET, G., 1975. Manuel sur les Paturages Tropicaux et les Cultures Fourragères 2. édition. Ministère de la Coopération, Paris.

212 BOULANGER, J., 1977. Classification des Malvales fibres jutières. Coton Fib. Trop. 32, 285-290.
213 BOULANGER, J., FOLLIN, J.C., et BOURELY, J., 1984. Les Hibiscus Textiles en Afrique Tropicale. 1ère Partie: Conditions Particulières de la Production du Kénaf et de la Roselle. Suppl. à Coton et Fibres Tropicales, Série Documents, Études et Synthèses No. 5.
214 BOURDEAUT, J., 1971. Le safoutier (*Pachylobus edulis*). Fruits 26, 663-666.
215 BOURDOUX, P., MAFUTA, M., HANSON, A., and ERMANS, A.M., 1980. Cassava toxicity: the role of linamarin. In: ERMANS, A. M., MBULAMOKO, N. M., DELANGE, F., and AHLUWALIA, R. (eds.): Role of Cassava in the Etiology of Endemic Goitre and Cretinism. Intern. Devel. Res. Centre, Ottawa, Canada.
216 BOURELY, J., 1987. Le cotonnier sans gossypol, une source de protéines alimentaires. Situation actuelle et perspectives d'avenir, après le colloque d'Abidjan. Cot. Fib. Trop. 42, 55-63.
217 BOURIQUET, G. (ed.), 1954. Le Vanillier et la Vanille dans le Monde. Lechevalier, Paris.
218 BOURKE, R.M., 1975. Evaluation of leguminous cover crops at Keravat, New Britain. Papua New Guinea Agric. J. 26, 1-9.
219 BOURKE, R.M., 1985. Sweet potato (*Ipomoea batatas*) production and research in Papua New Guinea. Papua New Guinea J. Agric., For., Fish. 33, 89-108.
220 BOUWERKAMP, J.C. (ed.), 1985. Sweet Potato Products: A Natural Resource for the Tropics. CRC Press, Boca Raton, Florida.
221 BOWEN, J.E., and KRATKY, B.A., 1980. Tropical weed control. World Farming 22 (3), 30-35.
222 BOWERS, W.S., OHTA, T., CLEERE, J.S., and MARSELLA, P.A., 1976. Discovery of insect anti-juvenile hormones in plants. Science 193, 542-547.
223 BRADBURY, J.H., and HOLLOWAY, W.D., 1988. Chemistry of Tropical Root Crops: Significance for Nutrition and Agriculture in the Pacific. Australian Centre Intern. Agric. Res., Monograph 6.
224 BRADDOCK, R.J., 1974. Citrus seeds: a potential food source. Span 17, 86-87.
225 BRADDOCK, R.J., and KESTERSON, J.W., 1973. Citrus Seed Oils. Agric. Exp. Sta. Bull. 756, Gainesville, Fla., USA.
226 BRANDLE, J.R., HINTZ, D.L., and STURROCK, J.W., (eds.), 1988. Windbreak Technology. Proc. Intern. Symp. Windbreak Technology, 23-27 June 1986, Lincoln, Nebraska. Elsevier Science Publ., Amsterdam.
227 BRANTON, R., and BLAKE, J., 1983. A lovely clone of coconuts. New Scientist 98, 554-557.
228 BRAUDEAU, J., 1969. Le Cacaoyer. Maisonneuve et Larose, Paris.
229 BRAY, R.A., HUTTON, E.M., and BEATIE, W.M., 1984. Breeding Leucaena for low-mimosine: field evaluation of selections. Trop. Grasslands 18, 194-198.
230 BREEMEN, N.VAN, HEYDENDAEL, A.J.F., LAMBERS, D.H.R., MOLSTER, H.C., OLDEMANN, L.R., PLANTINGA, W.J., and WIELEMAKER, 1970. Aspects of Rice Growing in Asia and the Americas. Veenman u. Zonen, Wageningen.
231 BREKKE, J.E., CHAN, H.T.Jr., and CAVALETTO, C.G., 1973. Papaya Purre and Nectar. Res. Bull. 170, Hawaii Agric. Exp. Sta., Honolulu.
232 BRESLIN, P., and JONES, A., 1973. The Structure of the Pepper Market in the Unite Kingdom, the Federal Republic of Germany, the Netherlands and France. G 8 Trop. Prod. Inst., London.
233 BRIGGS, D.E., 1978. Barley. Chapman and Hall, London.
234 BRINGI, N.V. (ed.), 1987. Non-Traditional Oilseeds and Oils of India. Oxford and IBH Publ., New Delhi.
235 BROADHEAD, D.M., and FREEMAN, K.C., 1980. Stalk and sugar yield of sweet sorghum as affected by spacing. Agron. J. 72, 523-525.
236 BROUGHTON, W.J., 1977. Effect of various covers on soil fertility under *Hevea brasiliensis* Muell. Arg. and on growth of the tree. Agro-Ecosystems 3, 147-170.

237 BROUGHTON, W.J. (ed.), 1981. Nitrogen Fixation. Clarendon Press, Oxford.

238 BROUGHTON, W.J., and TAN, G., 1979. Storage conditions and ripening of the custard apple *Annona squamosa* L. Scientia Hortic. 10, 73-82.

239 BROUK, B., 1975. Plants consumed by Man. Academic Press, London.

240 BROWN, C.H., 1955. Egyptian Cotton. Leonard Hill, London.

241 BROWN, W.H., 1951-58. Useful Plants of the Philippines. 3 Vols. Techn. Bull. 10, Rep. Phil. Agric. Comm., Philippines.

242 BRÜCHER, H., 1989. Useful Plants of Neotropical Origin and Their World Relatives. Springer, Berlin.

243 BRUHN, J.G., 1973. Ethnobotanical search for hallucinogenic cacti. Planta Medica 24, 315-319.

244 BRUIJN, G.H. DE, 1971. Étude du caractère cyanogénétique du manioc (*Manihot esculenta* Crantz). Meded. Landbouwhogeschool Wageningen, 71-13, 1-140.

245 BRUIJN, G.H. DE, and DHARMAPUTRA, T.S., 1974. The Mukibat system, a highyielding method of cassava production in Indonesia. Neth. J. Agric. Sci. 22, 89-100.

246 BRUNKEN, J, N, 1977. A systematic study of *Pennisetum* sect. *Pennisetum* (Gramineae). Amer. J. Bot. 64, 161-176.

247 BRUNKEN, J., WET, J.M.J. DE, and HARLAN, J.R., 1977. The morphology and domestication of pearl millet. Econ. Bot. 31, 163-174.

248 BUENO, C.R., e WEIGEL, P., 1983. Armazenamento de tubérculos frescos de ariá (*Calathea allouia* (Aubl.) Lindl.). Acta Amazonica 13, 7-14.

249 BUKENYA, Z.R., and HALL, J.B., 1987. Six cultivars of *Solanum macrocarpum* (Solanaceae) in Ghana. Bothalia 17, 91-95.

250 BULL, M.R., 1962. Oldest and newest crops in Tunisia. World Crops 14, 417-418.

251 BUNDESSTELLE FÜR ENTWICKLUNGSHILFE, 1974. Conference on Plant Protection in Tropical and Sub-Tropical Areas. Nov., 4-15, 1974, Manila, Philippines. BfE, Eschborn.

252 BUOL, S.W., HOLE, F.D., and MCCRACKEN, R.J., 1980. Soil Genesis and Classification. 2nd ed. Iowa Univ. Press, Ames.

253 BURDICK, E.M., 1971. Carpaine: An alkaloid of *Carica papaya*. Its chemistry and pharmacology. Econ. Bot. 25, 363-365.

254 BUREAU DE PROMOTION DE VARIETES FOURRAGERES, 1982. Latest Technical Information on *Bromus catharticus*. Bureau Prom. Var. Four., Paris.

255 BURGESS, A.H., 1964. Hops. Botany, Cultivation, and Utilization. Leonard Hill, London.

256 BURINGH, P., 1979. Introduction to the Study of Soils in Tropical and Subtropical Regions. 3rd ed. PUDOC Wageningen.

257 BURK, L.G., and HEGGESTAD, H.E., 1966. The genus *Nicotiana*: A source of resistance to diseases of cultivated tobacco. Econ. Bot. 20, 76-88.

258 BURKILL, H.M., 1985. The Useful Plants of West Tropical Africa. 2nd ed., Vol. 1, Families A-D. Univ. Virginia Press, Charlottesville, VA.

259 BURKILL, I.H., 1966. A Dictionary of the Economic Products of the Malay Peninsula. 2 Vols. Ministry Agric. Co-operatives, Kuala Lumpur.

260 BURTON, W.G., 1989. The Potato. 3rd ed. Longman, London.

261 BUSEY, P., BROSCHAT, T.K., and CENTER, B.J., 1982. Classification of St. Augustin grass. Crop Sci. 22, 469-473.

262 BUSSON, F., 1965. Plantes Alimentaires de l'Ouest Africain. Leconte, Marseille.

263 BUTANI, D.K., 1977. Insect pests of water nut in India and their control. Fruits 32, 569-571.

264 BUTLER, G.W., and BAILEY, R.W., (eds.), 1973. Chemistry and Biochemistry of Herbage. 3 Vols. Academic Press, London.

265 BUYCKX, E.I.E., 1962. Précis des Maladies et des Insects Nuisibles Rencontrés sur les Plantes Cultivées au Congo, au Rwanda et au Burundi. Publ. I.N.E.A.C., Bruxelles.

266 CAMERON, D.G., JONES, R.M., WILSON, G.P.M., BISHOP, H.G., COOK, B.G., LEE, G.R., and LOWE, K.F., 1989. Legumes for heavy grazing in costal subtropical Australia. Trop. Grasslands 23, 153-161.

267 CAMPAGNON, P., 1986. Le Caoutchouc Naturel: Biologie, Culture et Production. Maisonneuve et Larose, Paris.

268 CAMPELL, C.W., 1974. The Wampee, a fruit well adapted to southern Florida. Florida State Hort. Soc. 87, 390-393.

269 CANCEL, L.E., RIVERA-ORTIZ, J.M., and RUIZ DE MONTALVO, M.C., 1972. Separating and washing coffee harvested with plastic nets. J. Agric. Univ. Puerto Rico 56, 11-17.

270 CAPOT, J., et AKE ASSI, L., 1975. Un nouveau caféier hybride de Côte d'Ivoire. Café Cacao Thé 19, 3-4.

271 CARDENAS, J., REYES, C.E., and DOLL, J.D., 1986. Tropical Weeds, Malezas Tropicales. COMALFI, Bogotá, Colombia.

272 CARDENAS, M., 1969. Manual de Plantas Económicas de Bolivia. Imprenta Icthus, Cochabamba.

273 CARLOWITZ, P.G.V., 1986. Multipurpose tree yield data - their relevance to agroforestry research and development and the current state of knowledge. Agrofor. Systems 4, 291-314.

274 CARO-COSTAS, R., VICENTE-CHANDLER, J., and ABRUÑA, F., 1976. Comparison of heavily fertilized Congo, Star and Pangola grass pastures in the humid mountain region of Puerto Rico. J. Agric. Univ. Puerto Rico 60, 179-185.

275 CARR, A.R., 1974. *Duboisia* growing. Queensl. Agric. J. 100, 495-505.

276 CARTER, J.F. (ed.), 1978. Sunflower Science and Technology. Series Agronomy No. 19. Amer. Soc. Agr., Madison, Wisc.

277 CARVAJAL, J.F., 1984. Cafeto - Cultivo y Fertilización. Ed. 2a. Inst. Intern. Potasa, Bern.

278 CARVALHO, M.M. DE, ANDREW, C.S., EDWARDS, D.G., and ASHER, C.J., 1980. Comparative performance of six *Stylosanthes* species in three acid soils. Aust. J. Agric. Res. 31, 61-76.

279 CASIDA, J.E. (ed.), 1973. Pyrethrum: the Natural Insecticide. Academic Press, New York.

280 CASIMIR, D.J., KEFFORD, J.F., and WHITFIELD, F.B., 1981. Technology and flavor chemistry of passion fruit juices and concentrates. Adv. Food Res. 27, 243-295.

281 CAVALCANTE, P.B., 1988. Frutas Comestíveis da Amazônia. 2 Vols. 4th ed. Publicações Avulsas do Museu Goeldi, Belém.

282 CAYGILL, J.C., COOKE, R.D., MOORE, D.J., READ, S.J., and PASSAM, H.C., 1976. The Mango (*Mangifera indica* L.). Harvesting and Subsequent Handling and Processing: an Annotated Bibliography. G 107, Tropical Products Inst., London.

283 CENTRE D'ETUDE DE L'AZOTE, 1960. Progressive Wheat Production. Centre d'Etude de l'Azote, Genf.

284 CENTRE FOR OVERSEAS PEST RESEARCH, 1978. Pest Control in Tropical Root Crops. PANS Manual No. 4. Ministry Overseas Development, London.

285 CENTRE FOR OVERSEAS PEST RESEARCH, 1981. Pest Control in Tropical Grain Legumes. Centre Overseas Pest Res., London.

286 CENTRE FOR OVERSEAS PEST RESEARCH, 1983. Pest Control in Tropical Tomatoes. Centre Overseas Pest Res., London.

287 CENTRO AGRONOMICO TROPICAL DE INVESTIGACION Y ESEÑANZA, 1979. Die genetischen Ressourcen der Kulturpflanzen Zentralamerikas. Genbank CATIE/GTZ, Turrialba, Costa Rica.

288 CENTRO DE PESQUISAS DO CACAU, 1981. Fermentação de Cacau. Levantamento Bibliográfico. CEPEC-CEPLAC, Ilheus, Bahia, Brazil.

289 CENTRO INTERNACIONAL DE AGRICULTURA TROPICAL, 1979. Resúmenes Analíticos sobre Pastos Tropicales. CIAT, Cali, Colombia.

290 CENTRO INTERNACIONAL DE AGRICULTURA TROPICAL, 1987. Cassava Breeding. A Multidisciplinary Review. CIAT, Cali, Colombia.
291 CENTRO INTERNACIONAL DE AGRICULTURA TROPICAL, 1988. Fresh cassava conservation goes commercial. CIAT International 7 (1), 9-11.
292 CENTRO INTERNACIONAL DE LA PAPA, 1977. The Potato; Major Potato Diseases and Nematodes. CIP, Lima, Peru.
293 CENTRO INTERNACIONAL DE LA PAPA, 1980. An Interim Approach to Bacterial Wilt Control. CIP Circular 8 (12), 1-3.
294 CENTRO INTERNACIONAL DE LA PAPA, 1984. Potatoes for the Developing World. CIP, Lima, Peru.
295 CENTRO INTERNACIONAL DE MEJORAMIENTO DE MAIZ Y TRIGO, 1985. Wheats for More Tropical Environments. Proc. Intern. Symp. CIMMYT, México.
296 CHADHA, K.L., and RANDHAWA, G. S., 1974. Grape Varieties in India - Description and Classification. Indian Council Agric. Res., New Delhi.
297 CHAMPION, J., 1988. Les Bananiers et leur Culture. Tome I. Botanique et Génétique. Institut de Recherches sur les Fruits et Agrumes, Paris.
298 CHAN, H.T. (ed.), 1983. Handbook of Tropical Foods. Marcel Dekker, New York.
299 CHANDEL, K.P.S., 1980. Buckwheat, a neglected crop of hills. Indian Farming 30 (4), 13-14.
300 CHANDLER, R.F.Jr., 1979. Rice in the Tropics: A Guide to the Development of National Programs. Westview Press, Boulder, Colorado.
301 CHANDRA, S. (ED.), 1984. Edible Aroids. Oxford University Press, Oxford.
302 CHANDRASEKERA, L.B., 1980. Ground covers in hevea plantations in Sri Lanka. Rubber Res. Inst. Sri Lanka Bull. 15, 20-23.
303 CHANDRASEKERA, L.B., 1980. Tapping experiments. Rubber Res. Inst. Sri Lanka Bull. 15, 9-19.
304 CHANG, S.T., and HAYES, W.A., 1978. The Biology and Cultivation of Edible Mushrooms. Academic Press, New York.
305 CHANG, T.T., ADAIR, C.R., and JOHNSTON, T.H., 1982. The conservation and use of rice genetic resoures. Adv. Agron. 35, 37-91.
306 CHARLOT, G., GERMAIN, E., et PRUNET, J.P., 1988. Le noyer: nouvelles techniques. Centre Technique Interprofessionel des Fruits et Légumes, Paris.
307 CHARNEY, W., and HERZOG, H.L., 1967. Microbial Transformations of Steroids. Academic Press, New York.
308 CHARPENTIER, J.M., 1976. La culture bananière aux îles Canaries. Fruits 31, 569-585.
309 CHATFIELD, C., 1953. Food Composition Tables for International Use. 2nd ed. FAO, Rome.
310 CHATFIELD, C., 1954. Food Composition Tables for International Use - Minerals and Vitamins. FAO, Rome.
311 CHAUDHARY, S.A., and ZAWAWI, M.A., 1983. A Manual of Weeds of Central and Eastern Saudi Arabia. Ministry of Agriculture and Water, Riyadh, Saudi Arabia.
312 CHAUHAN, D.V.S., 1968. Vegetable Production in India. 2nd ed. Ram Prasad and Sons, Agra-3 (India).
313 CHAUHAN, J.S., VERGARA, B.S., and LOPEZ, F.S.S., 1985. Rice Ratooning. IRRI Res. Paper Ser. 102. IRRI, Manila.
314 CHAUHAN, Y.S., VENKATARATNAM, N., and SHELDRAKE, A.R., 1987. A perennial cropping system from pigeon pea grown in post-rainy season. Indian J. Agric. Sci. 57, 895-899.
315 CHAURASIA, B.D., SIROHI, S.S., and CHOHAN, J.S., 1972. Effect of harvesting period on the growth and yield of chicory (*Cichorium intybus* L.). Indian J. Agric. Sci. 42, 1132-1134.
316 CHAUX, C., 1972. Productions Légumières. Baillière, Paris.
317 CHAVAN, V.M., 1961. Niger and Safflower. Indian Central Oilseeds Committee, Hyderabad.

318 CHEE, K.H., 1980. Management of South American leaf blight. Planter (Kuala Lumpur) 56, 314-325.
319 CHENEY, R.H., and SCHOLTZ, E., 1963. Rooibos tea, a South African contribution to world beverages. Econ. Bot. 17, 186-194.
320 CHENON, R.D. DE, 1978. Protection des pépinières de palmiers à huile *Guineensis* contre le "Blast" en Afrique de l'Ouest. Oléagineux 33, 13-14.
321 CHEW, W.Y., CHEN, H.J., ABDUL RAZAK, H.A.H., ARIFFIN, B., and MALEK, M.A., 1980. Effects of close planting on soyabeans (*Glycine max*) cultivars in Malaysia. Exp. Agric. 16, 175-178.
322 CHHATTRAPATI, A.C., 1981. Tendances dans la production des graines oléagineuses en Inde. Oléagineux 36, 509-514.
323 CHI CHU WANG, 1968. Weeds Found on Cultivated Land in Taiwan. Coll. Agric. Nat. Taiwan Univ., Taipei.
324 CHILD, R., 1974. Coconuts. 2nd ed. Longman, London.
325 CHIN, C.W., and TANG, T.L., 1979. The oil palm - fertile *Pisifera*. Planter (Malaysia) 55 (635), 64-77.
326 CHITTENDEN, A.E., and PADDON, A.R., 1974. Cashew nut shell liquid as an adhesive in particle board manufacture. Trop. Sci. 15, 329-352.
327 CHOPRA, R.N., NAYAR, S.L., and CHJOPRA, I.C., 1956. Glossary of Indian Medicinal Plants. Council Sci. Ind. Res., New Delhi.
328 CHOUTEAU, J., and FAUCONNIER, D., 1989. Tobacco. Intern. Potash Inst., Bern, Switzerland.
329 CHRISTIAN, K.R., 1977. Effects of the environment on the growth of alfalfa. Adv. Agron. 29, 183-228.
330 CHUA, S.E., and HO, S.Y., 1973. Fruiting on sterile agar and cultivation of straw mushrooms on padi straw, banana leaves and sawdust. World Crops 25, 90-91.
331 CHUPP, CH., and SHERF, A.F., 1960. Vegetable Diseases and Their Control. Ronald Press, New York.
332 CIBA-GEIGY, 1975. Citrus. Techn. Monogr. 4. Ciba-Geigy Agrochemicals, Basel.
333 CLARKE, H.D., SEIGLER, D.S., and EBINGER, J.E., 1989. *Acacia farnesiana* (Fabaceae, Mimosoideae) and related species from Mexico, the southwestern U.S., and the Caribbean. System. Bot. 14, 549-564.
334 CLARKE, P.A., 1978. Rice Processing: a Check List of Commercially Available Machinery. G 114, Trop. Prod. Inst., London.
335 CLARKE, R.J., and MACRAE, R. (eds.), 1987/88. Coffee. 5 Vols. Elsevier Science Publ., Amsterdam.
336 CLEMENT, C.R., 1989. The potential use of the pejibaye palm in agroforestry systems. Agrofor. Systems 7, 201-212.
337 CLEMENT, C.R., MÜLLER, C.H., e FLORES, W.B.C., 1982. Recursos genéticos de especies frutiferas nativas da Amazônia Brasileira. Acta Amazonica 12, 677-695.
338 CLEMENTS, H.F., 1980. Sugarcane Crop Logging and Crop Control - Principles and Practices. Pitman, London.
339 CLIFFORD, M.N., and WILLSON, K.C. (eds.), 1985. Coffee: Botany, Biochemistry and Production of Beans and Beverage. Croom Helm, London.
340 CLYDESDALE, F.M., MAIN, J.H., and FRANCIS, F.J., 1979. Roselle (*Hibiscus sabdariffa* L.) anthocyanins as colorants for beverages and gelatin deserts. J. Food Protection 42, 204-207.
341 COBLEY, L.S., and STEELE , W.M., 1976. An Introduction to the Botany of Tropical Crops. 2nd ed. Longman, London.
342 COCHE, A.G., 1967. Fish culture in rice fields. A worldwide synthesis. Hydrobiologia 30, 1-44.
343 COCK, J.H., 1976. Characteristics of high yielding cassava varieties. Exp. Agric. 12, 135-143.
344 COCK, J.H., 1985. Cassava: New Potential for a Neglected Crop. Westview Press, Boulder, Colorado.

345 COCK, J.H., FRANKLIN, D., SANDOVAL, G., and JURI, P., 1979. Ideal cassava plant for maximum yield. Crop. Sci. 19, 271-279.
346 COCK, M.J.W., 1982. Potential biological control agents for *Mikania micrantha* H.B.K. from the neotropical region. Trop. Pest. Management 28, 242-254.
347 CODD, L.E., 1971. Generic limits in *Plectranthus*, *Coleus* and allied genera. Mitt. Bot. Staatssamml. München 10, 245-252.
348 COELHO, R.W., MOTT, G.O., OCUMPAUGH, W.R., and BROLMANN, J.B., 1981. Agronomic evaluation of some *Stylosanthes* species in North Florida, USA. Trop. Grasslands 15, 31-36.
349 COESTER, W.A., and OHLER, J.G., 1976. Cashew propagation by cuttings. Trop. Agric. (Trinidad) 53, 353-358.
350 COHEN, A., 1976. Citrus Fertilization. I. P. I. Bull. 4, Int. Potash Inst., Bern.
351 COKER, R.D., 1979. Aflatoxin: past, present and future. Trop. Sci. 21, 143-162.
352 COLLINS, J.L., 1960. The Pineapple - Botany, Cultivation and Utilization. Leonard Hill, London.
353 COMONT., G., and JACQUEMARD, J.C., 1977. Germination of oil palm seeds (*E. guineensis*) in polythene bags. Oléagineux 32, 151-153.
354 CONDIT, I.J., 1947. The Fig. Chronica Botanica, Waltham, Mass.
355 CONELLY, W.T., 1984. Copal and rattan collecting in the Philippines. Econ. Bot. 39, 39-46.
356 CONTICINI, L., 1986. Frutti esotici importati, coltivati o "coltivabili" in Italia. Riv. Agric. trop. e subtrop. 80, 43-69.
357 CONVERSE, R.H., 1981. The Israel strawberry industry. Hort Science 16, 19-22.
358 COOK, A.A., 1975. Diseases of Tropical and Subtropical Fruits and Nuts. Hafner Press, New York.
359 COOKE, R.D., RICKARD, J.E., and THOMPSON, A.K., 1988. The storage of tropical root and tuber crops - cassava, yam and edible aroids. Exp. Agric. 24, 457-470.
360 COOMARASANY, A., PERERA, P.P., and NADARAJAH, M., 1981. Preparation and use of cyclised rubber obtained from papain-coagulated natural rubber. J. Rub. Res. Inst. Sri Lanka 58, 46-57.
361 COONS, M.P., 1982. Relationships of *Amaranthus caudatus*. Econ. Bot. 36, 129-146.
362 COPPEN, J.J.W., 1979. Steroids: from plants to pills - the changing picture. Trop. Sci. 21, 125-141.
363 COPPER, W.C., and HENRY, W.H., 1974. Influence of rootstock on citrus fruit abscission response to cycloheximide treatment. Florida State Hort. Soc. Proc. 86, 52-55.
364 CORLEY, R.H.V., HARDON, J.J., and WOOD, B.J. (eds.), 1976. Oil Palm Research. Elsevier, Amsterdam.
365 CORLEY, R.H.V., WOOI, K.C., and WONG, C.Y, 1979. Progress with vegetative propagation of oil palm. Planter (Malaysia) 55 (641), 377-380.
366 CORNELIS, J., NUGTEREN, J.A., and WESTPHAL, E., 1985. Kangkong (*Ipomoea aquatica* Forsk.): an important leaf vegetable in Southeast Asia. Abstr. Trop. Agric. 10 (4), 9-21.
367 CORNELIUS, D.R., BHATT, B.N., and PATHAK, R.L., 1977. Windbreak plantation on sandy land in northern Gujarat. Indian Forester 103, 251-259.
368 CORNELIUS, J.A., 1980. Rice bran for edible purposes: a review. Trop. Sci. 22, 1-26.
369 CORNELIUS, J.A., and SIMMONS, E.A., 1969. *Crambe abyssinica* - a new commercial oilseed. Trop. Sci. 11, 17-22.
370 CORNER, E.J.H., 1966. The Natural History of Palms. Univ. Calif. Press, Berkeley.
371 CORNER, E.J.H., and WATANABE, K., 1969. Illustrated Guide to Tropical Plants. Hirokawa, Tokyo.
372 CORONEL, R.E., and ZUNO, J.C., 1980. Evaluation of fruit characters of some pili seedling trees in Calaunan and Los Baños, Laguna. Phil. Agric. 63, 166-173.
373 COSTA, J.A., OPLINGER, E.S., and PENDLETON, J.W., 1980. Response of soybean cultivars to planting patterns. Agron. J. 72, 153-156.

374 COSTANZA, S.H., WET, J.M.J. DE, and HARLAN, J.R., 1979. Literature review and numerical taxonomy of *Eragrostis tef* (T'ef). Econ. Bot. 33, 413-424.
375 COSTE, R., 1989. Caféiers et Cafés. Maisonneuve et Larose, Paris.
376 COUNCIL OF SCIENTIFIC AND INDUSTRIAL RESEARCH, 1948-1976. The Wealth of India. Raw Materials. 11 Vols. 2nd ed. 1985-. Publications Information Directorate, C.S.I.R., New Delhi.
377 COUNCIL OF SCIENTIFIC AND INDUSTRIAL RESEARCH, 1960. Coir, Its Extraction, Properties and Uses. C.S.I.R., New Delhi.
378 COURSEY, D.G., 1967. Yams. Longman, London.
379 COURSEY, D.G., and AIDOO, A., 1966. Ascorbic acid levels in Ghanaian yams. J. Sci. Food Agric. 17, 446-449.
380 COURSEY, D.G., and HAYNES, P.H., 1970. Root crops and their potential as food in the tropics. World Crops 22, 261-265.
381 COURTER, J.W., and RHODES, A.M., 1969. Historical notes on horseradish. Econ. Bot. 23, 156-164.
382 COX, P.A., 1980. Two Samoan technologies for breadfruit and banana preservation. Econ. Bot. 34, 181-185.
383 CRANE, E. (ed.), 1975. Honey. A Comprehensive Survey. Heinemann, London.
384 CRANE, E., 1978. Bibliography of Tropical Apiculture. Intern. Bee Res. Assoc., London.
385 CRANE, E. (ed.), 1978. Apiculture in Tropical Climates. Intern. Bee Res. Assoc., Bucks, England.
386 CRANE, E., WALKER, P., and DAY, R., 1984. Directory of Important World Honey Sources. Intern. Bee Research Association, London.
387 CRANE, J.C., and FORDE, H.I., 1976. Effects of four rootstocks on yield and quality of pistachio nuts. J. Amer. Soc. Hort. Sci. 101, 604-606.
388 CROSSA-RAYNAUD, P., et GERMAIN, E., 1982. Avenir de la culture des arbres fruitiers à fruits secs dans les pays méditerranéens: amandier, noyer, noisetier, pistachier. Fruits 37, 617-626.
389 CROSTA, G., e VECCHIO, V., 1979. Il fico d'India come fonte alimentare per il bestiame nelle zone aride. Riv. Agric. Subtrop. Trop. 73, 79-85.
390 CROWDER, L.V., and CHHEDA, H.R., 1982. Tropical Grassland Husbandry. Longman, London.
391 CURLEY, R.G., 1982. Long-term study reaffirms yield increase of narrow-row cotton. Calif. Agric. 36 (9/10), 8-10.
392 CURTIS, J.R., 1977. Prickly pear farming in the Santa Clara Valley, California. Econ. Bot. 31, 175-179.
393 CUTULI, G., DI MARTINO, E., LO GIUDICE, V., PENNISI, L., RACITI, G., RUSSO, F., SCUDERI, A., e SPINA, P., 1985. Trattado di Agrumicoltura. Edagricola, Bologna.
394 DAGUIN, F., et LETOUZE, R., 1988. Régénération du palmier dattier (*Phoenix dactylifera* L.) par embryogenèse somatique: amélioration de l'efficacité par passage en milieu liquide agité. Fruits 43, 191-194.
395 DAHLGREN, R., 1968. Revision of the genus *Aspalathus*. II. The species with ericoid and pinoid leaflets. 7. Sub-genus *Nortieria*. With remarks on rooibos-tea cultivation. Bot. Notiser 121, 165-208.
396 DALRYMPLE, D.G., 1976. Development and Spread of High-Yielding Varieties of Wheat and Rice in the Less Developed Nations. Foreign Devel. Di,v. - Econ. Res. Service, USDA, Washington, D. C.
397 DALZIEL, J.M., 1955. The Useful Plants of West Tropical Africa. 2nd ed. Crown Agents, London.
398 DAMME P.L. VAN, 1986. Comportement de deux aubergines africaines (*Solanum aethiopicum* et *Solanum macrocarpon*) dans la valleé du fleuve Sénégal. Agronomie Trop. 41, 218-230.
399 DANIEL, C., et OCHS, R., 1975. Amélioration de la production des jeunes palmiers à huile du Pérou par l'emploi d'engrais choré. Oléagineux 30, 295-298.

400 D'ARCY, W.G. (ed.), 1986. Solanaceae: Biology and Systematics. 2nd Int. Symp. Biol. Syst. Solanaceae, 3-6 Aug. 1983. Columbia University Press, New York, NY.
401 DARGRN, K.S., and CHILLAR, R.K., 1973. A technique for sugar beet and vegetables in saline alkali soils. Indian Farming 22 (12), 13-16.
402 DARLINGTON, C.D., 1973. Chromosome Botany and the Origins of Cultivated Plants. Allen and Unwin, London.
403 DARRAH, H.H., 1974. Investigation of the cultivars of the basils (*Ocimum*). Econ. Bot. 28, 63-67.
404 DAS GUPTA, D.K., 1971. Studies on the crop physiology and cultural practices of *Urena lobata* L. for fibre production in Sierra Leone. I. Crop physiology. Afr. Soils 16, 127-151.
405 DAS GUPTA, D.K., 1973. Studies on the crop physiology and cultural practices of *Urena lobata* L. for fibre production in Sierra Leone. II. Cultural practices. Afr. Soils 18, 33-46.
406 DASTUR, J.F., 1964. Useful Plants of India and Pakistan. Reprint 1985. Taraporevala, Bombay.
407 DAVIES, D.D., 1978. Hybrid cotton: specific problems and potentials. Adv. Agron. 30, 129-157.
408 DAVIES, W.N.L., 1970. The carob tree and its importance in the agricultural economy of Cyprus. Econ. Bot. 24, 460-470.
409 DAWIDAR, A.M., SALEH, A.A., and ELMOTEI, S.L., 1973. Steroid sapogenin constituents of fenugreek seed. Planta Medica 24, 367-370.
410 DAYANANDA, G.P., and RAO, K.B., 1980. Effect of soil nutrient status on yield of oil from *Bursera penicillata* Engl. Mysore J. Agric. Sci. 14, 505-507.
411 DEB ROY, A., KAUL, R.N., and GYANCHAND, 1973. Israeli babool, a promising tree for arid and semi-arid lands. Indian Farming 23 (8), 19-20.
412 DE-DATTA, S.K., 1981. Principles and Practices of Rice Production. Wiley, New York.
413 DEGRAS, L., 1986. L'Igname, Plante à Tubercule Tropicale. Maissoneuve et Larose, Paris.
414 DEINUM, B., and DIRVEN, J.G.P., 1972. Climate, nitrogen and grass. 5. Influence of age, light intensity and temperature on the production and chemical composition of Congo grass (*Brachiaria ruziziensis* Germain et Everard). Neth. J. Agric. Sci. 20, 125-132.
415 DEINUM, B., and DIRVEN, J.G.P., 1973. Preliminary investigations on the digestibility of some tropical grasses under different temperature regimes. Surinaamse Landbouw 21, 121-126.
416 DELABARRE, Y., 1989. Synthèse bibliographique sur le ramboutan ou litchi chevelu (*Nephelium lappaceum* L.). Fruits 44, 33-44, 91-98.
417 DELATTRE, R., 1973. Parasites et Maladies en Culture Cotonnière. Div. Docum. l'I. R. C. T., Paris.
418 DELFS-FRITZ, W., 1970. Citrus Cultivation and Fertilization. Ruhr-Stickstoff, Bochum.
419 DEMARNE, F., and WALT, J.J.A. VAN DER, 1989. Origin of the rose-scented *Pelargonium* cultivar grown on Réunion Island. S. Afr. J. Bot. 55, 184-191.
420 DEMPSEY, J.M., 1975. Fiber Crops. University Florida Presses, Gainesville.
421 DENNIS, F.G.Jr., HERNER, R.C., and CAMACHO, S., 1985. Naranjilla: A potential cash crop for the small farmer in Latin America. Acta Hortic. 158, 475-481.
422 DEODIKAR, G.B., PATIL, V.P., and RAO, V.S.P., 1979. Breeding of Indian *durum* and *dicoccum* wheats by interspecific hybridization with 4n *Triticum* species. Indian J. Genetics 39, 114-125.
423 DEPARTAMENTO DE DIVULGAÇAO ESTATISTICA, 1975. Anuario Estatístico do Brasil 1974, Rio de Janeiro.
424 DEUSE, J., et LAVABRE, E.M., 1979. Le Désherbage des Cultures sous les Tropiques. Maisonneuve et Larose, Paris.

425 DEVAUX, A., et HAVERKORT, A.J., 1986. Manuel de la Culture de la Pomme de Terre en Afrique Centrale. 2me éd. Programme Régional d'Amélioration de la Culture de la Pomme de Terre en Afrique Centrale. Ruhengeri, Rwanda.
426 DEWEY, D.R., 1962. Breeding crested wheatgrass for salt tolerance. Crop Sci. 2, 403-407.
427 DHAGAT, N.K., GOSWAMI, U., and NARSINGHANI, V.G., 1978. Genetic variability, character correlations and path analysis in barnyard-millet. Indian J. Agric. Sci. 48, 211-214.
428 DHERI, M., et GILLIER, P., 1971. Un nouveau pas dans la lutte contre la rosette de l'arachide. Résultats obtenus en Haute-Volta avec les nouveaux hybrides. Oléagineux 26, 243-251.
429 DIEHL, L., 1982. Smallholder Farming Systems with Yam in the Southern Guinea Savannah of Nigeria. GTZ, Eschborn.
430 DINESH, C., 1987. Coir: an industry in search of viability. Ceres No. 118 (Vol. 20/4), 32-34.
431 DIRVEN, J.G.P., 1973. Tropical roughage. Thai J. Agric. Sci. 6, 323-334.
432 DÖBEREINER, J., BURRIS, R.H., and HOLLAENDER, A. (eds.), 1978. Limitations and Potentials for Biological Nitrogen Fixation in the Tropics. Plenum Press, New York.
433 DOBKIN DE RIOS, M., 1970. *Banisteriopsis* in witchcraft and healing activities in Iquitos, Peru. Econ. Bot. 24, 296-300.
434 DOGGETT, H., 1988. Sorghum. 2nd ed. Longman, London.
435 DOKU, E.V., and KARIKARI, S.K., 1971. Bambarra groundnut. Econ. Bot. 25, 255-262.
436 DOLLET, M., et GIANNOTTI, J., 1976. Maladie de Kaincopé: présence de mycroplasmes dans le phloème des cocotiers malades. Oléagineux 31, 169-171.
437 DOVAL, S.L., MATHUR, R.S., and KACHHWAH, R.K., 1984. Cultivation of cumin in Rajasthan. Indian Farming 34 (1), 11-13.
438 DOWNES, R.W., and TONNET, M.L., 1985. Effect of environmental conditions on growth and rubber production of guayule (*Parthenium argentatum*). Aust. J. Agric. Res. 36, 285-294.
439 DOWSON, V.H.W., 1976. Bibliography of the Date Palm. Field Research Projects. Coconut Grove, Miami, Florida.
440 DOWSON, V.H.W., and ATEN, A., 1962. Dates: Handling, Processing and Packing. FAO Agric. Devel. Pap. 72, FAO, Rome.
441 DUKE, J., 1986. Dealers in Botanicals. ASP Newsletter 22 (1), 1-29.
442 DUKE, J.A., 1970. Ethnobotanical observations on the Chocó Indians. Econ. Bot. 24, 344-366.
443 DUKE, J.A., 1973. Utilization of papaver. Econ. Bot. 27, 390-400.
444 DUKE, J.A., 1981. Handbook of Legumes of World Economic Importance. Plenum Press, New York.
445 DUKE, J.A., 1985. Medicinal Herbs. CRC Press, Boca Raton, Florida.
446 DUKE, J.A., 1989. Handbook of Nuts. CRC Press, Boca Raton, Florida.
447 DUKE, J.A., and TERRELL, E.E., 1974. Crop diversification matrix introduction. Taxon 23, 759-799.
448 DUNCAN, R.R., and GARDNER, W.A., 1984. The influence of ratoon cropping on sweet sorghum yield, sugar production, and insect damage. Can. J. Plant Sci. 64, 261-273.
449 DUPAIGNE, P., 1973. Édulcorants synthétiques, édulcorants naturals. Perspectives d'avenir. Fruits 28, 51-64.
450 DUPAIGNE, P., 1974. Consommation des boissons de fruits dans le monde. Fruits 29, 619-624.
451 DUPAIGNE, P., 1974. Les colorants rouges d'origine naturelle. Fruits 29, 797-814.
452 DUPAIGNE, P., 1975. Effets biochimiques des bromélines. Leur utilisation en thérapeutique. Fruits 30, 545-567.

453 DUPAIGNE, P., 1977. Nouvelle mise au point sur la question des édulcorants. Fruits 32, 117-136.
454 DUPAIGNE, P., 1977. Nos connaissances actuelles sur les effets biochimiques des papaines. Fruits 32, 677-696.
455 DUTTA, P.K., CHOPRA, I.C., and KAPOOR, L.D., 1963. Cultivation of *Rauvolfia serpentina* in India. Econ. Bot. 17, 243-251.
456 DUVAL, Y., GASSELIN, T.D., KONAN, K., et PANNETIER, C., 1988. Multiplication végétative du palmier à huile (*Elaeis guineensis* Jacq.) par culture in vitro. Stratégie et résultats. Oléagineux 43, 39-44.
457 DUVERNEUIL, G., et HAENDLER, L., 1973. Évolution des méthodes de traitement des noix de cajou. Fruits 28, 561-581.
458 DUVERNEUIL, G., et HAENDLER, L., 1973. Organisation des usines de décorticage mécanique de noix de cajou. Fruits 28, 711-736.
459 EATON, F.M., 1955. Physiology of the cotton plant. Ann. Rev. Plant Physiol. 6, 299-328.
460 EDEN, T., 1976. Tea. 3rd ed. Longman, London.
461 EDMOND, J.B., and AMMERMAN, G.R., 1971. Sweet Potatoes: Production, Processing, Marketing. Avi Publ. Co., Westport, Conn.
462 EDMONDS, M.J., 1966. The Markets for Coir Fibre and Coir Yarn. G 18, Trop. Prod. Inst., London.
463 EDWARDS, D., 1974. The Industrial Manufacture of Cassava Products: An Economic Study. G 88, Trop. Prod. Inst., London.
464 EFFERSON, J.N., 1984. Growing potential of hybrid rice. World Farming and Agrimanagement 26 (3), 6, 22, 24.
465 EIJNATTEN, C.L.M. VAN, 1969. Kola: its Botany and Cultivation. Koninkl. Inst. Tropen, Amsterdam.
466 EIJNATTEN, C.L.M. VAN, 1973. Kola, a review of the literature. Trop. Abstr. 28 (8), 1-10.
467 EIJNATTEN, C.L.M. VAN, and ABUBAKER, A.S., 1983. New cultivation techniques for cashew (*Anacardium occidentale* L.). Neth. J. Agric. Sci. 31, 13-26.
468 EIMERN, J. VAN, KARSCHON, R., RAZUMOVA, L.A., and ROBERTSON, G.W., 1964. Windbreaks and Shelterbelts. Techn. Notes 59, World Meteorol. Org. Genf.
469 EL-BARADI, T.A., 1971. Onion growing in the tropics. Trop. Abstr. 26, 285-291.
470 EL-BARADI, T.A., 1972. Sesame. Trop. Abstr. 27, 153-160.
471 EL-BARADI, T.A., 1973. Sunflower. Trop. Abstr. 28, 309-316.
472 EL-BARADI, T.A., 1975. Guava. Abstr. Trop. Agric. 1 (3), 9-16.
473 ELLA, A., JACOBSEN, C., STÜR, W.W., and BLAIR, G., 1989. Effect of plant density and cutting frequency on the productivity of four tree legumes. Trop. Grasslands 23, 28-34.
474 ELLIOT, F.C., HOOVER, M., and PORTER, W.K.Jr., (eds.), 1968. Advances in Production and Utilization of Quality Cotton: Principles and Practices. Iowa State Univ. Press, Ames.
475 EL SHEIKH, M.O.A., EL HASSAN, G.M., HAFEEZ, A.-R. EL T., ABDOLLA, A.A., and ANTOUN, M.D., 1982. Studies on Sudanese medicinal plants III: Indigenous *Hyoscyamus muticus* as possible commercial source of hyoscyamin. Planta medica 45, 116-119.
476 EMBODEN, W.A., 1974. *Cannabis* - a polytypic genus. Econ. Bot. 28, 304-310.
477 EMPRESA BRASILEIRA DE PESQUISA AGROPECUARIA, 1975. Sistemas de Produção para Malva. EMBRAPA, Capanema, PA. Brazil.
478 ENDRES, H., HOWES, F.N., and REGEL, C. V., 1962. Gerbstoffe. Tanning Materials. REGEL, C. V. (ed.), 1962-1968. Wiesner, Rohstoffe des Pflanzenreiches. 5th ed. Cramer, Lehre.
479 ENTWISTLE, P.F., 1973. Pests of Cocoa. Longman Group, Harlow.
480 EPENHUIJSEN, C.W. VAN, 1974. Growing Native Vegetables in Nigeria. FAO, Rome.

481 EPENHUIJSEN, C.W. VAN, 1976. Deciduous Fruits in Tanzania. Ministry of Foreign Affairs, International Technical Assistance Department, The Hague.

482 EPSTEIN, E., and NORLYN, J.D., 1977. Seawater-based crop production: a feasibility study. Science 197, 249-251.

483 EPSTEIN, E., NORLYN, J.D., RUSH, D.W., KINGSBURY, R.W., KELLEY, D.B., CUNNINGHAM, G.A., and WRONA, A.F., 1980. Saline culture of crops: a genetic approach. Science 210, 399-404.

484 ERDMANN, M.D., and ERDMANN, B.A., 1984. Arrowroot (*Maranta arundinacea*), food, feed, fuel, and fibre resource. Econ. Bot. 38, 332-241.

485 ERDMANN, M.D., PHATAK, S.C., and HALL, H.S., 1985. Potential for production of arrowroot in the southern United States. J. Amer. Soc. Hort. Sci. 110, 403-406.

486 ERICKSON, H.T., CORREA, M.P.F., and ESCOBAR, J.R., 1984. Guaraná (*Paullinia cupana*) as a commercial crop in Brazilian Amazonia. Econ. Bot. 38, 273-286.

487 ERWIN, D.C., TSAI, S., and KHAN, R.A., 1979. Growth retardants mitigate *Verticillium* wilt and increase yield of cotton. Calif. Agric. 33 (4), 8-10.

488 ESBENSHADE, H.W., and WILSON, G., 1986. Growing carobs in Australia. Goddard and Dobson, Victoria, Australia.

489 ESEN, A., and SOOST, R.K., 1977. Adventive embryogenesis in citrus and its relation to pollination and fertilization. Amer. J. Bot. 64, 607-614.

490 ESPINOSA, A.J., BORROCAL, A.R., JARA, M., ZORILLA, G.C., ZANABRIA, P.C., et MEDINA, T.J., 1973. Quelques propriétés et essais prélimimares de conservation des fruits et du jus de figue de Barbarie (*Opuntia ficus-indica*). Fruits 28, 285-289.

491 ESPINOSA, C., 1972. Ensayo de un sistema rotativo en suelos de sabana: *Canavalia ensiformis* para abono verde, maíz fertilizado y maní. Agron. Trop. Venezuela 22, 133-148.

492 ESSIG, F.B., and YUN-FA DONG, 1987. The many uses of *Trachycarpus fortunei* (Aracaceae) in China. Econ. Bot. 41, 411-417.

493 ESTILAI, A., NAQVI, H., and WAINES, J.G., 1988. Developing guayule as a domestic rubber crop. Calif. Agric. 42 (5), 29-30.

494 EVANS, D.O., and ROTARS, P.P., 1987. *Sesbania* in Agriculture. Westview Press, Boulder, CO., USA.

495 EVANS, L.T., (ed.), 1975. Crop Physiology. Cambridge Univ. Press, London.

496 EYO, E.S., und ABEL, H., 1979. Untersuchungen zur Fettsäurezusammensetzung der Samen von *Irvingia gabonensis*, *Cucumeropsis mannii* und *Mucuna sloanei* aus Nigeria. Tropenlandwirt 80, 7-13.

497 EZUEH, M.I., 1984. African yam bean as a crop in Nigeria. World Crops 36, 199-200.

498 FABIANI, G., and LINTAS, C. (ed.), 1988. Durum wheat. Chemistry and Technology. Amer. Assoc. Cereal Chemists, St. Paul, Minnesota.

499 FAIRBAIRN, J.W., and WILLIAMSON, E.M., 1978. Anatomical studies on *Papaver bracteatum* (Lindley). Planta Medica 33, 34-45.

500 FALCAO, M. DE A., LLERAS, E., KERR, W.E., e CARREIRA, L.M.M., 1981. Aspectos fenológicos, ecológicos e de produtividade do biribá (*Rollinia mucosa* (Jacq.) Baill.). Acta Amazonica 11, 297-306.

501 FALCAO, M. DE A., LLERAS, E., e LEITE, A.M.C., 1982. Aspectos fenológicos, ecológicos e de produtividade da graviola (*Annona muricata* L.) na região de Manaus. Acta Amazonica 12, 27-32.

502 FAO, 1960. Soil Erosion by Wind and Measures for Its Control on Agricultural Land. FAO Agric. Devel. Paper 71, FAO, Rome.

503 FAO, 1962-. Sessions of the Intergovernmental Group on Jute, Kenaf and Allied Species. FAO, Rome.

504 FAO, 1969. The World Wine and Wine Products Economy. A Study of Trends and Problems. FAO Commodity Bull. Ser. 43, FAO, Rome.

505 FAO, 1970. Upland Rice. Better Farming Series 20, FAO, Rome.

506 FAO, 1970. Wet Paddy or Swamp Rice. Better Farming Series 21, FAO, Rome.

507 FAO, 1970. Amino-Acid Content of Foods and Biological Data on Proteins. FAO Nutritional Studies 24, FAO, Rome.
508 FAO, 1971. Approaches to International Action on World Trade in Oilseeds, Oils and Fats. FAO Commodity Policy Studies, FAO, Rome.
509 FAO, 1972. Processed Tropical Fruit. Trends and Outlook for Production and Trade of Canned Pineapple and Processed Tropical Exotic Fruit. FAO Commodity Bull. Ser. 51, FAO, Rome.
510 FAO, 1973. Guide for Planning Pulp and Paper Enterprises. FAO Forestry and Forest Prod. Studies 18, FAO, Rome.
511 FAO, 1974. Tree Planting Practices in African Savannas. FAO Forestry Devel. Paper 19, FAO, Rome.
512 FAO, 1979. Eucalypts for Planting. FAO Forestry Series No. 11. FAO, Rome.
513 FAO, 1980. Dietary Fats and Oils in Human Nutrition. FAO Food and Nutr. Ser. 20. FAO, Rome.
514 FAO, 1980-89. Production Yearbook, Vols 33-42. FAO, Rome.
515 FAO, 1982. Statistics on Sisal and Henequen. Intergov. Group on Hard Fibres, 17th Session. FAO, Rome.
516 FAO, 1982. Date Production and Protection. FAO Plant Prod. Prot. Paper 35. FAO, Rome.
517 FAO, 1982-90. Commodity Review and Outlook. FAO, Rome.
518 FAO, 1990. Trade Yearbook, Vol. 42. FAO, Rome.
519 FAO-UNESCO, 1971-1981. Soil Map of the World. UNESCO, Paris.
520 FATTEH, U.G., and PATEL, P.S., 1986. Package of practices for increasing castor production. Indian Farming 36 (1), 13-18.
521 FAUCONNIER, R., et BASSERAU, D., 1970. La Canne à Sucre. Maisonneuve et Larose, Paris.
522 FAURE, M., 1976. Hybrid sunflowers: they're here now. World Farming 18 (1), 24-27.
523 FAZLI, F.R.Y., and HARDMAN, R., 1968. The spice, fenugreek, (*Trigonella foenum-graecum* L.): its commercial varieties of seed as a source of diosgenin. Trop. Sci. 10, 66-78.
524 FEAKIN, S.D., 1972. Pest Control in Bananas. PANS Manual No. 1. 2nd ed. Centre Overseas Pest Res., London.
525 FEAKIN, S.D., 1973. Pest Control in Groundnuts. PANS Manual No. 2. 3rd ed. Centre Overseas Pest Res., London.
526 FEDERACION NACIONAL CAFETEROS DE COLOMBIA, n.d. El cultivo de la pitaya. Federacion Nacional de Cafeteros de Colombia. Bogotá, Colombia.
527 FELKER, P., 1981. Uses of tree legumes in semiarid regions. Econ. Bot. 35, 174-186.
528 FELKER, P., and BANDURSKI, R.S., 1979. Uses and potential uses of leguminous trees for minimal energy input agriculture. Econ. Bot. 33, 172-184.
529 FERGUSON, A.R., 1984. Kiwifruit - a botanical review. Hort. Reviews 6, 1-64.
530 FERNANDO, D.M., 1974. The selection of stocks in *Hevea*. Quart. J. Rubber Res. Inst. Sri Lanka 51, 28-30.
531 FERRARIS, R., 1973. Pearl Millet (*Pennisetum typhoides*). Commonwealth Agric. Bureaux, Farnham Royal, Hough, England.
532 FERRARIS, R., and STEWART, G.A., 1979. Agronomic assesment of *Pennisetum purpureum* cultivars for agro-industrial application. Field Crops Res. 2, 45-54.
533 FERRARIS, R., and STEWART, G.W., 1979. New options for sweet sorghum. J. Aust. Inst. Agric. Sci. 45, 156-164.
534 FERRIERE, J., 1988. La dormance des bourgeons floraux chez le jojoba: conséquences sur l'extension de la culture. Oléagineux 43, 343-353.
535 FERWERDA, F.P., and WIT, F., (eds.), 1969. Outlines of Perennial Crop Breeding in the Tropics. Veenman and Zonen, Wageningen.
536 FERY, R.L., 1980. Genetics of *Vigna*. Hort. Reviews 2, 311-394.

537 FEUELL, A.J., 1965. Insecticides. REGEL, C. v. (ed.), 1962-1968. Wiesner, Rohstoffe des Pflanzenreiches. 5th ed. Cramer, Lehre.
538 FIELD CROPS RESEARCH, 1985. Special Issue: Pearl millet. Field Crops Res. 11, 114-290.
539 FINKEL, H.J. (ed.), 1983. Handbook of Irrigation Technology, 2 vols. CRC Press, Boca Raton, Florida.
540 FIRMAN, I.D., and WALLER, J.M., 1977. Coffee Berry Disease and Other *Colletotrichum* Diseases of Coffee. CMI Phytopathol. Paper No. 20. Commonwealth Agric. Bureau, Farnham Royal, England.
541 FISCH, B.E., 1976. The roseapple. Calif. Rare Fruit Growers Yearbook 8, 100-111.
542 FISHER, H.H., 1973. Origin and uses of ipecac. Econ. Bot. 27, 231-234.
543 FLACH, M., 1983. The Sago Palm: Domestication, Exploitation and Products. FAO Plant Production and Protection Paper 47, FAO, Rome.
544 FLINN, J.C., et HOYOUX, J.M., 1976. Le bananier plantain en Afrique. Fruits 31, 520-530.
545 FLORES, F.A., and LEWIS, W.H., 1978. Drinking the South American hallucinogenic ayahuasca. Econ. Bot. 32, 154-156.
546 FLYNN, G., 1975. The Market Potential for Papain. G 99, Trop. Prod. Inst., London.
547 FLYNN, G., and CLARKE, P.A., 1980. An Industrial Profile of Rice Milling. G 148, Trop. Prod. Inst., London.
548 FOLLIN, J.C., et SCHWENDIMAN, J., 1974. La résistance du kénaf (*Hibiscus cannabinus* L.) à l'anthracnose (*Colletotrichum hibisci* Poll.). Cot. Fib. Trop. 29, 331-338.
549 FORSBERG , R.A. (ed.), 1985. Triticale. Crop Sci. Soc. Amer., Special Publ. No. 9, Madison, Wisconsin.
550 FOUQUE, A., 1976. Espèces Fruitières d'Amérique Tropicale. Inst. Français de Recherches Fruitières Outre-Mer, Paris.
551 FOUQUE, A., 1981. Les plantes médicinales présentes en forêt Guynaise. Fruits 36, 451-476.
552 FOURNIER, P., 1982. Expérimentation fraisier à l'île de la Réunion. Première partie: Essaix variétaux. Fruits 37, 365-379.
553 FOURNIER, P., 1982. Expérimentation fraisier à l'île de la Réunion. II. Etude de quelque techniques culturales. Fruits 37, 609-615.
554 FOURNIER, P., et HOAREAU, J., 1980. Les pêchers à la Réunion. Quatre ans d'observations sur la collections de Cilaos. Fruits 35, 537-549.
555 FRANCIS, CH.A. (ed.), 1986. Multiple Cropping Systems. Macmillan, New York.
556 FRANCOIS, R., 1974. Les Industries des Corps Gras. Techniques et Documentation, Paris.
557 FRANKLIN, F.W., and POLLACK, B.L., 1979. Vegetables for the hot humid tropics. Part 5. Eggplant, *Solanum melongena*. Sci. Educ. Adm., USDA, New Orleans.
558 FRANKS, H.D., 1968. Mechanized Harvesting of Sugar Cane. Univ. West Indies, Trinidad.
559 FREE, J.B., 1976. Beekeeping and pollination in developing countries. Span 19, 73-75.
560 FREMOND, Y., ZILLER, R., and NUCE DE LAMOTHE, M. DE, 1966. The Coconut Palm. Intern. Potash Inst., Bern.
561 FRENCH, CH.D., 1972. Papaya - the Melon of Health. Exposition Press, New York.
562 FRESCO, L.O., 1986. Cassava in Shifting Cultivation. A Systems Approach to Agricultural Technology Development in Africa. Royal Trop. Inst., Amsterdam, Netherlands.
563 FREY, K.J., (ed.), 1966. Plant Breeding. Iowa State U.P., AMES.
564 FREY-WYSSLING, A., 1935. Die Stoffausscheidung der höheren Pflanzen. Springer, Berlin.
565 FRIEDBERG, C., 1977. Les palmiers à sucre et à vin dans le sud-est asiatique et en Indonésie. J. Agric. Tradit. Bot. Appl. 24, 341-345.

566 FRIEND, J.C., 1981. Effect of night temperature on flowering and fruit size in pineapple (*Ananas comosus* (L.) Merrill). Bot. Gaz. 142, 188-190.
567 FRÖHLICH, G., and RODEWALD, W., 1970. Pests and Diseases of Tropical Crops and Their Control. Pergamon Press, London.
568 FUCHS, Y., ZAUBERMAN, G., YANKO, U., and HOMSKY, S., 1975. Ripening of mango fruits with ethylene. Trop. Sci. 17, 211-216.
569 FURIA, T.E. (ed.), 1977. Current Aspects of Food Colorants. CRC Press, West Palm Beach, Florida.
570 GABR, M.F., and TISSERAT, B., 1985. Propagating palms in vitro with special emphasis on the date palm (*Phoenix dactylifera* L.). Scientia Hortic. 25, 255-262.
571 GADE, D.W., 1966. Achira, the edible canna, its cultivation and use in the Peruvian Andes. Econ. Bot. 20, 407-415.
572 GADE, D.W., 1970. Ethnobotany of Canihua (*Chenopodium pallidicaule*), rustic seed crop of the Altiplano. Econ. Bot. 24, 55-61.
573 GAILLARD, J.P., 1974. Cycle de l'ananas en fonction de la position initiale du rejet. Fruits 29, 3-15.
574 GAILLARD, J.-P., 1987. L'Avocatier. Sa Culture, ses Produits. Maisonneuve et Larose, Paris.
575 GALANO, A.M., 1977. La noix babaçu. Economiste de Tiers Monde 20, 61-66.
576 GALIL, J., 1977. Fig biology. Endeavour, N. S. 1, 52-56.
577 GALLI, R., 1986. New crops for Semi-Arid Regions of Mediterranean European Countries. FAST Occasional Paper No. 88, Commision of the EC, Brussels, Belgium.
578 GALWEY, N.W., 1984. The *Chenopodium* grains of the Andes: Inca crops for modern agriculture. Adv. Appl. Biol. 10, 145-216.
579 GANDHI, S. R., n.d. The Papaya in India. Indian Council Agric. Res., New Delhi.
580 GANGULY, J.K., and KAUL, R.N., 1969. Wind Erosion Control. I.C.A.R. Technical Bull. (Agric.) No. 20, Indian Council Agric. Res., New Delhi.
581 GANGWAR, A.K., and RAMAKRISHNAN, P.S, 1988. Cultivation and Use of Lesser-Known Plants of Food Value by Tribals in North-East India. Agric., Ecosystems Env. 25, 253-267.
582 GARCIA, J.G., MACBRIDE, B., MOLINA, A.R., and HERRERA-MACBRIDE, O., 1975. Malezas prevalentes de America Central. Intern. Plant Protection Center, El Salvador, San Salvador.
583 GARIBOLDI, F., 1974. Rice Parboiling. FAO Agric. Devel. Paper 97, FAO, Rome.
584 GARNER, R.J., CHAUDRI, S.A., and STAFF OF THE COMMONWEALTH BUREAU OF HORTICULTURE and PLANTATION CROPS, 1976. The Propagation of Tropical Fruit Trees. FAO and Commonwealth Agric. Bureau, Farnham Royal.
585 GATES, P.J., and BROWN, K., 1988. *Acacia tortilis* and *Prosopis cineraria*: leguminous trees for arid areas. Outlook on Agriculture 17, 61-64.
586 GATTY, R., 1956. Kava - Polynesian beverage shrub. Econ. Bot. 10, 241-249.
587 GAUDY, M., 1959. Manuel d'Agriculture Tropicale. La Maison Rustique, Paris.
588 GAUTREAUT, J., GARET, B., et MAUBOUSSIN, J.C., 1980. Une nouvelle variété d'arachide sénégalaise adaptée à la secheresse: la 73-33. Oléagineux 35, 149-154.
589 GEER, T., 1970. The international market for natural rubber. Z. Ausl. Landw. 9, 211-244.
590 GENTRY, H.S., 1961. Buchu, a new cultivated crop in South Africa. Econ. Bot. 15, 326-331.
591 GENTY, PH., 1981. Entomological research on the oil palm in Latin America. Oil Palm News No. 25, 17-23.
592 GEPTS, P. (ed.), 1988. Genetic Resources of Phaseolus Beans. Their Maintenance, Domestication, Evolution, and Utilization. Kluwer Academic Publ., Dordrecht.
593 GERDTS, M., and CLARK, J.K., 1979. Caprification: a unique relationship between plant and insect. Calif. Agric. 33 (11/12), 12-14.

594 GETAHUN, A., and KRIKORIAN, A.D., 1973. Chat: coffee's rival from Harar, Ethiopia. I. Botany, cultivation and use. II. Chemical composition. Econ. Bot. 27, 353-389.
595 GEUS, J.G. DE, 1973. Fertilizer Guide for the Tropics and Subtropics. 2nd ed. Centre d'Étude de l'Azote, Zürich.
596 GHAI, S.K., RAO, D.L.N., and BATRA, L., 1985. Comparative study of the potential of sesbanias for green manuring. Trop. Agric. (Trinidad) 62, 52-56.
597 GHOSH, B.N., 1973. Drying cocoa beans by gas. World Crops 25, 232-237.
598 GHOSH, T., 1983. Handbook on Jute. FAO Plant Production and Protection Paper No. 51. FAO, Rome.
599 GIACOMELLI, E.J., et PY, C., 1981. L'ananas au Brésil. Fruits 36, 645-687.
600 GICHURU, M.P., and KANG, B.T., 1989. *Calliandra calothyrsus* Meissn. in an alley cropping system with sequentially cropped maize and cowpea in southwestern Nigeria. Agrofor. Systems 9, 191-203.
601 GIESBERGER, G., 1972. Climatic problems in growing deciduous fruit trees in the tropics. Trop. Abstr. 27, 1-8.
602 GILLIER, P., et SILVESTRE, P., 1969. L'Arachide. Maisonneuve et Larose, Paris.
603 GILMOUR, J.S.L., HORNE, F.R., LITTLE, E.L.JR., STAFLEU, F.A., and RICHENS, R.H., 1969. International Code of Nomenclature for Cultivated Plants. Intern. Bureau Plant Taxon. Nomenclat., Utrecht.
604 GIULIANI, F., 1982. La macadamia. Riv. Agric. Subtrop. Trop. 76, 103-161.
605 GLADSTONES, J.S., 1975. Lupin breeding in Western Australia. The narrow-leaf lupin (*Lupinus angustifolius*). J. Agric. W. Aust. 16, 44-49.
606 GLADSTONES, J.S., and MCKEOWN, N.R., 1977. Serradella - a pasture legume for sandy soils. J. Agric. W. Aust. 18, 11-14.
607 GLAS, K., 1988. Sunflower. Fertilizing for High Yield and Quality. Internat. Potash Institute, Worblaufen-Bern.
608 GLICKSMAN, M., 1969. Gum Technology in the Food Industry. Academic Press, New York.
609 GODIN, V.J., and SPENSLEY, P.C., 1971. Oils and Oilseeds. Trop. Prod. Inst., London.
610 GOERING, T.J., 1982. Natural Rubber (Sector Policy Paper). World Bank, Washington, D. C.
611 GOH, T.A., 1987. Rubberwood - from fuelwood to a major timber species. Planters' Bull. 191, 35-42.
612 GÖHL, B., 1981. Tropical Feeds. Feed Information Summaries and Nutritive Values. FAO Animal Prod. and Health Series No. 12, FAO, Rome.
613 GOLDBLATT, L.A., 1959. The tung industry. II. Processing and utilization. Econ. Bot. 13, 343-364.
614 GOLMIRZAIE, A.M., and MENDOZA, H.A., 1988. Breeding strategies for true potato seed production. CIP Circular 16 (4), 1-8.
615 GOMEZ, G., 1979. Cassava as a swine feed. World Animal Rev. No. 29, 13-20.
616 GOMEZ, G., CUESTA, D. DE LA, VALDIVIESO, M., y KAWANO, K., 1980. Contenido de cianuro total y libre en parénquima y cáscara de raíces de diez variedades promisorias de yuca. Turrialba 30, 361-365.
617 GOMEZ, P.C., FEREIRA, A.S., FIALHO, E.T., e BARBOSA, H.T., 1981. O cereal adlay na alimentação de suínos em crescimento e terminação. Pesq. agropec. bras. (Brasília) 16, 901-906.
618 GONÇALVES, J.R.C., 1971. A cultura do guaraná. Sér. Cult. Amazônica, Inst. Pesq. Exp. Agropec. Norte (IPEAN) 2 (1), 1-13.
619 GOODIN, J.R., and NORTHINGTON, D.K. (eds.), 1979. Arid Land Plant Resources. Intern. Center for Arid and Semi-Arid Land Studies, Lubbock, Texas.
620 GOODING, E.G.B., 1972. The production of instant yam in Barbados. I. Process development. Trop. Sci. 14, 323-333.
621 GOODLAND, R.J.A., WATSON, C., and LEDEC, G., 1984. Environmental Management in Tropical Agriculture. Westview Press, Boulder, Colorado.
622 GOODSPEED, T.H., 1954. The Genus *Nicotiana*. Chronica Botanica. Waltham, Mass.

623 GOODWIN, T.W. (ed.), 1988. Plant Pigments. Academic Press, London.
624 GOREN, R., and MENDEL, K., (eds.), 1989. Citriculture. Intern. Citrus Congr. Middle-East 1988. 4 vols. Margraf Scientific Publishers, Weikersheim.
625 GOUJON, P., LEFEBVRE, A., LETURCQ, P., MARCELLESI, A.P., et PRALORAN, J.C., 1973. Études sur l'anacardier. Régions écologiques favorables à la culture de l'anacardier en Afrique francophone de l'Ouest. Bois Forêts Trop. No. 151, 27-53.
626 GOULD, F.W., 1968. Grass Systematics. McGraw-Hill, New York.
627 GOULD, W.A., 1974. Tomato Production, Processing and Quality Evaluation. Avi Publ. Co., Westport, Conn.
628 GRACE, M.R., 1977. Cassava Processing. FAO Plant Production and Protection Series No. 3; FAO, Rome.
629 GRAHAM, K.M., 1971. Plant Diseases of Fiji. H. M. Sta. Off., London.
630 GRAHAM, P.H., 1981. Some problems of nodulation and symbiotic nitrogen fixation in *Phaseolus vulgaris* L.: a review. Field Crops Res. 4, 93-112.
631 GRAINGE, M., and AHMED, S., 1988. Handbook of Plants with Pest-Control Properties. Wiley, New York.
632 GRANT, J.A., and RYUGO, K., 1984. Influence of within-canopy shading on fruit size, shoot growth, and return bloom in kiwifruit. J. Amer, Soc. Hort. Sci. 109, 799-802.
633 GRAVES, W.L., WILLIAMS, W.A., WEGRZYN, V.A., CALDERON, D., GEORGE, M.R., and SULLINS, J.L., 1987. Berseem clover is getting a second chance. Calif. Agric. 41 (9/10), 15-18.
634 GRAW, D., und REHM, S., 1977. Vesikulär-arbuskuläre Mykorrhiza in den Fruchtträgern von *Arachis hypogaea* L. Z. Acker- und Pflanzenbau 144, 75-78.
635 GRAY, A.R., 1982. Taxonomy and evolution of broccoli (*Brassica oleracea* var. *italica*). Econ. Bot. 36, 397-410.
636 GRAY, J.C., KUNG, S.D., WILDMAN, S.G., and SHEEN, S.J., 1974. Origin of *Nicotiana tabacum* L. detected by polypeptide composition of Fraction I protein. Nature 252, 226-227.
637 GREATHOUSE, D.C., LAETSCH, W.M., and PHINNEY, B.O., 1971. The shootgrowth rhythm of a tropical tree, *Theobroma cacao*. Amer. J. Bot. 58, 281-286.
638 GREAVES, A., 1984. *Acacia nilotica*. Annotated Bibliography No. F 36. Commonwealth Agric. Bureaux, Oxford.
639 GREAVES, A., 1984. *Acacia tortilis*. Annotated Bibliography No. F 37. Commonwealth Agric. Bureaux, Oxford.
640 GREENHALGH, P., 1979. Ramie fibre: production, trade and markets. Trop. Sci. 21, 1-9.
641 GREENHALGH, P., 1979. The Market for Mint Oils and Menthol. G 126, Trop. Prod. Inst., London.
642 GREENHALGH, P., 1979. The Market for Culinary Herbs. G 121, Trop. Prod. Inst., London.
643 GREENHALGH, P., 1980. Production, trade and markets for culinary herbs. Trop. Sci. 22, 159-188.
644 GREENHALGH, P., 1982. The Production, Marketing and Utilisation of Naval Stores. G 170, Trop. Prod. Inst., London.
645 GREENHALGH, P., 1986. The World Market for Silk. G 195, Trop. Devel. Res. Inst., London.
646 GREGORIOU, C., and RAJ KUMAR, D., 1984. Effects of irrigation and mulching on shoot and root growth of avocado (*Persea americana* Mill.) and mango (*Mangifera indica* L.). J. Hort. Sci. 59, 109-117. .
647 GREGORY, P.H., (ed.), 1974. Phytophthora Disease of Cocoa. Longman, London.
648 GREGORY, R.S., 1975. The commercial production of triticale. Span 18, 65-66.
649 GRENBY, T.H. (ed.), 1988. Development in Sweeteners -3. Elsevier Applied Science Publ., Barking Essex, U.K.
650 GRIMWOOD, B.E., 1971. The Processing of Macadamia Nuts. G 66, Trop. Prod. Inst., London.

651 GRIMWOOD, B.E., 1975. Coconut Palm Products. Their Processing in Developing Countries. FAO Agric. Devel. Paper No. 99. FAO, Rome.
652 GRISI, B.M., 1976. Biodinâmica de solo cultivado com cacaueiros sombreados e ao sol. Rev. Theobroma 6, 87-99.
653 GRIST, D.H., 1986. Rice. 6th ed. Longman, London.
654 GRIST, D.H., and LEVER, R.J.A.W., 1969. Pests of Rice. Longman, London.
655 GROF, B., 1982. Breeding *Centrosema pubescens* in tropical South America. Trop. Grasslands 16, 80-83.
656 GROOM, Q.J., and REYNOLDS, T., 1987. Barbaloin in Aloe species. Planta medica 53, 345-348.
657 GROSS, R., BAER, E.V., and ROHRMOSER, K., 1983. The lupin - a new cultivated plant in the Andes. III. The output and quality of lupins (*Lupinus albus* and *Lupinus mutabilis*) in one South American and three European locations. J. Agron. Crop Sci. 152, 19-31.
658 GRUBBEN, G.J.H., 1976. The Cultivation of Amaranth as a Tropical Leaf Vegetable. Communication 67, Dep. Agric. Res., Royal Tropical Institute, Amsterdam.
659 GRUBBEN, G.J.H., 1977. Tropical Vegetables and Their Genetic Resources. FAO, Rome.
660 GRUY, I.V. DE, CARRA, J.H., and GOYNES, W.R., 1973. The Fine Structure of Cotton. An Atlas of Cotton Microscopy. Dekker, New York.
661 GUENTHER, E., 1948-52. The Essential Oils. 6 Vols. Van Nostrand, New York.
662 GUENTHER, E., 1954. The French lavender and lavandin industry. Econ. Bot. 8, 166-173.
663 GULERIA, W.S., and SINGH, C.M., 1983. Effects of cultural practices on growth, yield and economics of fibre flax production in the north western Himalayas. Exp. Agric. 19, 87-90.
664 GUNSTONE, F.D., 1987. Palm Oil. Critical Review on Applied Chemistry, Vol. 15. Wiley and Sons, Chichester.
665 GUPTA, P.K., and PRIYADARSHAN, P.M., 1982. Triticale - present status and future prospects. Adv. Genetics 21, 255-346.
666 GUPTA, U.S. (ed.), 1975. Physiological Aspects of Dryland Farming. Oxford and IBH Publishing Co., New Delhi.
667 GUPTA, Y.P., 1980. Improving nutritional quality of pulses. Indian Farming 30 (1), 9-11.
668 GUPTA, Y.P., 1980. Khesari dal consumption - a health hazard. Indian Farming 30 (2), 7-9.
669 GURNAH, A.M., and GACHANJA, S.P., 1984. Spacing and pruning of purple passion fruit. Trop. Agric. (Trinidad) 61, 143-147.
670 GUTTERIDGE, R.C., 1981. The productivity of a range of forage legumes oversown with and without Sabi grass into native grassland in northeast Thailand. Trop. Grasslands 15, 134-140.
671 GUYOT, A., PINON, A., et PY, C., 1974. L'ananas en Côte d'Ivoire. Fruits 29, 85-117.
672 GUYOT, A., et PY, C., 1970. La floraison controlée de l'ananas par l'éthrel, nouveau regulateur de croissance. Fruits 25, 427-445.
673 HABART, J.L., 1974. La baie de l'*Actinidia chinensis* Planch. var. *chinensis*. Fruits 29, 191-207.
674 HACKER, J.B., and JONES, R.J., 1969. The *Setaria sphacelata* complex - a review. Trop. Grasslands 3, 13-34.
675 HADFIELD, J., 1969. Vegetable Gardening in Central Africa. 2nd ed. Purnell, Cape Town.
676 HAGENMAIER, R., 1980. Coconut Aqueous Processing. 2nd ed. Univ. San Carlos, Cebu City, Philippines.
677 HAGIN, J., and TUCKER, B., 1982. Fertilization of Dryland and Irrigated Soils. Springer, Berlin.

678 HAHN, S.K., 1984. Tropical Root Crops. Their Improvement and Utilization. IITA, Ibadan, Nigeria.
679 HAHN, S.K., OSIVU, D.S.O., AKORODA, M.O., and OTOO, J.A., 1987. Yam production and its future prospects. Outlook on Agric. 16, 105-110.
680 HAIGH, J., 1951. A Manual on the Weeds of the Major Crops of Ceylon. Dept. Agric., Ceylon.
681 HAINSWORTH, E., 1952. Tea Pests and Diseases and Their Control. Heffer, Cambridge.
682 HALEVY, J., and BAZELET, M., 1989. Fertilizing for High Yield and Quality Cotton. IPI-Bulletin 2 (revised). International Potash Institute, Bern.
683 HALL, A.E., CANNEL, G.H., and LAWTON, H.W. (eds.), 1979. Agriculture in Semi-Arid Environments. Springer, Berlin.
684 HALL, C.J.J. VAN, en KOPPEL, C. VAN DE, (Uitg.), 1946-50. De Landbouw in den Indischen Archipel. 3 Vols. Van Hoeve, s'Gravenhage.
685 HALL, D.W., 1970. Handling and Storage of Food Grains in Tropical and Subtropical Areas. FAO Agric. Devel. Paper 90, FAO, Rome.
686 HALL, T.J., 1985. Adaptation and agronomy of *Clitoria ternatea* L. in northern Australia. Trop. Grasslands 19, 156-163.
687 HALLAUER, A.R., and FO MIRANDA, J.B., 1981. Quantitative Genetics in Maize Breeding. Iowa State Univ. Press, Ames, Iowa.
688 HAMILTON, L.S., and MURPHY, D.H., 1988. Use and management of nipa palm (*Nypa fruticans*, Arecaceae): a review. Econ. Bot. 42, 206-213.
689 HAMILTON, R.A., and FUKUNAGA, E.T., 1959. Growing *Macadamia* Nuts in Hawaii. Hawaii Agric. Exp. Sta. Bull. 121, Honolulu.
690 HAMPTON, M.G., 1981. Compare fuel requirement, then select tea process. World Crops 33, 85-86.
691 HAMPTON, R.E., and THOMPSON, P.G., 1974. Passionfruit production in Fiji. Fiji Agric. J. 36, 23-27.
692 HANAI, Y., 1974. Mat rush growing in Japan. World Crops 26, 122-126.
693 HANSON, A.A., BARNES, D.K., and HILL, R.R.Jr.(eds.), 1987. Alfalfa and Alfalfa Improvement. Agronomy No. 29. Amer. Soc. Agronomy, Madison.
694 HARBERD, D.J., 1972. A contribution to the cyto-taxonomy of *Brassica* (Cruciferae) and its allies. Bot. J. Linn. Soc. 65, 1-23.
695 HARDMAN, R., 1969. Pharmaceutical products from plant steroids. Trop. Sci. 11, 196-228.
696 HARDON, J.J., 1969. Interspecific hybrids in the genus *Elaeis*. II. Vegetative growth and yield of $F_1$ hybrids of *E. guineensis* x *E. oleifera*. Euphytica 18, 380-388.
697 HARDON, J.J., and TAN, G.Y., 1969. Interspecific hybrids in the genus *Elaeis*. I. Crossability, cytogenetics and fertility of $F_1$ hybrids of *E. guineensis* x *E. oleifera*. Euphytica 18, 372-379.
698 HARDY, F., 1961. Manual de Cacao. Instituto Interamer. Ciencias Agrícolas, Turrialba.
699 HARLAN, J.R., 1971. Agricultural origins: centers and noncenters. Science 174, 468-474.
700 HARLAN, J.R., and WET, J. DE, 1972. A simplified classification of cultivated sorghum. Crop. Sci. 12, 172-176.
701 HARLER, C.R., 1963. Tea Manufacture. Oxford Univ. Press, London.
702 HARLER, C.R., 1970. Propagation of the tea plant. World Crops 22, 375-377.
703 HARMAN, G.W., 1984. The World Market for Canned Pineapple and Pineapple Juice. G 186, Trop. Dev. Res. Inst., London.
704 HARPER, J.E., 1974. Soil and symbiotic nitrogen requirements for optimum soybean production. Crop Sci. 14, 255-260.
705 HARRIES, H.C., 1978. The evolution, dissemination and classification of *Cocos nucifera* L. Bot. Rev. 44, 265-320.

706 HARRIS, P.M. (ed.), 1978. The Potato Crop. The Scientific Basis for Improvement. Chapman, London.
707 HARRIS, W.V., 1969. Termites as Pests of Crops and Trees. Commonwealth Institute Entomology, London.
708 HARRISON, D.G., 1984. Sisal by-products as feed for ruminants. World Animal Rev. 49, 25-31.
709 HARTEN, A.M. VAN, 1970. Melegueta pepper. Econ. Bot. 24, 208-216.
710 HARTEN, A.M.VAN, 1974. Koriander: de geschiedenis van een oud gewas. Landbouwk. Tijdschr. 86, 58-64.
711 HARTLEY, C.W.S., 1989. The Oil Palm (*Elaeis guineensis* Jacq.) 3rd ed. Longman, London.
712 HARTLEY, W., 1979. A Checklist of Economic Plants in Australia. CSIRO, Melbourne.
713 HARTMANN, H.D., and KLAPPROTH, H., 1978. Cultivation of asparagus in the tropics. Plant Res. and Devel. 7, 67-77.
714 HARWOOD, C.E., 1989. *Grevillea robusta*: An Annotated Bibliography. ICRAF, Nairobi.
715 HASHIM, I., 1979. South American leaf blight - recent advances. Planters' Bull. No. 158, 20-24.
716 HASHIM, I.BIN, 1986. Micro-tapping of young rubber as a means of reducing immaturity period. Planters' Bull. 189, 138-158.
717 HATTIANGDI, G.S., 1958. The Vanaspati Industry. Indian Central Oilseeds Committee, Hyderabad.
718 HAVARD-DUCLOS, B., 1967. Les Plantes Fourragères Tropicales. Maisonneuve et Larose, Paris.
719 HAWTIN, G., and WEBB, C., 1982. Faba Bean Improvement. Martinus Nijhoff, The Hague.
720 HE GUITING, AMANDA TE, ZHU XIGANG, TRAVERS, S.L., LAI XIUFANG, and HERDT, R.W., 1984. The Economics of Hybrid Rice Production in China. IRRI Res. Paper, Ser. No. 101, IRRI, Manila.
721 HEATON, E.K., SHEWFELT, A.L., BADENHOP, A.E., and BEUCHAT, L.R., 1977. Pecans: Handling, Storage, Processing and Utilization. Georgia Exp. Sta. Res. Bull. 197, Georgia, USA.
722 HEBBELTHWAITE, P.D., DAWKINS, T.C.K., HEATH, M.C., and LOCKWOOD, G. (eds.), 1984. *Vicia faba*: Agronomy, Physiology and Breeding. Martinus Nijhoff/Dr. Junk, The Hague.
723 HEISER, C.B., 1979. The Gourd Book. Univ. Oklahoma Press, Norman, Oklahoma.
724 HEISER, C.B.Jr., 1981. The Sunflower. Univ. of Oklahoma Press, Norman, Oklahoma.
725 HEISER, C.B.Jr., 1985. Ethnobotany of the naranjilla (*Solanum quitoense*) and its relatives. Econ. Bot. 39, 4-11.
726 HEISER, C.B.Jr., and PICKERSGILL, B., 1969. Names for the cultivated *Capsicum* species (Solanaceae). Taxon 18, 277-283.
727 HENDERSON, M., and ANDERSON, J.G., 1966. Common Weeds in South Africa. Gov. Printer, Pretoria.
728 HENRY, B.S., 1979. Potential of plant material as a source of food colour. Trop. Sci. 21, 207-216.
729 HENTY, E.E., and PRITCHARD, G.H., 1975. Weeds of New Guinea and Their Control. Div. Botany, Lae, Papua New Guinea.
730 HEPPER, F.N., 1970. Bambara groundnut (*Voandzeia subterranea*). Field Crop Abstr. 23, 1-6.
731 HERKLOTS, G.A.C., 1972. Vegetables in South-East Asia. Allen & Unwin, London.

732 HERNANDEZ, J.E., CARPENA, A.L., LANTICAN, R.M., and NAVARRO, R.S., 1978. Yield performance and other agronomic characters of twenty-three varieties of adzuki bean (*Vigna angularis* (Willd.) Owhi and Ohashi) under Philippine conditions. Phil. J. Crop. Sci. 3, 13-18.
733 HERRERA, F., WYLLIE, D., and PRESTON, T.R., 1980. Fattening steers on a basal diet of ensiled sisal pulp and molasses urea supplemented with sunflower meal and *Leucaena* forage. Trop. Animal Prod. 5, 71-72.
734 HILL, D., 1987. Agricultural Insect Pests of the Tropics and Their Control. 2nd ed. Cambridge Univ. Press, London.
735 HILL, D.S., and WALLER, J.M., 1982/1989. Pests and Diseases of Tropical Crops. Vol. I: Principles and Methods of Control, Vol. II: Handbook of Pests and Diseases . Longman, London.
736 HILLS, F.J., LEWELLEN, R.T., and SKOYEN, I.O., 1990. Sweet sorghum cultivars for alcohol production. Calif. Agric. 44, 14-16.
737 HILTON, P.J., 1974. The effect of shade upon the chemical composition of the flush of tea (*Camellia sinensis* L.). Trop. Sci. 16, 15-22.
738 HILU, K.W., and WET, J.M.J. DE, 1976. Domestication of *Eleusine coracana*. Econ. Bot. 30, 199-208.
739 HINATA, K., and PRAKASH, S., 1984. Ethnobotany and evolutionary origin of Indian oleiferous Brassicae. Indian J. Genetics 44, 102-112.
740 HIRSINGER, F., 1986. Oleochemical raw materials and new oilseed crops. Oléagineux 41, 345-350.
741 HITCHCOCK, A.S., 1971. Manual of the Grasses of the United States. 2nd ed. rev. by A. CHASE. Dover Publications, New York.
742 HITIER, H., et SABOURIN, L., 1970. Le Tabac. Presses Universitaires de France, Paris.
743 HO, C.Y., 1979. Contribution to Improve the Effectiveness of Breeding, Selection and Planting Recommendations of *Hevea brasiliensis* Muell. Arg. Agr. Rijks Universiteit, Gent.
744 HODGE, W.H., 1953. The drug aloes of commerce with special reference to the Cape species. Econ. Bot. 7, 99-129.
745 HODGE, W.H., 1956. Chinese water chestnut or matai - a paddy crop of China. Econ. Bot. 10, 49-65.
746 HODGE, W.H., and SINEATH, H.H., 1956. The Mexican candelilla plant and its wax. Econ. Bot. 10, 134-154.
747 HOGAN, J.T., and HOUSTON, D.F., 1967. Rice By-products Utilization. Processing Agric. Products Informal Working Bull. 30, FAO, Rome.
748 HOLLIDAY, P., 1970. South American Leaf Blight (*Microcyclus ulei*) of *Hevea brasiliensis*. Phytopath.Papers No.12, Commonwealth Mycol. Inst., Kew.
749 HOLLIDAY, P., 1980. Fungus Diseases of Tropical Crops. Cambridge Univ. Press, London.
750 HOLM, L., 1969. Weed problems in developing countries. Weed Sci. 17, 113-118.
751 HOLM, L.G., PANCHO, J.V., HERBERGER, J.P., and PLUCKNETT, D.L., 1979. A Geographical Atlas of World Weeds. Wiley, New York.
752 HOLM, L.G., PLUCKNETT, D.L., PANCHO, J.V., and HERBERGER, J.P., 1977. The World's Worst Weeds. Distribution and Biology. Univ. Press Hawaii, Honolulu.
753 HOLMES, A.W., 1973. Substitute foods - a practical alternative?. Phil. Trans. R. Soc. Lond. B. 267, 157-166.
754 HOSAKA, E.Y., 1957. Common Weeds of Hawaii. Univ. Hawaii Ext. Serv., Honolulu.
755 HOSTETTMANN, K., 1984. On the use of plants and plant-derived compounds for the control of schistosomiasis. Naturwissenschaften 71, 247-251.
756 HOUCK, J.P., RYAN, M.E., and SUBOTNIK, A., 1972. Soybeans and Their Product. Markets, Models, and Policy. Univ. Minnesota Press, Minneapolis.
757 HOUSE, L.R., 1985. Guide to Sorghum Breeding. 2nd ed. ICRISAT, Patancheru, A. P., India.

758 HOWARD, G.M., 1974. W. A. Poucher's Perfumes, Cosmetics and Soaps. Vol. I: The Raw Materials of Perfumery. 7th ed. Chapman and Hall, London.
759 HOWELER, R.H., 1981. Mineral Nutrition and Fertilization of Cassava (*Manihot esculenta* Crantz). CIAT, Ser., 09EC-4. Cali, Colombia.
760 HOWES, F.N., 1949. Vegetable Gums and Resins. Chronica Botanica, Waltham, Mass.
761 HOWES, F.N., 1953. Vegetable Tanning Materials. Chronica Botanica, Waltham, Massachusetts.
762 HSIUNG, W., 1987. Bamboo in China: prospects for an ancient resource. Unasylva No. 156, Vol. 39, 42-49.
763 HU, SHIU YING, 1976. The genus *Panax* (ginseng) in Chinese medicine. Econ. Bot. 30, 11-28.
764 HUDSON, J.C., BOYCOTT, C.A., and SCOTT, D.A., 1975. A new method of sugarcane harvesting. World Crops 27, 164-169.
765 HUET, R., CASSIN, J., et TISSEAU, R., 1978. Citrons et limes acides. Les limes à gros fruits. Production d'avenir pour les Dom-Tom. Fruits 33, 701-714.
766 HUFFNAGEL, H.P., 1961. Agriculture in Ethiopia. FAO, Rome.
767 HUGOT, E., 1972. Handbook of Cane Sugar Engineering. 2nd ed. Elsevier, Amsterdam.
768 HULME, A.C., (ed.), 1970/71. The Biochemistry of Fruits and Their Products. 2 Vols. Academic Press, London.
769 HULSE, J.H., and LAING, E.M., 1974. Nutritive Value of Triticale Protein (and the Proteins of Wheat and Rye). Intern. Devel. Res. Centre, Ottawa.
770 HULSE, J.H., LAING, E.M., and PEARSON, O.E., 1981. Sorghum and the Millets: Their Composition and Nutritive Value. Academic Press, London.
771 HUMBERT, R.P., 1968. The Growing of Sugar Cane. 2nd ed. Elsevier, Amsterdam.
772 HUMBERT, R.P., 1978. Alcohol from sugar cane. Brazil "harvests" energy crop. World Farming 20, 30-31.
773 HUMPHREY, R.R., 1970. Arizona Range Grasses. Univ. Arizona Press, Tucson.
774 HUMPHREYS, L.R., 1979. Tropical Pasture Seed Production. 2nd ed. FAO, Rome.
775 HUMPHREYS, L.R., 1988. Tropical Pastures and Fodder Crops. 2nd ed. Longman, London.
776 HUNT, L.A., WHOLEY, D.W., and COCK, J.H., 1977. Growth physiology of cassava (*Manihot esculenta* Crantz). Field Crop Abstr. 30, 77-91.
777 HUNTER, R.B., HUNT, W.A., and KANNENBERG, L.W., 1974. Photoperiod and temperature effects on corn. Canad. J. Plant Sci. 54, 71-78.
778 HUSAIN, A., VIRMANI, O.P., SHARMA, A., KUMAR, A., and MISRA, L.N., 1988. Major Essential Oil-Bearing Plants of India. Central Inst. Medicinal and Aromatic Plants, Lucknow, India.
779 HUSZ, G.ST., 1972. Sugar Cane. Cultivation and Fertilization. Ruhr-Stickstoff, Bochum.
780 HUSZ, G.ST., 1986. Sugar cane factory efficiency as a function of agro-ecological growing conditions. Zuckerind. 111, 57-61.
781 HUTCHINSON, J.B., 1959. Genetics and Improvement of Tropical Crops. Cambridge Univ. Press, London.
782 HUTCHINSON, J.B., (ed.), 1974. Evolutionary Studies in World Crops. Cambridge Univ. Press, London.
783 HUTCHINSON, J.B., CLARK, J.G.G., JOPE, E.M., and RILEY, R. (ed.), 1977. The Early History of Agriculture. Oxford Univ. Press, Oxford.
784 HUTCHINSON, J.B., SILOW, R.A., and STEPHENS, S.G., 1947. The Evolution of *Gossypium* and the Differentiation of the Cultivated Cottons. Oxford Univ. Press, London.
785 HUTTON, E.M., 1970. Tropical pastures. Adv. Agron. 22, 1-73.
786 HUTTON, E.M., 1976. Selecting and breeding tropical pasture plants. Span 19, 21-22.

787 HUTTON, E.M., 1985. *Centrosema* breeding for acid tropical soils, with emphasis on efficient Ca absorption. Trop. Agric. (Trinidad) 62, 273-280.

788 HUTTON, E.M., and BEALL, L.B., 1977. Breeding of *Macroptilium atropurpureum*. Tropical Grasslands 11, 15-31.

789 HWANG, S.C., CHAN, C.L., LIN, J.C., and LIN, H.L., 1984. Cultivation of banana using plantlets from meristem culture. HortScience 19, 231-233.

790 HYMOWITZ, T., and BOYD, J., 1977. Origin, ethnobotany and agricultural potential of the winged bean - *Psophocarpus tetragonolobus*. Econ. Bot. 31, 180-188.

791 HYMOWITZ, T., and NEWELL, C.A., 1981. Taxonomy of the genus *Glycine*, domestication and uses of soybeans. Econ. Bot. 35, 272-288.

792 IBRAHIM, S.H., 1976. A list of pollen plants visited by honeybees in Egypt. Agric. Res. Rev. (Kairo) 54, 217-219.

793 IGWILO, N., and OKOLI, O., 1988. Evaluation of yam cultivars for seed yam production, using the minisett technique. Field Crops Res. 19, 81-89.

794 IIZUKA, H., and NAITO, A., 1967. Microbial Transformation of Steroids and Alkaloids. Univ. Tokyo Press, Tokyo.

795 IKENAGA, T., and OHASHI, H., 1978. Cultivation and harvest method of *Duboisia* by cultivators in Australia. Japan. J. Trop. Agric. 21, 221-222.

796 ILTIS, H.H., and DOEBLEY, J.F., 1980. Taxonomy of *Zea* (Gramineae). II. Subspecific categories in the *Zea mays* complex and generic synopsis. Amer. J. Bot. 67, 994-1004.

797 ILYAS, M., 1979. The spices of India II. Econ. Bot. 32, 238-263.

798 ILYAS, M., 1980. Spices in India III. Econ. Bot. 34, 236-259

799 INCH, A.J., 1978. Passion fruit diseases. Queensland Agric. J. 104, 479-484.

800 INDIAN CENTRAL TOBACCO INSTITUTE, 1960. Indian Tobacco. A Monograph. Indian Central Tob. Inst., Madras.

801 INDIAN COUNCIL OF AGRICULTURAL RESEARCH, 1971. Crop Diseases Calender. 2nd ed. Indian Council Agric. Res., New Delhi.

802 INDIAN COUNCIL OF AGRICULTURAL RESEARCH, 1988. Groundnut. Publications and Information Division, ICAR, New Delhi, India.

803 INDIAN FARMING, 1976. Special Issue on Lac. Indian Farming 27 (8), 3-35.

804 INDIAN FARMING, 1978. Special Issue on Wheat. Indian Farming 27 (11), 1-53.

805 INDIAN FARMING, 1979. Special Issue on Cashew. Indian Farming 28 (12), 3-32.

806 INDIAN FARMING, 1981. Special Issue on Grain Legumes. Indian Farming 31 (5), 3-87.

807 INDIAN FARMING, 1982. Special Issue on Oilseeds. Indian Farming 32 (8), 3-116.

808 INDIAN FARMING, 1982. Special Issue on Arecanut. Indian Farming 32 (9), 3-51.

809 INDIAN FARMING, 1983. Special Issue on Barley. Indian Farming 33 (7), 1-59.

810 INDIAN FARMING, 1987. Special Issue on Grasslands. Indian Farming 36 (8), 3-19.

811 INDIAN JOURNAL OF GENETICS AND PLANT BREEDING, 1975. Special Issue on Pulse Crops. Indian J. Genetics 35, 169-305.

812 INDIAN JOURNAL OF GENETICS AND PLANT BREEDING, 1979. Symposium on Use of Multilines for Reducing Disease Epidemics. Indian J. Genetics 39, 1-109.

813 INGLETT, G.E., (ed.), 1974. Symposium: Sweeteners. Avi Publ., Westport, Conn.

814 INGLETT, G.E. (ed.), 1982. Maize: Recent Progress in Chemistry and Technology. Academic Press, New York.

815 INGLETT, G.E., and CHARALAMBOUS, G. (eds.), 1979. Tropical Foods: Chemistry and Nutrition. 2 Vols. Academic Press, New York.

816 INGRAM, J.S., 1969. Saffron (*Crocus sativus* L.). Trop. Sci. 11, 177-184.

817 INGRAM, J.S., 1972. Cassava Processing: Commercially Available Machinery. G 75, Trop. Prod. Inst., London.

818 INGRAM, J.S., 1975. Standards, Specifications and Quality Requirements for Processed Cassava Products. G 102, Trop. Prod. Inst., London.

819 INGRAM, J.S., and FRANCIS, B.J., 1969. The annatto tree (*Bixa orellana* L.). A guide to its occurrence, cultivation, preparation and uses. Trop. Sci. 9, 97-103.

820 INSTITUT DE RECHERCHES POUR LES HUILES ET OLEAGINEUX, 1980. Fourth Meeting of the International Council on Lethal Yellowing. Oléagineux 35, 297-304.
821 INSTITUT NATIONAL DE LA RECHERCHE AGRONOMIQUE, 1962. Le Coton au Maroc. Inst. Nat. Rech. Agron., Rabat.
822 INSTITUT NATIONAL DE LA RECHERCHE AGRONOMIQUE, 1968. Les Agrumes au Maroc. Inst. Nat. Rech. Agron., Rabat.
823 INTERNATIONAL CROPS RESEARCH INSTITUTE FOR THE SEMI-ARID TROPICS, 1981. Proceedings of the International Workshop on Pigeonpeas, 2 vols. ICRISAT, Patancheru, India.
824 INTERNATIONAL CROPS RESEARCH INSTITUTE FOR THE SEMI-ARID TROPICS, 1983. Proceedings of the Symposium Sorghum in the Eighties. ICRISAT, Patancheru, India.
825 INTERNATIONAL CROPS RESEARCH INSTITUTE FOR THE SEMI-ARID TROPICS, 1987. Adaptation of Chickpea and Pigeonpea to Abiotic Stresses. ICRISAT, Patancheru, India.
826 INTERNATIONAL CROPS RESEARCH INSTITUTE FOR THE SEMI-ARID TROPICS, 1988. Summary and Recommendations of the International Workshop on Aflatoxin Contamination of Groundnut. ICRISAT, Patancheru, India.
827 INTERNATIONAL DEVELOPMENT RESEARCH CENTRE, 1980. Rattan. IDRC, Ottawa, Canada.
828 INTERNATIONAL LUPIN ASSOCIATION, 1984. Proceedings of the 3rd International Lupin Conference, La Rochelle (France). L'Union Nat. Interprof. des Protéagineux, Paris.
829 INTERNATIONAL RICE RESEARCH INSTITUTE, 1965. The Mineral Nutrition of the Rice Plant. Johns Hopkins Press, Baltimore.
830 INTERNATIONAL RICE RESEARCH INSTITUTE, 1965. The Rice Blast Disease. Johns Hopkins Press, Baltimore.
831 INTERNATIONAL RICE RESEARCH INSTITUTE, 1967. The Major Insect Pests of the Rice Plant. Johns Hopkins Press, Baltimore.
832 INTERNATIONAL RICE RESEARCH INSTITUTE, 1972. Rice Breeding. Intern. Rice Res. Inst., Los Baños, Philippines.
833 INTERNATIONAL RICE RESEARCH INSTITUTE, 1975. Major Research in Upland Rice. IRRI, Los Baños, Philippines.
834 INTERNATIONAL RICE RESEARCH INSTITUTE, 1976. Climate and Rice. IRRI, Los Baños, Philippines.
835 INTERNATIONAL RICE RESEARCH INSTITUTE, 1978. Soils and Rice. IRRI, Los Baños, Philippines.
836 INTERNATIONAL RICE RESEARCH INSTITUTE, 1980. Innovative Approaches to Rice Breeding. IRRI, Los Baños, Philippines.
837 INTERNATIONAL RICE RESEARCH INSTITUTE, 1988. Rice Ratooning. IRRI, Los Baños, Laguna, Philippines.
838 INTERNATIONAL RICE RESEARCH INSTITUTE, 1989. Progress in Irrigated Rice Research. IRRI, Manila, Philippines.
839 INTERNATIONAL SOCIETY FOR HORTICULTURAL SCIENCE, 1972. Symposium on Mango and Mango Culture (New Delhi 1969). Acta Horticulturae No.24.
840 INTERNATIONAL TRADE CENTRE, 1968. The Markets for Manioc as a Raw Material for Compound Animal Feeding Stuffs in the Federal Republic of Germany, the Netherlands, and Belgium. Intern. Trade Centre, Genf.
841 INTERNATIONAL TRADE CENTRE, 1977. Spices - A Survey of the World Markets. 2 Vols. International Trade Centre, Genf.
842 IRIKURA, Y., COOK, J.H., and KAWANO, K., 1979. The physiological basis of genotype-temperature interactions in cassava. Field Crops Res. 2, 227-239.
843 IRVINE, F.R., 1969. West African Crops. 3rd ed. Oxford Univ. Press, London.
844 ISHIZUKA, Y., 1971. Physiology of the rice plant. Adv. Agron. 23, 241-315.

845 ITOH, T., TAMURA, T., MATSUMOTO, T., et DUPAIGNE, P., 1975. Études sur l'huile d'avocat, en particulier sur la fraction stérolique de l'insaponifiable. Oléagineux 30, 687-695.
846 IVENS, G.W., 1967. East African Weeds and Their Control. Oxford Univ. Press, London.
847 IVENS, G.W., MOODY, K., and EGUNJOBI, J.K., 1978. West African Weeds. Oxford Univ. Press, Ibadan, Nigeria.
848 IVORY, D.A., 1982. Evaluation of temperate grass species for the eastern Darling Downs of Queensland. Trop. Grasslands 16, 63-72.
849 IVORY, D.A., 1982. Evaluation of five white clover (*Trifolium repens*) cultivars and Kenya white clover (*T. semipilosum*) on the eastern Darling Downs of Queensland. Trop. Grasslands 16, 72-75.
850 IWU, M.M., and COURT, W.E., 1977. Root alkaloids of *Rauvolfia vomitoria* Afz. Planta medica 32, 88-99.
851 JACKS, T.J., HENSARLING, T.P., and YATSU, L.Y., 1972. Cucurbit seeds: I. Characterizations and uses of oils and proteins. A review. Econ. Bot. 26, 135-141.
852 JACKSON, M.G., 1977. Rice straw as livestock feed. World Animal Review 23, 25-31.
853 JACOB, A., and UEXKÜLL, H.V. 1963. Fertilizer Use. Nutrition and Manuring of Tropical Crops. 3rd ed. Verlagsgesellschaft für Ackerbau, Hannover.
854 JACOBSON, M., and CROSBY, D.G., (eds.), 1971. Naturally Occurring Insecticides. Dekker, New York.
855 JACQUEMARD, J.C., et AHIZI, P., 1981. L'induction des inflorescences mâles chez les palmiers pisifera. Oléagineux 36, 51-58.
856 JACQUET, M., VINCENT, J.-C., HAHN, J., et LOTODE, R., 1980. Le séchage artificiel des fèves de cacao. Café Cacao Thé 24, 43-56.
857 JACQUOT, M., et COURTOIS, B., 1983. Le Riz Pluvial. Maisonneuve et Larose, Paris.
858 JACQUY, P., 1973. La Culture du Pistachier en Tunisie. Ministère de l'Agriculture, Tunis.
859 JAGADISHCHANDRA, K.S., 1975. Recent studies on *Cymbopogon* Spreng. (aromatic grasses) with special reference to Indian taxa: taxonomy, cytogenetics, chemistry, and scope. J. Plantation Crops 3, 43-57.
860 JAGTIANI, J., CHAN, H.T., and SAKAI, W.S., 1988. Tropical Fruit Processing. Academic Press, San Diego, CA.
861 JAGWITZ-BIEGNITZ, F. VON, (ed.) 1984. Anbau der Zuckerrübe in wärmeren Ländern. In: Institut für Zuckerrübenforschung (Hrsg.). Geschichte der Zuckerrübe, 175-187. Bartens, Berlin.
862 JAHN, S. AL A., MUSNAD, H.A., and BURGSTALLER, H., 1986. The tree that purifies water. Cultivating multipurpose Moringaceae in the Sudan. Unasylva 152, Vol. 38 (2), 23-28.
863 JANARDHANAN, K.K., GUPTA, M.L., and HUSAIN, A., 1982. A new commercial strain of ergot adapted from a wild grass. Planta medica 44, 166-167.
864 JANGPO, B., and SHARMA, K.D., 1984. Cultivation of kala zira in Himachal Pradesh. Indian Farming 35 (5), 9, 14.
865 JANICK, J., 1974. The apple in Java. HortScience 9, 13-15.
866 JANICK, J., WOODS, F.W., SCHERY, R.W., and RUTTAN, V.W., 1974. Plant Science. An Introduction to World Crops. 2nd ed. Freeman, San Francisco.
867 JANIYA, J.D., and MOODY, K., 1984. Use of azolla to suppress weeds in transplanted rice. Trop. Pest Management 30, 1-6.
868 JANSEN, R.K., 1981. Systematics of *Spilanthes* (Compositae: Heliantheae). Systematic Bot. 6, 231-257.
869 JANZEN, D.H., 1973. Tropical agrosystems. Science 182, 1212-1219.
870 JARITZ, G., 1982. Amélioration des Herbages et Cultures Fourragères dans le Nord-Ouest de la Tunesie: Étude Particulière des Prairies de Tréfles-Graminées avec *Trifolium subterraneum*. Schriftenreihe GTZ No. 119, GTZ, Eschborn.

871 JARITZ, G., et SCHÜLKE, E., 1972. Premières Expériences sur les Pâturages à Base de Trèfle Souterrain en Grande Culture dans le Nord-Ouest de la Tunisie. Doc. Tech. 63, INRAT, Tunis.
872 JARMAN, C.G., CANNING, A.J., and MYKOLUK, S., 1978. Cultivation, extractation and processing of ramie fibre: a review. Trop. Sci. 20, 91-116.
873 JARMAN, C.G., and JAYASUNDERA, D.S., 1975. The Extraction and Processing of Coconut Fibre. G 94, Trop. Prod. Inst., London.
874 JARMAN, C.G., and ROBBINS, S.R.J., 1986. An Industrial Profile of Coconut Fibre Extraction and Processing. G 189, Trop. Devel. Res. Inst., London.
875 JAUHAR, P.P., 1981. Cytogenetics and Breeding of Pearl Millet and Related Species. Progress and Topics in Cytogenetics, Vol. I. Liss, New York.
876 JAW-KAI WANG (ed.), 1983. Taro. A Review of *Colocasia esculenta* and its Potentials. University of Hawaii Press, Honolulu.
877 JAYASANKAR, N.P., and BAVAPPA, K.V.A., 1986. Coconut root (wilt) disease. Past studies, present status and future strategy. Indian J. Agric. Sci. 56, 309-328.
878 JEFFREY, C., 1973. Biological Nomenclature. Edward Arnold, London.
879 JEFFREY, C., 1978. Further notes on Cucurbitaceae. IV. Some New-World taxa. Kew Bull. 33, 347-380.
880 JEFFREYS, M.D.W., 1967. Pre-Columbian maize in Southern Africa. Nature 215, 695-697.
881 JENKINS, G.H., 1966. Introduction to Cane Sugar Technology. Elsevier, Amsterdam.
882 JENNINGS, D.L., 1970. Cassava in Africa. Field Crop Abstr. 23, 271-278.
883 JENSMA, J.R., 1973. The sunflower has emigrated to the tropics and is doing well. Riv. Agric. Subtrop. Trop. 67, 89-92.
884 JEYASINGHAM, T., 1974. Rubberwood is abundant and accessible; will it ever be successfully exploited?. Quart. J. Rubber Res. Inst. Sri Lanka 51, 13-15.
885 JHORAR, B.S., SOLANKI, K.R., JATASRA, D.S., and GREWAL, R.P.S., 1985. Inheritance of gum content in clusterbean under different environments. Indian J. Genetics 45, 16-20.
886 JIKA, N.I., ST.-PIERRE, C.A., et DENIS, J.C., 1980. L'adaptation de cultivars de sorgho-grain à différent régimes hydriques. Can. J. Plant Sci. 60, 233-239.
887 JIMEMENZ, M.F., and ULRICH, D., 1982. Calixim, a powerful weapon against black sigatoka in bananas in Latin America. BASF Agric. News 3/82, 19-21.
888 JOAS, J., 1982. Les mombins: Des possibilités technologiques intéressantes. Fruits 37, 727-729.
889 JOHANNESSEN, C.L., and PARKER, A.Z., 1989. Maize ears sculptured in 12th and 13th century A.D. India as indicators of Pre-Columbian diffusion. Econ. Bot. 43, 164-180.
890 JOHN, C.K., and NEWSAM, A., 1969. Continuous coagulation of hevea latex. Planters' Bull. R.R.I.M. 105, 289-293.
891 JOHNSON, D., 1972. The carnauba wax palm (*Copernicia prunifera*). III. Exploitation and plantation growth. Principes 16, 111-114.
892 JOHNSON, D., 1973. The botany, origin, and spread of the cashew, *Anacardium occidentale* L. J. Plantation Crops 1, 1-7.
893 JOHNSON, E.J., and JOHNSON, T.J., 1976. Economic plants in a rural nigerian market. Econ. Bot. 30, 375-381.
894 JOHNSON, J.D., and HINMAN, C.W., 1980. Oils and rubber from arid land plants. Science 208, 460-464.
895 JOHNSON, L.A., SULEIMAN, T.M., and LUSAS, E.W., 1979. Sesame protein: a review and prospectus. J. Amer. Oil Chem. Soc. 56, 463-468.
896 JOHNSON, R.T., ALEXANDER, J.T., RUSH, G.E., and HAWKES, G.R. (eds.), 1971. Advances in Sugarbeet Production: Principles and Practices. Iowa State Univ. Press, Ames, Iowa.
897 JOHNSON, S.S., and ROGERS, R.T., 1974. Progress in mechanization of wine grapes. Calif. Agric. 28 (8), 4-6.

898 JOHNSON, V.A., SCHMIDT, J.W., and MATTERN, P.J., 1968. Cereal breeding for better protein impact. Econ. Bot. 22, 16-25.

899 JOLLY, M.S., SEN, S.K., and DAS, M.G., 1976. Silk from the forest. Unasylva 28 (No. 114), 20-23.

900 JONES, A., 1973. The Market for Mango Products with Particular Reference to the United Kingdom. G 74, Trop. Prod. Inst., London.

901 JONES, A., 1974. World Protein Resources. MTP Publ. Co., St. Leonardgate, Lancaster, England.

902 JONES, C.A., 1979. The potential of *Andropogon gayanus* Kunth in the oxisol and ultisol savannas of tropical America. Herbage Abstracts 49, 1-8.

903 JONES, H.A., and MANN, L.K., 1963. Onions and Their Allies. Leonard Hill, London.

904 JONES, R.J., 1979. The value of *Leucena leucocephala* as a feed for ruminants in the tropics. World Animal Rev. No. 31, 13-23.

905 JONES, S.F., 1980. The World Market for Desiccated Coconut. Trop. Sci. 22, 277-285.

906 JORDAN, D.C., 1984. Rhizobiaceae. In: KRIEG, N.R. (ed.): Bergey's Manual of Systematic Bacteriology, Vol. I, 234-256. Williams and Wilkins, Baltimore.

907 JOSEPH, J., 1980. The nutmeg - its botany, agronomy, production, composition and uses. J. Plantation Crops 8, 61-72.

908 JOSEPH, J., and SAHA, A., 1978. Photoperiod insensitivity in jute. Indian J. Genetics 38, 313-317.

909 JOSEPH, R., 1971. The economic significance of *Cannabis sativa* in the Moroccan Rif. Econ. Bot. 27, 235-240.

910 JOUGHIN, J., 1986. The Market for Processed Tropical Fruit. G 196, Trop. Devel. Res. Inst., London.

911 JOURDAIN, J.M., 1989. Le Kiwi. Tome 2. Techiniques de production. Centre Technique Interprofessionel des Fruits et Légumes, Paris.

912 JOURNAL OF PLANTATION CROPS, 1980. National seminar on ginger and turmeric, 8-9 April, 1980, Calicut, Kerala, India. J. Plantation Crops 8, 48-52.

913 JOURNAL OF THE AMERICAN OIL CHEMISTS SOCIETY, 1975. Symposium: Soy Proteins. 52, 237A-282A.

914 JOY, C., 1987. Selected European Markets for Speciality and Tropical Fruit and Vegetables. G 201, Trop. Devel. Res. Inst., London.

915 JUAREZ, J., NAVARRO, L., et GUARDIOLA, J.L., 1976. Obtention de plants nucellaires de divers cultivars de clémentiers au moyen de la culture de nucelle in vitro. Fruits 31, 751-762.

916 JUD, B., und LÖSSL, U., 1986. Tarakernmehl - ein Verdickungsmittel mit Zukunft. Intern. Z. Lebensmittel-Technologie Verfahrenstechnik 37 (1).

917 JUDD, B.I., 1979. Handbook of Tropical Forage Grasses. Garland STPM Press, New York.

918 JUGENHEIMER, R.W., 1976. Corn: Improvement, Seed Production and Uses. John Wiley, Chichester.

919 KADER, A.A., CHORDAS, A., and YU LI, 1984. Harvest and postharvest handling of Chinese date. Calif. Agric. 38 (1/2), 8-9.

920 KALDY, M.S., 1972. Protein yield of various crops as related to protein value. Econ. Bot. 26, 142-144.

921 KALI UND SALZ AG, n.d. German Potash for World Agriculture. Kali und Salz AG, Hannover.

922 KALIDURAI, M., and KANNAYAN, S., 1989. Effect of *Sesbania rostrata* on nitrogen uptake and yield of lowland rice. J. Agron. Crop Sci. 163, 284-288.

923 KALMBACHER, R.S., HODGES, E.M., and MARTIN, F.G., 1980. Effect of plant height and cutting height on yield and quality of *Indigofera hirsuta*. Trop. Grasslands 14, 14-18.

924 KALPAGE, F.S.C.P., 1976. Tropical Soils. Classification, Fertility and Management. Macmillan, New Delhi.

925 KALSHOVEN, L.G.E., 1981. Pests of Crops in Indonesia. Elsevier Science Publ., Amsterdam.

926 KALU, B.A., 1989. Seed yam production by minisett technique: evaluation of three *Dioscorea* species in the Guinea and derived savanna zone of Nigeria. Trop. Agric. (Trinidad) 66, 83-86.

927 KALU, B.A., NORMAN, J.C., PAL, U.R., and ADEDZWA, D.K., 1989. Seed yam multiplication by the mini-sett technique in three yam species in a tropical Guinea savanna location. Exp. Agric. 25, 181-188.

928 KAMAL, R., YADAV, R., and SHARMA, G.L., 1987. Diosgenin content in fenugreek collected from different geographical regions in south India. Indian J. Agric. Sci. 57, 674-676.

929 KAMATH, J., 1973. The Small Scale Manufacture of Soluble Coffee. G 82, Trop. Prod. Inst., London.

930 KANG, B.T., 1979. Soil fertility management for maize production in the humid and sub-humid regions of West Africa. IITA 1st Ann. Res. Conf., IITA, Ibadan, Nigeria.

931 KANG, B.T., ISLAM, R., SANDERS, F.E., and AYANABA, A., 1980. Effect of phosphate fertilization and inoculation with VA- mycorrhizal fungi on performance of cassava (*Manihot esculenta* Crantz) grown on an alfisol. Field Crops Res. 3, 83-94.

932 KANWAR, J.S., 1969. Sugarbeet for saline and alkali soils of Northern India. Indian Farming 19 (2), 5-6.

933 KAR, A., and MITAL, H.C., 1981. The study of shea butter. 6. The extraction of shea butter. Qualitas Plantarum 31, 67-70.

934 KARAWYA, M.S., WAHAB, S.M.A., and HIFNAWY, M.S., 1979. Essential oil of *Xylopia aethiopica* fruit. Planta medica 37, 57-59.

935 KARAWYA, M.S., WASSEL, G.M., BAGHDADI, H.H., and AMMAR, N.M., 1980. Mucilagenous contents of certain Egyptian plants. Planta medica 38, 73-78.

936 KARIKARI, S.K., 1971. Economic importance of bambarra groundnut. World Crops 23, 195-196.

937 KASASIAN, L., 1971. Weed Control in the Tropics. Leonard Hill, London.

938 KASASIAN, R., 1982. Bibliography of Rice Parboiling. G 158, Trop. Prod. Inst., London.

939 KATIYAR, R.K., SARAN, G., and GIRI, G., 1986. Evaluation of *Brassica carinata* as a new oilseed crop in India. Exp. Agric. 22, 67-70.

940 KATS, K., 1965. Some achievements in the cultivation and production of tea in the USSR. World Crops 17, 57-61.

941 KAUR, M., DASS, H.C., and RANDHAWA, G.S., 1973. Systematic status of round melon (*Citrullus vulgaris* var. *fistulosus*) as studied by leaf phenolics. Curr. Sci. 42, 730-731.

942 KAWAGUCHI, K., 1966. Tropical paddy soils. Jap. Agr. Res. Quart. 1, 7-11.

943 KAWATANI, T., KANEKI, Y., TANABE, T., and TAKAHASHI, T., 1978. Cultivation of kaa he-e (*Stevia rebaudiana* Bertoni). 3. Response of kaa he-e to fertilizer application amount and to nitrogen fertilization rates. Japan. J. Trop. Agric. 21, 165-178.

944 KAY, D.E., 1967. Banana Products. G 32, Trop. Prod. Inst., London.

945 KAY, D.E., 1970. The production and marketing of pepper. Trop. Sci. 12, 201-218.

946 KAY, D.E., 1973. Root Crops. Trop. Prod. Inst., London.

947 KAY, D.E., 1979. Food Legumes. TPI Crop Digest No. 3. Trop. Products Inst., London.

948 KAYONGO-MALE, H., THOMAS, J.N., ULLREY, D.E., DEANS, R.J., and ARROYA AGUILU, J.A., 1976. Chemical composition and digestibility of tropical grasses. J. Agric. Univ. Puerto Rico 60, 186-200.

949 KAYS, S.J., 1975. Production of New Zealand spinach (*Tetragonia expansa* Murr.) at high plant densities. J. Hort. Sci. 50, 135-141.

950 KAYS, S.J., GAINES, T.P., and KAYS, W.R., 1979. Changes in the composition of the tuber crop *Oxalis tuberosa* Molina during storage. Scientia Hortic. 11, 45-50.

951 KEARNS, H.G.H., 1963. Crop protection in the tropics. Ann. Appl. Biol. 51, 353-360.

952 KEELER, J.T., and FUKUNAGA, 1968. The Economic and Horticultural Aspects of Growing *Macadamia* Nuts Commercially in Hawaii. Agric. Econ. Bull. 27, Hawaii Agric. Exp. Sta., Honolulu.

953 KEMMLER, G., 1974. Modern Aspects of Wheat Manuring. Intern. Potash Inst., Bern.

954 KEMPANNA, C., 1974. Prospects for medicinal plants in Indian agriculture. World Crops 26, 166-168.

955 KENT, N.L., 1975. Technology of Cereals. 2nd ed. Pergamon Press, Oxford.

956 KERR, R.W., 1968. Chemistry and Industry of Starch. 2nd ed. Academic Press, New York.

957 KESDEN, D., and WILL, A.A.JR., 1988. Purslane: A ubiquitous garden weed with nutritional potential. Proc. Florida State Hort. Soc. 100, 195-197.

958 KESTER, E.B., 1951. Minor oil-producing crops of the United States. Econ. Bot. 5, 38-59.

959 KHAN, H.H., BHASKARA RAO, E.V.V., and NAYAR, N.M., 1979. International Cashew Symposium, 12-15 March, 1979, Cochin (Kerala, India): summary report and recommendations. J. Plantation Crops 7, 54-60.

960 KHANDUJA, S.D., 1972. India: growing grapes in a difficult climate. Span 15, 95-96.

961 KHELIL, A., et KELLAL, A., 1980. Possibilités de culture et délimitation des zones à vocation pistachier en Algérie. Fruits 35, 177-185.

962 KHOSHOO, T.N., 1955. Cyto-taxonomy of Indian species of *Citrullus*. Curr. Sci. 24, 377-378.

963 KHUSH, G.S., 1977. Disease and insect resistance in rice. Adv. Agron. 29, 265-343.

964 KHYBRI, M.L., 1975. Fuel and fodder from ravine lands. Indian Farming 24 (11), 21-23.

965 KIMBER, A.J., 1972. The sweet potato in subsistence agriculture. Papua New Guinea Agric. J. 23, 80-102.

966 KINCH, D.M., WANG, J.K., and STROHMAN, R.E., 1961. Equipment for Husking *Macadamia* Nuts. Hawaii Agric. Exp. Sta. Bull. 126, Honolulu.

967 KING, K.W., 1971. The place of vegetables in meeting the food needs in emerging nations. Econ. Bot. 25, 6-11.

968 KING, P., 1975. The Market for French Beans in Selected Western European Countries. G 92, Trop. Prod. Inst., London.

969 KING, P.I., 1979. Selected European markets for dehydrated vegetables - the potential for a developing country. Trop. Sci. 21, 231-247.

970 KING, S.R., and GERSHOFF, S.N., 1987. Nutritional evaluation of three underexploited Andean tubers: *Oxalis tuberosa* (Oxalidaceae), *Ullucus tuberosus* (Basellaceae), and *Tropaeolum tuberosum* (Tropaeolaceae). Econ. Bot. 41, 503-511.

971 KINGSHORN, A.D., and SOEJARTO, D.D., 1986. Sweetening agents of plant origin. CRC Critical Rev. Plant Sci. 4, 79-120.

972 KIRBY, R.H., 1963. Vegetable Fibres. Leonard Hill, London.

973 KIRK, J.T.O., and ORAM, R.N., 1981. Isolation of erucic acid-free lines of *Brassica juncea*: Indian mustard now a potential oilseed crop in Australia. J. Aust. Inst. Agric. Sci. 47, 51-52.

974 KITAGAWA, H., and GLUCINA, P.G., 1984. Persimmon. Culture in New Zealand. DSIR Information Ser. Nr. 159, Wellington, N. Z.

975 KLATT, R.C. (ed.), 1988. Wheat Production Constraints in Tropical Environments. CIMMYT, México.

976 KLOOS, H., and MCCULLOUGH, F.S., 1982. Plant molluscicides. Planta medica 46, 195-209.

977 KNORR, L.C., 1973. Citrus Diseases and Disorders. Univ. of Florida Press, Gainesville, Florida.

978 KNOTT, J.E., and DEANON, J.R.Jr., (eds.), 1967. Vegetable Production in Southeast Asia. Univ. Philipp., Los Baños.

979 KNOWLES, P.F., 1975. Recent research on safflower, sunflower, and cotton. J. Amer. Oil Chemists' Soc. 52, 374-376.

980 KOHEL, R.J., and LEWIS, C.F. (eds.), 1984. Cotton. Amer. Soc. Agron., Washington, D. C.
981 KRAMER, F.L., 1957. The pepper tree, *Schinus molle* L. Econ. Bot. 11, 322-326.
982 KRANZ, J., SCHMUTTERER, H., and KOCH, W., (eds.) 1977. Diseases, Pests and Weeds in Tropical Crops. Parey, Berlin.
983 KRAUSS, B.H., and HAMILTON, R.A., 1970. Bibliography of Macadamia. II Parts 1970-1972. Res. Rep. 176, Hawaii Agric. Exp. Sta., Honolulu.
984 KRIKORIAN, A.D., 1985. Growth mode and leaf arrangement in *Catha edulis* (Kat). Econ. Bot. 39, 514-521.
985 KRIKORIAN, A.D., and KANN, R.P., 1986. Oil palm improvement via tissue culture. Plant Breeding Reviews 4, 175-202.
986 KRIKORIAN, A.D., and LEDBETTER, M.C., 1975. Some observations on the cultivation of opium poppy (*Papaver somniferum* L.) for its latex. Bot. Rev. 41, 30-103.
987 KRISHNAMURTHY, N., MATHEW, A.G., NAMBUDIRI, E.S., SHIVASHANKAR, S., LEWIS, Y.S., and NATARAJAN, C.P., 1976. Oil and oleoresin of turmeric. Trop. Sci. 18, 37-45.
988 KROCHMAL, A., and KROCHMAL, C., 1974. The complete Illustrated Book of Dyes from Natural Sources. Doubleday & Co., Garden City, N. Y.
989 KRUKOFF, B.A., and BARNEBY, R.C., 1974. Conspectus of the genus *Erythrina*. Lloydia 37, 332-459.
990 KUHNE, F.A., and LOGIE, J.M., 1977. Granadilla longevity improved by grafting. Citrus and Subtropical Fruit J. No. 524, 13-14.
991 KUIJT, J., 1969. The Biology of Parasitic Flowering Plants. Univ. California Press, Berkeley.
992 KUMAR, A., ABROL, I.P., and DARGAN, K.S., 1981. Shaftal - a promising fodder for alkali soils. Indian Farming 31 (4), 18-19.
993 KUMAR, D., and WAREING, P.F., 1973. Studies on tuberization in *Solanum andigena*. I. Evidence for the existence and movement of a specific tuberization stimulus. New Phytol. 72, 283-287.
994 KUMAR, D., and WAREING, P.F., 1974. Studies on tuberization of *Solanum andigena*. II. Growth hormones and tuberization. New Phytol. 73, 833-840.
995 KUMAR, S., 1973. Famous plants: isubgol. Botanica (India) 24, 39-42.
996 KUNDU, B.C., 1956. Jute, world's foremost fibre. Econ. Bot. 10, 103-133, 203-240.
997 KUNDU, B.C., BASAK, K.C., and SARCAR, P.B., 1959. Jute in India. Indian Central Jute Committee, Calcutta.
998 KUNG, P., 1975. Farm Crops of China. 1. Distribution and major crops. World Crops 27, 55-64.
999 KURUP, C.G.R., (ed.), 1967. Handbook of Agriculture. 3rd ed. Indian Council Agric. Res., New Delhi.
1000 KUSHARI, D.P., 1987. Development of inoculum banks of Azolla for its large-scale utilization as biofertilizer. Indian Farming 37 (2), 22-23, 35.
1001 LAFLECHE, D., et BOVE, J.M., 1970. Mycoplasmes dans les agrumes atteints de "greening", de "stubborn" ou de maladies similaires. Fruits 25, 455-465.
1002 LAGIERE, R., 1966. Le Cotonnier, Techniques Agricoles et Productions Tropicales. Maisonneuve et Larose, Paris.
1003 LAL, R., 1987. Tropical Ecology and Physical Edaphology. John Wiley, New York.
1004 LAL, R., and GREENLAND, D.J. (eds.), 1979. Soil Physical Properties and Crop Production in the Tropics. Wiley, Chichester.
1005 LAL, S., 1981. Production technology for urd. Indian Farming 31 (5), 81-83.
1006 LAL, S., and YADAV, D.S., 1980. Pulses and oilseeds ideally suited for drylands. Indian Farming 30 (6), 3-6.
1007 LAMB, K.P., 1974. Economic Entomology in the Tropics. Academic Press, London.
1008 LAMBERTI, F., and TAYLOR, C.E. (eds.), 1979. Root-Knot Nematodes (*Meloidogyne* Species). Systematics, Biology and Control. Academic Press, London.

1009 LANCASTER, P.A., INGRAM, J.S., LIM, M.Y., and COURSEY, D.G., 1982. Traditional cassava-based foods: survey of processing. Econ. Bot. 36, 12-45.
1010 LANDSBERG, H.E., LIPPMANN, H., PAFFEN, K.H., and TROLL, C., 1965. World Maps of Climatology. 2nd ed. Springer, Berlin.
1011 LANE, E.V., 1957. Piquí-á - potential source of vegetable oil for an oil-starving world. Econ. Bot. 11, 187-207.
1012 LAROUSSILHE, F. DE, 1980. Le Manguier. Maisonneuve et Larose, Paris.
1013 LARUE, M., 1975. L'*Actinidia sinensis* et sa culture. Fruits 30, 45-50.
1014 LARUE, T.A., and PATTERSON, T.G., 1981. How much nitrogen do legumes fix? Adv. Agron. 34, 15-38.
1015 LASSOUDIERE, A., 1969. La papaye. Récolte, conditionnement, exportation, produits transformés. Fruits 24, 491-502.
1016 LAURENCE, R.C.N., 1973. Improvement of groundnut pod-filling by lime and gypsum application to some soils in northern Malawi. Exp. Agric. 9, 353-360.
1017 LAVABRE, E.M., 1970. Insectes Nuisibles aux Cultures Tropicales. Maisonneuve et Larose, Paris.
1018 LAVABRE, E.M., 1977. Les Mirides du Cacaoyer. Maisonneuve et Larose, Paris.
1019 LAVILLE, E., 1979. Utilisation d'un nouveau fongicide systématique: l'Aliette, dans la lutte contre la gommose à *Phytophthora* des agrumes. Fruits 34, 35-41.
1020 LAWANI, S.M., DEVOS, P., and ODUBANJO, M.O., 1977. A bibliography of plantains and other cooking bananas. Paradisiaca (IITA), No. 2, 1-95.
1021 LAWRENCE, A.A., 1973. Natural Gums for Edible Purposes. Noyes Data Corporation, Park Ridge, New Jersey.
1022 LAWRIE, R.A., 1970. Proteins as Human Foods. Butterworths, London.
1023 LEAKEY, C.L.A., and WILLS, J.B. (eds.), 1977. Food Crops of the Lowland Tropics. Oxford Univ. Press, Oxford.
1024 LEAL, F., 1989. On the history, origin and taxonomy of the pineapple. Interciencia 14, 235-241.
1025 LECLERCQ, P., 1984. France's contribution to sunflower breeding. Span 27, 61-63.
1026 LE DIVIDICH, J., GEOFFROY, CANOPE, I., and CHENOST, M., 1976. Using waste bananas as animal feed. World Animal Review No. 20, 22-30.
1027 LEE, C.W., and PALZKILL, D.A., 1984. Propagation of jojoba by single node cuttings. HortScience 19, 841-842.
1028 LEE, G.R., PROCTOR, F., and THOMPSON, A.K., 1973. Transport of papaya fruits from Trinidad to Britain. Trop. Agric. 50, 303-306.
1029 LEE, S.H., 1982. Vegetable crops growing in China. Scientia Horticulturae 17, 201-209.
1030 LEENHOUTS, P.W., 1971. A revision of *Dimocarpus* (Sapindaceae). Blumea 19, 113-131.
1031 LEES, P., 1979. Triticale. Man-made grain of the future. World Farming 21 (6), 40-44.
1032 LEES, P., 1981. Fenugreek: a crop that could bring a rise in food supply and a fall in population growth. World Farming 23 (5), 14-18.
1033 LEES, P., 1981. Sainfoin and forage peas, two legumes with great potential. World Farming 23 (5), 22, 46-47.
1034 LEES, P., 1983. The rediscovery of amaranth. World Farming Agrimanagement 25 (6), 6, 24-25.
1035 LE HOUEROU, H.N. (ed.), 1980. Browse in Africa. The Current Stage of Knowledge. ILCA, Addis Abeba.
1036 LEITAO, E.L., 1969. O mate. 3. - Tecnologia do mate. Lavoura 72, Nov. - Dec., 16-20.
1037 LENT, R., 1966. The origin of the cauliflorous inflorescence of *Theobroma cacao*. Turrialba 16, 352-358.
1038 LEON, J., 1964. The "maca" (*Lepidium meyenii*), a little known food plant of Peru. Econ. Bot. 18, 122-127.

1039 LEON, J., 1987. Botánica de los Cultivos Tropicales. Inst. Interamer. Cien. Agric., San José.
1040 LEON, J., and SGARAVATTI, E., 1971. Tropical Pastures: Grasses and Legumes. FAO, Rome.
1041 LEONARD, W.H., and MARTIN, J.H., 1963. Cereal Crops. Macmillan, New York.
1042 LEONG, S.K., and YOON, P.K., 1985. Young budding in hevea cultivation. Planters' Bull. 184, 59-67.
1043 LE PELLEY, R.H., 1968. Pests of Coffee. Longmans, London.
1044 LE SAINT, J.P., et NUCE DE LAMOTHE, M. DE, 1987. Les hybrides de cocotiers Nains: performances et intérêt. Oléagineux 42, 353-359.
1045 LESINS, K.A., and LESINS, I., 1979. Genus *Medicago* (Leguminosae). A Taxogenetic Study. W. Junk, The Hague.
1046 LESLEIGHTER, L.C., and SHELTON, H.M., 1986. A method for enhancing establishment of *Leucaena leucocephala* (Lam.) de Wit in infertile acid soils. Trop. Grasslands 20, 36-39.
1047 LEVER, R.J.A.W., 1969. Pests of the Coconut Palm. FAO Agric. Studies 77, FAO, Rome.
1048 LEVY, A., PALEVITCH, D., and LAVIE, D., 1981. Genetic improvement of *Papaver bracteatum*: heritability and selection response of thebaine and seed yield. Planta medica 43, 71-76.
1049 LEVY, D., 1984 Cultivated *Solanum tuberosum* L. as a source for the selection of cultivars adapted to hot climates. Trop. Agric. (Trinidad) 61, 167-170.
1050 LI CHENKANG, 1985. China wax and the China wax scale insect. World Animal Rev. 55, 26-33.
1051 LI, HUI-LIN, 1970. The origin of cultivated plants in Southeast Asia. Econ. Bot. 24, 3-19.
1052 LI, HUI-LIN, 1974. The origin and use of *Cannabis* in Eastern Asia. Linguistic-cultural implications. Econ. Bot. 28, 293-301.
1053 LI, P.H. (ed.), 1985. Potato Physiology. Academic Press, Orlando, Florida.
1054 LIANG, J., 1981. Le litchi, origine, utilisation et développement de sa culture. J. Agric. tradit. Bot. appl. 28, 259-270.
1055 LICHOU, J., 1980. Quelques méthodes de conduites palissées des fruitiers témpérés expérimentées à île de La Réunion. Fruits 35, 369-377.
1056 LICHOU, J., 1980. La multiplication du pommier en zone tropicale. Fruits 35, 769-777.
1057 LICHOU, J., 1981. La conduite du pêcher. Fruits 36, 37-46.
1058 LIENER, I.E., (ed.), 1969. Toxic Constituents of Plant Foodstuffs. Academic Press, New York.
1059 LILLEY, J.M., FUKAI, S., and HICKS, L.N., 1988. Growth and yield of perennial cassava in the subhumid tropics. Field Crops Res. 18, 45-56.
1060 LIM, G., TEE, L.L., and AVADHANI, P.N., 1982. Cultivating padi-straw mushrooms on industrial wastes. Phil. Agric. 65, 209-214.
1061 LINDNER, M.W., 1974. A monograph on the kaki-fig. Z. Lebensm. Unters. -Forsch. 155, 65-71.
1062 LING, K.C., 1972. Rice Virus Diseases. Intern. Rice Res. Inst., Los Baños.
1063 LIPPMANN, D., 1978. Cultivation of *Passiflora edulis* S. - General Information on Passion Fruit Growing in Kenya. Schriftenreihe GTZ No. 62, Eschborn.
1064 LIU YUXIANG, and DENG YUSHENG, 1984. Development of cotton science and technology in China. Cot. Fib. Trop. 39, 61-68.
1065 LLERAS, E., 1985. *Acrocomia*, um gênero com grande potential. Newsletter, Useful Palms of Tropical America No 1, 3-5.
1066 LLOYD, D.L., 1981. Makarikari grass - (*Panicum coloratum* var. *makarikariense*) - a review with particular reference to Australia. Trop. Grasslands 15, 44-52.
1067 LOCH, D.S., 1977. *Brachiaria decumbens* (signal grass) - A review with particular reference to Australia. Trop. Grasslands 11, 141-157.

1068 LOCK, C.S., 1977. Leguminous cover crops for rubber smallholdings. Planters' Bull. 150, 83-97.

1069 LOCK, G.W., 1969. Sisal. 2nd ed. Longmans, London.

1070 LOCK, J.M., HALL, J.B., and ABBIW, D.K., 1977. The cultivation of melegueta pepper (*Aframomum melegueta*) in Ghana. Econ. Bot. 31, 321-330.

1071 LOCKHART-SMITH, C.J., and ELLIOT, R.G.H., 1974. Tanning of Hides and Skins. G 86, Trop. Prod. Inst., London.

1072 LONG, W.H., and HENSLEY, S.D., 1972. Insect pests of sugar cane. Ann. Rev. Entom. 17, 149-176.

1073 LONGERI, S.L., y HERRERA, O.A., 1972. Inoculación del poroto soya (*Glycine max* Merr.). Efecto de la inoculación en el rendimiento y en los contenidos de proteína y aceite del grano. Agric. Técn. (Chile) 32, 132-137.

1074 LOO, T.G., 1971. Small-scale Processing of Soybeans and Some Applications. Publication 294, Royal Trop. Inst., Amsterdam.

1075 LOOY, R. VAN, 1981. Les mini-huileries de palme. Oléagineux 36, 147-151.

1076 LOPEZ, A.S., 1979. Fermentation and organoleptic quality of cacao as affected by partial removal of pulp juices from the beans prior to curing. Rev. Theobroma 9, 25-37.

1077 LORENZI, H., 1982. Plantas Daninhas do Brasil. Lorenzi, Nova Odessa, S. P., Brasil.

1078 LOUSSERT, R., et BROUSSE, G., 1978. L'Olivier. Maisonneuve et Larose, Paris.

1079 LOVETT, J.V., and LAZENBY, A. (eds.), 1979. Australian Field Crops. Vol. 2. Tropical Cereals, Oilseeds, Grain Legumes and Other Crops. Angus and Robertson, Sydney, Australia.

1080 LOWE, S.B., MAHON, J.D., and HUNT, L.A., 1976. The effect of daylength on shoot growth and formation of root tubers in young plants of cassava (*Manihot esculenta* Crantz). Plant Sci. Letters 6, 57-62.

1081 LOWRY, J.B., 1989. Agronomy and forage quality of *Albizia lebbek* in the semi-arid tropics. Trop. Grasslands 23, 84-91.

1082 LOWRY, J.B., and CHEW, L., 1974. On the use of extracted anthocyanin as a food dye. Econ. Bot. 28, 61-62.

1083 LOZANO, J.C., BELLOTTI, A., REYES, J.A., HOWELER, R., LEIHNER, D.E., and DOLL, J., 1981. Field Problems in Cassava. 2nd ed. Centro Internacional de Agricultura Tropical, Cali, Colombia.

1084 LOZANO, J.C., and TERRY, E.R., 1977. Cassava Diseases and Their Control. Intern. Devel. Center Publ. No. 080, Ottawa, Canada.

1085 LUBEIGT, G., 1977. Le palmier à sucre, *Borassus flabellifer*, ses differents produits et la technologie associée en Birmanie. J. Agric. Tradit. Bot. Appl. 24, 311-340.

1086 LUCAS, G.B., 1965. Diseases of Tobacco. 2nd ed. Scarecrow Press, New York.

1087 LUCKWILL, L.C., and CUTTING, C.V., (eds.), 1970. Physiology of Tree Crops. Academic Press, London.

1088 LUH, B.S., 1980. Rice: Production and Utilization. AVI Publ. Co. Westport, Connecticut.

1089 LUMPKIN, T.A., and PLUCKNETT, D.L., 1982. *Azolla* as a Green Manure. Use and Management in Crop Production. Westview Press/ Boulder, Colorado.

1090 LUNTUNGAN, H.T., 1982. Progress report on coconut breeding in Indonesia. Yearly Progr. Rep. Coconut Res. Devel. FAO, Rome.

1091 LYND, J.O., and AUSMAN, T.R., 1989. Effects of P, Ca with four K levels on nodule histology, nitrogenase activity and improved 'Spanco' peanut yield. J. Plant Nutr. 12, 65-84.

1092 MAC FARLANE, N., 1975. The castor oil industry: a comparison of lubricants derived from castor oil, mineral oil and synthetics. Trop. Sci. 17, 217-218.

1093 MACHIN, D.H., 1977. Ramie as an animal feed: a review. Trop. Sci. 19, 187-195.

1094 MAC KEY, J., 1966. Species Relationship in *Triticum*. Hereditas Suppl. 2, 237-276.

1095 MACMILLAN, H.F., 1962. Tropical Planting and Gardening. Macmillan, London.

1096 MADAN, C.L., KAPUR, B.M., and GUPTA, U.S., 1966. Saffron. Econ. Bot. 20, 377-385.

1097 MADRID, M.T.JR., PUNZALAN, F.L., and LUBIGAN, R.T., 1972. Some Common Weeds and Their Control. Weed Sci. Soc. Philippines. Laguna.

1098 MAEDA, E., 1982. Cultivation and morphology of sago palm (*Metroxylon sagu*). Jap. J. Trop. Agric. 26, 169-176.

1099 MAESEN, L.J.G., VAN DER, 1972. *Cicer* L., a monograph of the genus, with special reference to the chickpea (*Cicer arietinum* L.), its ecology and cultivation. Mededel. Landbouwhogeschool Wageningen, 72-10.

1100 MAESEN, L.J.G., VAN DER, 1985. *Cajanus* DC. and *Atylosia* W. et A. (Leguminosae). Pudoc, Wageningen.

1101 MAESEN, L.J.G. VAN DER, and SOMAATMADJA, S. (eds.), 1989. Pulses. Plant Resources of South-East Asia (Prosea) No. 1. Pudoc, Wageningen.

1102 MAGAT, S.S., MARGATE, R.Z., and HABANA, J.A., 1988. Effects of increasing rates of sodium chloride (common salt) fertilization on coconut palms grown under an inland soil (Tropudalfs) of Mindanao, Philippines. Oléagineux 43, 13-17.

1103 MAGGS, D.H., 1973. Genetic resources in pistachio. Plant Genetic Resourc. Newsletter 29, 7-15.

1104 MAHESHWARI, S.K., DAHATONDE, B.N., YADAV, S., and GANGRADE, S.K., 1985. Intercropping of *Rauvolfia serpentina* for higher monetary returns. Indian J. Agric. Sci. 55, 332-334.

1105 MAHESHWARI, S.K., YADAV, S., GANGRADE, S.K., and TRIVEDI, K.C., 1988. Effect of fertilizers on growth, root and alkaloid yield of rauvolfia (*Rauvolfia serpentina*). Indian J. Agric. Sci. 58, 487-488.

1106 MAIDEN, J.H., 1889. The Useful Native Plants of Australia (Including Tasmania). Reprint 1975. Compendium, Melbourne, Australia.

1107 MAISTRE, J., 1964. Les Plantes à Épices. Maisonneuve et Larose, Paris.

1108 MAITI, S.N., NEOGI, A.K., and SEN, S., 1975. Effect of crop age on fibre quality (tenacity) in jute (*Corchorus* sp.). Current Sci. 44, 274-276.

1109 MAITI, S., RAOOF, M.A., SASTRY, K.S., and YADAVA, T.P., 1985. A review of sesamum diseases in India. Trop. Pest Management 31, 317-323.

1110 MAKKY, A.M., YOUSSEF, M.S.S., HATHOUT, M.K., and AL-NOUBY, H.M., 1978. Economics of introducing elephant grass to the Egyptian agriculture by cultivation in one-fourth of the maize and sorghum area. Agric. Res. Rev. 56 (7), 1-9.

1111 MALAGAMBA, P., and MONARES, A., 1988. True Potato Seed. Past and Present Uses. CIP, Lima, Peru.

1112 MALAN, E.F., MEULEN, A. VAN DER, LOEST, F.C., and STOFBERG, F.J., 1955. Avocado Culture in South Africa. Bull. 342, Dep. Agric. Techn. Serv., Pretoria.

1113 MALO, S.E., 1972. Mango and avocado: emerging fruits in world horticulture and trade. Florida State Hort. Soc. Proc. 84, 311-313.

1114 MALO, S.E., and MARTIN, F.W., 1979. Cultivation of neglected tropical fruits with promise. Part 7. The durian. S.E.A., USDA, New Orleans.

1115 MALYON, T., and HENMAN, A., 1980. No marihuana but plenty of hemp. Hemp fibre is back in favour. France is cashing in, so why doesn't Britain ? New Scientist 88, No. 1227, 433-443.

1116 MANCIOT, R., OLLAGNIER, M., et OCHS, R., 1979/1980. Nutrition minérale et fertilisation du cocotier dans le monde. Oléagineux 34, 499-515, 563-580; 35, 13-27.

1117 MANGELSDORF, P.C., 1974. Corn: Its Origin, Evolution, and Improvement. Harvard Univ. Press, Cambridge.

1118 MANICA, I., 1981. Fruticultura Tropical. 1. Maracujá. Editoria Agronômica Ceres, São Paulo, Brazil.

1119 MANICA, I., 1981. Fruticultura Tropica. 2. Manga. Editoria Agronômica Ceres, São Paulo, Brazil.

1120 MANN, J.D., 1978. Production of solasodine for the pharmaceutical industry. Adv. Agron. 30, 207-245.

1121 MANN, L.K., and STEARN, W.T., 1960. Rakkyo or ch'iao t'ou (*Allium chinense* G.Don, syn. *A. bakeri* Regel) a little known vegetable crop. Econ. Bot. 14, 69-83.
1122 MANNING, C.E.F., 1969. The Market for Vanilla Beans. G 42, Trop. Prod. Inst., London.
1123 MANNING, C.E.F., 1970. The Market for Cinnamon and Cassia and Their Essential Oils. G 44, Trop. Prod. Inst., London.
1124 MANNING, C.E.F., 1970. The Market for Natural Rubber with Particular Reference to the Competitive Status of Synthetic Rubber. G 47, Trop. Prod. Inst., London.
1125 MANTELL, C.L., 1949. The water-soluble gums - their botany, sources and utilization. Econ. Bot. 3, 3-31.
1126 MARECHAL, R., MASCHERPA, J.M., et STAINIER, F., 1978. Combinaisons et noms nouveaux dans les genres *Phaseolus*, *Minkelersia*, *Macroptilium*, *Ramirezella* et *Vigna*. Taxon 27, 199-202.
1127 MARGARIS, N., KOEDAM, A., and VOKOE, D. (eds.), 1982. Aromatic Plants. Martinus Nijhoff, The Hague.
1128 MARKLEY, K.S., 1955. Caranday - a source of palm wax. Econ. Bot. 9, 39-52.
1129 MARKLEY, K.S., 1956. Mbocayá or Paraguay cocopalm - an important source of oil. Econ. Bot. 10, 3-32.
1130 MARKLEY, K.S., 1971. The babassú oil palm of Brazil. Econ. Bot. 25, 267-304.
1131 MARQUIE, C., 1987. Utilisation alimentaire des dérivés des cotonniers sans gossypol. Cot. Fib. Trop. 42, 65-73.
1132 MARS, P.A., 1970. An Inquiry into the Feasibility of Producing Particle Board from Groundnut Husks in India. G 55, Trop. Prod. Inst., London.
1133 MARS, P.A., 1971. The Manufacture of Orange Squash in Developing Countries. G 53, Trop. Prod. Inst., London.
1134 MARSTON, A., and HOSTETTMANN, K., 1985. Plant molluscicides. Phytochemistry 24, 639-652.
1135 MARTER, A.D., 1981. Castor: Markets, Utilisation and Prospects. G 152. Trop. Prod. Inst., London.
1136 MARTIN, F.W., 1979. Vegetables for the hot humid tropics. Part 4. Sponge and bottle gourds, *Luffa* and *Lagenaria*. S.E.A., USDA, New Orleans.
1137 MARTIN, F.W., 1982. Okra, potential multiple-purpose crop for the temperate zone and tropics. Econ. Bot. 36, 340-345.
1138 MARTIN, F.W., (ed.) 1984. Handbook of Tropical Food Crops. CRC Press, Boca Raton, Florida.
1139 MARTIN, F.W., and CABANILLAS, E., 1976. Leren (*Calathea allouia*), a little known tuberous crop of the Caribbean. Econ. Bot. 30, 249-256.
1140 MARTIN, F.W., and DELPIN, H., 1978. Vegetables of the Hot Humid Tropics. Part 1. The Winged Bean, *Psophocarpus tetragonolobus*. S.E.A., USDA, New Orleans.
1141 MARTIN, F.W., and JONES, A., 1986. Breeding sweet potatoes. Plant Breeding Rev. 4, 313-346.
1142 MARTIN, F.W., and NAKASONE, H.Y., 1970. The edible species of *Passiflora*. Econ. Bot. 24, 333-343.
1143 MARTIN, F.W., and RUBERTE, R.M., 1975. Edible Leaves of the Tropics. Agency for Intern. Devel., Mayagüez, Puerto Rico.
1144 MARTIN, F.W., and TELEK, L., 1979. Vegetables for the Hot Humid Tropics. Part 6. Amaranth and Celosia, *Amaranthus* and *Celosia*. S.E.A., USDA, New Orleans.
1145 MARTIN, G., et GUICHARD, P.H., 1979. A propos de quatre palmiers spontanés d'Amérique Latin. Oléagineux 34, 375-381.
1146 MARTIN, J.H., LEONARD, W.H., and STAMP, D.L., 1976. Principles of Field Crop Production. 3rd ed. Macmillan, New York.
1147 MARTIN, P.J., RILEY, J., and DABEK, A.J., 1987. Clove tree yields in the islands of Zanzibar and Pemba. Exp. Agric. 23, 293-303.

1148 MARTIN, P.M.D., and GILMAN, G.A., 1976. A Consideration of the Mycotoxin Hypothesis with Special Reference to the Mycoflora of Maize, Sorghum, Wheat and Groundnuts. G 105, Trop. Prod. Inst., London.

1149 MARTIN, R.J., 1975. A review of carpet grass (*Axonopus affinis*) in relation to the improvement of carpet grass based pasture. Trop. Grasslands 9, 9-19.

1150 MARTIN, R.T., 1970. The role of coca in the history, religion, and medicine of South American Indians. Econ. Bot. 24, 422-437.

1151 MARTINEZ, C.P., 1972. Comportamiento de las nuevas variedades mejorades de arroz en varios países latino-americanos. Arroz (Colombia), 21, 6, 8-10, 12.

1152 MÄRZ, U., 1989. The Economics of Neem Production and Its Use in Pest Control. Farming Systems and Resource Economics in the Tropics, Vol. 5, Kiel.

1153 MARZOLA, D.L., and BARTHOLOMEW, D.P., 1979. Photosynthetic pathway and biomass energy production. Science 205, 555-559.

1154 MASSERON, A., THIBAULT, B., DECCENE, C., HILAIRE, C., et DALLE, E., 1988. Le Nashi. Centre Technique Interprofessionnel des Fruits et Légumes, Paris.

1155 MATTESON, P.C. (ed.), 1984. Proceedings of the International Workshop in Integrated Pest Control for Grain Legumes. Dep. de Difusão de Tecnologia, EMBRAPA, Brasília.

1156 MATTHEW, A.G., 1984. Recent trends in post harvest research on spices and plantation products. J. Plantation Crops 12, 94-97.

1157 MATTHEWS, G.A., 1989. Cotton Insect Pests and Their Management. Longman, London.

1158 MATTOS, M.D.L. DE, e MATTOS, C.C.L.V., 1976. Palmito juçara, *Euterpe edulis* Mart. (Palmae), uma espécie a plantar, manejar e proteger. Brasil Florestal 7 (27), 9-20.

1159 MAY, P.H., ANDERSON, A.B., FRAZÃO, J.M.F., and BALIK, M.J., 1985. Babassu palm in the agroforestry systems of Brazil's Mid-North. Agrofor. Systems 3, 275-295.

1160 MAYDELL, H.-J. VON, 1990. Trees and Shrubs of the Sahel. Margraf, Weikersheim.

1161 MAZUMDAR, B.C., 1979. Cape-gooseberry - the jam fruit of India. World Crops 31, 19, 23.

1162 MAZUMDAR, B.C., 1985. Water chestnut - the aquatic fruit: cultivation in India. World Crops 37, 42-44.

1163 MCCALL, W.W., SHIGEURA, G.T., and TAMIMI, Y.N., 1970. Windbreaks for Hawaii. Coop. Ext. Service, Circ. 438, Honolulu.

1164 MCCOSKER, T.H., and TEITZEL, J.K., 1975. A review of Guinea grass (*Panicum maximum*) for the wet tropics of Australia. Trop. Grasslands 9, 177-190.

1165 MCDAVID, C.R., and ALAMU, S., 1980. Effect of daylength on the growth and development of whole plants and rooted leaves of sweet potato (*Ipomoea batatas*). Trop. Agric. (Trinidad) 57, 111-119.

1166 MCDONALD, D., and MEHAN, V.K., 1989. Aflatoxin Contamination of Groundnut. Proc. Intern. Workshop, 6-9 Oct. 1987. ICRISAT, Patancheru, India.

1167 MCDOWELL, L.R., CONRAD, J.H., THOMAS, J.E., and HARRIS, L.E., 1974. Latin American Tables of Feed Composition. Univ. Florida, Gainesville, Florida.

1168 MCKAY, J.W., and CRANE, H.L., 1953. Chinese chestnut - a promising new orchard crop. Econ. Bot. 7, 228-242.

1169 MCLEOD, M.J., GUTTMAN, SH.I., and ESHBAUGH, W.H., 1982. Early evolution of chili pepper (*Capsicum*). Econ. Bot. 36, 361-368.

1170 MCNEIL, D.L., 1989. Factors affecting the field establishment of *Plantago ovata* Forsk. in northern Australia. Trop. Agric. (Trinidad) 66, 61-64.

1171 MEADOWS, D.J., SULE, V.K., PALOMAR, R., et JENSEN, P., 1980. L'utilisation du bois de cocotier. Oléagineux 35, 365-369.

1172 MEEUSE, A.D.J., 1962. The Cucurbitaceae of southern Africa. Bothalia 8, 1-111.

1173 MEHAN, V.K., MCDONALD, D., NIGAM, S.N., and LALITHA, B., 1981. Groundnut cultivars with seed resistant to invasion by *Aspergillus flavus*. Oléagineux 36, 501-507.

1174 MEJIA, M., 1984. *Andropogon gayanus* Kunth. Bibliografía Analítica. CIAT, Cali, Colombia.
1175 MENDEL, K., 1989. Zitrusfrüchte. In: REHM S., (Hrsg.) 1989. Spezieller Pflanzenbau in den Tropen und Subtropen. Handbuch der Landwirtschaft und Ernährung in den Entwicklungsländern, 2. Aufl. Bd. 4, 310-328. Ulmer, Stuttgart.
1176 MENNINGER, E.A., 1977. Edible Nuts of the World. Horticultural Books, Stuart, Florida.
1177 MENZEL, C.M., 1983. The control of floral initiation in lychee: a review. Scientia Hortic. 21, 201-215.
1178 MENZEL, C.M., 1984. The pattern and control of reproductive development in lychee: a review. Scientia Hortic. 22, 333-345.
1179 MENZEL, C.M., 1984. Potato as a potential crop for the lowland tropics. Trop. Agric. (Trinidad) 61, 162-166.
1180 MEREDITH, D. (ed.), 1955. The Grasses and Pastures of South Africa. Central News Agency, Cape Town.
1181 METWALLY, A.M., and KHAFAGY, S.M., 1975. Fixed oils from seeds of certain *Citrus* plants. Planta medica 27, 242-246.
1182 MEUNIER, J., 1975. Le "palmier à huile" américain *Elaeis melanococca*. Oléagineux 30, 52-61.
1183 MEYER, L.H., 1960. Food Chemistry. Reinhold, New York.
1184 MICHELLON, R., 1978. *Geranium rosat* in Réunion: rise in cropping intensity and genetic improvement prospects. Agron. Trop. 33, 80-89.
1185 MIDGLEY, S.J., TURNBULL, J.W., and JOHNSTON, R.D. (eds.), 1983. Casuarina Ecology, Management and Utilization. Proc. Intern. Workshop, Canberra, 17-21 August 1981. CSIRO, Melbourne.
1186 MIEGE, J., and LYONGA, S.N., 1982. Yams/ Ignames. Oxford Univ. Press, Oxford.
1187 MIKOLA, P. (ed.), 1980. Tropical Mycorrhiza Research. Clarendon Press, Oxford.
1188 MILLER, D.A., 1984. Forage Crops. McGraver-Hill, New York.
1189 MILLER, J.M., and BACKHAUS, R.A., 1986. Rubber content in diploid guayule (*Parthenium argentatum*): chromosomes, rubber variation, and implications for economic use. Econ. Bot. 40, 366-374.
1190 MILNER, M. (ed.), 1975. Nutritional Improvement of Food Legumes by Breeding. Wiley, New York.
1191 MIRACLE, M.P., 1966. Maize in Tropical Africa. Wisconsin Press, Madison.
1192 MISLEVY, P., KALMBACHER, R.S., and MARTIN, F.G., 1981. Cutting management of the tropical legume American jointvetch. Agron. J. 73, 771-775.
1193 MITRA, C.R., 1963. Neem. Indian Central Oilseeds Committee, Hyderabad.
1194 MODI, J.M., MEHTA, K.G., and RAJENDRA GUPTA, 1974. Isabgol, a dollar earner of North Gujarat. Indian Farming 23 (10), 17-19.
1195 MOHAMMED, SH., and WILSON, L.A., 1984. Modern systems of fruit growing and their application for the improvement of tropical fruit production. Trop. Agric. (Trinidad) 61, 137-142.
1196 MOHR, E.C.J., BAREN, F.A. VAN, and SCHUYLENBORGH, J. VAN, 1973. Tropical Soils: A Comprehensive Study of Their Genesis. 3rd ed. Van Goor en Zoon, den Haag.
1197 MÖHR, P.J., and DICKINSON, E.B., 1979. Mineral nutrition of maize. In: HÄFLINGER, E. (ed.). Maize. CIBA-Geigy, Basel.
1198 MOLL, H.A.J., 1987. The Economics of Oil Palm. Pudoc, Wageningen.
1199 MONSELISE, S.P., (ed.) 1986. Handbook of Fruit Set and Development. CRC Press, Boca Raton, Florida.
1200 MONTAGUT, G., 1972. Essais de culture du *Synsepalum dulcificum* au Dahomey. Fruits 27, 219-221.
1201 MONTALDO, A., 1972. Cultivo de raíces y tubérculos tropicales. Inst. Interamer. Cienc. Agric. OEA, Lima.
1202 MONTALDO, A., 1979. La Yuca o Manioca. Inst. Interam. Ciencias Agric., San José, Costa Rica.

1203 MOODY, K., 1986. Weed Control in Tropical Crops, 2 Vols. Weed Sci. Soc. Philippines, Laguna, Philippines.
1204 MOORE, D.J., 1980. A Simple Method of Collecting and Drying Papaya (Papaw) Latex to Produce Crude Papain. Rural Technology Guide No. 8; Trop. Prod. Inst., London.
1205 MOORE, D.J., and CAYGILL, J.C., 1979. Proteolytic activity of Malaysian pineapples. Trop. Sci. 21, 97-102.
1206 MOORE, R.M., (ed.), 1970. Australian Grasslands. Australian National Univ. Press, Canberra.
1207 MORAKINYO, J.A., and OLORODE, O., 1984. Cytogenetic and morphological studies on *Cola acuminata* (P. Beauv.) Schott et Endl., *Cola nitida* (Vent.) Schott et Endl. and the *C. acuminata* x *C. nitida* $F_1$ hybrid. Café Cacao Thé 28, 251-256.
1208 MORALLO-REJESUS, B., 1987. Botanical pest control research in the Philippines. Philip. Entomol. 7, 1-30.
1209 MORETTINI, A., 1977. L'Olivicoltura, sec. ed. Rama Editoriale degli Agricoltura, Roma.
1210 MOREUIL, C., et HUET, R., 1973. Le combava: culture et débouchés à Madagascar. Fruits 28, 703-708.
1211 MORTENSEN, E., and BULLARD, E.T., 1970. Handbook of Tropical and Subtropical Horticulture. Agency Intern. Devel., Washington.
1212 MORTIMER, W.G., 1901. Golden History of Coca, the Divine Plant of the Incas. Reprint 1974. Vail, New York.
1213 MORTON, J.F., 1960. The emblic (*Phyllanthus emblica* L.). Econ. Bot. 14, 119-128.
1214 MORTON, J.F., 1971. The wax gourd, a year-round Florida vegetable with unusual keeping quality. Florida State Hort. Soc. Proc. 84, 104-109.
1215 MORTON, J.F., 1972. Cocoyams (*Xanthosoma caracu*, *X. atrovirens* and *X. nigrum*), ancient root- and leaf-vegetables, gaining in economic importance. Florida State Hort. Soc. Proc. 85, 85-94.
1216 MORTON, J.F., 1977. Major Medicinal Plants. Botany, Culture and Uses. Thomas, Springfield, Illinois.
1217 MORTON, J.F., 1983. Rooibos tea, *Aspalathus linearis*, a caffeine-less, low-tannin beverage. Econ. Bot. 37, 164-173.
1218 MORTON, J.F., 1985. Indian almond (*Terminalia catappa*), salt-tolerant, useful, tropical tree with "nut" worthy of improvement. Econ. Bot. 39, 101-112.
1219 MORTON, J.F., 1988. Notes on distribution, propagation, and products of *Borassus* palms (Arecaceae). Econ. Bot. 42, 420-441.
1220 MORTON, J.F., SMITH, R.E., LUGO-LOPEZ, M.A., and ABRAMS, R., 1982. Pigeonpeas (*Cajanus cajan* Millsp.): A Valuable Crop for the Tropics. College Agric. Sci., Univ. Puerto Rico, Mayagüez, P.R.
1221 MORTON, J.K., 1962. Cytogenetic studies on West African Labiatae. J. Linn. Soc. Bot. 58, 231-283.
1222 MORTON, M., 1973. Rubber Technology. 2nd ed. Van Nostrand Reinhold, London.
1223 MOST, B.H., SUMMERFIELD, R.J., and BOXALL, M., 1979. Tropical plants with sweetening properties. Physiological and agronomic problems of protected cropping. 2. *Thaumatococeus daniellii*. Econ. Bot. 32, 321-335.
1224 MUIRHEAD, W.A., BLACKERELL, J., HUMPHREYS, E., and WHITE, R.J.G., 1989. The growth and nitrogen economy of rice under sprinkler and flood irrigation in South East Australia. I. Crop response and N uptake. Irrig. Sci. 10, 183-199.
1225 MULDOON, D.K., PEARSON, C.J., and WHEELER, J.L., 1982. The effect of temperature on growth and development of *Echinochloa* millets. Ann. Bot. 50, 665-672.
1226 MÜLLER, A.V., und WINNER, C., 1976. Wirkung der Stickstoffdüngung auf Ertrag und Qualität von Zuckerrüben bei unterschiedlicher Bestandesdichte. Zucker 29, 243-251.
1227 MUNIER, P., 1973. Le Palmier-Dattier. Maisonneuve et Larose, Paris.
1228 MUNIER, P., 1973. Le jujubier et sa culture. Fruits 28, 377-388.

1229 MUNIER, P., 1974. Le problème de l'origine du palmier-dattier et l'atlantide. Fruits 29, 235-240.
1230 MUNIER, P., 1981. Origine de la culture du palmier-dattier et sa propagation en Afrique. Fruits 36, 437-450, 531-556, 615-631, 689-706.
1231 MUNRO, J.M., 1987. Cotton. 2nd ed. Longman, London.
1232 MUNSHI, A.M., SINDHU, J.S., and BABA, G.H., 1989. Improved cultivation practices for saffron. Indian Farming 39 (3), 27-30.
1233 MÜNTZING, A., 1979. Triticale. Results and Problems. Fortschritte der Pflanzenzüchtung. Beiheft zur Zeitschrift für Pflanzenzüchtung No. 10. Parey, Berlin.
1234 MURALIDHARAN, A., and BABU, B., 1974. *Eucalyptus citriodora*, lucrative on plantation scale. Indian Farming 24 (2), 21-23.
1235 MURALIDHARAN, A., and NAIR, E.V.G., 1973. Response of *Eucalyptus citriodora* Hook. to different levels of nitrogen and pruning. Indian J. Agric. Sci. 43, 868-871.
1236 MURTHY, B.N.S., and SWAMY, P.N., 1989. Coriander, a multipurpose aromatic herb. Indian Farming 39 (2), 13-15.
1237 MURTY, U.R., RAO, N.G.P., KIRTI, P.B., and BHARATI, M., 1981. Fertilization in groundnut, *Arachis hypogaea* L. Oléagineux 36, 73-76.
1238 MUSSELMAN, L.J. (ed.), 1987. Parasitic Weeds in Agriculture. CRC Press, Boca Raton, Florida.
1239 MUTURI, P., and ARUNGA, R.O., 1988. Cashewnut shell liquid: a review of production and research in Kenya. Trop. Sci. 28, 201-218.
1240 NADAL, A.M., and CARANGAL, V.R., 1979. Performance of the main and ratoon crops of thirteen advanced rice selections under dry-seeded rainfed bunded conditions. Phil. J. Crop Sci. 4, 95-101.
1241 NAGABHUSHANAN, S., and NAIR, R.V., 1988. Asexual propagation of cocoa (*Theobroma cacao* L.). J. Plantation Crops 16, 143-145.
1242 NAGPAL, R.L., 1966. Fig Cultivation in India. Farm Bull. (N. S.) 42, Indian Council Agric. Res., New Delhi.
1243 NAGY, S., and SHAW, P.E. (eds.), 1980. Tropical and Subtropical Fruits. Composition, Properties and Uses. AVI Publ. Co., Westport, Connecticut.
1244 NAGY, S., SHAW, P.E., and VELDHUIS, M.K. (eds.), 1977. Citrus Science and Technology. 2 vols. AVI Publ. Co., Westport, Conn.
1245 NAIR, M.K., BHASKARA, RAO, E.V.V., NAMBIAR, K.K.N., and NAMBIAR, M.C., 1979. Cashew (*Anarcardium occidentale* L.). Monograph on Plantation Crops Nr. 1. Central Plantation Crops Research Institute, Kasaragod, Kerala, India.
1246 NAIR, P.K.R., 1979. Intensive Multiple Cropping with Coconuts in India. Adv. Agron. Crop Sci. 6, Suppl. J. Agron. Crop Sci. Parey, Berlin.
1247 NAIR, P.K.R., FERNANDES, E.C.M., and WAMBUGU, P.N., 1984. Multipurpose leguminous trees and shrubs for agroforestry. Agrofor. Systems 2, 145-163.
1248 NAKASU, B.H., BASSOLS, M. DO C., and FELICIANO, A.J., 1981. Temperate fruit breeding in Brazil. Fruit Var. J. 35, 114-122.
1249 NAKATA, S., and WATANABE, Y., 1966. Effects of photoperiod and night temperature on the flowering of *Litchi chinensis*. Bot. Gaz. 127, 146-152.
1250 NAQVI, H.H., HASHEMI, A., DAVEY, J.R., and WAINES, J.G., 1987. Morphological, chemical, and cytogenetic characters of $F_1$ hybrids between *Parthenium argentatum* (guayule) and *P. fruticosum* var. *fruticosum* (Asteraceae) and their potential in rubber improvement. Econ. Bot. 41, 66-77.
1251 NARAIN, P., 1976. Buckwheat. A potential crop for Uttar Pradesh. Indian Farming 26 (2), 17-18.
1252 NARAYANAN, T.R., and DABADGHAO, P.M., 1972. Forage Crops of India. Indian Council Agric. Res., New Delhi.
1253 NASSAR, N.M.A., 1978. Wild *Manihot* species of Central Brazil for cassava breeding. Can. J. Plant Sci. 58, 257-261.

1254 NASSAR, N.M.A., 1986. Genetic variation of wild *Manihot* species native to Brazil and its potential for cassava improvement. Field Crops Res. 13, 177-184.
1255 NATARAJAN, K.R., 1980. Peanut protein ingredients: preparation, properties, and food uses. Adv. Food Res. 26, 215-273.
1256 NATIONAL ACADEMY OF SCIENCES,1972. Soils of the Humid Tropics. Nat. Acad. Sci., Washington, D. C.
1257 NATIONAL ACADEMY OF SCIENCES,1973. Toxicants Occurring Naturally in Foods. 2nd ed. Nat. Acad. Sci., Washington, D. C.
1258 NATIONAL ACADEMY OF SCIENCES,1975. Underexploited Tropical Plants with Promising Economic Value. Nat. Acad. Sci., Washington, D. C.
1259 NATIONAL ACADEMY OF SCIENCES,1977. Guayule: An Alternative Source of Natural Rubber. Nat. Acad. Sci., Washington, D.C.
1260 NATIONAL ACADEMY OF SCIENCES,1978. Postharvest Food Losses in Developing Countries. Nat. Acad. Sci., Washington, D. C.
1261 NATIONAL ACADEMY OF SCIENCES,1979. Tropical Legumes: Resources for the Future. Nat. Acad. Sci., Washington, D. C.
1262 NATIONAL ACADEMY OF SCIENCES,1980/83. Firewood Crops. Shrubs and Trees for Energy Production. 2 Vols. National Academy Press, Washington, D. C.
1263 NATIONAL ACADEMY OF SCIENCES,1981. The Winged Bean. A High-Protein Crop for the Tropics. 2nd ed. Nat. Acad. Sci., Washington, D. C.
1264 NATIONAL ACADEMY OF SCIENCES,1982. Marijuana and Health. Nat. Acad. Sci. Press, Washington, D. C.
1265 NATIONAL ACADEMY OF SCIENCES,1984. *Leucaena* - Promising Forage and Tree Crop for the Tropics. 2nd ed. Nat. Acad. Sci. Washington, D.C..
1266 NATIONAL ACADEMY OF SCIENCES,1984. Casuarinas: Nitrogen-Fixing Trees for Adverse Sites. National Academy Press, Washington, D. C.
1267 NATIONAL RESEARCH COUNCIL, 1983. Calliandra: a Versatile Small Tree for the Humid Tropics. National Academy Press, Washington, D. C.
1268 NATIONAL RESEARCH COUNCIL, 1984. Amaranth: Modern Prospects for an Ancient Crop. National Academy Press, Washington, D. C.
1269 NATIONAL RESEARCH COUNCIL, 1985. Jojoba: New Crop for Arid Lands, New Raw Material for Industry. National Academy Press, Washington, D. C.
1270 NAVILLE, R., 1987. Commerce mondial de l'ananas. Fruits 42, 25-41.
1271 NAVOT, NIR, and ZAMIR, D., 1987. Isozyme and seed protein phylogeny of the genus *Citrullus* (Cucurbitaceae). Pl. Syst. Evol. 156, 61-67.
1272 NAYAGAM, J., and SHAHABUDDIN, M.Y., 1987. Trend in labour utilization in the estates of peninsular Malaysia - a review. Planters' Bull. 190, 14-19.
1273 NAYAR, N.M., 1973. Origin and cytogenetics of rice. Adv. Genet. 17:, 153-292.
1274 NAYAR, N.M. (ed.), 1983. Coconut Research and Development. Wiley Eastern Ltd., New Delhi.
1275 NAYAR, N.M., and MEHRA, K.L., 1970. Sesame: its uses, botany, cytogenetics, and origin. Econ. Bot. 24, 20-31.
1276 NEALES, T.F., 1973. Effect of night temperature on the assimilation of carbon dioxide by mature pineapple plants, *Ananas comosus* (L.) Merr. Austr. J. Biol. Sci. 26, 539-546.
1277 NEGI, S.S., PAL, R.N., and EHRLICH, C., 1980. Tree Fodders in Himachal Pradesh (India). GTZ, Eschborn.
1278 NERKAR, Y.S., 1972. Induced variation and response to selection for low neurotoxin content in *Lathyrus sativus*. Indian J. Genetics 32, 175-180.
1279 NESTEL, B., and GRAHAM, M. (eds.), 1977. Cassava as Animal Feed. Intern. Devel. Res. Centre, Ottawa.
1280 NESTEL, B., and MACINTYRE, R., (eds.), 1973. Chronic Cassava Toxicity. Intern. Dev. Res. Centre, Ottawa.
1281 NG, S.K., n.d. The Oil Palm, its Culture, Manuring and Utilization. Intern. Potash Inst., Bern.

1282 NICKELL, L.G., 1979. Uses of plant growth substances in the production of sugarcane: a practical case history. In: SKOOG, F. (ed.), Plant Growth Substances, 419-425. Springer, Berlin.
1283 NICKELL, L.G., 1982. Plant growth regulators in the sugarcane industry. In: MCLAREN, J. S. (ed.): Chemical Manipulation of Crop Growth and Development, 167-189. Butterworth, London.
1284 NICKELL, L.G., 1982. Plant Growth Regulators. Agricultural Uses. Springer, Berlin.
1285 NIENHAUS, F., 1981. Virus and Similar Diseases in Tropical and Subtropical Areas. GTZ, Eschborn.
1286 NIHOUL, E., 1976. Le Yang Tao (*Actinidia chinensis* Planchon). Fruits 31, 97-109.
1287 NIKAM, S.M., PATIL, N.Y., and DEOKAR, A.B., 1985. Performance of safflower-based double-cropping sequences under rainfed conditions. Indian J. Agric. Sci. 55, 160-166.
1288 NISHIDA, T., 1963. Ecology of the Pollinators of Passion Fruit. Techn. Bull. 55, Hawaii Agric. Exp. Sta., Honolulu.
1289 NISHIYAMA, I., 1971. Evolution and domestication of the sweet potato. Bot. Mag. (Tokyo) 84, 377-387.
1290 NOBLE, B.F., 1978. The growing demand for kapok. Canopy (Forest Res. Inst., Philippines) 8 (11), 8.
1291 NORMAN, A.G., (ed.) 1978. Soybean Physiology, Agronomy, and Utilization. Academ. Press, New York.
1292 NORMAN, M.J.T., PEARSON, C.J., and SEARLE, P.G.E., 1984. The Ecology of Tropical Food Crops. Cambridge University Press, Cambridge.
1293 NORTON, G. (ed.), 1978. Plant Proteins. Butterworth, London.
1294 NWOKOLO, E., BRAGG, D., and SIM, J., 1987. Nutritional assessment of rubber seed meal with broiler chicks. Trop. Sci. 27, 195-204.
1295 NYMAN, U., and BRUHN, J.G., 1979. *Papaver bracteatum* - a summary of current knowledge. Planta medica 35, 97-117.
1296 OCHSE, J.J., and BAKHUIZEN VAN DEN BRINK, R.C., 1977. Vegetables of the Dutch East Indies. English edition of "Indische Groenten" (1931). Asher, Amsterdam.
1297 OCHSE, J.J., SOULE, M.J., DIJKMAN, M.J., and WEHLBURG, C., 1961. Tropical and Subtropical Agriculture. 2 Vols. Macmillan, New York.
1298 OGBE, F.M.D., and WILLIAMS, J.T., 1978. Evolution in indigenous West African rice. Econ. Bot. 32, 59-64.
1299 OGUTUGA, D.B.A., 1975. Chemical composition and potential commercial uses of kola nut, *Cola nitida* (Vent.) Schott and Endlicher. Ghana J. Agric. Sci. 8, 121-125.
1300 OHAIR, S.K., and ASOKAN, M.P., 1986. Edible aroids - botany and horticulture. Hort. Rev. 8, 43-100.
1301 OHLER, J.G., 1968. Annatto (*Bixa orellana* L.). Trop. Abstr. 23, 409-413.
1302 OHLER, J.G., 1979. Cashew. Communication 71, Dep. Agric. Res., Royal Tropical Inst. Amsterdam.
1303 OHLER, J.G., 1984. Coconut, Tree of Life. FAO Plant Production and Protection paper No. 57, FAO, Rome.
1304 OJO, A.A., and OLANIRAN, Y.A.O., 1982. Evaluation of selected *Cola nitida* clones and their crosses. II. Heterosis. Café Cacao Thé 26, 48-52.
1305 OKA, H.I., 1988. Origin of Cultivated Rice. Elsevier, Amsterdam.
1306 OKIGBO, B.N., 1986. Broadening the food base in Africa: The potential of traditional food plants. Food and Nutrition 12 (1), 4-17.
1307 OKOLI, B.E., and MGBEOGU, C.M., 1983. Fluted pumpkin, *Telfairia occidentalis*: West African vegetable crop. Econ. Bot. 37, 145-149.
1308 OLEAGINEUX, 1979-1988. L'évolution du marché des oléagineux. Oléagineux, 34-47.
1309 OLEAGINEUX, 1981. La culture in vitro de tissus chez le palmier à huile. Oléagineux 36, 111-126.
1310 OLEAGINEUX, 1981. Les ravageurs du palmier à huile et du cocotier en Afrique occidentale. Oléagineux 36, 169-228.

1311 Olmo, H.P., 1971. *Vinifera* x *rotundifolia* hybrids as wine grapes. Amer. J. Enol. Vitic. 22, 87-91.
1312 Olmo, H.P., 1979. Disease and insect resistance in cultivated grapes. World Farming 21 (3), 14-15.
1313 Oludemokun, A.A., 1983. Processing, storage and utilization of kola nuts, *Cola nitida* and *C. acuminata*. Trop. Sci. 24, 111-117.
1314 Olufowote, J.O., 1979. The potentials of some *Oryza glaberrima* cultivars in the germplasm of the National Cereals Research Institute, Ibadan, Nigeria. Intern. Rice Comm. Newsletter 28 (2), 23-25.
1315 Onwueme, I.C., 1978. The Tropical Tuber Crops: Yams, Cassava, Sweet Potato, Cocoyams. Wiley, New York.
1316 Onwueme, I.C., Onochie, B.E., and Sofowora, E.A., 1979. Cultivation of *T. daniellii* - the sweetener. World Crops 31, 106-111.
1317 Oomen, H.A.P.C., and Grubben, G.J.H., 1978. Tropical Leaf Vegetables in Human Nutrition. Comm. 69, Dep. Agric. Res., Royal Trop. Inst., Amsterdam.
1318 Oomen, H.A.P.C., Spoon, W., Heesterman, J.E., Ruinard, J., Luyken, R., and Slump, P., 1961. The sweet potato as the staff of life of the Highland Papuan. Trop. Geogr. Med. 13, 55-66.
1319 Opeke, L.K., 1982. Tropical Tree Crops. John Wiley, Chichester.
1320 Opena, R.T., Kuo, C.G., and Yoon, J.Y., (eds.), 1988. Breeding and Seed Production of Chinese Cabbage in the Tropics and Subtropics. Asian Vegetable Research and Development Centre, Shanhua, Tainan, Taiwan.
1321 Opitz, K.W., 1972. The Pistachio Nut. Agric. Ext., Univ. California. AXT-315, Riverside.
1322 Orr, E., 1972. The Use of Protein-Rich Foods for the Relief of Malnutrition in Developing Countries: an Analysis of Experience. G 73, Trop. Prod. Inst., London.
1323 Orsi, M.A., e Pagani, L., 1972. L'ananas alle Azzorre. Osservazioni e rilievi sulla coltura protetta. Riv. Agric. Subtrop. Trop. 66, 145-208.
1324 Ortiz, A.J., y Cordero, O.L., 1984. El rambután (*Nephelium lappaceum*) ; composición química del fruto y su conservación. Turrialba 34, 243-246.
1325 Osborne, J.F., and Singh, D.P., 1980. Sisal and Other Long Fibre Agaves. Hybridization of Crop Plants. Amer. Soc. Agron., Madison, USA.
1326 Ou, S.H., 1986. Rice Diseases. 2nd ed. Commonwealth Mycol. Inst., Kew.
1327 Oyebade, T., 1972. Some aspects of developmental physiology of the Nigerian kola (*Cola nitida*) fruit. Econ. Bot. 27, 417-422.
1328 Page, P.E. (ed.), 1984. Tropical Tree Fruits for Australia. Queensl. Dep. Primary Ind., Inf. Ser. Q 183/018, Brisbane, Australia.
1329 Pagsuberon, I.C., 1969. Better rice varieties deserve better management. World Farming 11 (10), 12-16.
1330 Pahlen, A. von der, 1977. Cubiu (*Solanum topiro* Humb. and Bonpl.), uma fruteira da Amazônia. Acta Amazonica 7, 301-307.
1331 Paiva, W.O. de, 1982. Comportamento de cultivares de cebola (*Allium cepa* L.), caracterizado para dias curtos, em Manaus, AM. Acta Amazonica 12, 263-269.
1332 Paiva, W.O. de, 1983. Eficiência da seleção de plantas de bertalha (*Basella alba* L. syn. *B. rubra*) para resistência a nematóides. Acta Amazonica 13, 217-226.
1333 Pakianathan, S.W., and Tharmalingam, C., 1982. A technique for improved field planting of hevea budded stumps for smallholdings. Planters' Bull. 172, 79-84.
1334 Pal, B.P., 1984. Environmental conservation and agricultural production. Indian J. Agric. Sci. 54, 233-250.
1335 Palgrave, K.C., 1977. Trees of Southern Africa. C. Struik, Cape Town.
1336 Palmer, E., and Pitman, N., 1972. Trees of South Africa. 3 vols. Balkema, Amsterdam.
1337 Palmer-jones, R.W., 1985. Harvesting policies for tea in Malawi. Exp. Agric. 21, 357-368.

1338 PANDEY, R.M., and SHARMA, H.C., 1989. The Litchi. Indian Council Agric, Res., New Dehli.
1339 PANDITA, P.N., BHAN, M.K., and KAUL, A., 1987. Asparagus - a cash crop of Kashmir. Indian Farming 36 (11), 3-4.
1340 PANIGRAHI, U.C., PATRO, G.K., and MOHANTI, G.C., 1987. Package of practices for turmeric cultivation in Orissa. Indian Farming 37 (4), 4-6.
1341 PANNETIER, C., et BUFFARD-MOREL, J., 1982. Premiers résultats concernant la production d'embryons somatiques à partir de tissus foliaires de cocotier, *Cocos nucifera* L. Oléagineux 37, 349-354.
1342 PANS, 1970. Pest Control in Rice. Manual 3, Ministry Overseas Development, London.
1343 PANSE, V.G., ABRAHAM, T.P., and LEELAVATHI, C.R., 1965. Green Manuring of Crops. I.C.A.R. Techn. Bull. (Agric.) 2, Indian Council Agric. Res., New Delhi.
1344 PANTASTICO, E.B. (ed.), 1975. Postharvest Physiology, Handling and Utilization of Tropical and Subtropical Fruits and Vegetables. AVI Publ. Co., Westport, Connecticut.
1345 PANTULU, J.V., and RAO, M.K., 1982. Cytogenetics of pearl millet. Theor. Appl. Genetics 61, 1-18.
1346 PARAMESWAR, N.S., 1973. Floral biology of cardamom (*Elettaria cardamomum* Maton). Mysore J. Agric. Sci. 7, 205-213.
1347 PARDO, E.M., y GARCIA, C.R., 1984. Praderas y Forrajes. Mundi-Prensa, Madrid.
1348 PAREEK, O.P., 1983. The Ber. Indian Council Agric. Res., New Delhi.
1349 PAREEK, S.K., MAHESHWARI, M.L., and GUPTA, R., 1981. Cultivation of palmarosa oil grasses. Indian Farming 31 (4), 22-25.
1350 PAREEK, S.K., SINGH, S., SRIVASTAVA, V.K., MANDAL, S., MAHESHWARI, M.L., and GUPTA, R., 1981. Advances in periwinkle cultivation. Indian Farming 31 (6), 18-21.
1351 PAREEK, S.K., SRIVASTAVA, V.K., and GUPTA, R., 1989. Effect of source and mode of nitrogen application on senna (*Cassia angustifolia* Vahl). Trop. Agric. (Trinidad) 66, 69-72.
1352 PAREEK, S.K., SRIVASTAVA, V.K., MAHESHWARI, M.L., SINGH, S., and GUPTA, R., 1980. Grow senna in North India. Indian Farming 30 (9), 15-17.
1353 PARFIT, D.E., 1984. Relationship of morphological plant characteristics of sunflower to bird feeding. Can. J. Plant Sci. 64, 37-42.
1354 PARISER, E.R., 1973. Proteins of aquatic origin as foods for human consumption. Adv. Food Res. 20, 189-216.
1355 PARKER, C., 1972. The role of weed science in developing countries. Weed Science 20, 408-413.
1356 PARKER, C., and WILSON, A.K., 1986. Parasitic weeds and their control in the Near East. FAO Plant Prot. Bull. 34, 83-98.
1357 PARRY, G., 1982. Le Cotonnier et Ses Produits. Maisonneuve et Larose, Paris.
1358 PARRY, J.W., 1969. Spices. 2 Vols. Chemical Publ., New York.
1359 PASSMORE, R., NICOL, B.M., and NARAYANA RAO, M., 1974. Handbook of Human Nutritional Requirements. FAO Nutritional Studies 28, FAO, Rome.
1360 PATTERSON, D.T., 1980. Shading effects on growth and partitioning of plant biomass in cogongrass (*Imperata cylindrica*) from shaded and exposed habitats. Weed Sci. 28, 735-739.
1361 PATTISON, E.S., (ed.), 1968. Fatty Acids and Their Industrial Uses. Dekker, New York.
1362 PATURAU, J.M., 1982. By-products of the Cane Sugar Industry. 2nd ed. Elsevier, Amsterdam.
1363 PAULIN, D., 1981. Contribution à l'étude de la biologie florale du cacaoyer. Bilan de pollinisations artificielles. Café Cacao Thé 25, 105-112.
1364 PAULY, G., 1979. Les glandes à pigment du cotonnier: aspect génétique et sélection des variétés "glandless" et "high gossypol". Cot. Fib. Trop. 34, 379-402.

1365 PAYENS, J.P.D.W., 1967. A monograph of the genus *Barringtonia* (Lecythidaceae). Blumea 15, 157-263.
1366 PAYNE, J.A., JAYNES, R.A., and KAYS, S.J., 1983. Chinese chestnut production in the United States: practice, problems, and possible solutions. Econ. Bot. 37, 187-200.
1367 PEACHY, J.E., (ed.), 1969. Nematodes of Tropical Crops. Commonwealth Agric. Bureaux, Farnham Royal.
1368 PEARSON, E.O., and DARLING, M.R.S., 1958. The Insect Pests of Cotton in Tropical Africa. E.C.G.C. and Commonwealth Inst. Entomol., London.
1369 PEE, T.Y., and ANI BIN AROPE, 1976. Rubber Owners' Manual. Rubb. Res. Inst. Malaysia, Kuala Lumpur.
1370 PENDSE, G.S., DANGE, P.S., and SURANGE, S.R., 1974. The present and future senna cultivation in India. J. Univ. Poona 46, 151-162.
1371 PENFOLD, A.R., and WILLIS, J.L., 1961. The Eucalypts. Leonard Hill, London.
1372 PENG, S.Y., 1984. The Biology and Control of Weeds in Sugarcane. Elsevier, Amsterdam.
1373 PERALTA, C.G. DE, e ABBATE, M.L.E., 1981. Caratteristiche anatomiche ed usi di 25 specie legnose provenienti dalla Republica di Panama. Rev. Agric. Subtrop. Trop. 75, 325-379.
1374 PERDUE, R.E.Jr., 1958. *Arundo donax* - source of musical reeds and industrial cellulose. Econ. Bot. 12, 368-404.
1375 PERDUE, R.E.Jr., and KRAEBEL, C.J., 1961. The rice-paper plant - *Tetrapanax papyriferum* (Hook.) Koch. Econ. Bot. 15, 165-179.
1376 PERDUE, R.E.Jr., KRAEBEL, C.J., and YANG, CHI-WEI, 1961. Brush and filling fibers from bamboo. Econ. Bot. 15, 156-160.
1377 PEREZ, J.A., 1974. El Fique. Compañia de Empaques, Medellin, Colombia.
1378 PEREZ-ARBELAEZ, E., 1978. Plantas Útiles de Colombia. 4th ed. Roldan, Bogotá, Colombia.
1379 PEREZ-BUVIEL, J., y WATTS, A.B., 1973. Avances sobre la inactivación del gossipol en la harina de algodón. Agron. Trop. (Venezuela) 23, 323-331.
1380 PERIES, O.S., (ed.), 1970. A Handbook of Rubber Culture and Processing. Rubber Res. Inst. Ceylon, Angalawatta.
1381 PERKINS, R.M., and BURKNER, P.F., 1973. Mechanical pollination of date palms. Ann. Date Grower's Inst. 50, 4-7.
1382 PERROT, É., et PARIS, R., 1971. Les Plantes Médicinales. 2 Vols. Presses Universitaires de France, Vendome.
1383 PERRY, L.M., 1980. Medicinal Plants of East and Southeast Asia: Attributed Properties and Uses. MIT Press, Cambridge, Massachusetts.
1384 PERSINOS, G.J., QUIMBY, M.W., and SCHERMERHORN, J.W., 1964. A preliminary pharmacognostical study of ten Nigerian plants. Econ. Bot. 18, 329-341.
1385 PERSLEY, G.J., and DE LANGHE, E.A., (eds.) 1987. Banana and Plantain Breeding Strategies. Australian Centre Intern. Agric. Res.
1386 PERSLEY, G., TERRRY, E.R., and MACINTYRE, R., (eds.), 1977. Cassava Bacterial Blight. Intern. Developm. Res. Centre, Ottawa.
1387 PESCE, C., 1985. Oil Palms and Other Oilseeds of the Amazon (Transl. by JOHNSON, D. V. of "Oleaginosas da Amazônia" 1941). Reference Publications, Michigan.
1388 PETERS, R.E., and LEE, T.H., 1977. Composition and physiology of *Monstera deliciosa* fruit and juice. J. Food Sci. 42, 1132-1133.
1389 PETERSON, R.F., 1965. Wheat. Botany, Cultivation, and Utilization. Leonard Hill, London.
1390 PETHIYAGODA, U., 1979. Coconut inflorescence sap. Planter (Malaysia) 55 (641), 390-397.
1391 PHADNIS, N.A., and SHINDE, V.K., 1969. Studies in the development of new varieties of grape (*Vitis vinifera* L.) by seedling selection. Trop. Sci. 11, 286-297.
1392 PHILIP, J., 1981. Curry leaf - and its uses. World Crops 33, 125-127.

1393 PHILIP, J., 1983. Studies on growth, yield and quality components in different turmeric types. Indian Cocoa, Arecanut and Spices J. 6, 93-97.
1394 PHILIPPINE AGRICULTURIST, 1986. Special Azolla Issue. Phil. Agric. 69, 449-542.
1395 PHILIPPINE COCONUT AUTHORITY, 1979. Technical Data Handbook on Coconut: Its Products and By-Products. Res. Coord. Docum. Center, Phil. Coconut Authority, Quezon, Philippines.
1396 PHILLIPS, L.L., 1966. The cytology and phylogenetics of the diploid species of *Gossypium*. Amer. J. Bot. 53, 328-335.
1397 PHILLIPS, S.H., and YOUNG, H.M.Jr., 1973. No-tillage Farming. Reiman Associates, Milwaukee.
1398 PHILLIPS, T.P., 1974. Cassava Utilization and Potential Markets. Intern. Dev. Res. Cent., Ottawa, Canada.
1399 PHILLIPS, T.P., 1979. Prospects for processing and marketing cassava products as animal feed. World Animal Rev. No. 31, 36-40.
1400 PICKERSGILL, B., 1988. The genus *Capsicum*: a multidisciplinary approach to the taxonomy of cultivated and wild plants. Biol. Zentralbl. 107, 381-389.
1401 PIENIAZEK, S.A., 1976. Fruit growing in China: changes during two decades. Span 19, 61-63.
1402 PILCHER, J.R., and MERWE, G. VAN DER, 1976. The development of a simple cane cutter. S. Afr. Sugar Technol. Assoc. Proc., June 1976, 1-5.
1403 PILI, E.C., 1968. Dormancy of rice seeds - its causes and methods of breaking. Philippine J. Plant Ind. 33, 127-136.
1404 PILLAI, M.D., and HAVERI, R.R., 1979. Bibliography on Cashew (*Anacardium occidentale* L.). Central Plantation Crops Res. Inst., Kasaragod, India.
1405 PILLAY, P.N.R. (ed.), 1980. Handbook of Natural Rubber Production in India. Rubber Research Institute of India, Kottayam, India.
1406 PIRIE, N.W. (ed.), 1975. Food Protein Sources. Cambridge Univ. Press, Cambridge.
1407 PIRINGER, A.A., 1962. Photoperiodic responses of vegetable plants. In: Proc. Plant Sci. Symposium, Camden, New Jersey, Campbell Soup Co., Camden, 173-185.
1408 PLATT, B.S., 1962. Tables of Representative Values of Foods Commonly Used in Tropical Countries. Series No. 302, Medical Res. Council, London.
1409 PLUCKNETT, D.L., 1977. Giant swamp taro, a little-known Asian-Pacific food crop. Proc. 4th Symp. Intern. Soc. Trop. Root Crops, IDRC - 080e, 36-40.
1410 PLUCKNETT, D.L., 1979. Managing Pastures and Cattle under Coconuts. Westview Press, Boulder, Colorado.
1411 PLUCKNETT, D.L., and BEEMER, H.L.Jr. (eds.), 1981. Vegetable Farming Systems in China. Frances Pinter, London.
1412 PLUCKNETT, D.L., PEÑA, R.S. DE LA, and OBRERO, F., 1970. Taro (*Colocasia esculenta*). Field Crop Abstr. 23, 413-426.
1413 POEHLMAN, J.M., 1979. Breeding Field Crops. AVI Publ. Co., Westport, Connecticut.
1414 POEHLMAN, J.M., and BORTHAKUR, D., 1969. Breeding Asian Field Crops. Oxford and IBH Publ., Calcutta.
1415 POLI, M., 1979. Étude bibliographique de la physiologie de l'alternance de production chez l'olivier (*Olea europaea* L.). Fruits 34, 687-695.
1416 POLLMER, W.G., and PHIPPS, R.H. (eds.), 1980. Improvement of Quality Traits of Maize for Grain and Silage Use. Martinus Nijhoff, The Hague.
1417 POMERANZ, Y., (ed.), 1971. Wheat: Chemistry and Technology. 2nd ed. Amer. Assoc. Cereal Chemists, St. Paul, Minn.
1418 PONGRACZ, D.P., 1978. Practical Viticulture. Philip Publ., Cape Town, South Africa.
1419 PONNAMPERUMA, F.N., 1979. IR42: A Rice Type for Smaller Farmers of South and Southeast Asia. IRRI Res. Paper Series No. 44, IRRI, Manila.
1420 POPENOE, J., 1979. The genus *Spondias* in Florida. Proc. Fla. State Hort. Soc. 92, 277-279.
1421 POPENOE, P., 1973. The Date Palm. Coconut Grove, Miami, Florida.

1422 PORTER, J.W.G., and ROLLS, B.A., (eds.), 1973. Proteins in Human Nutrition. Academic Press, London.
1423 PORTER, R.H., 1950. Maté - South American or Paraguay tea. Econ. Bot. 4, 37-51.
1424 PORTERFIELD, W.M.Jr., 1951. The principal Chinese vegetable foods and food plants of Chinatown markets. Econ. Bot. 5, 3-37.
1425 POSEY, D.A., 1985. Indigenous management of tropical forest ecosystems: the case of the Kayapó Indians of the Brazilian Amazon. Agrofor. Systems 3, 139-158.
1426 POSTGATE, J.R., (ed.), 1971. The Chemistry and Biochemistry of Nitrogen Fixation. Plenum Press, London.
1427 POULTER, N.H., and CAYGILL, J.C., 1985. Production and utilization of papain - a proteolytic enzyme from *Carica papaya* L. Trop. Sci. 25, 123-137.
1428 POULTER, N., and DENCH, J.E., 1981. The Winged Bean (*Psophocarpus tetragonolobus* (L.) DC.). An Annotated Bibliography. Trop. Prod. Inst., London.
1429 PRAKASA RAO, E.V.S., SINGH, M., and GANESHA RAO, R.S., 1988. Intercropping studies in Java citronella (*Cymbopogon winterianus*). Field Crops Res. 18, 279-286.
1430 PRALORAN, J.-C., (ed.), 1971. Les Agrumes. Maisonneuve et Larose, Paris.
1431 PRAN VOHRA, H.R.B., and WILSON, W.O., 1972. The Use of ipil-ipil (*Leucaena leucocephala*) in the diets of laying chickens and laying quail. Philip. Agric. 16, 104-113.
1432 PRENTICE, A.N., 1972. Cotton, With Special Reference to Africa. Longmans, London.
1433 PRESCOTT, J.M., BURNETT, P.A., SAARI, E.E., et al., 1986. Wheat Diseases and Pests: A Guide for Field Identification. CIMMYT, México.
1434 PRESTON, T.R., and LENG, R.A., 1978. Sugar cane as cattle feed. Part I: Nutrition constraints and perspectives. World Animal Rev. 27, 7-12.
1435 PRESTON, T.R., and LENG, R.A., 1978. Sugar cane as cattle feed. Part II: Commercial application and economics. World Animal Rev. 28, 44-48.
1436 PRICE, S., 1963. Cytogenetics of modern sugar canes. Econ. Bot. 17, 97-106.
1437 PRINE, G.M., 1973. Perennial peanuts for forage. Soil Crop Sci. Soc. Flor. Proc. 32, 33-35.
1438 PRINSEN, J.H., 1986. Potential of *Albizia lebbeck* (Mimosaceae) as a tropical fodder tree. A review of literature. Trop. Grasslands 20, 78-83.
1439 PRINSLEY, R.T., and TUCKER, G. (eds.), 1987. Mangoes - A Review. Commonwealth Secretariat, London.
1440 PRINZ, D., 1989. Solanaceae. In: REHM S., (Hrsg.) 1989. Spezieller Pflanzenbau in den Tropen und Subtropen. Handbuch der Landwirtschaft und Ernährung in den Entwicklungsländern, 2. Aufl. Bd. 4, 278-294. Ulmer, Stuttgart.
1441 PROCTOR, J.T.A., and BAILEY, W.G., 1987. Ginseng - industry, botany, and culture. Hort. Reviews 9, 187-236.
1442 PRUTHI, J.S., 1980. Spices and Condiments: Chemistry, Microbiology, Technology. Academic Press, New York.
1443 PUNIA, M.S., SHARMA, G.D., and VERMA, P.K., 1988. Technology for growing liquorice, a medicinal herb. Indian Farming 38 (4), 20-21.
1444 PURSEGLOVE, J.W., 1968. Tropical Crops. Dicotyledons, 2 Vols. Longmans, London.
1445 PURSEGLOVE, J.W., 1972. Tropical Crops. Monocotyledons. 2 Vols. Longmans, London.
1446 PURSEGLOVE, J.W., BROWN, E.G., GREEN, C.L., and ROBBINS, S.R.J., 1981. Spices. 2 Vols. Longman, London.
1447 PUSHKARNATH, 1959. Production of healthy seed potatoes in plains: A new approach. Indian Potato J. 1, 63-72.
1448 PUSHPARAJAH, E., and CHEW, P.S. (eds.), 1982. The Oil Palm in Agriculture in the Eighties. Incorporated Society of Planters, Kuala Lumpur.
1449 PUSHPARAJAH, E., and RAJADURAI, M. (eds.), 1983. Palm Oil Product Technology in the Eighties. Palm Oil Research Institute of Malaysia, Kuala Lumpur, Malaysia.
1450 PY, C., 1969. Problèmes de mécanisation en culture d'ananas. Fruits 24, 5-19.

1451 PY, C., LACOEUILHE, J.-J., et TEISSON, C., 1984. L'Ananas, sa Culture, ses Produit. Maisonneuve et Larose, Paris.
1452 QUIAMCO, M.B., 1979. Philippine rattan: industrial potentials and problems. Monitor (Philippines), 7 (12), 8-9, 14.
1453 QUIMBY, M.W., 1953. *Ammi visnaga* Lam. - a medicinal plant. Econ. Bot. 7, 89-92.
1454 QUIMBY, M.W., DOORENBOS, N.J., TURNER, C.E., and MASOUD, A., 1973. Mississippi-grown marihuana - *Cannabis sativa*. Cultivation and observed morphological variations. Econ. Bot. 27, 117-126.
1455 RACHIE, K.O., 1975. The Millets. Importance, Utilization, and Outlook. ICRISAT, Hyderabad.
1456 RACHIE, K.O., and MAJMUDAR, J.V., 1980. Pearl Millet. Pennsylvania State University Press, University Park, Pennsylvania.
1457 RACHIE, K.O., and PETERS, L.V., 1977. The Eleusines. A Review of the World Literature. ICRISAT, Begrumpet A. P. India.
1458 RACHIE, K.O., and RAWAL, K.M., 1976. Integrated Approaches to Improving Cowpeas, *Vigna unguiculata* (L.) Walp. Techn. Bull. No. 5, IITA. Ibadan, Nigeria.
1459 RACHIE, K.O., and ROBERTS, L.M., 1974. Grain legumes of the lowland tropics. Adv. Agron. 26, 1-132.
1460 RADLEY, J.A., 1968. Starch and Its Derivatives. 4th ed. Chapman and Hall, London.
1461 RAHIMAN, B.A., and NAIR, M.K., 1983. Morphology, cytology and chemical constituents of some *Piper* species from India. J. Plantation Crops 11, 72-90.
1462 RAI, R.S., NEWATIA, R.K., CHAUDHARI, L.B., and RAI, B., 1980. How to have two tonnes of "sawan" from a hectare. Indian Farming 30 (6), 16-18.
1463 RAIMONDO, A., 1981. Osservazioni sul compartamento di alcune cultivar di papaya in serra in Sicilia. Riv. Agric. Subtrop. Trop. 75, 267-276.
1464 RAJAGURU, A.S.B., and WETTIMUNY, S.G. DE S., 1971. Rubber seed meal as a protein supplement in poultry feeding. J. Nat. Agric. Soc. Ceylon 8, 1-12.
1465 RALIARISON, A.V., 1972. Actions nematicide et physiologique de l'extrait aqueux des racines de *Derris elliptica* Benth. Rev. Minist. Dévelop. Rur. 6 (12), 61-108.
1466 RAM, C.S.V., 1980. Calixin in blister blight control. BASF Agric. News 4/80, 6-7.
1467 RAM, M., 1980. High-yielding, huskless and nutritionally rich barley variations. Indian Farming 30 (8), 21-26.
1468 RAMACHANDRAN, C., PETER, K.V., and GOPALAKRISHNAN, P.K., 1980. Drumstick (*Moringa oleifera*): a multipurpose Indian vegetable. Econ. Bot. 34, 276-283.
1469 RANDHAWA, G.S., and BAMMI, R.K., 1971. Grape culture under tropical conditions. Trop. Sci. 13, 137-142.
1470 RANDHAWA, G.S.SAHOTA, T.S.BAINS, D.S., and MAHAJAN, V.P., 1978. The effect of sowing date, seed rate and nitrogen fertilizer on the growth and yield of isabgol (*Plantago ovata* F.). J. Agric. Sci. (Cambridge) 90, 341-343.
1471 RANJHAN, S.K., 1978. Use of agro-industrial by-products in feeding ruminants in India. World Animal Rev. 28, 31-37.
1472 RAO, G.M., SURYANARAYANA, M.C., and THAKER, C.V., 1980. Bees can boost oilseed production. Indian Farming 29 (11), 25-26.
1473 RAO, M.V., and PRASAD, M., 1975. You can have jute and rice too. Indian Farming 24 (12), 22-25.
1474 RAO, N.G.P., 1980. New technology for higher sorghum production. Indian Farming 30 (7), 31-35.
1475 RAO, N.G.P., 1981. Genotype alteration: the basis for agricultural transformation in tropical drylands. Indian J. Genetics 41, 384-388.
1476 RAO, S.A., MENGESHA, M.H., and SUBRAMANIAN, V., 1982. Collection and preliminary evaluation of sweet-stalk pearl millet (*Pennisetum*). Econ. Bot. 36, 286-290.
1477 RAPONDA-WALKER, A., and SILLANS, R., 1961. Les Plantes Utiles du Gabon. Lechevalier, Paris.

1478 RARIVOSON, G., et SCHRAMM, M., 1988. Utilisation de *Sesbania* et *Aeschynomene* comme engrais vert en riziculture. Beitr. trop. Landwirtsch. Veterinärmed. 26, 353-358.

1479 RASHAP, A.W., BRALY, B.A., and STONE, J.T. (eds.), 1984. The Ginseng Research Institute's Indexed Bibliography. Ginseng Res. Inst., Roxbury, NY.

1480 RASMUSSON, D.C., 1985. Barley. Agronomy Series Monograph No. 26. Amer. Soc. Agron., Madison Wisconsin.

1481 RATNAMBAL, M.J., and NAIR, M.K., 1986. High yielding turmeric selection PCT-8. J. Plantation Crops 14, 94-98.

1482 RATTAN, P.S., and PAWSEY, R.G., 1981. Death of tea in Malawi caused by *Pseudophaeolus baudonii*. Trop. Pest Management 27, 225-229.

1483 RAUT, R.N., and KAUL, T., 1982. Synthesis of new genotypes of *Brassica napus* suitable for cultivation in India. Current Sci. 51, 838-839.

1484 RAY, P.K., MISHRA, S., and MISHRA, S.S., 1985. Better elephant-foot yams for Bihar. Indian Farming 34 (11), 11-13, 36.

1485 RAYNAL-ROQUES, A., 1978. Les plantes aquatiques alimentaires. Adansonia, Sér. 2, 18, 327-343.

1486 REBOUR, H., 1966. Les Agrumes. Manual de Culture des Citrus pour le Bassin Méditerranéen. Maillière et Fils, Paris.

1487 REDDEN, R., DRABO, I., and AGGARWAL, V., 1984. IITA programme of breeding cowpeas with acceptable seed types and disease resistance for West Africa. Field Crops Res. 8, 35-48.

1488 REDDY, D.B., 1968. Plant Protection in India. Allied Publ., Bombay.

1489 REDDY, P.V., REDDY, K.B., and RAO, G.N.S.N., 1989. Influence of soil moisture content of pod zone temperature in groundnut. Intern. Arachis Newsl. 6, 9-10.

1490 REE, J.H., 1966. Hemp growing in the Republic of Korea. Econ. Bot. 20, 176-186.

1491 REED, W., LATEEF, S.S., SITHANANTHAM, S., and PAWAR, C.S., 1989. Pigeonpea and Chickpea Insect Identification Handbook. ICRISAT, Patancheru, India.

1492 REHM, S. (Hrsg.), 1986. Grundlagen des Pflanzenbaues in den Tropen und Subtropen. Handb. der Landwirtschaft und Ernährung in den Entwicklungsländern, 2. Aufl., Bd.3. Ulmer, Stuttgart.

1493 REHM S., (Hrsg.) 1989. Spezieller Pflanzenbau in den Tropen und Subtropen. Handbuch der Landwirtschaft und Ernährung in den Entwicklungsländern, 2. Aufl. Bd. 4. Ulmer, Stuttgart.

1494 REHM, S., ENSLIN, P.R., MEEUSE, A.D.J., and WESSELS, J.H., 1957. Bitter principles of the Cucurbitaceae. VII. Distribution of bitter principles in this plant family. J. Sci. Food Agric. 8, 679-686.

1495 REICHEL-DOLMATOFF, G., 1969. Notes on the cultural extent of the use of yajé (*Banisteriopsis caapi*) among the Indians of the Vaupés, Colombia. Econ. Bot. 24, 32-33.

1496 RENARD, J.L., 1979. La fusariose du palmier à huile. Diagnostic en plantation. Méthodes de lutte. Oléagineux 34, 59-63.

1497 RENARD, J.-L., et QUILLEC, G., 1984. Les maladies graves du palmier à huile en Afrique et en Amérique du Sud. Oléagineux 39, 57-64.

1498 RESEARCH CENTER OF AZOLLA HANOI, 1985. Propagation and agricultural use of Azolla in Vietnam. Intern. Rice Comm. Newsletter 34 (1), 79-86.

1499 REUTHER, W. (ed.), 1973. The Citrus Industry. Vol. 3. Production Technology. 2nd ed. Univ. of California, Div. Agric. Sci., Riverside.

1500 REUTHER, W., BATCHELOR, L.D., and WEBBER, H.J. (eds.), 1968. The Citrus Industry. Vol. 2. Anatomy, Physiology, Genetics and Reproduction. 2nd ed. Univ. of California, Div. Agric. Sci., Riverside.

1501 REUTHER, W., CARMAN, E.C., and CARMAN, G.E. (eds.), 1978. The Citrus Industry. Vol. 4. Crop Protection. 2nd ed. Univ. California, Div. Agric. Sci., Riverside, California.

1502 REUTHER, W., WEBBER, H.J., and BATCHELOR, L.D., (eds.), 1967. The Citrus Industry. Vol. 1. History, World Distribution, Botany and Varieties. 2nd ed. Univ. California, Div. Agric. Sci., Riverside.
1503 REVRI, R., 1983. *Catha edulis* Forsk. Geographical Dispersal, Botanical, Ecological and Agronomical Aspects with Special Reference to Yemen Arabic Republic. Göttinger Beitr. Land- und Forstwirtschaft Trop. und Subtrop. Heft 1. Göttingen.
1504 REYES, G.F., 1967. El cultivo de lulo en la zona cafetera Colombiana. Rev. Caf. Colombia 17, 75-77.
1505 REYNOLDS, S.G., 1980. Grazing cattle under coconuts. World Animal Rev. 35, 40-45.
1506 RIBEREAU-GAYON, J., et PEYNAUD, E., (eds.), 1971. Sciences et Techniques de la Vigne. 2 Vols. Dunod, Paris.
1507 RICAUD, B.C., EGAN, B.T., GILLASPIE, A.G.Jr., and HUGHES, G.C. (eds.), 1988. Diseases of Sugarcane. Major Diseases. Elsevier, Amsterdam.
1508 RICCI, P., TORREGROSSA, J.P., and ARNOLIN, R., 1979. Storage problems in the cush-cush yam. I. Post-harvest decay. Trop. Agric. (Trinidad) 56, 41-48.
1509 RICH, A.E., 1983. Potato Diseases. Academic Press, New York.
1510 RICHHARIA, R.H., and GOVINDASHWAMI, S., 1966. Rices of India. Scientific Book, Patna-4, India.
1511 RIJN, P.J.VAN, 1974. The production of pyrethrum. Trop. Abstr. 29, 237-244.
1512 RIPPER, W.E., and GEORGE, L., 1965. Cotton Pests in the Sudan. Blackwell, Oxford.
1513 RITCHIE, G.A. (ed.), 1979. New Agricultural Crops. Amer. Ass. Adv. Sci., Washington, D. C.
1514 RIVALS, P., et ASSAF, R., 1977. Modalités de croissance et systèmes de reproduction du néflier du Japon (*Eriobotrya japonica* Lindl.). Fruits 32, 105-115.
1515 RIVALS, P., et MANSOUR, A.H., 1974. Sur les cardamomes de Malabar (*Elettaria cardamomum* Maton). J. Agric. Trop. Bot. Appl. 21, 37-43.
1516 RIVIER, L., and LINDGREN, J.-E., 1971. "Ayahuasca", the South American hallucinogenic drink: an ethnobotanical and chemical investigation. Econ. Bot. 26, 101-129.
1517 RIVNAY, E., 1962. Field Crop Pests in the Near East. Junk, den Haag.
1518 RIZZINI, C.T., e MORS, W.B., 1976. Botânica Econômica Brasileira. Editora Pedagôgica e Universitária, São Paulo.
1519 RÖBBELEN, G., DOWNEY, R.K., and ASHRI, A., 1989. Oil Crops of the World. Their Breeding and Utilization. McGraw-Hill, New York.
1520 ROBBINS, S.R.J., 1982. Selected Markets for the Essential Oils of Patchouli and Vetiver. G 167, Trop. Prod. Inst., London.
1521 ROBBINS, S.R.J., 1983. Natural essential oils: current trends in production, marketing and demand. Trop. Sci. 24, 65-75.
1522 ROBBINS, S.R.J., 1983. Selected Markets for the Essential Oils of Lemongrass, Citronella and Eucalyptus. G 171, Trop. Prod. Inst., London.
1523 ROBBINS, S.R.J., 1983. Selected Markets for the Essential Oils of Lime, Lemon and Orange. G 172, Trop. Prod. Inst., London.
1524 ROBBINS, S.R.J., 1984. Pyrethrum: A Review of Market Trends and Prospects in Selected Countries. G 185, Trop. Devel. Res. Inst., London.
1525 ROBBINS, S.R.J., 1985. Geranium oil: market trends and prospects. Trop. Sci. 25, 189-196.
1526 ROBBINS, S.R.J., and GREENHALGH, P., 1979. The Markets for Selected Herbaceous Essential Oils. G 120, Trop. Prod. Inst., London.
1527 ROBINSON, F.E., 1988. Kenaf: a new fiber crop for paper production. Calif. Agric. 42 (5), 31-32.
1528 ROCHA, H.M., e DANTAS MACHADO, A., 1972. Fatores ambientes associados com a prodridão do cacaueiro. Rev. Theobroma 2, 26-34.
1529 RODRIGUEZ, F.R., y GARAYAR, M.H., 1969. Cultivo de Cocona, Maracuya y Naranjilla. Informe Expec. S.I.P.A. 25, Minist. Agric., Lima.

1530 RODRIGUEZ, O., e VIEJAS, F. (eds.), 1980. Citricultura Brasileira. Fundação Cargill, Campinas, São Paulo.

1531 ROGER, P.A., and KULASOORIYA, S.A., 1980. Blue-Green Algae and Rice. IRRI, Los Baños, Philippines.

1532 ROGER, P.A., and WATANABE, I., 1982. Research on Algae, Blue-Green Algae, and Phototrophic Nitrogen Fixation at the International Rice Research Institute (1963-81). Summarization, Problems, and Prospects. IRRI Research Paper Series No. 78. IRRI, Manila.

1533 ROGERS, D.J., and APPAN, S.G., 1973. Manihot and Manihotoides. Hafner Press, New York.

1534 ROGERS, D.J., and FLEMING, H.S., 1973. A monograph of *Manihot esculenta* with an explanation of the taximetric methods used. Econ. Bot. 27, 1-113.

1535 ROGERS, D.J., and MORTENSEN, J.A., 1979. The native grape species of Florida. Proc. Fla. State Hort. Soc. 92, 286-289.

1536 ROHAN, T.A., 1963. Processing of Raw Cocoa for the Market. FAO Agric. Studies 60, FAO, Rome.

1537 ROSE INNES, R., 1977. A Manual of Ghana Grasses. Land Resources Div., Min. Overs. Dev., Tolworth Tower, Surbiton, England.

1538 ROSENGARTEN, F.JR., 1973. The Book of Spices. 2nd ed. Pyramid Books, New York.

1539 ROSENGARTEN, F.JR., 1984. The Book of Edible Nuts. Walker, New York.

1540 ROSS, W.M., and EASTIN, J.D., 1972. Grain sorghum in the USA. Field Crop Abstr. 25, 169-174.

1541 ROSSITER, R.C., 1974. The relative success of strains of *Trifolium subterraneum* L. in binary mixtures under field conditions. Austr. J. Agric. Res. 25, 757-766.

1542 ROSTRON, H., 1974. The effect of chemical ripeners on the growth, yield and quality of sugarcane in South Africa and Swaziland. S. Afr. Sugar J. 58, 74-77, 79, 81-83, 85.

1543 ROUGHLEY, R.L., 1970. Preparation and use of legume seed inoculants. Plant Soil 32, 675-701.

1544 ROUSI, A., JOKELA, P., KALLIOLA, R., PIETILÄ, L., SALO, J., and YLI-REKOLA, M., 1989. Morphological variation among clones of ulluco (*Ullucus tuberosus*, Basellaceae) collected in southern Peru. Econ. Bot. 43, 58-72.

1545 ROWE, P.R., and RICHARDSON, D.L., 1975. Breeding Bananas for Disease Resistance, Fruit Quality, and Yield. Trop. Agric. Res. Services, La Lima, Honduras.

1546 ROY, A.B., and MANDEL, A.K., 1967. Retting and quality of jute fibre. Jute Bull. 30, 1-10.

1547 ROY, S.K., and SINGH, R.N., 1979. Bael fruit (*Aegle marmelos*) - a potential fruit for processing. Econ. Bot. 33, 203-212.

1548 RUBBER RESEARCH INSTITUTE OF MALAYSIA, 1973. R.R.I.M. clonal seedlings trials: final report. Planters' Bull. No. 127, 115-126.

1549 RUBBER RESEARCH INSTITUTE OF MALAYSIA, 1974. Manuring of hevea under Ethrel stimulation. Planters' Bull. No. 133, 131-138.

1550 RUBBER RESEARCH INSTITUTE OF MALAYSIA, 1976. Nursery practices and planting techniques. Planters' Bull. No. 143, 25-49.

1551 RUBBER RESEARCH INSTITUTE OF MALAYSIA, 1980. Tapping tools of tomorrow; Puncture tapping - an overview; The micro-tapping system; Implements for micro-tapping. Planters' Bull. No. 164, 99-135.

1552 RUBBER RESEARCH INSTITUTE OF MALAYSIA, 1985. Crop protection in rubber. Planters' Bull. No. 182, 1-28.

1553 RUBBER RESEARCH INSTITUTE OF MALAYSIA, 1988. Reducing immaturity. Planters' Bull. No. 195, 25-63.

1554 RUCK, H.C., 1975. Deciduous Fruit Tree Cultivars for Tropical and Subtropical Regions. Hort. Rev. 3. Commonwealth Agric. Bureaux, Farnham Royal, England.

1555 RUEHLE, G.D., and LEDIN, R.B., 1960. Mango Growing in Florida. Bull. No. 174, Agric. Ext. Serv., Gainsville, Fla.

1556 RUSCHEL, A.P., and VOSE, P.B., 1984. Biological nitrogen fixation in sugar cane. In: SUBBA RAO, N. S. (ed.): Current Developments in Biological Nitrogen Fixation, 219-235. Arnold, London.
1557 RUSSELL, C.E., and FELKER, P., 1987. The prickly-pears (*Opuntia* spp., Cactaceae): a source of human and animal food in semiarid regions. Econ. Bot. 41, 433-445.
1558 RUSSEL G.E., (ed.) 1985. Progress in Plant Breeding - 1. Butterworths, London.
1559 RUSSEL, T.A., 1965. The raphia palms of West Africa. Kew Bull. 19, 173-196.
1560 RUTHENBERG, H., 1980. Farming Systems in the Tropics. 3rd ed. Clarendon Press, Oxford.
1561 SABIITI, E.N., 1980. Dry matter production and nutritive value of *Indigofera hirsuta* L. in Uganda. E. Afr. agric. For. J. 45, 296-303.
1562 SACKS, F.M., 1977. A literature review of *Phaseolus angularis* - the adsuki bean. Econ. Bot. 31, 9-15.
1563 SAHU, B.N., 1979/83. *Rauvolfia serpentina* (Sarpagandha). 2 Vols. Today & Tomorrow's, New Delhi.
1564 SALAMA, R.B., 1973. Sterols in the seed oil of *Nigella sativa*. Planta medica 24, 375-377.
1565 SALGADO, A.L.B., LOVADINI, L.A.C., e PIMENTEL, J.M., 1980. Instruções para a cultura da *Crotalaria juncea*. 2a ed. Bol. Inst. Agron. (Brasil) No. 198.
1566 SALIH, F.A., 1984. Effect of sowing date and row width on the kenaf dry ribbon production in Central Sudan. Beitr. trop. Landwirtsch. Veterinärmedizin 22, 377-382.
1567 SAMARAWIRA, I., 1972. Cardamom. World Crops 24, 76-78.
1568 SAMARAWIRA, I., 1979. A classification of the stages in the growth cycle of the cultivated paddy straw mushroom (*Volvariella volvacea* Singer) and its commercial importance. Econ. Bot. 33, 163-171.
1569 SAMARAWIRA, I., 1983. Date palm, potential source for refined sugar. Econ. Bot. 37, 181-186.
1570 SAMSON, B.T., ZANDSTRA, H.G., and MABBAYAD, B.B., 1980. The effect of different temperature regimes on the ratoonability of two IRRI rice varieties. Phil. J. Crop. Sci. 5, 15-18.
1571 SAMSON, B.T., ZANDSTRA, H.G., and MABBAYAD, B.B., 1980. The effect of clipping portion of the rice panicles/leaves as a method of inducing ratoonability in rice. Phil. J. Crop. Sci. 5, 19-21.
1572 SAMSON, J.A., 1970. Rootstocks for tropical fruit trees. Trop. Abstr. 25, 145-151.
1573 SAMSON, J.A., 1986. Tropical Fruits. 2nd ed. Longman, London.
1574 SAMUELS, A., and ARIAS, L.F., 1979. Agronomic observation on ackee (*Blighia sapida* L.) and preliminary tests on industrial processing. Agron. Costarricense 3, 79-88.
1575 SAMUELS, G., 1986. Growing sugarcane as a renewable energy crop. Soil and Crop Sci. Florida Proc. 45, 103-105.
1576 SAN ANTONIO, J.P., 1981. Cultivation of the shiitake mushroom. HortScience 16, 151-156.
1577 SANCHEZ, P.A., 1976. Properties and Management of Soils in the Tropics. Wiley, New York.
1578 SANCHEZ, P.A., and BUOL, S.W., 1975. Soils of the tropics and the world food crisis. Science 188, 598-603.
1579 SANCHEZ, P.A., and TERGAS, L.E. (eds.), 1979. Pasture Production in Acid Soils of the Tropics. CIAT, Cali, Colombia.
1580 SANCHEZ, P.C., 1979. The prospects of fruit wine production in the Philippines. Phil. J. Crop Sci. 4, 183-190.
1581 SANCHEZ-MONGE y PARELLADA, E., 1981. Diccionario de Plantas Agrícolas. Ministerio de Agricultura, Madrid.
1582 SANGAKKARA, U.R., 1990. Relationship between storage and germinability of cardamom (*Elettaria cardamomum* Maton). J. Agron. Crop Sci. 164, 16-19.

1583 SANGARE, A., et ROGNON, F., 1980. Production de l'hybride Port-Bouet 121. Oléagineeux 35, 79-83.
1584 SANTA CRUZ, H.G., 1976. The genetics and origin of quinua (*Chenopodium quinoa*). Bol. Genético (Argentina) 9, 3-14.
1585 SANTA CRUZ, H.G., y CANEDO, G.E., 1981. Relación entre el rendimiento y la forma de la panoja en la quinua. Turrialba 31, 385-388.
1586 SANTOS, G.A., CARPIO, C.B., ILAGAN, M.C., CANO, S.B., and DELA CRUZ, B.V., 1982. Flowering and early yield performance of four I.R.H.O. coconut hybrids in the Philippines. Oléagineux 37, 571-582.
1587 SARMA, M.S., 1969. Jute. Field Crop Abstr. 22, 323-336.
1588 SARMA, P.V., 1979. Possibilities for integrated control of major pests of tea in India. PANS 25, 237-245.
1589 SATYABALAN, K., 1982. The present status of coconut breeding in India. J. Plantation Crops 10, 67-80.
1590 SATYAVATI, G.V., RAINA, M.K., and SHARMA, M., 1976. Medicinal Plants of India, Vol. I. Indian Council Medical Res., New Delhi.
1591 SAUER, J.D., 1979. Living fences in Costa Rican agriculture. Turrialba 29, 255-261.
1592 SAURE, M., 1973. Successful apple growing in the tropical Indonesia. Fruit Var. J. 27, 44-45.
1593 SAURE, M., 1985. Dormancy release in deciduous fruit trees. Hortic. Rev. 7, 239-300.
1594 SAURE, M., 1989. Obstarten der gemäßigten Breiten. In: REHM S., (Hrsg.) 1989. Spezieller Pflanzenbau in den Tropen und Subtropen. Handbuch der Landwirtschaft und Ernährung in den Entwicklungsländern, 2. Aufl. Bd. 4, 400-414. Ulmer, Stuttgart.
1595 SAXENA, M.C., and SINGH, K.B., 1987. The Chickpea. C.A.B. International, Wallingford, Oxen.
1596 SAXENA, M.C., and VARMA, S., 1985. Faba Beans, Kabali Chickpeas and Lentils in the 1980s. ICARDA, Aleppo, Syria.
1597 SAXON, E.C., 1981. Tuberous legumes: preliminary evaluation of tropical Australian and introduced species as fuel crops. Econ. Bot. 35, 163-173.
1598 SAY MEZA, C.R., y MIRANCA COLIN, S., 1986. El corte de precosecha (calentamiento) en el rendimiento y precocida del tomate de cáscara *Physalis ixocarpa* Brot. Agric. Téc. Méx. 12, 159-171.
1599 SCHALL, S., 1987. La multiplication de l'avocatier (*Persea americana* Mill. cv. Fuerte) par microbouturage in vitro. Fruits 42, 171-176.
1600 SCHELD, H.W., and COWLES, J.R., 1981. Woody biomass potential of the Chinese tallow tree. Econ. Bot. 35, 391-397.
1601 SCHMIDT, G., und HESSE, F.-W., 1975. Einführung der Zuckerrübe in Marokko. GTZ, Eschborn.
1602 SCHMIDT, G., KOCH, W., und ROHRMOSER, K., 1969. Zuckerrübenerzeugung in Marokko. Z. Zuckerindustrie 19, 137-141.
1603 SCHMIDT, G. und WINNER, CH., 1989. Zuckerrübe. In: REHM S., (Hrsg.) 1989. Spezieller Pflanzenbau in den Tropen und Subtropen. Handbuch der Landwirtschaft und Ernährung in den Entwicklungsländern, 2. Aufl. Bd. 4, 156-161. Ulmer, Stuttgart.
1604 SCHMUTTERER, H., 1969. Pests of Crops in Northeast and Central Africa. Gustav Fischer, Stuttgart.
1605 SCHMUTTERER, H., 1977. Plagas y Enfermedades del Algodón en Centroamérica. Deutsche Gesellschaft für Technische Zusammenarbeit (GTZ), Eschborn, W Germany.
1606 SCHMUTTERER, H., and ASCHER, K.R.S. (eds.), 1984. Natural Pesticides from the Neem Tree (*Azadirachta indica* A. Juss.) and Other Tropical Plants. Deutsche Gesellschaft für Technische Zusammenarbeit (GTZ), Eschborn, W Germany.

1607 SCHULTES, R.E., 1955. El guaraná, su historia y su uso. Revista Agricult. Trop. 11, 131-140.

1608 SCHULTES, R.E., 1970. The botanical and chemical distribution of hallucinogens. Ann. Rev. Plant Physiol. 21, 571-598.

1609 SCHULTES, R.E., 1979. The Amazonia as a source of new economic plants. Econ. Bot. 33, 259-266.

1610 SCHULTES, R.E., 1987. Studies in the genus *Hevea*. VIII. Notes on infraspecific variants of *Hevea brasiliensis* (Euphorbiaceae). Econ. Bot. 41, 125-147.

1611 SCHULTZE-MOTEL, J. (ed.), 1986. Rudolf Mansfeld, Verzeichnis landwirtschaftlicher und gärtnerischer Kulturpflanzen. 4 Bde. 2. Aufl. Springer, Berlin.

1612 SCHWARTZ, H.F., and GALVEZ, G.E. (eds.), 1980. Bean Production Problems: Disease, Insect, Soil and Climatic Constraints of *Phaseolus vulgaris*. CIAT, Cali, Colombia.

1613 SCHWARZ, R.E., and KNORR, L.C., 1974. Presence of citrus greening and its psylla vector in Thailand. FAO Plant Prot. Bull. 21, 132-138.

1614 SCHWITZER, M.K., 1980. Utilisations non-alimentaires de l'huile de palme. Oléagineux 35, 261-267.

1615 SCORA, R.W., KUMAMOTO, J., SOOST, R.K., and NAUER, E.M., 1982. Contribution to the origin of the grapefruit, *Citrus paradisii* (Rutaceae). Systematic Botany 7, 170-177.

1616 SCOTT, F.S., and SHORAKA, R., 1974. Economic Analysis of the Market for Guava Nectar. Hawaii Agricult. Experiment Station, Research Rep. 230. Univ. of Hawaii, Honolulu.

1617 SCOTT, W.O., and ALDRICH, S.R., 1970. Modern Soybean Production. S & A Publications, Illinois.

1618 SECOY, D.M., and SMITH, A.E., 1983. Use of plants in control of agricultural and domestic pests. Econ. Bot. 37, 28-57.

1619 SEDDIGH, M., JOLLIFF, G.D., CALHOUN, W., and CRANE, J.M., 1982. *Papaver bracteatum*, potential commercial source of codeine. Econ. Bot. 36, 433-441.

1620 SEEGELER, C.J.P., 1983. Oil Plants in Ethiopia, Their Taxonomy and Agricultural Significance. Centre Agric. Publ. Documentation, Wageningen.

1621 SEEGELER, C.J.P., 1989. *Sesamum orientale* L. (Pedaliaceae): Sesame's correct name. Taxon 38, 657-659.

1622 SEESCHAAF, K.W., 1971. Changes in the amounts of total nitrogen, carbon, and reducing sugars in normal young cocoa pods and those affected by physiological cherelle wilting. Angew. Botanik 45, 285-297.

1623 SEETHARAM, A., 1981. Sunflower cultivation. Indian Farming 30 (10), 13-17.

1624 SEETHARAMAN, R., 1981. Rice improvement in India. Current Sci. 50, 517-522.

1625 SEIGLER, D.S. (ed.), 1977. Crop Resources. Academic Press, New York.

1626 SEMPLE, A.T., 1970. Grassland Improvement. Leonard Hill, London.

1627 SENGUPTA, S.C., 1972. Improved lac cultivation. Indian Farming 22, 11-14, 23.

1628 SESHADRI, R., NAGALAKSHMI, S., MADHUSUDHANA RAO, J., and NATARAJAN, C.P., 1986. Utilization of by-products of the tea plant: a review. Trop. Agric. 63, 2-6.

1629 SETHI, B.L., SIKKA, S.M., DASTUR, R.H., GADKARI, P.D., BALASUBRAHMANYAN, R., MAHESHWARI, P., RANGA SWAMI, and JOSHI, A.B., 1960-61. Cotton in India. 4 Vols. Indian Central Cotton Committee, Bombay.

1630 SETHI, G.S., and KAUR SINGH, 1976. Kuth - a cash crop of Himachal Pradesh. Indian Farming 27 (5), 13-14, 21.

1631 SETHI, S.C., BYTH, D.E., GOWDA, C.L.L., and GREEN, J.M., 1981. Photoperiodic response and accelerated generation turnover in chickpea. Field Crops Res. 4, 215-225.

1632 SHAH, M.H., 1989. Sainfoin, an ideal forage legume for dry, temperate areas of Kashmir. Indian Farming 38 (11), 31-32.

1633 SHALITIN, G., 1974. Nouvelle approche en viticulture tropicale. Études de conduite et de taille au Kenya. Fruits 29, 375-383.

1634 SHAMI, A., 1974. Sélection de cotonniers sans gossypol en Syrie. Cot. Fib. Trop. 29, 269-276.
1635 SHANKARACHARY, N.B., and NATARAJAN, C.P., 1971. Coriander - chemistry, technology and uses. Indian Spices 8 (2), 4-13.
1636 SHANKARIKUTTY, B., NARAYANAN, C.S., KAJARAMAN, K., SUMATHIKUTTY, M.A., OMANSKUTTY, M., and MATHEW, A.G., 1982. Oils and oleoresins from major spices. J. Plantation Crops 10, 1-20.
1637 SHAW, H.K.A., 1973. Willis, A Dictionary of the Flowering Plants and Ferns. 8th ed. Cambridge Univ. Press, New York.
1638 SHAW, N.H., and BRYAN, W.W. (eds.), 1976. Tropical Pasture Research, Principles and Methods. Bull. 51, Commonwealth Agric. Bureaux, Farnham Royal, England.
1639 SHAW, N.H., and WHITEMAN, P.C., 1977. Siratro - a success story in breeding a tropical pasture legume. Tropical Grasslands 11, 7-14.
1640 SHENK, M.D., 1971. Weed control in cacao. World Farming 13 (9), 12-13.
1641 SHERMAN, W.B., and LYRENE, P.M., 1984. Biannual peaches in the tropics. Fruit Var. J. 38, 37-38.
1642 SHERRY, S.P., 1971. The Black Wattle (*Acacia mearnsii* de Wild.). Univ. Natal Press, Pietermaritzburg.
1643 SHIGEURA, G.T., and BULLOCK, R.M., 1983. Guava (*Psidium guajava* L.) in Hawaii - History and Production. Res. Extension Series 035, Hawaii Inst. Trop. Agric. Human Resources, University of Hawaii, Manoa.
1644 SHIGEURA, G.T., and MCCALL, W.W., 1970. The Use of Wild Cane, *Saccharum* Hybrid Clone Moentai, for Windbreak in Hawaii. Circ. 445, Univ. Hawaii, Honolulu.
1645 SHIGEURA, G.T., and OOKA, H., 1984. *Macadamia* nuts in Hawaii: History and Production. Res. Ext. Ser. 039, College of Trop. Agric., Univ. of Hawaii.
1646 SHOCK, C.C., 1982. Rebaudi's stevia: natural noncaloric sweetener. Calif. Agric. 36 (9/10), 4-5.
1647 SHOCK, C.C., 1982. Experimental Cultivation of Rebaudi's Stevia in California. Agron. Progr. Rep. No. 122, Univ. Calif., Davis.
1648 SHOLTON, E.J., 1968. Kenaf in Thailand. USOM/Thailand and Royal Thai Government, Cecchi, Bangkok.
1649 SIDHOM, M.Z., et GEERTS, S., 1983. *Ambrosia maritima* L., molluscicide végétal prometteur. Tropicultura 1, 136-141.
1650 SIEGENTHALER, I.E., n.d. Useful Plants of Ethiopia. Exp. Sta. Bull. 14, Imperial Ethiopian College Agric. Mechan. Arts, Jima Exp. Sta., USAID Contract Publication.
1651 SIEVERDING, E., 1990. Vesicular Arbuscular Mycorrhiza Management in Tropical Agrosystems. Deutsche Gesellschaft für Technische Zusammenarbeit (GTZ), Eschborn, W Germany.
1652 SILSBURY, J.H., 1979. Growth, maintenance and nitrogen fixation of nodulated plants of subterranean clover (*Trifolium subterraneum* L.). Aust. J. Plant Physiol. 6, 165-176.
1653 SILVESTRE, P., et ARRADEAU, M., 1983. Le Manioc. Maisonneuve et Larose, Paris.
1654 SIMMONDS, N.W., 1962. The Evolution of the Bananas. Longmans, London.
1655 SIMMONDS, N.W., 1965. The grain chenopods of the tropical American highlands. Econ. Bot. 19, 223-235.
1656 SIMMONDS, N.W., 1971. The potential of potatoes in the tropics. Trop. Agric. (Trinidad) 48, 291-299.
1657 SIMMONDS, N.W. (ed.), 1976. Evolution of Crop Plants. Longman, London.
1658 SIMPSON, B.B. (ed.), 1977. Mesquite. Its Biology in Two Desert Scrub Ecosystems. US/IBP Synthesis Series, Vol. 4; Dowden, Hutchinson & Ross, Stroudsburg, Pennsylvania.
1659 SIMPSON, M.J., and MCGIBBON, R., 1982. White lupin in cultivation in Iberia. Econ. Bot. 36, 442-446.

1660 SINCLAIR, J.B., 1977. Infectious soybean diseases of world importance. PANS 23, 49-57.
1661 SINCLAIR, J.B., and SHURTLEFF, M.C., 1975. Compendium of Soybean Diseases. American Phytopath. Soc., Minnesota.
1662 SINCLAIR, W.B., 1984. The Biochemistry and Physiology of the Lemon ond Other Citrus Fruits. Univ. of California, Div. of Agriculture and Natural Resources, Oakland, CA.
1663 SINGER, R., 1961. Mushrooms and Truffles. Leonard Hill, London.
1664 SINGH, A.L., and SINGH, P.K., 1989. A comparison of the use of azolla and blue-green algal bio-fertilizers with green manuring, organic manuring and urea in transplanted and direct-seeded rice. Exp. Agric. 25, 485-491.
1665 SINGH, G., 1973. Black zira, a new cash crop. Indian Farming 23 (5), 27-28.
1666 SINGH, G.B., PANT, H.G., and GUPTA, P.N., 1978. Large cardamom. A foreign exchange earner from Sikkim. Indian Farming 27 (12), 3-7.
1667 SINGH, H., and THOMAS, T.A., 1978. Grain Amaranths, Buckwheat and Chenopods. Indian Council Agric. Res., New Delhi.
1668 SINGH, H.B., and ARORA, R.K., 1972. Raishan (*Digitaria* sp.) - a minor millet of the Khasi Hills, India. Econ. Bot. 26, 376-380.
1669 SINGH, H.B., and ARORA, R.K., 1973. Soh-phlong, *Moghania vestita*, a leguminous root crop of India. Econ. Bot. 27, 332-338.
1670 SINGH, J.A., 1978. Cultivation of matting reed in Manipur. Indian Farming 28 (6), 19-21.
1671 SINGH, K.A., RAI, R.N., PATIRAM, and BHUTIA, D.T., 1989. Large cardamom (*Amomum subulatum* Roxb.) plantation - an age old agroforestry system in Eastern Himalayas. Agrofor. Systems 9, 241-257.
1672 SINGH, K.P., SINGH, J.R.P., and RAY, P.K., 1981. 'Rajendra Mishrikand A-1', a promising yam bean. Indian Farming 31 (9), 19-21.
1673 SINGH, L.B., 1970. Utilization of saline-alkali soils without prior reclamation - *Rosa damascena*, its botany, cultivation and utilization. Econ. Bot. 24, 175-179.
1674 SINGH, R.A., DAS, B., AHMED, K.M., and PAL, V., 1980. Chemical control of bacterial leaf blight of rice. Trop. Pest Management 26, 21-25.
1675 SINGH, R.N., 1978. Mango. Indian Council Agric. Res., New Delhi.
1676 SINGH, S.D., and YUSUF, M., 1981. Effect of water, nitrogen and row spacing on the yield and oil content of safflower. Indian J. agric. Sci. 51, 38-43.
1677 SINGH, S.R., EMDEN, H.F. VAN, and TAYLOR, T.A. (eds.), 1978. Pests of Grain Legumes: Ecology and Control. Academic Press, London.
1678 SINGH, S.R., and RACHIE, K.O. (eds.), 1985. Cowpea. Research, Production and Utilization. John Wiley, Chichester, England.
1679 SINGH, S.R., RACHIE, K.O., and DASHIELL, K.S., 1987. Soyabeans for the Tropics. Research, Production and Utilization. John Wiley, New York.
1680 SINNADURAI, S., 1973. Shallot farming in Ghana. Econ. Bot. 27, 438-441.
1681 SINNADURAI, S., and ABU, J.F., 1977. Onion farming in Ghana. Econ. Bot. 31, 312-314.
1682 SIVANESAN, A., and WALKER, J.M., 1986. Sugarcane Diseases. CAB International, Farnham Royal, Hough, England.
1683 SIVETZ, M., and DESROSIER, N.W., 1979. Coffee Technology. Avi Publ., Westport, Connecticut.
1684 SKERMAN, P.J., CAMERON, D.G., and RIVEROS, F., 1988. Tropical Forage Legumes. 2nd. ed. FAO Plant Production and Protection Series No. 2. FAO, Rome.
1684a SKERMAN, P.J., and RIVEROS, F., 1990. Tropical Grasses. FAO Plant Production and Protection Series No. 23. FAO, Rome.
1685 SMARTT, J., 1977. Tropical Pulses. Longman, London.
1686 SMITH, A., 1982. Selected Markets for Chillies and Paprika. G 155. Trop. Prod. Inst., London.

1687 SMITH, A., 1982. Selected Markets for Tumeric, Coriander Seed, Cumin Seed, Fenugreek Seed and Curry Powder. G 165, Trop. Prod. Inst., London.
1688 SMITH, A.E., 1986. International Trade in Cloves, Nutmeg, Mace, Cinnamon, Cassia and Their Derivatives. G 193, Trop. Devel. Res. Inst., London.
1689 SMITH, A.K., 1972. Soybeans. Chemistry and Technology. Vol. 1. Proteins. Avi Publ., Westport, Conn.
1690 SMITH, G.A., BAGBY, M.O., LEWELLAN, R.T., et al., 1987. Evaluation of sweet sorghum for fermentable sugar production potential. Crop Sci. 27, 788-793.
1691 SMITH, H.P., 1964. Farm Machinery and Equipment. McGraw-Hill, New York.
1692 SMITH, J.H.E., 1972. The White Sapote. Leaflet No. 74, Dep. Agric. Techn. Serv., Pretoria.
1693 SMITH, J.H.E., 1980. Propagation of Mangoes. C. 3, Farming in South Africa. Div. Agric. Inform., Pretoria.
1694 SMITH, M.A., and WHITEMAN, P.C., 1983. Evaluation of tropical grasses in increasing shade under coconut canopies. Exp. Agric. 19, 153-161.
1695 SMITH, O.B., and VAN HOUTERT, M.F.J., 1987. The feeding value of *Gliricidia sepium*. A review. World Animal Rev. 62, 57-68.
1696 SMITHSON, J.B., 1985. Breeding advances in chickpeas at ICRISAT. 4587, 223-237.
1697 SMOLENSKI, S.J., KINGHORN, A.D., and BALANDRIN, M.F., 1981. Toxic constituents of legume forage plants. Econ. Bot. 35, 321-355.
1698 SNEEP, J., and HENDRIKSON, A.J.T. (eds.), 1979. Plant Breeding Perspectives. Pudoc, Wageningen.
1699 SNOOK, L.C., 1982. Tagasaste (tree lucerne) *Chamaecytisus palmensis*: a shrub with high potential as a productive fodder crop. J. Austr. Inst. Agric. Sci. 48, 209-213.
1700 SNOWDEN, J.D., 1936. The Cultivated Races of Sorghum. Adlard and Son, London.
1701 SOEJARTO, D.D., COMPADRE, C.M., MEDON, P.J., KAMATH, S.K., and KINGHORN, A.D., 1983. Potential sweetening agents of plant origin. II. Field search for sweet-tasting *Stevia* species. Econ. Bot. 37, 71-79.
1702 SOFO-KANTANKA, O., and LAWSON, N.C., 1980. The effect of different row spacings and plant arrangements on soybeans. Can. J. Plant Sci. 60, 227-231.
1703 SOFOWORA, A., 1982. Medicinal Plants and Traditional Medicine in Africa. John Wiley, Chichester, England.
1704 SOH, A.C., 1987. Abnormal oil palm clones. Possible causes and implications: further discussions. Planter (Kuala Lumpur) 63, 59-63.
1705 SOIL SURVEY STAFF, 1975. Soil Taxonomy. USDA, Handbook No. 436, US Gov. Print. Office, Washington, D. C.
1706 SOLLENBERGER, L.E., PRINE, G.M., OCUMPAUGH, W.R., SCHANK, S.C., KALMBACHER, R.S., and JONES, C.S.Jr., 1987. Dwarf elephantgrass: a high quality forage with potential in Florida and the tropics. Soil Crop Sci. Soc. Florida Proc. 46, 42-46.
1707 SOMARRIA, E., 1988. Guava (*Psidium guajava* L.) trees in a pasture: population model, sensitivity analysis, and applications. Agrofor. Systems 6, 3-17.
1708 SORENSEN, R.C., and PENAS, E.J., 1978. Nitrogen fertilization of soybeans. Agron. J. 70, 213-216.
1709 SORIA, M.V., 1983. Métodos de multiplicación del babaco (*Carica pentagona* L.) por injertos. Turrialba 33, 215-217.
1710 SORIA, S. DE J., GAREIA, J.R., and TREVIZAN, S., 1980. Mechanical pollination of cacao using motorised knapsack sprayers in Brazil, agro-economical assessment. Rev. Theobroma 10, 149-155.
1711 SORIA, S. DE J., and WIRTH, W.W., 1979. Ceratopogonid midged (Diptera: Nematocera) collected from cacao flowers in Palmira, Colombia: an account of their pollinating abilities. Rev. Theobroma 9, 77-84.
1712 SORKAR, G.K., 1972. The World Tea Economy. Oxford Univ. Press, London.
1713 SOTO, M., 1985. Bananas. Cultivo y Comercialization. Litografía e Imprenta LIL, San José, Costa Rica.

1714 SPENCER, N.R., 1984. Velvetleaf, *Abutilon theophrasti* (Malvaceae), history and economic impact in the United States. Econ. Bot. 38, 407-416.
1715 SPLITTSTOESSER, W.E., MARTIN, F.W., and RHODES, A.M., 1973. The amino acid composition of five species of yams (*Dioscorea*). J. Amer. Soc. Hort. Sci. 98, 563-567.
1716 SPOON, W., 1959. Het tropische plantaardige insekticide ryania. Landbouwk. Tijdschrift 71, 369-373.
1717 SPRAGUE, G.F., and DUDLEY, J.W. (eds.), 1988. Corn and Corn Improvement. 3rd ed. Agronomy Monograph No. 18. Amer. Soc. Agron., Madison, USA.
1718 SPRAGUE, H., 1981. Management of Rangelands and Other Grazing Lands of the Tropics and Subtropics for Support of Livestock Production. US Agency Intern. Devel., Washington, D. C.
1719 SPURLING, A.T., and SPURLING, D., 1974. Effect of various organic and inorganic fertilizers on the yield of Montana tung (*Aleurites montana*) in Malawi. Trop. Agric. 51, 1-12.
1720 SPURLING, D., and SPURLING, A.T., 1972. A close planting trial with tung (*Aleurites montana*). Trop. Agric. 50, 347-348.
1721 SREEKANTARADHYA, R., SHETTAR, B.I., CHANDRAPPA, H.M., and SHIVASHANKAR, G., 1975. Genetic variability in horsegram (*Macrotyloma uniflorum*, syn. *Dolichos uniflorus*). Mysore J. Agric. Sci. 9, 361-363.
1722 SREERAMULU, C., 1981. Higher yields of horsegram. Indian Farming 31 (5), 80, 83.
1723 SRIPATHI RAO, B., 1965. Pests of Hevea Plantations in Malaya. Rub. Res. Inst. Malaysia, Kuala Lumpur.
1724 SRIPATHI RAO, B., 1975. Maladies of Hevea in Malaya. Rubber Res. Inst. Malaysia, Kuala Lumpur.
1725 SRZEDNICKI, Z., 1978. Winterkultur der Zuckerrüben in warmen Ländern. Zuckerind. 103, 565-570.
1726 STACE, H.M., and CAMERON, D.F., 1987. Cytogenetic review of taxa in *Stylosanthes hamata* sensu lato. Trop. Grasslands 21, 182-188.
1727 STACE, H.M., and EDYE, L.A. (eds.), 1984. The Biology and Agronomy of *Stylosanthes*. Academic Press, North Ryde, Australia.
1728 STAFLEU, F.A., (ed.), 1972. International Code of Botanical Nomenclature. Oosthoek, Utrecht.
1729 STALKER, H.T., and MOSS, J.P., 1987. Speciation, cytogenetics, and utilization of *Arachis* species. Adv. Agron. 41, 1-40.
1730 STANDAL, B.R., STREET, J.M., and WARNER, R.M., 1974. Tasty, protein rich, and easy to grow - the new edible 'Sunset' hibiscus. Hawaii Farm Sci., 2nd Quartal 1974, 2-3.
1731 STANTON, W.R., 1966. Grain Legumes in Africa. FAO, Rome.
1732 STANTON, W.R., and FLACH, M. (ed.), 1981. Sago. The Equitorial Swamp as a Natural Resource. Martinus Nijhoff, The Hague.
1733 STAPF, O., and HUBBARD, C.E., 1934. *Pennisetum*. In: PRAIN, D., (ed.). Flora of Tropical Africa, Vol. 9, 954-1070.
1734 STAVELY, J.R., PITTARELLI, G.W., and BURK, L.G., 1973. *Nicotiana repanda* as a potential source of disease resistance in *N. tabacum*. J. Heredity 64, 265-272.
1735 STEGEMANN, H., MAJINO, S., and SCHMIEDICKE, P., 1988. Biochemical differentiation of clones of oca (*Oxalis tuberosa*, Oxalidaceae) by their tuber proteins and the properties of these proteins. Econ. Bot. 41, 37-44.
1736 STEMMER, W.P.C., ADRICHEM, J.C.J. VAN, and ROORDA, F.A., 1982. Inducing orthotropic shoots in coffee with the morphactin chlorflurenolmethylester. Exp. Agric. 18, 29-35.
1737 STEVENSON, G.C., 1965. Genetics and Breeding of Sugar Cane. Longmans, London.
1738 STEWART, G.A., 1971. Photoperiod characteristics of Bluebonnet rice. J. Austr. Inst. Agric. Sci. 37, 246-249.

1739 STIGTER, C.J., and MWAMPAJA, A.R., 1984. An interpretation of temperature patterns under mulched tea at Kericho, Kenya. Agric. Forest Meteorology 31, 231-239.

1740 STOBART, T., 1973. Herbs, Spices and Flavorings. McGraw-Hill, New York.

1741 STOECKELLER, J.H., 1965. The design of shelterbelts in relation to crop yield improvement. World Crops 17, 27-32.

1742 STONE, B.C., 1973. The Wild and Cultivated *Pandanus* of the Marshall Islands. Cramer, Lehre.

1743 STONE, J.F., and WILLIS, W.O. (eds.), 1983. Plant Production and Management under Drought Conditions. Elsevier Science Publ., Amsterdam.

1744 STOSIC, D.D., and KAYKAY, J.M., 1981. Rubber seeds as animal feed in Liberia. World Animal Rev. 39, 29-39.

1745 STOTHER, J., 1971. The Market for Fresh Mangoes in Selected Western European Countries. G 59, Trop. Prod. Inst., London.

1746 STOTHER, J., 1971. The Market for Avocados in Selected Western European Countries. G 60, Trop. Prod. Inst., London.

1747 STOUT, B.A., 1966. Equipment for Rice Production. FAO Agric. Devel. Paper 84, FAO, Rome.

1748 STOUT, D.G., KANNANGARA, T., and SIMPSON, G.M., 1978. Drought resistance of *Sorghum bicolor* 2. Water stress effects on growth. Can. J. Plant Sci. 58, 225-233.

1749 STOUT, D.G., and SIMPSON, G.M., 1978. Drought resistance of *Sorghum bicolor* 1. Drought avoidance mechanisms related to leaf water status. Can. J. Plant Sci. 58, 213-224.

1750 STOVER, R.H., 1972. Banana, Plantain and Abaca Diseases. Commonwealth Mycol. Inst., Kew.

1751 STOVER, R.H., and SIMMONDS, N.W., 1987. Bananas. 3rd ed. Longman, London.

1752 STRAUSS, M.S., STEPHENS, G.C., GONZALES, C.J., and ARDITTI, J., 1980. Genetic variability in taro, *Colocasia esculenta* (L.) Schott (Araceae). Annals Bot. 45, 429-438.

1753 STRICKLAND, R.W., 1973. Dry matter production, digestibility and mineral content of *Eragrostis superba* Peyr. and *E. curvula* (Schrad.) Nees at Stamford, south-eastern Queensland. Trop. Grasslands 7, 233-242.

1754 STRYDOM, E., and HYMAN, L.G., 1965. The Production and Marketing of Sweet Potatoes. Bull. 382. Dep. Agric. Techn. Serv., Pretoria.

1755 STUBBS, R.W., PRESCOTT, J.M., SAARI, E.E., and DUBIN, H.J., 1986. Cereal Disease Methodology Manual. CIMMYT, México.

1756 STURROCK, J.W., 1977. Shelterbelts in New Zealand. Experience and innovation. Span 20, 118-120.

1757 SUBBA RAO, A., and SAMPATH, S.R., 1979. Chemical composition and nutritive value of horsegram (*Dolichos biflorus*). Mysore J. Agric. Sci. 13, 198-202.

1758 SUBBA RAO, N.S. (ed.), 1988. Biological Nitrogen Fixation. Recent Developments. Gordon and Breach, New York.

1759 SUBRAMANYAM, H., KRISHNAMURTHY, S., and PARPIA, H.A.B., 1975. Physiology and Biochemistry of Mango Fruit. Academic Press, New York.

1760 SUKARTAATMADJA, K., and SIREGAR, O., 1971. Control of alang-alang by combination of shading with *Gliricidia maculata* H.B.K. and dalapon application. Contr. Weed Sci. Soc. Indonesia 1, 167-172.

1761 SULZBERGER, E.W., and MCLEAN, B.T. (eds.), 1986. Soybean in Tropical and Subtropical Cropping Systems. AVRDC, Tainan, Taiwan.

1762 SUMMERFIELD, R.J., and BUNTING, A.H., (eds.) 1980. Advances in Legume Science. Royal Botanic Gardens, Kew.

1763 SUMMERFIELD, R.J., HUXLEY, P.A., and STEELE, W., 1974. Cowpea (*Vigna unguiculata* (L.) Walp.). Field Crop Abstr. 27, 301-312.

1764 SUMMERFIELD, R.J., MOST, B.H., and BOXALL, M., 1977. Tropical plants with sweetening properties: Physiological and agronomic problems of protected cropping I. *Dioscoreophyllum cumminsii*. Econ. Bot. 31, 331-339.

1765 SUMMERFIELD, R.J., and ROBERTS, E.H. (eds.), 1985. Grain Legume Crops. Collins, London.
1766 SVENDSEN, A.B., and SCHEFFER, J.J.C. (eds.), 1985. Essential Oil and Aromatic Plants. Nijhoff/Junk, Alt Dordrecht, Netherlands.
1767 SWAMINATHAN, M.S., 1975. Breeding for yield and quality in *Hevea brasiliensis*. Planters' Bull. No. 141, 136-144.
1768 SWAMINATHAN, M.S., 1985. Genetics and some emerging trends in agriculture. Indian J. Genetics 45, 1-11.
1769 SWARBRICK, J.T., 1974. The Australian Weed Control Handbook. 2nd ed. Herbicide Recommendations, Toowoomba.
1770 SWARUP, V., and CHATTERJEE, S.S., 1972. Origin and genetic improvement of Indian cauliflower. Econ. Bot. 26, 381-393.
1771 SWERN, D., 1964. Bailey's Industrial Oil and Fat Products. 3rd ed. Interscience, New York.
1772 SWINCER, G.D., OADES, J.M., and GREENLAND, D.J., 1969. The extraction, characterization, and significance of soil polysaccharides. Adv. Agron. 21, 195-235.
1773 SYED, R.A., 1981. Insect pollination of oil palm: feasibility of introducting *Elaeeidobius* spp. into Malaysia. Oil Palm News No. 25, 2-16.
1774 SZOLNOKI, T.W., 1985. Food and Fruit Trees of the Gambia. Stiftung Walderhaltung in Afrika, und Bundesforschungsanstalt für Forst- und Holzwirtschaft, Hamburg.
1775 TAFFIN, G. DE, et QUENCEZ, P., 1980. Aspect de la nutrition anionique chez le palmier à huile et le cocotier. Problème du chlore. Oléagineux 35, 539-546.
1776 TAINTON, M.M. (ed.), 1981. Veld and Pasture Management in South Africa. Shuter and Shooter, Pietermaritzburg, South Africa.
1777 TAN, A.G., NAMBIAR, J., and SUJAN, A., 1981. Rubber wood for SMR pallet manufacture. Planters' Bull. No. 167, 76-83.
1778 TAN, H.T., 1982. Sago palm - a review. Abstr. Trop. Agric. 8 (9), 9-23.
1779 TANAKA, T., 1975. Tanaka's Cyclopedia of Edible Plants of the World. Keigaku Publ. Co., Tokyo.
1780 TANNER, R.D., HUSSAIN, S.S., HAMILTON, L.A., and WOLF, F.T., 1979. Kudzu (*Pueraria lobata*): potential agricultural and industrial resource. Econ. Bot. 33, 400-412.
1781 TANTON, T.W., 1981. The banjhi (dormancy) cycle in tea (*Camellia sinensis*). Exp. Agric. 17, 149-156.
1782 TARR, S.A.J., 1962. Diseases of Sorghum, Sudan Grass and Broom Corn. Commonwealth Mycol. Inst., Kew.
1783 TAUBE, E., 1952. Carnauba wax - product of a Brazilian palm. Econ. Bot. 6, 379-401.
1784 TAYLOR, W.L., and FARMSWORTH, N.R. (eds.), 1973. The *Catharanthus* alkaloids: botany, chemistry, pharmacology and clinical uses. Marcel Decker, New York.
1785 TEIWES, G., and GRÜNEBERG, F., 1967. Science and Practice in the Manuring of Pineapples. 3rd ed. Verlagsgesellschaft für Ackerbau, Hannover.
1786 TENGA, A.Z., and ORMROD, D.P., 1985. Reponses of okra (*Hibiscus esculentus* L.) cultivars to photoperiod and temperature. Scientia Hortic. 27, 177-187.
1787 TEOH, C.H., TOH, P.Y., CHONG, C.F., and EVANS, R.C., 1979. Recent development in the use of herbicides in rubber and oil palm. PANS 24, 503-513.
1788 TERRA, G.J.A., 1966. Tropical Vegetables. Koninkl. Inst. Trop., Amsterdam.
1789 TERRELL, E.E., HILL, S.R., WIERSEMA, J.H., and RICE, W.E., 1986. A Checklist of Names for 3000 Vascular Plants of Economic Importance. USDA, Agriculture Handbook No. 505, Washington, D. C.
1790 TERRY, P.J., and MICHIEKA, R.W., 1987. Common Weeds of East Africa. FAO, Rome.
1791 TEZENAS DU MONTCEL, H., 1989. Le Bananier Plantain. Maisonneuve et Larose, Paris.

1792 THAMPAN, P.K., 1981. Handbook on Coconut Palm. Oxford and IBH Publishing Co., New Delhi.

1793 THARP, W.H., 1979. The Cotton Plant. Agricultural Handbook, USDA, Washington, D. C.

1794 THEBERGE, R.L. (ed.), 1985. Common African Pests and Diseases of Cassava, Yam, Sweet Potato and Cocoyam. IITA, Ibadan, Nigeria.

1795 THEODOSE, R., 1973. Traditional methods of vanilla preparation and their improvement. Trop. Sci. 15, 47-57.

1796 THEUNS, H.G., THEUNS, H.L., and LOUSBERG, R.J.J.CH., 1986. Search for new natural sources of morphinans. Econ. Bot. 40, 485-497.

1797 THIEME, J.G., 1968. Coconut Oil Processing. FAO Agric. Devel. Paper 89, FAO, Rome.

1798 THOMAS, C.A., 1980. Jackfruit, *Artocarpus heterophyllus* (Moraceae) as source of food and income. Econ. Bot. 34, 154-159.

1799 THOMAS, D.G., 1970. Finger millet (*Eleusine coracana* (L.) Gaertn.). In: JAMESON, J. D. (ed.), 1970. Agriculture in Uganda. Oxford Univ. Press, London.

1800 THOMAS, D.W., 1980. Toward more rational policy. In: STAPLES, R. C., and KUHR, R.J. (eds.): Linking Research to Crop Production, 199-208. Plenum Press, New York.

1801 THOMAS, G.V., SUNDARAJU, P., ALI, S.S., and GHAI, S.K., 1989. Individual and interactive effects of VA mycorrhizal fungi and root-knot nematode, *Meloidogyne incognita*, on cardamom. Trop. Agric. (Trinidad) 66, 21-24.

1802 THOMAS, P.R., and TAYLOR, C.E., 1968. Plant Nematology in Africa South of the Sahara. Commonwealth Agric. Bureaux, Farnham Royal.

1803 THOMAS, W.T., 1979. Ericulture: handicraft for the poor. Appropriate Technology 6, 18-19.

1804 THOMPSON, A.E., and RAY, D.T., 1989. Breeding guayule. Plant Breeding Reviews 6, 93-166.

1805 THOMPSON, A.K., 1982. The Storage and Handling of Kiwifruit. G 159, Trop. Prod. Inst., London.

1806 THOMPSON, H.C., and KELLY, W.C., 1957. Vegetable Crops. 5th ed. McGraw-Hill, New York.

1807 THOMSON, P.H., 1982. Jojoba Handbook. Bonsall Publ., Bonsall, California.

1808 THOMSON, R. (ed.), 1981. *Vicia faba*, Physiology and Breeding. Martinus Nijhoff, The Hague.

1809 THOROLD, C.A., 1975. Diseases of Cocoa. Clarendon Press, Oxford.

1810 THURSTON, H.D., 1984. Tropical Plant Diseases. Amer. Phytopathol. Soc., St. Paul, Minnesota.

1811 TINDALL, H.D., 1974. Vegetable production and research in tropical Africa. Scientia Hortic. 2, 199-207.

1812 TINDALL, H.D., 1983. Vegetables in the Tropics. Macmillan, London.

1813 TINGWA, P.O., and YOUNG, R.E., 1975. Studies on the inhibition of ripening in attached avocado (*Persea americana* Mill.) fruits. J. Amer. Soc. Hort. Sci. 100, 447-449.

1814 TINSLEY, A.M., SCHEERENS, J.C., ALEGBEJO, J.O., ADAN, F.H., KRUMHAR, K.C., BUTLER, L.E., and KOPPLIN, M.J., 1985. Tepary beans (*Phaseolus acutifolius* var. *latifolius*) - a potential food source for African and Middle Eastern cultures. Qualitas Plantarum 35, 87-102.

1815 TISSEAU, R., 1977. Activité protéolytique de l'ananas en conserverie et des ses déchets. Recherche d'une technique simplifiée d'extraction de la broméline. Fruits 32, 87-92.

1816 TISSERAT, B., ESAN, E.B., and MURASHIGE, T., 1979. Somatic embryogenesis in angiosperms. Hort. Rev. 1, 1-78.

1817 TONGDEE, S.C., SCOTT, K.J., and MCGLASSON, W.B., 1982. Packaging and cool storage of litchi fruit. CBIRO Food Res. Quarterly 42, 25-28.

1818 TORRES, F., 1983. Role of woody perennials in animal agroforestry. Agrofor. Systems 1, 131-163.
1819 TRAN VAN DAT, 1985. Upland rice review. Intern. Rice Comm. Newsletter 34 (1), 1-17.
1820 TRINH TON THAT, 1982. Potentialities and constraints of rainfed lowland rice development in tropical Africa. Intern. Rice Comm. Newsletter 31 (1), 1-6.
1821 TROPICAL DEVELOPMENT and RESEARCH INSTITUTE, 1986. Pest Control in Tropical Onions. TDRI, London.
1822 TROPICAL PRODUCTS INSTITUTE, 1968. Essential Oils Production in Developing Countries. Trop. Prod. Inst., London.
1823 TROPICAL PRODUCTS INSTITUTE, 1970. Proceedings of the Conference on Tropical and Subtropical Fruits. Trop. Prod. Inst., London.
1824 TROPICAL PRODUCTS INSTITUTE, 1970. The Preparation of Coir or Coconut Fibre by Traditional Methods. G 52, Trop. Prod. Inst., London.
1825 TROPICAL PRODUCTS INSTITUTE, 1973. Proceedings of the Conference on Spices, 1972. Trop. Prod. Inst., London.
1826 TROPICAL PRODUCTS INSTITUTE, 1975. Proceedings of the Conference on Animal Feeds of Tropical and Subtropical Origin. Trop. Prod. Inst., London.
1827 TROPICAL PRODUCTS INSTITUTE, 1975. Bibliography of Insecticide Materials of Vegetable Origin, No. 125. Trop. Prod. Inst., London.
1828 TROPICAL PRODUCTS INSTITUTE, 1980. Abaca for Papermaking: an Atlas of Micrographs. G 136. Trop. Prod. Inst., London.
1829 TROPICAL PRODUCTS INSTITUTE, 1980. A Hand-Operated Bar Mill Sunflower Seed Decorticator. Rural Techn. Guide 9. Trop. Prod. Inst., London.
1830 TROPICAL PRODUCTS INSTITUTE, 1980. A Hand-Operated Disc Mill Sunflower Seed Decorticator. Rural Techn. Guide 10. Trop. Prod. Inst., London.
1831 TRUEBLOOD, E.W.E., 1973. Omixochitl - the tuberose (*Polianthes tuberosa*). Econ. Bot. 27, 157-173.
1832 TSO, T.C., 1972. Physiology and Biochemistry of Tobacco Plants. Dowden, Hutchinson and Ross, Stroudsburg, Pa.
1833 TSUNO, Y., n.d. Sweet Potato. Nutrient Physiology and Cultivation. Intern. Potash Inst., Bern.
1834 TSUNODA, S., HINATA, K., and GOMES-CAMPO, G., 1980. Brassica Crops and Wild Allies. Japan Scientific Societies Press, Tokyo.
1835 TSUNODA, S., and TAKAHASHI, N. (eds.), 1984. Biology of Rice. Elsevier, Amsterdam.
1836 TUCKER, A.O., 1986. Frankincense and myrrh. Econ. Bot. 40, 425-433.
1837 TUCKER, A.O., MACIARELLO, M.J., and HOWELL, J.T., 1980. Botanical aspects of commercial sage. Econ. Bot. 34, 16-19.
1838 TURCHI, F., 1987. L'amaranto: una coltura poco nota ricca di interessanti prospettive. Riv. Agric. subtrop. trop. 81, 89-116.
1839 TURNER, P.D., 1981. Oil Palm Diseases and Disorders. Oxford Univ. Press, Oxford.
1840 TURNER, P.D., and BULL, R.A., 1967. Diseases and Disorders of the Oil Palm in Malaysia. Incorp. Soc. Planters, Kuala Lumpur.
1841 TURNER, P.D., and GILLBANKS, R.A., 1974. Oil Palm Cultivation and Management. Incorp. Soc. Planters, Kuala Lumpur.
1842 TYLER, V.T., 1986. Plant drugs in the twenty-first century. Econ. Bot. 40, 279-288.
1843 UEXKÜLL, H.R. VON, 1976. Aspects of Fertilizer Use in Modern, High-Yield Rice Culture. Bull. 3, Intern. Potash Inst., Bern.
1844 UEXKÜLL, H.R. VON, 1985. Chlorine in the nutrition of palm trees. Oléagineux 40, 67-71.
1845 UPADHYA, M.D., PUROHIT, A.N., and SHARDA, R.T., 1972. Breeding potatoes for subtropical and tropical areas. World Crops 24, 314-316.
1846 UPHOF, J.C.T., 1968. Dictionary of Economic Plants. 2nd ed. Cramer, Lehre.

1847 US DEPARTMENT OF AGRICULTURE, 1970. Selected Weeds of the United States. Agriculture Handbook 366, US Gov. Print. Office, Washington D. C.
1848 US DEPARTMENT OF AGRICULTURE, 1971. World Supply and Demand Prospects for Oilseeds and Oilseed Products in 1980. Foreign Agric. Econ. Rep. 71, US Dep. Agric., Washington D. C.
1849 VALDES III, L.J., HATFIELD, G.M., KOREEDA, M., and PAUL, A.G., 1987. Studies of *Salvia divinorum* (Lamiaceae), an hallucinogenic mint from the Sierra Mazateca in Oaxaca, Central Mexico. Econ. Bot. 41, 283-291.
1850 VALLEAU, W.D., 1952. Breeding tobacco for disease resistance. Econ. Bot. 6, 69-102.
1851 VANAGA, V., SHAHABUDDIN, N.Y., KOW, P.A., and AHMAD, M.R., 1987. Controlled upward tapping with stimulation: preliminary results. Planters' Bull. 191, 29-34.
1852 VARUGHESE, G., BARKER, T., and SAARI, E., 1987. Triticale. CIMMYT, México, D. F.
1853 VASANIYA, P.C., 1966. Palm sugar - a plantation industry in India. Econ. Bot. 20, 40-45.
1854 VASTANO, B.Jr., e BARBOSA, A.P., 1983. Propagação vegetativa do piquiá (*Caryocar villosum* Pers.) por estaquia. Acta Amazonica 13, 143-148.
1855 VAUGHAN, D.A., 1989. The Genus *Oryza* L. Current Status of Taxonomy. IRRI Res. Paper Ser. 138. IRRI, Manila, Philippines.
1856 VAUGHAN, J.G., 1970. The Structure and Utilization of Oil Seeds. Chapman and Hall, London.
1857 VAUGHAN, J.G., MACLEOD, A.J., and JONES, B.M.G. (eds.), 1976. The Biology and Chemistry of the Cruciferae. Academic Press, New York.
1858 VELTKAMP, J.H., 1986. Physiological Causes of Yield Variation in Cassava (*Manihot esculenta* Crantz). Pudoc, Wageningen.
1859 VENUGOPAL, G., KRISHNA DOSS, C., VISWANADHAM, R.K., THIRUMALA RAO, S.D., and REDDY, B.R., 1972. Processing of Indian tea seed. Oléagineux 27, 605-609.
1860 VERBOOM, W., FANSHAWE, D., WILD, H., and NEALE, K., 1973. Common Weeds of Arable Land of Zambia. Ministry Rural Devel., Lusaka.
1861 VERGARA, B.S., and CHANG, T.T., 1976. The Flowering Response of the Rice Plant to Photoperiod: A Review of the Literature. IRRI Techn. Bull. 8. IRRI, Los Baños, Philippines.
1862 VERMA, S.K., 1980. Field pests of pearl millet in India. Trop. Pest Management 26, 13-20.
1863 VIAUD, P., and TEISSEIRE, D., 1979. Euphorbia latex, a possible source of hydrocarbons and rubber. Rev. Génér. Caout. et Plast. No. 593, 181-185.
1864 VIETMEYER, N., 1984. The lost crops of the Incas. Ceres 17 (3), 37-40.
1865 VILJOEN, L., 1971. Mechanical cultivation of denuded areas to encourage biological reclamation. Dep. Agric. Techn. Services, Leaflet 66, Div. Agric. Engineering, Pretoria.
1866 VILLALOBOS CRUZ, M., 1978. El borojó y sus posibilidades de industrialización. Technología No. 113, 8-22.
1867 VILLAREAL, R.L., 1980. Tomatoes in the Tropics. Westview Press, Boulder, Colorado.
1868 VILLAREAL, R.L., 1981. Tomatoes for the Humid Tropics. Span 24, 72-74.
1869 VILLAREAL, R.L., and GRIGGS, T.D., 1982. Sweet Potato. Proc. 1st Intern. Symp., AVRDC, Tainan, Taiwan.
1870 VILLAREAL, R.L., LIN, S.K., CHANG, L.S., and LAI, S.L., 1979. Use of sweet potato (*Ipomoea batatas*) leaf tips as vegetables. Part I-III. Exp. Agric. 15, 113-127.
1871 VILLAX, E.J., 1963. La Culture des Plantes Fourragères dans la Région Méditerranéenne Occidentale. Inst. Nat. Rech. Agron., Rabat.
1872 VINCENTE-CHANDLER, J., CARO-COSTAS, R., PEARSON, R.W., ABRUNA, F., FIGARELLA, J., and SILVA, S., 1964. The Intensive Management of Tropical Forages in Puerto Rico. Univ. Puerto Rico, Agric. Exp. Sta., Bull. 187, Puerto Rico.

1873 VIRMANI, S.S., and EDWARDS, I.B., 1983. Current status and future prospects for breeding hybrid rice and wheat. Adv. Agron. 36, 145-214.
1874 VIVIEN, J., and FAURE, J.J., 1988. Fruitiers sauvages du Cameroun. Fruits 43, 375-516, 465-471, 507-516, 585-601; 44, 155-163.
1875 VOKOU, D., KOKKINI, S., and BESSIERE, J.-M., 1988. *Origanum onites* (Lamiaceae) in Greece: distribution, volatile oil yield, and composition. Econ. Bot. 42, 407-412.
1876 VREEN, G. VAN, and AHMADZABILI, A.L., 1986. Pests of Rice and Their Natural Enemies in Peninsular Malaysia. Pudoc, Wageningen.
1877 VRIES, C.A. DE, 1973. Mushroom cultivation. Trop. Abstr. 28, 849-857.
1878 VRIES, C.A. DE, 1974. Sericulture. Trop. Abstr. 29, 633-642.
1879 VRIES, C.A. DE, 1975. Bee-keeping in the tropics. Abstr. Trop. Agric. 1 (6), 9-23.
1880 VRIES, C.A. DE, 1985. Optimum harvest time of cassava (*Manihot esculenta*). Abstr. Trop. Agric. 10 (1), 9-14.
1881 VRIES, C.A. DE, FERWERDA, J.D., and FLACH, M., 1967. Choice of food crops in relation to actual and potential production in the tropics. Netherl. J. Agric. Sci. 15, 241-248.
1882 VUYLSTEKE, D., SWENNEN, R., WILSON, G.F., and LANGHE, E. DE, 1988. Phenotypic variation among in vitro propagated plantain (*Musa* sp. cultivar 'AAB'). Scientia Hortic. 36, 79-88.
1883 VYAS, S.C., 1981. Diseases of sesamum and niger in India and their control. Pesticides (India) 15 (9), 10-15.
1884 WAGNER, H., and WOLFF, P., 1977. New Natural Products and Plant Drugs with Pharmacological, Biological or Therapeutical Activity. Springer, Berlin.
1885 WAINES, J.G., 1978. Protein contents, grain weight, and breeding potential of wild and domesticated tepary beans. Crop. Sci. 18, 587-590.
1886 WALDEN, D.B. (ed), 1979. Maize Breeding and Genetics. John Wiley and Sons, Chichester.
1887 WALKER, H., 1966. The Market for Sandalwood Oil. G 22, Trop. Prod. Inst., London.
1888 WALKER, J.T., 1973. Breeding cereals for protein. Span 16, 7-9.
1889 WALL, D., (ed.), 1973. Potentials of Field Beans and Other Food Legumes in Latin America. CIAT, Cali, Colombia.
1890 WALL, J.S., and ROSS, W.M., (eds.), 1970. Sorghum Production and Utilization. Avi Publ., Westport.
1891 WALMSLEY, D., and TWYFORD, I.T., 1968. The zone of nutrient uptake by the Robusta banana. Trop. Agric. 45, 113-118.
1892 WALTER, H., HARNICKELL, E., and MUELLER-DOMBOIS, D., 1975. Climate-Diagram Maps of the Individual Continents and the Ecological Regions of the Earth. Springer, Berlin.
1893 WALTER, H., und LIETH, H., 1960-67. Klimadiagramm-Weltatlas. Gustav Fischer, Jena.
1894 WALTERS, P.R., MACFARLANE, N., and SPENSLEY, P.C., 1979. Jojoba: an Assessment of Prospects. G 128, Trop. Prod. Inst., London.
1895 WALTERS, T.W., 1989. Historical overview on domesticated plants in China with special emphasis on the Cucurbitaceae. Econ. Bot. 43, 297-313.
1896 WANDERS, A.A., and MOENS, A., 1982. Present status of mechanisation of rice production and further needs, especially in small-scale farming. Intern. Congr. 12th Agric. Mech. Exhib. Landbouw RAI 82, Amsterdam, 107-141.
1897 WANG, H., 1975. The Breeding and Cultivation of Sweet Potatoes. Techn. Bull. No. 26. Food and Fertilizer Technology Center Taiwan.
1898 WANG, SHIH-CHI, and HUFFMAN, J.B., 1981. Botanochemicals: supplements to petrochemicals. Econ. Bot. 35, 369-382.
1899 WARNOCK, S.J., 1988. A Review of Taxonomy and Phylogeny of the Genus *Lycopersicon*. HortScience 23, 669-673.
1900 WASTIE, R.L., 1986. Disease resistance in rubber. FAO Plant Prot. Bull. 34, 193-199.

1901 WATKINS, G.M. (ed.), 1981. Compendium of Cotton Diseases. Amer. Phytopath. Soc., St. Paul.
1902 WATSON, G.A., 1957. Cover plants in rubber cultivation. J. Rub. Res. Inst. Malaysia 15, 2-18.
1903 WATSON, G.A., WONG, P.W., and NARAYANAN, R., 1964. Effect of cover plants on soil nutrient status and growth of Hevea. IV. Leguminous creepers compared with grasses, *Mikania cordata* and mixed indigenous covers on four soil types. J. Rubber Res. Inst. Malaya 18, 123-145.
1904 WATSON, W. (ed.), 1982. Lacquerwork in Asia and Beyond. Redwood Burn, Trowbridge, England.
1905 WATT, J.M., and BREYER-BRANDWIJK, M.G., 1962. The Medicinal and Poisonous Plants of Southern and Eastern Africa. Livingstone, Edinburgh.
1906 WEBB, C., and HAWTIN, G. (eds.), 1981. Lentils. Commonwealth Agricultural Bureaux, Farnham Royal, Slough, U.K.
1907 WEBER, E., NESTEL, B., and CAMPBEL, M. (eds.), 1979. Intercropping with Cassava. Proc. Intern. Workshop Trivandrum, India. Intern. Devel. Res. Centre, Ottawa, Canada.
1908 WEBER, H.CH., and FORSTREUTER, W. (eds.), 1987. Parasitic Flowering Plants. 4th Intern. Symp. Parasitic Flowering Plants. Phillipps-Universität Marburg, W Germany.
1909 WEBER, H.W., and ROOYEN, P.C.VAN, 1971. Polysaccharides in molasses meal as an ameliorant for saline-sodic soils compared to other reclamation agents. Geoderma 6, 233-253.
1910 WEBSTER, C.C., and BAULKWILL, W.J., (eds.) 1989. Rubber. Longman, London.
1911 WEBSTER, C.C., and WILSON, P.N., 1980. Agriculture in the Tropics. 2nd ed. Longman, London.
1912 WEE, Y.C., 1974. The Masmerah pineapple: a new cultivar for the Malaysian pineapple industry. World Crops 26, 64-67.
1913 WEIR, B.L., and GAGGERO, J.M., 1982. Ethephon may hasten cotton boll opening, increase yield. Calif. Agric. 36 (9/10), 29.
1914 WEISS, E.A., 1971. Castor, Sesame and Safflower. Leonard Hill, London.
1915 WEISS, E.A., 1983. Oilseed Crops. Longman, London.
1916 WELLMAN, F.L., 1972. Tropical American Plant Disease (Neotropical Phytopathology Problems). Scarecrow Press, Metuchen, N.J.
1917 WELLMAN, F.L., 1977. Dictionary of Tropical American Crops and Their Diseases. Scarecrow, Metuchen, N.J.
1918 WENT, F.W., 1957. The Experimental Control of Plant Growth. Ronald Press, New York.
1919 WESTWOOD, M.N., and CHESTNUT, N.E., 1964. Rest period chilling requirement of Bartlett pear as related to *Pyrus calleryana* and *P. communis* rootstocks. Amer. Soc. Hort. Sci. Proc. 84, 82-87.
1920 WET, J.M.J. DE, 1978. Systematics and evolution of *Sorghum* sect. *Sorghum* (Gramineae). Amer. J. Bot. 65, 477-484.
1921 WET, J.M.J. DE, 1987. Pearl Millet (*Pennisetum glaucum*) in Africa and India. Proc. Intern. Pearl Millet Workshop, 3-4. ICRISAT, Patancheru, India.
1922 WET, J.M.J. DE, GRAY, J.R., and HARLAN, J.R., 1976. Systematics of *Tripsacum* (Gramineae). Phytologia 33, 203-227.
1923 WET, J.M.J. DE, and HARLAN, J.R., 1971. The origin and domestication of *Sorghum bicolor*. Econ. Bot. 25, 128-135.
1924 WET, J.M.J. DE, PRASADO RAO, K.E., BRINK, D.E., and MENGESHA, M.H., 1984. Systematics and evolution of *Eleusine coracana* (Gramineae). 'Amer. J. Bot. 71, 550-557.
1925 WET, J.M.J. DE, PRASADO RAO, K.E., MENGASHA, M.H., and BRINK, D.E., 1983. Diversity in kodo millet, *Paspalum scrobiculatum*. Econ. Bot. 37, 159-163.

1926 WHEAT, D., 1979. Branch formation in cocoa (*Theobroma cacao* L., Sterculiaceae). Turrialba 29, 275-284.
1927 WHEELER, L.C., 1978. Hevea (rubber) seeds for human food. Rubber Res. Inst. Sri Lanka Bull. 13, 17-21.
1928 WHISTLER, R.L., (ed.), 1973. Industrial Gums, Polysaccharides and Their Derivatives. 2nd ed. Academic Press, New York.
1929 WHISTLER, R.L., 1982. Industrial gums from plants: guar and chia. Econ. Bot. 36, 195-202.
1930 WHISTLER, R.L., and HYMOWITZ, T., 1979. Guar: Agronomy, Production, Industrial Use, and Nutrition. Purdue Univ. Press, West Lafayette, Indiana.
1931 WHITAKER, T.W., and CUTLER, H.C., 1965. Cucurbits and cultures in the Americas. Econ. Bot. 19, 344-349.
1932 WHITAKER, T.W., and DAVIS, G.N., 1962. Cucurbits. Leonard Hill, London.
1933 WHITE, G.A., and HAUN, J.R., 1965. Growing *Crotalaria juncea*, a multi-purpose legume, for paper pulp. Econ. Bot. 19, 175-183.
1934 WHITEHOUSE, W.E., 1957. The pistachio nut - a new crop for the western United States. Econ. Bot. 11, 281-321.
1935 WHITEMAN, P.C., 1980. Tropical Pasture Science. Oxford Univ. Press, London.
1936 WHITMAN, W.F., 1974. The camu camu, the 'wan' maprang and the 'manila' santol. Proc. Fla. State Hort. Soc. 87, 375-379.
1937 WHITMORE, T.C., 1980. Utilization, potential, and conservation of *Agathis*, a genus of tropical Asian conifers. Econ. Bot. 34, 1-12.
1938 WHITTAKER, D.E., 1972. Passion fruit: agronomy, processing and marketing. Trop. Sci. 14, 59-77.
1939 WHYTE, R.O., 1974. Tropical Grazing Lands. Communities and Constituent Species. Junk, den Haag.
1940 WICKREMASINGHE, R.L., 1978. Tea. Adv. Food Res. 24, 229-286.
1941 WIDRLECHNER, M.P., 1981. History and utilization of *Rosa damascena*. Econ. Bot. 35, 42-58.
1942 WILCOX, J.R. (ed.), 1987. Soybeans: Improvement, Production, and Uses. Amer. Soc. Agron., Washington, D. C.
1943 WILD, A. (ed.), 1988. Russel's Soil Conditions and Plant Growth. 11th ed. Longman, London.
1944 WILDING, J.L., BARNETT, A.G., and AMOR, R.L., 1986. Crop Weeds. Inkata Press, North Clayton, Australia.
1945 WILES, T.L., and HAYWARD, D.M., 1981. The principles and practice of weed control for no-tillage soyabean in southern Brazil using bipyridyl herbicides. Trop. Pest Management 27, 388-400.
1946 WILLEY, R.W., 1975. The use of shade in coffee, cocoa and tea. Hort. Abstr. 45, 791-198.
1947 WILLIAMS, C.N., 1975. The Agronomy of the Major Tropical Crops. Oxford Univ. Press, Kuala Lumpur.
1948 WILLIAMS, C.N., CHEW, W.Y., and RAJARATNAM, J.H., 1980. Tree and Field Crops of the Wetter Regions of the Tropics. Longman, London.
1949 WILLIAMS, C.N., and JOSEPH, K.T., 1970. Climate, Soil and Crop Production in the Humid Tropics. Oxford Univ. Press, Kuala Lumpur.
1950 WILLIAMS, J., 1978. Progrès récents de la recherche sur les acides gras essentiels. Oléagineux 35, 457-459.
1951 WILLIAMS, J.R., METCALFE, J.R., MUNGOMERY, R.W., and MATHES, R., 1969. Pests of Sugar Cane. Elsevier, Amsterdam.
1952 WILLIAMS, L., 1962. Laticiferous plants of economic importance. III. *Couma* species. Econ. Bot. 16, 251-263.
1953 WILLIAMS, L., 1963. Laticiferous plants of economic importance. IV. Jelutong (*Dyera* spp.). Econ. Bot. 17, 110-126.

1954 WILLIAMS, L., 1964. Laticiferous plants of economic importance. V. - Resources of gutta-percha, *Palaquium* species (Sapotaceae). Econ. Bot. 18, 5-26.
1955 WILLIAMS, L.O., 1957. Ginseng. Econ. Bot. 11, 344-348.
1956 WILLIAMS, L.O., 1958. Bayberry wax and bayberries. Econ. Bot. 12, 103-107.
1957 WILLIAMS, L.O., 1970. Jalap or Veracruz jalap and its allies. Econ. Bot. 24, 399-401.
1958 WILLIAMS, L.O., 1970. A yellow food dye - *Escobedia*. Econ. Bot. 24, 459.
1959 WILLIAMS, L.O., 1977. The avocados, a synopsis of the genus *Persea*, subg. *Persea*. Econ. Bot. 31, 315-320.
1960 WILLIAMS, L.O., 1981. The Useful Plants of Central America. Ceiba 24 (1-4), 1-381.
1961 WILLIAMS, R.J., FREDERIKSEN, R.A., and GIRARD, J.-C., 1978. Sorghum and Pearl Millet Disease Identification Handbook. Information Bull. No. 2, ICRISAT, Hyderabad.
1962 WILLIAMSON, E.M., and EVANS, F.J., 1988. Potter's New Cyclopaedia of Botanical Drugs and Preparations. C.W. Daniel Co., Saffron Walden, Essex.
1963 WILSON, F.D., and MENZEL, M.Y., 1967. Interspecific hybrids between kenaf (*Hibiscus cannabinus*) and roselle (*H. sabdariffa*). Euphytica 16, 33-44.
1964 WILSON, H.D., and HEISER, C.B., 1979. Origin and evolutionary relationships of huauzontle (*Chenopodium nuttalliae* Safford), domesticated chenopod of Mexico. Am. J. Bot. 66, 198-206.
1965 WILSON, H.K., 1955. Grain Crops. 2nd ed. McGraw-Hill, New York.
1966 WILSON, R.J., 1975. The Market for Cashew-Nut Kernels and Cashew-Nut Shell Liquid. G 91, Trop. Prod. Inst., London.
1967 WILSON, R.J., 1975. The Market for Edible Groundnuts. G 96, Trop. Prod. Inst., London.
1968 WILSON, R.J., 1975. The International Market for Banana Products for Food Use. G 103, Trop. Prod. Inst., London.
1969 WINDER, J.A., 1978. Cocoa flower Diptera, their identity, pollinating activity and breeding sites. PANS 24, 5-18.
1970 WINKLER, A.J., COOK, J.A., KLIEWER, W.M., and LIDER, L.A, 1974. General Viticulture. 2nd ed. Univ. Calif. Press, Berkeley.
1971 WINNER, C., 1981. Zuckerrübenbau. DLG-Verlag, Frankfurt.
1972 WINTER, J.D., 1968. The Market for Edible Groundnuts. G 36 Trop. Prod. Inst., London.
1973 WINTER, J.D., and KING, P., 1973. The market for exotic and out-of-season fruits and vegetables in the United Kingdom. Trop. Sci. 15, 59-75.
1974 WINTERS, H.F., and MISKIMEN, G.W., 1967. Vegetable Gardening in the Caribbean Area. Agric. Handb. No. 323, Washington D. C.
1975 WINTERS, W.D., BENAVIDES, R., and CLOUSE, W.J., 1981. Effects of aloë extracts on human normal and tumor cells in vitro. Econ. Bot. 35, 89-95.
1976 WISNIAK, J., and ZABICKY, K. (eds.), 1985. Jojoba. Proc. 6th Intern. Conf. on Jojoba and Its Uses. Ben-Gurion Univ. Negev, Beer-Sheva, Israel.
1977 WITCOMBE, J.R., and ERSKINE, W. (eds.), 1984. Genetic Resources and Their Exploitation. Chickpeas, Faba Beans and Lentils. Martinus Nijhoff, The Hague.
1978 WITTWER, S.H., 1975. Food production: technology and the resource base. Science 188, 579-584.
1979 WODAGENCH, A., 1985. Cassava and cassava pests in Africa. FAO Plant Prot. Bull. 33, 101-108.
1980 WOLVERTON, B.C., 1982. Hybrid wastewater treatment system using anaerobic microorganisms and reed (*Phragmites communis*). Econ. Bot. 36, 373-380.
1981 WOOD, B.J., 1968. Pests of Oil Palms in Malaysia and Their Control. Incorp. Soc. Planters, Kuala Lumpur.
1982 WOOD, B.J., 1977. The economics of crop protection in oil palms. PANS 23, 253-267.
1983 WOOD, G.A.R., and LASS, R.A., 1985. Cocoa. 4th ed. Longman, London.

1984 WOOD, I.M., QUICK, D.J., STIFF, R.A., and ADAMS, N.H., 1978. Harvesting kenaf with sugar cane harvester. World Crops 30, 200-201, 204-205.

1985 WOODARD, K.R., PRINE, G.M., and OCUMPAUGH, W.R., 1985. Techniques in the establishment of elephantgrass (*Pennisetum purpureum* Schum.). Soil and Crop Sci. Soc. Florida Proc. 44, 216-221.

1986 WOODROOF, J.G., 1979. Tree Nuts. Production, Processing, Products. 2 Vols. 2nd ed. Avi Publ., Westport, Conn.

1987 WOODROOF, J.G., 1979. Coconuts. 2nd ed. Avi Publ., Westport, Conn.

1988 WOODROOF, J.G., 1983. Peanuts. Production, Processing, Products. 3rd ed. Avi Publ., Westport, Conn.

1989 WOODSON, R., YOUNGKEN, H.W., SCHLITTLER, E., and SCHNEIDER, J.A., 1957. *Rauwolfia*: Botany, Pharmacognosy, Chemistry and Pharmacology. Little, Brown, Boston.

1990 WORLD AGRICULTURAL OUTLOOK BOARD, 1987. Major World Crop Areas and Climatic Profiles. Agric. Handbook No. 664, USDA, Washington, D.C.

1991 WORLD BANK, 1988. Vetiver grass (*Vetiveria zizanioides*). A method of vegetative soil and moisture conservation. 2nd ed. World Bank, New Delhi.

1992 WORMER, T.M., and GITUANJA, J., 1972. Seasonal patterns of growth and development of arabica coffee in Kenya. Part II. Kenya Coffee 35, 270-277.

1993 WRIGHT, M.J. (ed.), 1976. Plant Adaptation to Mineral Stress in Problem Soils. Cornell Univ., Ithaka, N.Y.

1994 WRIGLEY, G., 1988. Coffee. Longman, London.

1995 WU, K.K., and JAIN, S.K., 1977. A note on germ plasm diversity in the world collections of safflower. Econ. Bot. 31, 72-75.

1996 WUEST, P.J., ROYSE, D.J., and BEELMAN, R.B. (ed.), 1988. Cultivating Edible Fungi. Elsevier, Amsterdam.

1997 WUIDART, W., 1981. Production de matériel végétal cocotier. Tenue d'un germoir. Oléagineux 36, 305-309.

1998 WUIDART, W., 1981. Production de matériel végétal cocotier. Pépinière en sacs de plastique. Oléagineux 36, 367-376.

1999 WUIDART, W., 1981. Production de matériel végétal cocotier. Sélection des hybrides en germoir. Oléagineux 36, 497-500.

2000 WYCHERLEY, P.R., 1963. The range of cover plants. Planters' Bull., Rubber Res. Inst. Malaya 68, 117-122.

2001 WYNIGER, R., 1962. Pests of Crops in Warm Climates and Their Control. Verlag für Recht und Gesellschaft, Basel.

2002 WYNNE, J.C., and GREGORY, W.C., 1981. Peanut breeding. Adv. Agron. 34, 39-72.

2003 XIAOSHAN, S., 1978. Le coton en Chine: surproduction et réajustement. Cot. Fib. Trop. 42, 173-177.

2004 YABUNO, T., 1968. Biosystematic studies of the genus *Echinochloa*. Proc. 12th Intern. Congr. Genetics I, 184.

2005 YADURAJU, N.T., and HOSMANI, M.M., 1979. *Striga asiatica* control in sorghum. PANS 25, 163-167.

2006 YAMADA, N., and KAINUMA, K. (eds.), 1986. Sago - 85. Intern. Sago Symp., Tokyo. Sago Palm Research Fund, Trop. Agric. Res. Centre, Yatabe, Tzukuba, Itaraki, Japan.

2007 YAMAGUCHI, M., 1983. World Vegetables. Principles, Production and Nutritive Values. AVI Publ. Co., Westport, Connecticut.

2008 YAMASAKI, T., 1971. Rice Cultivation in Japan, an Example for Intensive Agriculture. Intern. Potash Inst., Bern.

2009 YAPA, P.A.J., 1983. Yield stimulation in Hevea. Rub. Res. Inst. Sri Lanka Bull. 17, 24-31.

2010 YAPA, P.A.J., and SENANAYAKE, S.I., 1983. Use of papain treatment in the manufacture of deproteinized natural rubber. RRISL Bull. 18, 29-32.

2011 YAWALKAR, K.S., 1969. Vegetable Crops of India. Agric. Hortic. Publ. House. Nagpur 1.

2012 YAYOCK, J.Y., and QUINN, J.G., 1977. Agronomy of linseed oil cultivars (*Linum usitatissimum*) in northern Nigeria. Exp. Agric. 13, 93-100.

2013 YAZDISAMADI, B., 1979. Evaluation of safflower cultivars and lines for agronomic traits. Crop Sci. 19, 327-329.

2014 YAZDISAMADI, B., and ZALI, A.A., 1979. Comparison of winter-type and spring-type safflower. Crop Sci. 19, 783-785.

2015 YEN, D.E., 1974. Arboriculture in the subsistence of Santa Cruz, Solomon Islands. Econ. Bot. 28, 247-284.

2016 YEN, D.E., 1974. The Sweet Potato and Oceania. Bishop Museum Press, Honolulu, Hawaii.

2017 YEOH CHONG HOE, 1979. Propagation of legume cover crops in rubber plantations. Planters' Bull. No. 159, 54-64.

2018 YERMANOS, D.M., 1974. Agronomic survey of jojoba in California. Econ. Bot. 28, 160-174.

2019 YERMANOS, D.M., 1982. Jojoba: out of the Ivory Tower and into the Real World of Agriculture. Univ. California, Riverside, CA.

2020 YOON, P.K., LEONG, S.K., LEONG, H.T., and PHUN, H.K., 1988. Preparation of crown budding in nursery. Planters' Bull. 195, 64-70.

2021 YOUMBI, E., CLAIR-MACZULAJTYS, D., et BORY, G., 1989. Variations de la composition chimique des fruits de *Dacryodes edulis* (Don) Lam. Fruits 44, 149-154.

2022 YOUNG, A., 1976. Tropical Soils and Soil Survey. Cambridge Univ. Press, Cambridge.

2023 YOUNGE, O.R., and PLUCKNETT, D.L., 1981. Papaya Fruit Yield and Quality as Influenced by Crop Rotation, Cover Cropping, Liming, and Soil Fumigation in Hawaii. Res. Bull. 155, Hawaii Agric. Exp. Sta., University of Hawaii, Honolulu.

2024 ZAPATA, A., 1972. Pejibaye palm from the Pacific coast of Colombia (a detailed chemical analysis). Econ. Bot. 26, 156-159.

2025 ZENTMYER, G.A., 1984. Avocado diseases. Trop. Pest Management 30, 388-400.

2026 ZEVEN, A.C., and WET, J.M.J. DE, 1982. Dictionary of Cultivated Plants and their Regions of Diversity. 2nd ed. Centre Agric. Publ. Docum., Wageningen.

2027 ZHU WENZHI, and HOU DAN, 1987. Hybrid varieties boost yield in Chinese cotton. Ceres 119 (Vol. 20, No. 5), 10-11.

2028 ZILLINSKY, F., SKOVMAND, B., and AMAYA, A., 1980. Triticale: Adaption, production and uses. Span 23, 83-84.

2029 ZIMDAHL, R.L., 1984. Improving chemical systems of weed control. FAO Plant Prot. Bull. 32, 105-109.

2030 ZOHARY, D., and BASNIZKY, J., 1975. The cultivated artichocke - *Cynara scolymus*. Its probable wild ancestors. Econ. Bot. 29, 233-235.

2031 ZON, A.P.M. VAN DER, et GRUBBEN, G.J.H., 1976. Les Légumes-Feuilles Spontanés et Cultivés du Sud-Dahomey. Communication 65, Dep. Agric. Res., Royal Tropical Institute, Amsterdam.

2032 ZUBAIR, M.U., and ZAHEER, Z., 1978. New possible indigenous source of fat. Pakistan J. Sci. Ind. Res. 21, 136-137.

2033 ZYL, H.J. VAN, and JOUBERT, A.J., 1979. Cultivation of kiwifruit. Deciduous Fruit Grower 29, 18-24.

# Index